OXFORD LOGIC GUIDES: 18

General Editors

ANGUS MACINTYRE
JOHN SHEPHERDSON
DANA SCOTT

OXFORD LOGIC GUIDES

1. Jane Bridge: *Beginning model theory: the completeness theorem and some consequences*
2. Michael Dummett: *Elements of intuitionism*
3. A. S. Troelstra: *Choice sequences: a chapter of intuitionistic mathematics*
4. J. L. Bell: *Boolean-valued models and independence proofs in set theory* (1st edition)
5. Krister Segerberg: *Classical propositional operators: an exercise in the foundation of logic*
6. G. C. Smith: *The Boole–De Morgan correspondence 1842–1864*
7. Alec Fisher: *Formal number theory and computability: a work book*
8. Anand Pillay: *An introduction to stability theory*
9. H. E. Rose: *Subrecursion: functions and hierarchies*
10. Michael Hallett: *Cantorian set theory and limitation of size*
11. R. Mansfield and G. Weitkamp: *Recursive aspects of descriptive set theory*
12. J. L. Bell: *Boolean-valued models and independence proofs in set theory* (2nd edition)
13. Melvin Fitting: *Computability theory: semantics and logic programming*
14. J. L. Bell: *Toposes and local set theories: an introduction*
15. R. Kaye: *Models of Peano arithmetic*
16. J. Chapman and F. Rowbottom: *Relative category theory and geometric morphisms: a logical approach*
17. Stewart Shapiro: *Foundations without foundationalism: a case for second-order logic*
18. John P. Cleave: *A study of logics*
19. T. E. Forster: *Set theory with a universal set*

A Study of Logics

JOHN P. CLEAVE
Formerly
Reader in Mathematics
University of Bristol

CLARENDON PRESS · OXFORD
1991

Oxford University Press, Walton Street, Oxford OX2 6DP
Oxford New York Toronto
Delhi Bombay Calcutta Madras Karachi
Petaling Jaya Singapore Hong Kong Tokyo
Nairobi Dar es Salaam Cape Town
Melbourne Auckland
and associated companies in
Berlin Ibadan

Oxford is a trade mark of Oxford University Press

Published in the United States
by Oxford University Press, New York

© John P. Cleave, 1991

All rights reserved. No part of this publication may be reproduced,
stored in a retrieval system, or transmitted, in any form or by any means,
electronic, mechanical, photocopying, recording, or otherwise, without
the prior permission of Oxford University Press

A catalogue record for this book is available from the British Library

Library of Congress Cataloging in Publication Data
Cleave, J. P. (John P.)
A study of logics/John P. Cleave
(Oxford logic guides: 18)
1. Logic, Symbolic and mathematical. I. Title. II. Series.
BC 135.C47 1992   160—dc20   91-24813
ISBN 0-19-853211-3

Typeset by Integral Typesetting, Gorleston, Norfolk
Printed in Great Britain by
Bookcraft (Bath) Ltd
Midsomer Norton, Avon

Preface

It is a fact of modern scientific culture that there is a great diversity of logics—classical, intuitionistic, many-valued, temporal, modal, the logic of this, that, and the other—which are embraced for a variety of practical, theoretical, or metaphysical reasons. This book seeks the unity in this diversity by asking what properties a structure should have to qualify it to be a logic. It does not provide a complete atlas of the logic manifold but propounds a principle which, though it makes neither the crooked straight nor the rough places plane, facilitates the drawing of local charts of the fortuitous crookedness of real logical highways and byways.

The traditional approach to formal logic proceeds by defining a set of formulas which are then methodically assigned the truth-values *true* or *false* via truth-tables for the connectives. The tautologies—formulas which are true whatever the assignment of truth-values to atomic formulas—are then the centre of further development of the logic. Thus the initial orientation of the conventional study of logic is via the classical (two-valued) logic, so that non-classical systems appear as pathological or special cases.

It is proposed here, however, to invert this traditional order of development by using Tarski's well-known axiomatization of *logical consequence* as a definition of a logic. The functional parts of a logic—connectives, quantifiers, etc.—can then be identified as such by virtue of certain relations with the logical consequence with which the logic is equipped. Thus, the approach here is consequence-oriented rather than theorem-oriented. It sets the background for the classification of logics and investigation of principle types of construction.

In pursuing this objective it is acknowledged that logic is a living growing domain with regions where strata of philosophical dispute are clearly exposed to the discerning eye. No investigation into logic, even if it follows conventional procedures, can avoid philosophical premises. Thus the author espouses the social construction of knowledge (e.g. Bloor 1983) and sees logic as reasoning according to (public) conventions. As a species of regulated linguistic activity, the rules, like those of other institutions, are subject to historical variability. In order to communicate rationally, however, there is no way of standing outside our present, socially given, canons of validity and objectivity. These criteria appear in reified form as constructive and classical logics, the logics of the player and the spectator respectively. This study of logics must therefore be conducted with current mathematical and logical methods.

The metaphysical assumptions behind the formal development of logics usually go unrecorded. This is particularly so with classical logic. Neither is the philosophical basis of intuitionistic logic very well known. Accordingly, Chapter 10 contains a large section devoted to a sketch of Brouwer's philosophy drawn mostly from the thesis of van Stigt (1971). There are several historical notes, most of which are taken from Bocheński (1961) and Kneale (1962). Some quotations from the public domain have been made (provocatively) which express certain philosophical and methodological points with admirable succinctness and lucidity. In some cases they are made to support the present author's own prejudices. Apologies are offered to those quoted authors who might think that they have thereby been misrepresented or their words misused.

As to the technical qualities of this book, it is assumed that the reader has a modest acquaintance with mathematical symbolism and methods and, in particular, with the classical propositional and predicate calculus such as is given in a first undergraduate course in formal logic. There is nothing here beyond the grasp of a philosophically minded mathematics or computer science undergraduate or a mathematically inclined philosophy student. Certain passages are used as exercises. These are marked 'Example' or elsewhere by the traditional invocations to cortical dexterity with pencil and paper: 'it can easily be seen that...', '...similarly...', etc. Occasionally it has been necessary to refer to results which are well known in certain areas outside logic, such as universal algebra and topology. Such results are recorded in the Appendix.

The author is beholden to many logicians, particularly Harvey Rose and John Shepherdson, and others who have been at the School of Mathematics of the University of Bristol at some time or other during the last 25 years. But a debt beyond words is owed to Stephan Körner whose collaboration and advice over 20 years created the inspiration for this book, its plan, and the substance of several chapters. Needless to say, neither Stephan Körner nor any of the authors quoted in the text are responsible for the views expressed herein.

Finally, gratitude for the preparation of the manuscript must be expressed to Joyce, without whose utilities the task would never have been completed.

Bristol J.P.C.
October 1991

Contents

Glossary of notation		xi
1.	The multiplicity of logics	1
	1.1 The province of logic	1
	1.2 The development of the concept of deductive science	5
	1.3 Criticisms of the positivist view	8
	1.4 The science of logics	12
	1.5 Logics	14
2.	Classical logic	16
	2.1 Introduction	16
	2.2 Propositional logic	19
	2.3 Predicate logic	38
3.	Abstract logics	58
	3.1 Introduction	58
	3.2 Prelogics	59
	3.3 Conditions on prelogics	65
	3.4 Transfer principles	69
	3.5 Logics	75
	3.6 Closed sets	79
	3.7 Logical operations	82
	3.8 Compact logics	89
	3.9 Formal systems and combinatorial systems	93
	3.10 Finite logics	96
	3.11 Gentzen systems	98
	3.12 Comparison of logics	102
4.	Logical operations	105
	4.1 Connectives	105
	4.2 The historical connectives	106
	4.3 Relations between connectives	125
	4.4 Quantifiers	129
	4.5 Quantifiers and connectives	132

viii   Contents

5. Order and lattices   136

    5.1 Introduction   136
    5.2 Relations   136
    5.3 Equivalence relations   140
    5.4 Abstraction   142
    5.5 Order   145
    5.6 Upper and lower bounds   151
    5.7 Lattices   155
    5.8 Complete lattices   161
    5.9 Various kinds of lattices   163
    5.10 Complemented lattices   175

6. Constructing logics   193

    6.1 Introduction   193
    6.2 Many-valued logics   193
    6.3 Logics from truth-values   195
    6.4 Logics from lattices   199
    6.5 Logical consequence in classical logic   202
    6.6 Logical consequence in semantic logics   204
    6.7 Functionally free algebras   208
    6.8 The logic of quantum mechanics   209
    6.9 Predicate logics   212
    6.10 Temporal logics   218
    6.11 Model structures   225

7. Quasi-Boolean algebras and empirical continuity   228

    7.1 Introduction   228
    7.2 Quasi-Boolean algebras   228
    7.3 Constructing quasi-Boolean algebras   230
    7.4 The boundary elements of a quasi-Boolean algebra   233
    7.5 Quasi-Boolean algebras in relation to Boolean algebras   235
    7.6 The structure of quasi-Boolean algebras   238
    7.7 Inexact classes and predicates   243
    7.8 Empirical continuity   247

8. Three-valued logic   253

    8.1 Introduction   253
    8.2 The classical imperative   253
    8.3 Sources of the third truth-value   255
    8.4 The logics of truth-values   257
    8.5 The continuity principle for truth-values   258

		Contents	ix
	8.6	N_3 propositional calculus	263
	8.7	Normal forms and the analysis of implications	268
	8.8	The nucleus of $\mathfrak{L}[N_3]\P$	272
	8.9	Connectivity and continuity in $\mathfrak{L}[N_3]\P$	274
	8.10	Three-valued predicate calculus	275
9.	Relevance		286
	9.1	Introduction	286
	9.2	Rigid propositions	288
	9.3	Minimal implications	298
	9.4	A semantic method	305
10.	The calculus of logics: effective logic		309
	10.1	Introduction	309
	10.2	Effective extension of a logic	312
	10.3	The cut condition	316
	10.4	The triviality condition	324
	10.5	Regular junctors and effective positive logic	326
	10.6	Negation in effective logic	328
	10.7	Effective quantifier logics	330
	10.8	Effective, intuitionistic, and classical logics	339
	10.9	Brouwer's intuitionism	341
	10.10	Bishop's constructivism	353
	10.11	Other kinds of constructivism	356
11.	Modal logics		360
	11.1	Introduction	360
	11.2	Necessity and logic	363
	11.3	Relative necessity	365
	11.4	Constructing a classical modal logic	367
	11.5	Axiomatizability of theories of necessity	372
	11.6	Axiomatization of the general theory of necessity	374
	11.7	Effective modal logics	377
	11.8	Deontic logic	385

Appendix   390

References   397

Index   411

Glossary of notation

Abbreviations in text

iff	if, and only if
AC	axiom of choice
ACC	ascending chain condition
DCC	descending chain condition
DL	distributive lattice
NQBA	normal quasi-Boolean algebra
OCL	orthocomplemented lattice
OML	orthomodular lattice
QBA	quasi-Boolean algebra
SDPC	simple definite program clause
◆	end of example
■	end of proof

Non-formal logical abbreviations

&, $\vee$, $\sim$, $\Rightarrow$, $\Leftrightarrow$ 'and', 'or', 'not', 'implies' and 'if and only if' (or 'is equivalent to') respectively. The symbol '$\Rightarrow$' should not be confused with the logical consequence relation, $\rightarrow$, between elements of a given logic.

$(En)A(n)$ there exists an n such that $A(n)$ holds.

$(En, m, \ldots)A(n, m, \ldots)$ there exist $n, m, \ldots$ such that $A(n, m, \ldots)$ holds.

$(n)A(n)$ for all n, $A(n)$ holds.

$(n, m, \ldots)A(n, m, \ldots)$ for all $n, m, \ldots$, $A(n, m, \ldots)$ holds.

$(\bar{x})A(\bar{x})$ abbreviation of $(n, m, \ldots)A(n, m, \ldots)$ where $\bar{x}$ stands for the sequence $n, m, \ldots$.

$(En \in X)A(n)$ there exists an n in X such that $A(n)$ holds.

$(n \in X)A(n)$ for all n in X, $A(n)$ holds.

$(\mu n)A(n)$ the least n such that $A(n)$.

$(\iota x)A(x)$ the unique x such that $A(x)$.

$\bigwedge, \bigvee$ arbitrary conjunction and disjunction respectively as in $\bigwedge \Delta$ where Δ is a set of propositions. $\bigwedge_{1 \leq i \leq m} A_i$ is an alternative notation for $\bigwedge \{A_i; 1 \leq i \leq m\}$.

Definitions

$A =_{def} B$ A is defined to equal B

$A = B$ $(x, y \in \Sigma, X, Y, \ldots \subseteq \Sigma)$ $A = B$ for all elements $x, y, \ldots$ of Σ and all subsets $X, Y, \ldots$ of Σ.

$A \Leftrightarrow B$ $(x, y \in \Sigma, X, Y, \ldots \subseteq \Sigma)$ the conditions A and B are equivalent for the range of variables stated in brackets.

Sets, sequences, and functions

$(\ ,\ )$ round brackets used in normal mathematical usage and in logics. Thus, in a logic equipped with connectives $\supset$ and $\vee$, if a, b, c are elements of the logic then so is $(a \supset (b \vee c))$.

$a \equiv b$ is an abbreviation of the expression $(a \supset b) \wedge (b \supset a)$. Thus, for a logic equipped with connectives $\supset$ and $\wedge$, if a, b, c are elements of the logic then so is $a \equiv b$. The symbol $\equiv$ is also used as an abbreviation in certain formal languages (see Formulas).

$\{A, B, \ldots\}$ set whose members are $A, B, \ldots$.

$x \in X$ x is a member of X.

$\{x; A(x)\}$ the set of x's which satisfy $A(x)$.

$A \supseteq B, B \subseteq A$ A includes B.

X^c complement of X with respect to a fixed universe S which includes X. Thus $X^c = S - X$.

$X + Y, X \cdot Y$ union and intersection respectively of sets X, Y.

$\sum K, \prod K$ union and intersection respectively of the family K of sets. Thus

$$\sum K = \{x; (EX \in K)(x \in X)\},$$
$$\prod K = \{x; (X \in K)(x \in X)\}.$$

(This notation for union and intersection departs from current convention and reverts to an earlier practise (Tarski 1956, 1965). It has the advantage of distinguishing between operation on sets and lattice operations.)

$\Delta, \Gamma, \ldots, A, B, \ldots$ abbreviation of $\Delta + \Gamma + \cdots \{A, B, \ldots\}$ where $\Delta, \Gamma, \ldots$ are sets of elements (formulas) of a logic and $A, B, \ldots$ are elements of the logic. Thus $\Delta, A \to B$ means that B is a logical consequence of $\Delta + \{A\}$.

$\mathbb{P}(X)$ power set of the set X, i.e. the set of subsets of X.

$\mathbb{P}_\omega(X)$ the set of finite subsets of X.

$\mathring{\mathbb{P}}_\omega(X)$ the set of non-empty finite subsets of X.

$\mathbb{N}$ the set of natural numbers $0, 1, 2, \ldots$.

Glossary of notation   xiii

$\langle a_1, a_2, \ldots, a_n \rangle$ the sequence whose first member is a_1, second member is $a_2, \ldots,$ nth member is a_n.
$U \cap V$ concatenation of sequences U, V. If $U = \langle a_1, a_2, \ldots, a_n \rangle$ and $V = \langle b_1, \ldots, b_m \rangle$, then $U \cap V = \langle a_1, \ldots, a_n, b_1, \ldots, b_m \rangle$.
λ null sequence.
$f: X \to Y$ f maps X to Y.
$f \upharpoonright_A$ restriction of function f to domain A.
id_X identity relation on the set X, i.e. $\{(x, x); x \in X\}$.

Algebras, structures, and logics

$|\mathfrak{A}|$ the domain of an algebra, structure, or logic $\mathfrak{A}$ (Appendix, §§A, R).
$\vdash_H$ provability in the system H.
Θ^e an equivalence relation constructed (in context) from Θ. The expressions $x \Theta^e y$, $(x, y) \in \Theta^e$ and $x = y \bmod(\Theta^e)$ have the same meaning.
$\mathfrak{A}/\rho$ algebra $\mathfrak{A}$ reduced by congruence ρ (Definition 5.3(i)).
(J) ideal (of lattice) generated by J (Definition 5.12).

Lattices

xy or $x \cap y$ *meet* or *greatest lower bound* of x and y.
$x \cup y$ *join* or *least upper bound* of x and y

Formulas

$\wedge, \vee, \neg, \supset$ symbols used as junctors in the construction of formulas of a formal language. At a later stage in the construction of a logic from the set of formulas, these become connectives— usually some form of conjunction, disjunction, negation, and implication respectively.
$[\,,\,]$ square brackets used in the formation of formulas in a formal language, e.g. $[A \supset [B \vee C]]$.
$A \equiv B$ Abbreviation of the formula $[[A \supset B] \wedge [B \supset A]]$ of a formal language containing junctors $\wedge$ and $\supset$.
$|F|$ logical length of formula F equals the number of occurrences of logical symbols.
$A(a_1, \ldots, a_n/x_1, \ldots, x_n)$ result of simultaneously substituting a_1 for x_1, a_2 for $x_2, \ldots, a_n$ for x_n in A (Appendix, §W).

1

The multiplicity of logics

1.1 The province of logic

Logics, in the plural, are the subject of this book. Why should there be more than one logic? Philosophical reasons have been advanced for the multiplicity of logics:

> *In logic there are no morals.* Everyone is at liberty to build up his own logic, i.e. his own form of language, as he wishes. All that is required of him is that, if he wishes to discuss it he must state his methods clearly and give syntactical rules instead of philosophical arguments.
>
> Carnap 1949, p. 52

This is Carnap's *principle of tolerance*. But philosophical arguments are not needed to draw this conclusion—it is an unavoidable sociological fact that there are many distinct structures having diverse aims and applications which can claim the title 'logic'. How this state of affairs came about is a matter of history.

Logic has been conceived as the science of valid inference. It was so studied in antiquity by Aristotle. It was developed by medieval scholars. Mathematical treatments of various aspects of logic in the nineteenth century blossomed through its association with the axiomatic method and the foundations of mathematics. Controversies over the foundations resulted in the boom growth in the twentieth century. Today, mathematical logic, as an institution, is a major sector of mathematics.

The formal character of implication and valid inference was recognized early, though it was many centuries before logic was purged of psychological elements and transcendent metaphysics:

> The essential purpose of logic is attained if we can analyse the various forms of inference and arrive at a systematic way of discriminating the valid from the invalid forms. Writers on logic, however, have not generally been content to restrict themselves to this, especially since the days of Locke they have engaged in a good deal of speculative discussion as to the general nature of knowledge and the operations by which the human mind attains truth as to the external world, . . .
>
> Cohen and Nagel 1934, p. 20

Traditional logicians applied logic to contextual reasoning. Though deeply concerned with fallacies, they remained bounded by the Aristotelian framework and did not succeed in formulating a general theory. However, there were a few traditional logicians who prefigured the modern formal outlook by maintaining that logic is a science of correct forms in which the study of fallacies is irrelevant. The formal character of logic was clearly expressed by Albert of Saxony (1295–1366).

Early Indian logic, which developed independently of the West, originated in disputation. Classical texts and commentaries over ten centuries elaborated the classification of fallacies. With the development in the fourteenth century of the New Nyāya school, logic became more formal, and disputation and fallacies were no longer considered with logic. The New Nyāya was responsible for an original method of universal quantification by abstraction (Bocheński 1961), the discovery of some laws related to propositional logic, and a study of the definitions of relations and their use in complex operations. The creativeness of the New Nyāya school diminished in the eighteenth century. Meanwhile, Western logic remained trapped in Aristotle's subject–predicate form.

The logical ideas of Aristotle (384–322 BC) stem from a theory of the subject–predicate relation based on Plato, who maintained that the simplest form of proposition is composed of two essentially different kinds of element: a noun and a verb. Aristotle dropped Plato's type-distinction:

> ... instead he treats predication as an attachment of one term to another term ... it is impossible on the new doctrine for any term to be essentially predicative; on the contrary, any term that occurs in a proposition predicatively may be made into the subject–term of another predication. I call this 'Aristotle's thesis of interchangeability'. His adoption of it marks a transition from the original name-and-predicable theory to a *two-term* theory ... Aristotle's going over to the two-term theory was a disaster.... The restitution of genuine logic is due to two men above all: Bertrand Russell and Gottlob Frege. To Frege we owe it that modern logicians almost universally accept absolute category-difference between names and predicables.
>
> Geach 1968

It is well known that Aristotle's doctrines in physics and biology persisted, independently of evidence, for more than a thousand years. Aristotle's subject–predicate logic lasted even longer. It became a complete self-contained grotesquely overrated corner of thought whose claim to universality was contradicted by the relational arguments used to sustain it (see Chapter 5, §5.2). Thus, Kant (1787) said of logic that:

> Since Aristotle it has not had to retrace a single step unless we choose to consider as improvements the removal of some unnecessary subtleties, or the clearer definition of its matter, both of which refer to the elegance rather than the solidity of the science. It is remarkable also, that to the present day, it has not been able to make one step in advance, so that, to all appearance, it may be considered as completed and perfect.

The 'disaster', which was propagated in textbooks well into the twentieth century, can hardly be attributed to Aristotle himself—responsibility (in a causative, not moral, sense) lies with the social institutions which vested these ideas with such cultural momentum. A history of logic which relates ideas to society has yet to be written.

From the Megarians, and later the Stoics, there developed some basic principles of propositional logic concerning the logical particles such as conjunction, disjunction, and conditionals. Against the background of the ascendancy of mathematical method, Boole's rediscovery of this logic and its treatment as a calculus marks the modern era. de Morgan and Boole elaborated an algebra of logic, so contributing to the mathematical development of the subject. Schröder and Peirce liberated logic from its subject–predicate fixation. Meanwhile the Weierstrass–Dedekind movement in mathematics reduced the continuum to rational numbers and thence to integers. Frege attempted the final reduction of mathematics to logic by a philosophical analysis of the concept of *number*. Russell and Whitehead achieved a synthesis of these movements in their *Principia Mathematica*.

In Frege's *Begriffschrift* (Frege 1879) the *predicate logic*—a theory of all logical particles (connectives and quantifiers)—was given for the first time. This 'classical' logic, like the subject–predicate logic, was intended to be universally applicable—it was *the* logic. Later, the universality was challenged principally by the ferment in the 'foundations' of mathematics, particularly in the controversy over the Frege–Russell project of reducing all mathematics to logic—the 'logistic thesis'. From this and the controversies over 'completed infinities' in the classical theories of Weierstrass, Dedekind, and Cantor, there arose Brouwer's intuitionism (Chapter 10) whose logic is remarkable for its rejection of the principle of the excluded middle.

Classical mathematics treated the infinite as *actual* or *completed*. An infinite set was regarded as existing as a complete whole, independent of any process of generation. But Brouwer treated the infinite as *potential* or *becoming*.

> Brouwer made it clear, as I think beyond any doubt, that there is no evidence supporting the belief in the existential character of the totality of all natural numbers... The sequence of numbers which grows beyond any stage already reached by passing to the next number, is the manifold of possibilities open towards infinity: it remains forever in the state of creation but is not a closed realm of things existing in themselves. That we blindly converted one into the other is the true source of our difficulties, including the antinomies—a source of more fundamental nature than Russell's vicious principle indicated. Brouwer opened our eyes and made us see how far classical mathematics, nourished by a belief in the 'absolute' that transcends all possibilities of realization, goes beyond such statements as can claim real meaning and truth founded on evidence.
>
> Weyl 1946

Hilbert accepted that the areas of classical mathematics which use completed infinities lacked evidential support. But rather than following Brouwer's radical rejection of classical logic, he proposed to formulate classical mathematics up to an appropriate level as a formal axiomatic theory and then to prove this theory consistent. The only way to achieve this aim is to reconstruct the informal theory as a *formal system* or *logistic system*, a structure in which the assumptions of the theory together with its primitive propositions and rules of inference of the underlying logic are listed. Hilbert proposed to make such objects the study of *metamathematics*.

Prior to Hilbert's plan, consistency proofs had been given for axiomatic theories by the construction of *models*. Thus the consistency of geometries is reduced to that of analysis via co-ordinate geometry (Descartes). But the method of models offers no hope for a consistency proof of analysis itself. Hilbert's proposal required that a consistency proof be based *directly* on the meaning of consistency as freedom from contradiction, i.e. no proof in the formal system under consideration terminates in a contradiction. Metamathematics should deal with the formal system by *finitary* methods, i.e. using performable processes in intuitively conceivable objects. The interpretation of the formal theory should play no part in metamathematics—the meanings of the undefined terms of the theory are omitted.

Investigations of the consistency of formal number theory by Gödel forced a clarification of the methods which must be counted as finitary (transfinite induction (Gentzen), functionals of finite type (Gödel, Kreisel), etc.). Hilbert's finitist programme prompted the study of arbitrary formal systems (Post's canonical systems and so on) and the formulation of a viable concept of provability.

The proof-theoretical wing of classical logic was established by Hilbert, Herbrand, and Gentzen. The semantic wing, which is not confined to finitist methods, was established above all by Tarski. It blossomed under his influence into a major branch of mathematical logic—model theory—which has many fruitful applications to algebra. Many of these rest on the *compactness theorem* (Tarski 1930). (For a history of model theory, see Chang (1974) and Vaught (1974).)

The study of formal logic now excludes inductive logic and probable inference; *degree of confirmation* belongs to inductive logic, which is considered as another subject. But it embraces all areas concerned with the 'foundations of mathematics'. As a subject (i.e. institution) it now covers several areas, each with their own distinctive concepts and methods. Sacks' (1972) estimate of the subject was:

> The subject of mathematical logic splits fourfold into: recursive functions, the heart of the subject, proof theory which includes the best theorem in the subject, sets and classes whose romantic appeal far outweigh their mathematical substance, and model theory, whose value is its applicability to, and roots in, algebra..

These areas interpenetrate but form a more or less coherent whole. One ought nowadays to include in the subject certain areas of computer science (automatic theorem-proving, complexity, program specification and verification, logic programming, etc.)—an important growth area (see, for instance, Börger 1986).

The logic at the root of the subject is primarily classical logic, particularly in the applications to set theory, model theory, and computer science. Classical logic sets the benchmarks against which other logics are measured: non-classical logics have their identity by virtue of the classical paradigm. For this reason, Chapter 2 is a brief summary of some elementary facts about it. Intuitionist logic is secondary in the institutional sense that there is virtually no overt use of intuitionist mathematics in undergraduate pure mathematics, and none whatsoever in applied mathematics, but it has fundamental importance for the constructive aspects of the foundations of mathematics. Further, in a certain sense (Chapter 10) Heyting's formal intuitionistic logic is the 'maximal' logic. It is not our concern to argue which of these is the 'correct' logic, for it is by no means clear what constitutes correctness. It is preferable to take them as facts of social life and try to elucidate the connection between them and their relation to other logics, to mathematics, and to the other institutions of knowledge.

1.2 The development of the concept of deductive science

An adequate account of logic cannot avoid the philosophical problems surrounding the applications, particularly in the study of scientific method (e.g. Nagel 1972). Indeed, a critical study of the positivist theory of scientific method supports the case for a multiplicity of logics.

A characteristic tendency of science is codification and unification of the subject domain into a hypothetico-deductive system. This method originated with Aristotle. A record of Aristotle's theory of science and its later developments has been given by Beth (1959, Chapter II). In Aristotle's theory a deductive science is a set Σ of sentences which satisfies the following four conditions.

1. *Reality postulate.* The sentences of Σ are about a particular subject matter, a specific area of real entities (Aristotle 1949, p. 513 §75^{a}38, p. 597 §87^{a}38).

2. *Truth postulate.* Every sentence of Σ is true.

3. *Deductive postulate.* Everything that can be deduced from a subset of Σ is also a member of Σ—Σ is closed under logical consequence (Aristotle 1949, p. 597 §72^{a}25).

4. *Evidence postulate.* This has two parts.
   (a) *Definability.* The sentences of Σ are built from a finite set T of terms, the meaning of which is immediately obvious without further explanation and such that every other term in Σ can be defined from T (Aristotle 1949, p. 597 §87^{a}38–9).
   (b) *Axiomatizability.* The truth of every sentence of Σ can be inferred logically from a certain finite subset Π of Σ whose truth is so obvious that no further proof is needed (Aristotle 1949, p. 507 §71^{b}19, p. 512 §72^{b}5, p. 536 §76^{a}16).

Thus, according to Aristotle, every science has a deductive structure based on self-evident axioms and accepted principles, and has an empirical foundation. Rival non-Aristotelian conceptions of science crystallized by the seventeenth century. One conformed to Aristotle's deductive and evidence postulates; the other (typically Locke) emphasized the reality postulate. The Cartesian tendency eventually moved further away from the Aristotelian model by dropping the evidence postulate. A decisive step in this direction was the development of non-Euclidean geometry. Recent research into logic and the foundations of mathematics have modified or abandoned one or other of the postulates of evidence or deductivity.

A typical defence of the modern 'positivist' view of scientific theories is given by Nagel (1972). Here it is recognized that not all deductive inference is syllogistic. Thus the deductive scaffolding of a scientific theory is replaced by an abstract calculus or 'deductive system'. In his examination of the character and cognitive status of theories, Nagel outlines the tripartite structure of theories as follows:

> (1) an abstract calculus that is the logical skeleton of the explanatory, and that 'implicitly defines' the basic notions of the system,
> (2) a set of rules that in effect assign an empirical content to the abstract calculus by relating it to the concrete materials of observation and experiment, and
> (3) an interpretation or model for the abstract calculus, which supplies flesh for the skeletal structure in terms of more or less familiar conceptual or visualizable materials.
>
> Nagel 1972, p. 90

Nagel has replaced Aristotle's set Σ of sentences by an abstract calculus together with a model. Nagel's condition (2) then covers the reality and truth postulates. The deductive postulate is included in the notion of 'abstract calculus' in (1). The evidence postulate then says something about the structure of the abstract calculus. It is noteworthy that these conditions are dropped in the modern view. All that is now required of the abstract calculus (axiomatic system) is that the relation of logical consequence be a transmitter of truth—self-evidence of axioms is abandoned, as has been explicitly stated by Carnap (see also Cohen and Nagel 1934, ch. 7, §1).

The development of the concept of deductive science

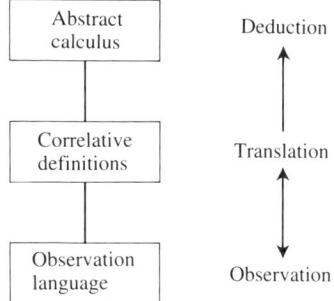

Fig. 1.1. Structure 1 of a scientific theory.

> There is the traditional view of an axiomatic system—current in Euclid's time, and continuing into our own—that requires its axioms to be self-evident (even today common usage tends to attribute this meaning to the word 'axiom'). The modern conception of axiomatic system does not include this requirement; arbitrary sentences may be selected as axioms.
>
> Carnap 1958, p. 171

Some philosophers of science (e.g. Campbell 1920, Reichenbach 1935) conceive the rules which 'assign an empirical content to the abstract calculus' as being effected by means of a dictionary of *correlative definitions* which translate empirical terms, predicates, and propositions of an 'observation language' into theoretical ones. According to this view, an application of the scientific theory to a concrete situation involves (Fig. 1.1):

(1) empirical determination of conditions expressed by an empirical sentence (i.e. in the observation language);

(2) its translation via the correlative definitions into a theoretical sentence (i.e. in the abstract calculus);

(3) deduction of another theoretical sentence using the concept of logical consequence in the calculus;

(4) reverse translation of the derived theoretical sentence via the correlative definitions into a corresponding (true) sentence of the observation language.

The dictionary of correlative definitions is thus a transmitter of truth to and from the truth-preserving operations of the abstract calculus.

1.3 Criticisms of the positivist view

The inadequacies of the treatment of the sciences as applied axiomatic theories have been delineated by many philosophers and scientists (e.g. Duhem 1954, Toulmin 1961, Kuhn 1962). For instance, such a treatment of a science concentrates only on theories that have achieved such a high degree of development as to be ready for axiomatization. Many of the sound reasons which have led to such maturity are ignored. Moreover, this logical treatment of scientific theories and their components regards them as frozen in time.

Criticizing the positivist view of science is a popular pastime. But the point of doing so here is that such criticism points to the need for different kinds of logics—modal logics at the metalevel, non-classical logics, possibly, in the internal structure of a scientific theory, and a three-valued logic at the interface of theory and observation.

1.3.1 *On reductionism*

Attempts to reduce theoretical concepts of science to observational terms have consistently failed to produce the 'detailed' reduction procedures for any of the most important concepts actually used—though there is no lack of proposals 'in principle'—and have been unable to provide a convincing theory of the nature and status of scientific theories.

The positivist view usually seems to assume that the abstract calculus is the classical (two-valued) logic. Further, by being an abstraction, it ignores the circumstances and conditions of scientific investigation. Science is an activity undertaken within a framework of scientific *institutions* (e.g. Bloor 1983) which sanction particular theoretical and practical standards governing the gathering of 'facts' and theorizing. The making of observations is already theory-laden. For example, the 'objects' considered in physics are not the material bodies of physical experience as they present themselves to our senses, though it is useful for many reasons to make the *identification*. But the identification is limited by purpose and extent. Further, to apply physics, the *significant* data must be extracted from the totality of an empirical situation—e.g. the velocity of a body at a certain place and time, rather than smell or temperature. Theoretical notions are the instruments of this selection.

General criticisms of the positivist conception of scientific theory, particularly relating to the methodological problems of the social sciences, have been formulated by Keat and Urry (1975) and by Hindess (1977), for instance, though Nagel tries to show that social-scientific explanation need be no different from explanation in the natural sciences.

The highly schematized relation of truth to theory in the positivist account of scientific theory is particularly vulnerable to criticism (Quine 1961, p. 42):

A dogma of reductionism survives in the supposition that each statement, taken in isolation from its fellows can admit of confirmation or infirmation at all. My countersuggestion, issuing from Carnap's doctrine of the physical world in the Aufbau (Carnap 1928) is that our statements about the external world face the tribunal of sense experience not individually but only as a corporate body.

Quine requires a more flexible concept of scientific theories in which perturbations of the structure at one level can be compensated at another level:

> ...total science is like a field of force whose boundary conditions are experience. A conflict with experience at the periphery occasions readjustments in the interior of the field. Truth-values have to be redistributed over some of our statements... But the total field is so undetermined by its boundary conditions, experience, that there is much latitude of choice as to what statements to re-evaluate in the light of any single contrary experience... even a statement very close to the periphery can be held true in the face of recalcitrant experience by pleading hallucination or amending certain statements of the kind called logical laws.

There are two points of special interest here: Quine allows a multiplicity of logics, but still makes the classical presupposition that statements are truth-definite (i.e. have a definite truth-value) and are either *true* or *false*. Note also that Carnap (1949, p. 180) allows logics with 'extra-logical' rules of inference, embodying physical laws say.

1.3.2 *Truth-values in science*

The preceding remarks of Quine raise the question of how truth is ascertained in scientific theories. Carnap (1949, p. 45) advocated the use of a 'co-ordinate language' as observation language—a language which denotes the objects belonging to its domain by positional designations:

> 'In such and such a place is a horse' means such and such a space–time domain has such and such a property.

Co-ordinate languages are highly theoretical, for the use of numerical co-ordinates requires a certain level of mathematical analysis—not trivial for the real continuum—and the sophisticated practice of setting a standard co-ordinate system and the determination of locations relative to it. The whole concept of co-ordinate language is too philosophically loaded to serve as a starting point for the explanation of the relation between theory and experiment. Furthermore, even if we were to grant the validity of the co-ordinate language enterprise, there is a serious problem of determining the truth-value of such sentences as: 'At the point k_1, k_2, k_3 at time k_4 the temperature is k_5' (Carnap 1949, p. 150). How could one know that the temperature at that space–time location was k_5 (exactly)? The problem is one of extracting the 'numerical' quintuple $(k_1, k_2, k_3, k_4, k_5)$ from the non-numerical 'action of measuring.' Co-ordinates are determined via measurements, but are not identical with the action of measuring. It is not a question of accuracy. Replacing k by $k + 10^{-5}$ simply removes the problem one stage further, but does not abolish the theory-laden problem of determining a sequence of real numbers by measurement.

The statement that k_1, k_2, k_3 at time k_4 has temperature k_5 is judged to be true or false by *decisions* made according to socially accepted norms of spatial, temporal and thermal measurements. The 'assumption' that every action of measurement or, more generally, of observation will enable the decision between true or false to be made is surely a heavy philosophical commitment. Empirical verification of this statement is an operation followed by the observation of a property. It is usually assumed that there are two such 'witnesses' P and N, say. The presence of P signifies the truth of the statement; the presence of N indicates its falsehood. But sometimes the methods and norms of measurement and observation fail to produce such a witness, so that the decision between true and false cannot be given. This is the *neutral* case, which we might later class as true or false for certain purposes.

Let us assume the structure of scientific theories illustrated in Fig. 1.1 and study further the connection between the observation language and experience. The problem of determining the truth of observation sentences cannot be avoided by taking the observation language to be a *name language* (Carnap 1949, pp. 12, 189) in which *objects* are designated by proper names. This device again imports a heavy theoretical load into the foundations of scientific theory which ought to be made explicit (Quine 1961):

> ... I see all objects as theoretical, ... Physical objects are conceptually imported into the situation as convenient intermediaries—not by definition in terms of experience but simply as irreducible posits comparable epistemologically to the gods of Homer.

Further, there is the inescapable problem of vagueness in naming and predicating. Any vagueness in the truth of 'A has the property P' comes not from there 'really' being an object A' closely resembling A and possessing a property P' much like P, but from the inadequacy in some way of the norms of observation, in the particular case, to yield a 'decision'. Truth is not something pre-existing, ascertained by passive reflection, but extracted by active intervention and decision-making. Sometimes the criteria for deciding truth or falsehood of a given empirical sentence S fail to provide an answer. S then has no truth-value. An abbreviated way of acknowledging this condition is to say that S is *neutral*. But saying this could be misconstrued as asserting that even before the observation S was neutral and the observation merely discovered this 'fact'. On the contrary, we mean that the neutrality of S was *created* by the decisions incurred in the observation. If an observation sentence S of a scientific theory is neutral, the correlative

Criticisms of the positivist view

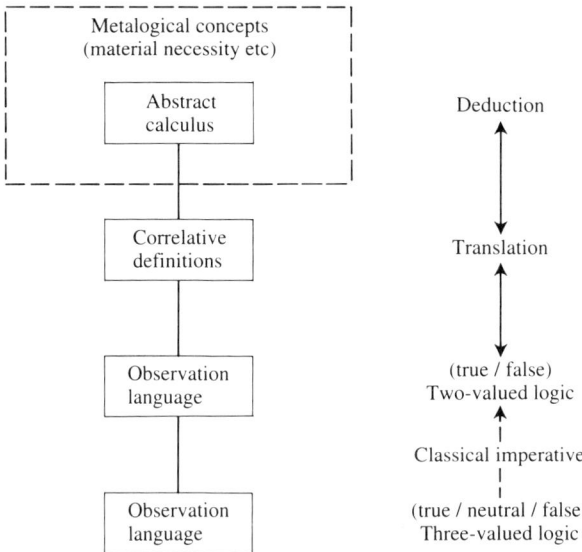

Fig. 1.2. Structure 2 of a scientific theory.

definitions cannot translate S into a true theoretical sentence for processing by the deductive machinery. Thus, the theory is only applicable to *idealized* states of affairs described by sentences with a definite truth-value. To pass from the raw material of experience to the clear sharp truth-definite empirical propositions of science is a process of *idealization* which can justly be called the *classical imperative*. It involves sharpening the truth-criteria so that neutral empirical propositions become truth-definite. Between observation language and experience there must be interposed a stage at which the process of observation and experiment assigns to observation sentences the judgements *true/false/neutral*. The structure of a scientific theory now looks like the building shown in Fig. 1.1 but with a basement corresponding to the three-valued observation language (Fig. 1.2). The classical imperative sharpens these decisions to the classical values *true/false* which are then transmitted to the theory as in Fig. 1.1. Thus all the propositions of a science—the idealized classically evaluated empirical sentences of the observation language and the deductive superstructure—are theoretical. Their link with experience is via the process of *idealization* which identifies the inexact and practical with the exact and theoretical for certain purposes and within certain limits and constraints.

1.3.3 Metalogical constraints

The positivist account of science takes a very unhistorical view—it does not tell how a science comes be formalized in a particular way, nor what cultural and metaphysical assumptions motivate the particular formalization. Thus logicians such as Duns Scotus in medieval times and J. S. Mill in the modern era have urged that a principle of uniformity in nature must function axiomatically in science. Uniformity cannot be verified by any finite number of experiments—it is an a priori assumption which is very difficult to do without.

In the realm of mathematical-physical theories certain propositions occur as explicitly stated or tacitly assumed qualitative restrictions on the class of models to be used. For instance, it is customary to formulate a physical law by a set of ordinary or partial differential equations (Newton's equations of motions, Maxwell's equations, etc.), on the assumption that the differential calculus is the 'correct' mathematical tool. One could externalize this procedural decision by expressing it as an axiom: 'the universe is a differentiable manifold'. This is a *materially necessary* proposition—it is not forced on one by logical considerations only; no amount of empirical evidence can verify or refute it. It performs a regulatory function in theory construction which can be justified on pragmatic grounds and by a priori conceptions of space and time. Similarly, the 'dogma of structural stability' (Abraham 1967, Thom 1967), which requires that models of a physical theory shall have certain qualitative features, finds its justification in certain metaphysical conceptions of the nature of physical enquiry. Dirac (1939, 1963) pointed out that there could be aesthetic criteria in the selection of an axiomatic foundation of physics:

> What makes the theory of relativity so acceptable to physicists ... is its great *mathematical beauty* ...
>
> The trend of mathematics and physics towards unification provides the physicist with a powerful new method of research into the foundations of his subject.... The method is to begin by choosing that branch of mathematics which one thinks will form the basis of the new theory. One should be influenced very much in this choice by considerations of mathematical beauty.

1.4 The science of logics

The remarks in the preceding sections indicate that the philosophy of science allow more than one logic. From studies of the foundations of mathematics there have emerged two logics with totally different philosophical bases. Order can be brought to this multiplicity of logics by treating logics as a deductive science (§1.2). Accordingly, logics will be defined in Chapter 3 as certain determinate kinds of structure which embody a notion of *logical consequence* following a well-known definition of Tarski (1956).

The science of logics

There is considerable systematic advantage in such an axiomatic organization, much as group theory as a mathematical institution is related to the axioms for groups. Group theory has its specialist branches—finite groups, infinite groups, topological groups, etc. One could also include various modifications of the group concept (e.g. semi-groups, loops, etc.) under the heading of group theory because, as a sociological fact, a specialist in semi-groups would surely be competent to teach group theory to undergraduates and might be expected to contribute to research in that area. Likewise, the various areas of mathematical logic can be considered as branches of the theory and application of deductive systems.

To use a definition of logics in this way, however, does not enable one to define exactly the province of logic. The logistic thesis generated an unresolved debate over which a priori elements—geometry, set theory, mathematical induction, for instance—should be called 'logical' (Lakatoś 1962). Some of these elements could also be classified as metalogical constraints.

Frege espoused a Kantian view of Euclidean geometry, that it is synthetic and a priori, but originally believed that the arithmetic of the natural and real numbers could be reduced to logic (Körner 1981). He failed to complete the task and finally concluded that:

> arithmetic and geometry have sprung from the same ground, namely the geometrical... The activity of counting, which has its psychological origin in a requirement of practical life, has misled the scholars...
>
> Frege 1969, p. 297

Zermelo (1930), in his axiomatization of cumulative type structures in set theory, counted the axiom of choice as a 'general principle of logic' as 'it has a different character from the others'. It seems that any grandiose task for logic can succeed only by the intervention of a priori elements via procedural and regulative decisions of organizing the 'logic'. These elements—e.g. the axiom of infinity in type theory, the axiom of choice in set theory—cannot always easily be counted as 'logical'.

Peano included the principle of mathematical induction as one of his five postulates of arithmetic. Many mathematicians agree with Peano in regarding this principle as just one of the postulates characterizing a particular mathematical discipline and as not being fundamentally different from the other postulates of arithmetic or other branches of mathematics. Poincaré maintained that mathematical induction is synthetic and a priori, i.e. neither reducible to a principle of logic or having a purely logical proof nor known independently of experience or observation. Thus mathematical induction has a special role in the constitution of mathematical reasoning: it allows the deduction from premisses of genuinely new results, something which is supposed not to be possible by logic alone. This doctrine was followed by intuitionism, which again treats mathematical induction as synthetic and a priori in the sense of Kant. Most constructivists admit some form of induction as a principle of constructive reasoning.

From the point of view espoused in this book, illuminating (or not) as such discussions of the domain of the purely logical might be, the relativity and flexibility of the chosen definition of *logics* and the absence of a precise definition of 'logical' considerably diminishes the urgency of deciding on an absolute distinction between the logical and the non-logical.

1.5 Logics

The subject of this book, then, is the science of logics. It is not a complete survey of all (existing) logics but a discussion of some of the principles of construction and classification of them together with hints, frequently unorthodox, on the non-formal-logical aspects—philosophical, practical, representational, etc.—particularly on the movements of thought which influence and shape the various logics.

Because of the supreme importance of classical logic—both theoretically and as a cultural fact—Chapter 2 is devoted to a brief survey of some elementary facts about it. The methods and results are typical, and in many respects they set a standard for comparison and contrast. In this book we shall regard higher-order logics—set theory, type theory, theory of choice sequences, etc.—as applications of 'first-order' logics and confine attention to the latter.

Chapter 3 defines the terms of the general discussion of logics. The notion of a *logic* is defined following Tarski in terms of *logical consequence*. This enables the notion of a *connective* to be defined and hence, in Chapter 4, provides a basis for an axiomatic classification of logics.

There are several reasons why the notion of *order* is of particular importance in analysing the structure of logics. Therefore Chapter 5 gives a brief elementary exposition of such topics in lattice theory as are necessary for this task.

There are two main classes of logics based upon a genetic classification as opposed to the axiomatic method founded in Chapter 4. These are *formal systems* and *semantic systems*. The two concepts are independent and can be developed separately.

In formal systems the centre of interest is the transformation rules (rules of inference) themselves. Derivation from premisses gives the notion of logical consequence. Effective logic (Chapter 10) falls under this heading, as do Smullyan's *elementary formal systems* (Chapter 2, §2.3.7) and numerous axiomatizations of classical and intuitionist logic which are used as examples at various places in this book.

Semantic systems are based on the treatment of truth via truth-values as objects and a definition of truth. The preceding discussion of truth-values in science (§1.3.2) provides the motive for a thorough examination of many-valued logic. The main problem here is how to pass from truth-values to a logic. The most general method, defined in Chapter 6, is to organize the truth-values themselves as a lattice, convert the lattice to a logic of truth-values in the abstract sense of Chapter 3 and then construct the logic by a systematic procedure from the logic of truth-values.

The two classes overlap in a certain sense because of completeness theorems, but it is quite possible to develop semantics without recourse to the notions of proof theory (e.g. Kreisel and Krivine 1967)—logical consequence can be defined, as in Chapter 6, in purely semantical terms.

In §1.3.2 it was indicated that three truth-values—*true/neutral/false*—stood between theory and observation. How can these be made into a logic? In accordance with the method of Chapter 6, the three truth-values are first made into a lattice. There are various ways of doing this: the one chosen here gives a *quasi-Boolean algebra*. Quasi-Boolean algebras bear the same relation to three-valued logic as Boolean algebras bear to classical two-valued logic; they are examined in Chapter 7, and the corresponding three-valued logic is constructed in Chapter 8.

The remaining three chapters are concerned with the extra-logical issues that enter into theory construction as was indicated in §1.3.3. The treatment in Chapter 9 of relevance logic, which avoids the postulational approach, also concerns a certain four-valued logic which is closely connected with the quasi-Boolean algebras of Chapter 7 and the three-valued logic of Chapter 8.

The problem broached in Chapter 10 is how to regularize the discourse about particular sciences considered as deductive systems. As a result of the definition of *logics* in Chapter 3, it is a question of the systematic construction of the logic of a given system (i.e. the logic of the given logic) without introducing metaphysical assumptions. This is achieved by pursuing an idea due to Lorenzen (1955). The logical structure thus obtained is formally identical with Lorenzen's effective logic and Heyting's formalization of intuitionistic logic. This chapter reveals a peculiarity of the science of logics—it includes its own metalevel judgements in the sense that once discourse about a particular deductive science (i.e. a logic) has been regularized, it becomes a logic itself.

Modal logic and the logic of material necessity are considered in Chapter 11. Our development of it follows ideas of Quine and Körner. It avoids the postulational approach—so running counter to conventional other-worldly wisdom—by extending any given logic so as to incorporate some of its metalogical concepts such as provability and so on.

2

Classical logic

2.1 Introduction

For the established naive realist view permeating mathematics and science the *classical two-valued logic* is *the* logic, which is naturally and unquestionably supreme. Other conceptions of logic or mathematics are quaintly ideological; *other* people's ideas are distinctly metaphysical. Nevertheless, the usually unspoken classical view is a definite philosophical stance whose metaphysical presuppositions are the legacy of history. Mathematical and logical truths are a reflection of an objective external structure—they are seen as analogous to a universe of clear-cut objects and relations between them essentially set apart from the knowing subject, the community, and institutions of mathematics and science. Truths, like supermarket commodities, are discovered, not invented. It is assumed that each well-formed formula of logic has one of the truth-values *truth* or *falsehood* independently of proof and the way in which it is determined (hence the Aristotelian principle of the excluded middle), i.e. it is truth-definite. This realist view was explicitly stated by the mathematician G. H. Hardy:

> I believe that mathematical reality lies outside us, that our function is to discover or *observe* it, and that the theorems which we prove, and which we describe grandiloquently as our 'creations', are simply our notes of our observations.
> Hardy 1969, pp. 123–4

Wittgenstein (1956, I, §8) characterized this view of logic as a 'kind of ultraphysics'.

There are several ways of criticizing this classical view, (for instance, Pivčević (1974) and various views of Kantian descent (see also Chapter 10)). But it is a remarkable fact that, despite these criticisms, no one has successfully maintained that there is a contradiction in classical logic and no bridge has accidently fallen down because of correct use of classical mathematics in its design.

Classical logic has two kinds of limit. First, its use presupposes that the relevant relations and individuals of the theoretical discourse are exact. Relations and individuals of empirical discourse are, however, usually inexact and so cannot be included in the theoretical domain. A correspondence with theoretical entities is achieved by idealization (Chapter 1, §1.3.2). Second, there is some difficulty in maintaining (with Frege) the universality of the predicate logic (i.e. *everything* can be equivalently expressed by sentences of the predicate logic). It is certainly not clear what a proof of this universality would look like. Besides, there are probably some facts about the logic itself which cannot be expressed within the logic. Whilst it is an empirical fact that the predicate logic has proved to be extensively applicable, the analysis of elementary sentences by predicate logic is not always very helpful and, indeed, may introduce confusions (Thiel 1981). The success of an application of predicate logic in empirical discourse is dependent on the process of idealization (Chapter 1, §1.3.2) which replaces empirical inexactness with exact structures.

Although there is a constructive tendency within the classical tradition based on the theory of recursive functions for constructions and the use of classical reasoning for proofs, the main opponents of classical hegemony are constructive groupings based on various philosophies which offer a constructive theory of proof itself. The chief contender is intuitionism, first expounded by L. E. J. Brouwer and popularly known for its denial of the principle of the excluded middle. The first attempt at the formalization of intuitionistic mathematics and logic was made in 1928 by Heyting (unpublished) in response to a challenge by Mannoury. This was the basis of the well-known 1930 publication. We shall refer to this formalization simply as 'intuitionistic logic'. Brouwerian solipsism is not the only philosophical basis for this formal logic, as will be seen later (Chapter 10).

Classical formal logic was brought to fruition by Frege in his *Begriffschrift* (Frege 1879). The propositional calculus appeared there in its modern form for the first time. The modern use of quantifiers was defined and there was the logical analysis of *proof* by mathematical induction in the form of the notion of *hereditary property* or *hereditary class*. The use of quantifiers was suggested independently of Frege by Mitchell, to whom the idea is credited by Peirce (1885).

Frege's predicate calculus provided a structure for the *propositions* which were unanalysed in Boole's algebras. Frege frequently derided Boole's work as 'abstract' logic in which propositions are reduced to mere truth-values. The predicate calculus provided a *language* in addition to a calculus. According to Frege, this language is *universal*: the ontological furniture of the universe divides exclusively into *objects* (but not necessarily of the material sort) and *functions*. For Boole the universe of discourse is determined by a temporary local convention, but Frege's universe was fixed and consisted of all that there is so that nothing can be said outside the system. Thus Frege never considered questions *about* the system such as consistency or completeness. The departure from universality was later accomplished by Löwenheim and Skølem whose first-order predicate logic with identity was based on set theory and shifted attention away from *proof* to *validity* in arbitrary domains (van Heijenoort 1967a). Löwenheim (1915) and Skølem (1920) considered the validity of formulas in *arbitrary* domains and proved that if a first-order theory has a model then it has a denumerable model.

Because of its dominant position classical logic has set the standards of formulation and structure of development for other logics. Therefore in the following sections we shall give a cursory sketch of the basic steps in the development of classical propositional and predicate logic which will be used in later sections.

We shall not be specifically concerned with the predicate logic with equality in this chapter, as one way of dealing with equality is to count it as a non-logical binary relation and add axioms requiring it to be an equivalence relation with substitutivity conditions as in Robinson (1963, §2.1). However, it will be used in several places later.

There are two traditional ways of developing formal logic. In the first method a propositional or predicate language is used to analyse ordinary language. It is assumed that substitution of statements or predicates of ordinary language for the propositional symbols and predicate letters respectively of the formal language gives true or false sentences. Heuristically valuable as this approach might be as a first encounter with logic, anyone who has experienced it will recall on the one hand the doubts generated by the lack of clarity of the ordinary language, the vagueness of its predicates and range of quantification, etc., and on the other hand the clarity achieved by illustrating the logic with examples in arithmetic.

The second approach proceeds by the valuation of logical formulas in set-theoretical structures. This enables rigorous and powerful development of the subject to proceed unhindered by worries over the range of quantification or whether the instances of the formulas 'really' express what is meant by the intended ordinary language interpretation. By defining *truth* in structures the question of correctness of ordinary language interpretation—which involves the problem of vagueness (Chapter 1, §1.3.2)—is sharpened to the problem of whether the meaning of ordinary language sentences can be adequately represented by a structure.

The treatment of classical logic given below is a cursory sketch of material which, for the most part, is already contained in first courses in predicate logic. Our exposition, which owes much to Kreisel and Krivine (1967), is designed to show the deep dependence of the predicate logic (§2.3) on the propositional logic (§2.2) and to allow generalization to some many-valued logics in Chapter 6.

We begin with a development of the semantics of the propositional logic, the central idea being the notion of logical consequence in Definition 2.4. Some basic notions of logic programming (§2.2.5) and compactness (§2.2.6) can be formulated in this context. A *very* quick sketch of appropriate syntactical concepts follows in §2.2.7. The idea of *resolution*, the fundamental concept of modern automated theorem-proving, can also be framed at the propositional logic level.

2.2 Propositional logic

A typical development of a classical propositional logic starts with the definition of the class of *formulas*, and then establishes the proof theory (around the ideas of *rule of inference, proof* and *theorem*) and the semantics (around the ideas of *validity* and *tautology*). The two branches then meet in a *completeness theorem* which establishes the identity of *theorem* and *tautology* ideas. The development is theorem–tautology oriented. There is no unique way of defining propositional logic. For alternatives and history see Church (1956) and numerous excellent treatises on the subject.

2.2.1 Formation rules

The set of formulas is defined by a uniform inductive definition. At various places in this text the set of propositional letters and the set of junctors will be changed, but a fixed set G of separators—square brackets, [,], and a comma , —will always be used. Thus the primitive symbols of the language L are the pairwise disjoint alphabets P, J, G where the following hold.

(i) P is a set of objects

$$p_1, p_2, p_3, \ldots$$

called *proposition letters*. (When the predicate calculus is considered the proposition letters themselves will be structured entities.)

(ii) J is the list of symbols

$$t, f, \neg, \wedge, \vee, \supset$$

of *junctors* or *connectives*: these are the *logical symbols*.
  (a) t, f are zero-place junctors, i.e. propositional constants, t is *truth* and f is *falsehood, contradiction,* or *absurdity*.
  (b) $\neg$ is a one-place junctor, the sign for *negation*. $\wedge, \vee, \supset$ are two-place junctors, the signs for *conjunction, disjunction* and *implication* respectively. It is customary to read informally '$\neg A$' as 'not A', '$A \wedge B$' as 'A and B', '$A \vee B$' as 'A or B' and '$A \supset B$' as 'A implies B' or 'if A then B' (if one takes this custom too seriously, then one is committed in classical logic to maintaining of any two propositions that one implies the other).

The class of formulas is a certain subset of L^* (the set of words on L) defined as follows.

DEFINITION 2.1.

(i) The class $\mathbb{F}(J, P)$ of *formulas on P* is defined inductively by the following rules.

(F1) t, f and all proposition letters are in $\mathbb{F}(J, P)$.
(F2) If $A, B \in \mathbb{F}(J, P)$, then $\neg A, [A \wedge B], [A \vee B], [A \supset B] \in \mathbb{F}(J, P)$.
(F3) A word on the alphabet L is a formula iff it is required to be so by (F1) and (F2).

Where the sets P and J are fixed, reference to them will be omitted and we shall simply write $\mathbb{F}_{prop}$ instead of $\mathbb{F}(J, P)$.

(ii) A *positive literal* is a proposition letter. A *negative literal* is the negation of a proposition letter. A *literal* is a positive or a negative literal.
(iii) Each formula A is a word on L: a subword B of A is a *subformula* if B is also a formula.
(iv) if W is a set of words of an alphabet which does not include $J + G$, then the set $\Lambda(W)$ of *logical combinations* of W is defined by analogy with $\mathbb{F}(J, P)$.
(F'1) $\{t, f\} + W \subseteq \Lambda(W)$.
(F'2) If $A, B \in \Lambda(W)$ then $\neg A, [A \wedge B], [A \vee B], [A \supset B] \in \Lambda(W)$.
(F'3) A word is a logical combination of W iff it is required to be so by (F'1) and (F'2).
(v) The *length* $|A|$ of a formula A is the number of occurrences of logical symbols in A.

It follows immediately from this definition that $\Lambda(P) = \mathbb{F}(J, P)$, so that formulas are logical combinations of proposition letters. Further, if $W \subseteq \mathbb{F}(J, P)$ then $\Lambda(W) \subseteq \mathbb{F}(J, P)$, i.e. the logical combinations of any set of formulas are again formulas. Of course, Λ depends upon the set J of symbols which have been selected to be 'logical'. The length of a formula is a measure of its logical complexity relative to J. Thus, if $|A| = 0$ then $A \in P$ and if $|A| > 0$ then A is a logical combination of formulas of shorter length.

The role of junctors in the construction of formulas is brought out by making functions of them. Thus, associated with the junctor $\neg$ is a one-place function $\neg: L^* \to L^*$ (denoted by the junctor itself) defined by

$$\neg(A) := \neg A \qquad (A \in L^*).$$

Similarly, associated with the junctors $\vee, \wedge,$ and $\supset$ are binary functions $\vee, \wedge, \supset: L^* \to L^*$ (*junctor operations*) defined respectively by

$$\vee(U, V) = [U \vee V] \qquad (U, V \in L^*)$$
$$\wedge(U, V) = [U \wedge V] \qquad (U, V \in L^*)$$
$$\supset(U, V) = [U \supset V] \qquad (U, V \in L^*).$$

$\mathbb{F}_{prop}$ is the smallest subclass of L^* which includes P and is closed under the junctor operations $\neg, \vee, \wedge,$ and $\supset$. Further, the formulas equipped with the junctor operations, i.e. the structure $\langle \mathbb{F}(J, P), \neg, \vee, \wedge, \supset \rangle$, form an algebra called the *algebra of formulas*. There should be no confusion in using $\mathbb{F}(J, P)$ for both the *set* and the *algebra* of formulas.

Propositional logic

It follows from the formation rules that the operations of substitution for proposition letters commute with the junctor operations. Thus, if $W_1, \ldots, W_n \in L^*$ and $q_1, \ldots, q_n$ are distinct proposition letters and $\sigma = \sigma(W_1, \ldots, W_n)$ is the operation of simultaneously substituting W_1 for $q_1, \ldots, W_n$ for q_n (see Appendix, §W), then for any junctor operation c

$$c(U, V, \ldots)\sigma = c(U\sigma, V\sigma, \ldots) \qquad (U, V \in L^*).$$

Clearly, $\mathbb{F}_{\text{prop}}$ is closed under substitution of formulas for proposition letters. If $W_1, \ldots, W_n$ (above) and A are formulas, then

$$A\sigma \in \mathbb{F}_{\text{prop}}.$$

2.2.2 Semantical concepts

The following definition of *truth* and *logical consequence* gives the connection between the formal language and the truth-values. It embodies the idea that the proposition letters stand for truth-definite propositions, i.e. propositions which have a determinate truth-value.

DEFINITION 2.2.

(i) The *algebra of truth-values* corresponding to L is the algebra $\mathbb{T}v = \langle \{0, 1\}, ', \cap, \cup, \supset, 0, 1 \rangle$ of type $\langle 1, 2, 2, 2, 0, 0 \rangle$ where 0, 1 are the truth-values *falsehood* and *truth* respectively and the operations are defined by Table 2.1. $\mathbb{T}v$ is a *Boolean algebra* (see Chapter 4). We shall also use $\mathbb{T}v$ to denote the set $\{0, 1\}$.

Table 2.1 Truth tables

t	t'		s	t	$s \cup t$	$s \cap t$	$s \supset t$
0	1		0	0	0	0	1
1	0		0	1	1	0	1
			1	0	1	0	0
			1	1	1	1	1

(ii) An ordering on the truth-values is defined by $0 \leqslant 1$. For $X \subseteq \mathbb{T}v$, $\text{lub}(X) =_{\text{def}} 0$ if $X = \emptyset$ and is the greatest element of X in this ordering otherwise; $\text{glb}(X) =_{\text{def}} 1$ if $X = \emptyset$ and is the least element of X in this ordering otherwise. (The significance of this notation will emerge in Chapter 5, §5.6.)

The first step in developing the semantics is to define the truth-values of formulas.

DEFINITION 2.3. A *valuation* (in $\mathbb{T}v$) is a function which assigns truth-values to proposition letters. $\mathbb{V}$ is the set of all valuations. A valuation v can be extended to an assignment of truth-values in $\mathbb{T}v$ to all formulas by the following rules:

(v1) If $A \in P$ then $v(A)$ has already been defined.

(v2) $$v(\neg A) =_{\text{def}} v(A)'$$
$$v((A \vee B)) =_{\text{def}} v(A) \cup v(B)$$
$$v((A \wedge B)) =_{\text{def}} v(A) \cap v(B)$$
$$v((A \supset B)) =_{\text{def}} v(A) \supset v(B).$$

(The valuations are exactly the homomorphisms of the algebra of formulas to the algebra of truth-values). v extends to sets of formulas in the obvious way:

$$v(\Delta) =_{\text{def}} \{v(A);\ A \in \Delta\} \qquad (v \in \mathbb{V},\ \Delta \subseteq \mathbb{F}_{\text{prop}}).$$

It is sometimes expedient to consider a valuation $v \in \mathbb{V}$ as the infinite sequence $v(p_1), v(p_2), \ldots, v(p_n), \ldots$ of its values (cf. Theorem 2.2).

DEFINITION 2.4.

(i) A valuation v is a *model* of a formula A if $v(I) = 1$; it is a model of a set Δ if it is a model of every member of Δ. We write $v \vDash A$ when v is a model of A and similarly for sets of formulas. Δ^* denotes the set of models of the set Δ of formulas. Similarly, for a class K of valuations, K^* is the class of formulas A such that $v \vDash A$ for all v in K. (Thus, if Δ is a set of formulas then Δ^* is a set of valuations, and if K is a set of valuations then K^* is a set of formulas.)

(ii) A formula A is said to be *valid* or to be a *tautology* if $\{A\}^* = \mathbb{V}$, to be *satisfiable* if it has a model, and to be a *contradiction* if it has no model. A set Δ of formulas is said to be *inconsistent* if $\Delta^* = \emptyset$; otherwise Δ is *consistent*.

(iii) A formula A is a *logical consequence* of the set Δ of formulas (written $\Delta \to A$, or $\Delta \to_{\text{prop}} A$ when proposition logic is to be stressed) if every model of Δ is a model of A, i.e. $\Delta^* \subseteq \{A\}^*$. An equivalent way of saying this, which facilitates a generalization to many-valued logics in Chapter 6, is

$$\text{glb}(v(\Delta)) \leqslant v(A) \qquad (v \in \mathbb{V}). \qquad (2.1)$$

$(v(\Delta)) \subseteq \mathbb{T}v$ so $\text{glb}(\Delta)$ is defined as in Definition 2.2(b).) If $\Delta = \{B_1, \ldots, B_k\}$ we shall sometimes write $B_1, \ldots, B_k \to A$ instead of $\Delta \to A$.

(iv) Two formulas A, B are *logically equivalent* (written $A \leftrightarrow B$) if each is a logical consequence of the other. A and B are *logically equivalent relative to the set Δ of formulas* (written $A \leftrightarrow_\Delta B$) if $\Delta, A \to B$ and $\Delta, B \to A$.

Propositional logic

Some philosophers, perhaps influenced by the Wittgenstein of the *Tractatus* (Wittgenstein 1922, 4.461)

> Propositions show what they say: tautologies and contradictions show that they say nothing!.

have urged that tautologies and mathematical truths are vacuous. There is, of course, a sense in which this is true. But the notion of truth cannot easily be divorced from the way that it is ascertained (Hintikka 1973). Thus the general problem of deciding of any given formula whether or not it is satisfiable is one of considerable computational complexity (Börger 1989, Chapter F, Part III): to possess a tautology is one thing, but to *know* that one has a tautology is quite another.

The following results are easy, but important, consequences of Definition 2.4. They are recorded here for later use.

EXAMPLE 2.1.

(i) For a collection Δ of formulas,
$$\Delta^* = \prod \{k^*; k \in \Delta\}.$$
Δ^{**} is the set of logical consequences of Δ, $\varnothing^{**}$ is the set of tautologies, and $f^{**} = \mathbb{F}_{\text{prop}}$.

(ii)
$$A \to A$$
$$\Gamma \to A \quad \Rightarrow \quad \Gamma, \Delta \to A.$$

If $\Delta \to A$ and $\Gamma \to B$ for all $B \in \Delta$, then $\Gamma \to A$. Hence if $\Delta, A \to B$ and $\Delta, B \to C$, then $\Delta, A \to C$.

(iii)
$$\Delta, A \wedge B \to A$$
$$\Delta, A \wedge B \to B$$
$$\Delta, A \to A \vee B$$
$$\Delta, B \to A \vee B$$
$$\Delta, A \to C \quad \& \quad \Delta, B \to C \quad \Rightarrow \quad \Delta, A \vee B \to C$$
$$\Delta, B \to C \quad \Leftrightarrow \quad \Delta \to B \supset C$$
$$\Delta \to A \quad \& \quad \Delta \to B \quad \Rightarrow \quad \Delta \to A \wedge B.$$

From the latter we have $B, B \supset C \to C$.

(iv) $\quad\quad \Delta, A \to B \quad \Leftrightarrow \quad \Delta, \neg B \to \neg A.$

(v) $\quad\quad \Delta, B \to A \quad \& \quad \Delta, \neg B \to A \quad \Rightarrow \quad \Delta \to A.$

(vi) $\Delta \to A$ iff $\Delta + \{\neg A\}$ is inconsistent.

(vii) For any set Δ of formulas, if $A \leftrightarrow_\Delta B$ and $C \leftrightarrow_\Delta D$ then $A \vee C \leftrightarrow_\Delta B \vee D$, $A \wedge C \leftrightarrow_\Delta B \wedge D$, $A \supset C \leftrightarrow_\Delta B \supset D$, and $\neg A \leftrightarrow_\Delta \neg B$. This shows that logical equivalence (see Chapter 3) is a congruence on the algebra of formulas.

(viii)
$$[A \wedge B] \leftrightarrow [B \wedge A] \qquad [A \vee B] \leftrightarrow [B \vee A]$$
$$[A \vee [B \vee C]] \leftrightarrow [[A \vee B] \vee C] \qquad [A \wedge [B \wedge C]] \leftrightarrow [[A \wedge B] \wedge C]$$
$$[A \wedge B] \vee A \leftrightarrow A \qquad [A \vee B] \wedge A \leftrightarrow A$$
$$[A \wedge [B \wedge C]] \leftrightarrow [[A \wedge B] \wedge [A \wedge C]]$$
$$[A \vee [B \wedge C]] \leftrightarrow [[A \vee B] \wedge [A \vee C]]$$
$$\neg \neg A \leftrightarrow A$$
$$\neg A \wedge \neg B \leftrightarrow \neg [A \vee B] \qquad \neg A \vee \neg B \leftrightarrow \neg [A \wedge B]$$
$$A \wedge t \leftrightarrow A \qquad A \wedge f \leftrightarrow f \qquad A \vee f \leftrightarrow A \qquad A \vee t \leftrightarrow t$$
$$A \wedge \neg A \leftrightarrow f \qquad A \vee \neg A \leftrightarrow t.$$

These equivalences show that $\mathbb{F}(J, P)$ reduced by logical equivalence is Boolean algebra (Chapter 5, §5.10.6).

(ix)
$$A \supset B \leftrightarrow \neg A \vee B \qquad A \supset f \leftrightarrow \neg A$$
$$[A_1 \supset [A_2 \supset \cdots [[A_n \supset B] \cdots]] \leftrightarrow \neg A_1 \vee \cdots \vee \neg A_n \vee B$$
$$\leftrightarrow [A_1 \wedge \cdots \wedge A_n] \supset B.$$

Thus the connective $\supset$ can be defined in terms of disjunction and negation. Although the semantic side of classical propositional logic is adequately serviced by negation, conjunction, and disjunction, there is considerable advantage in retaining $\supset$ for the syntactical aspect as it enables an equivalent system to be constructed with just one rule of inference, *modus ponens* (§2.2.7).

(x) $A \supset B$ is a tautology iff $A \to B$. A is a tautology iff $A \leftrightarrow t$ iff $\emptyset \leftrightarrow B$. $A \vee [A \supset B]$ and $[[[A \supset B] \supset A] \supset A]$ are tautologies for all $A, B \Vdash_{\text{prop}}$. Any formula of one of the forms (A0)–(A10) of §2.2.7 is a tautology. ◆

CONVENTIONS For the purposes of exposition in this book certain abbreviations of the 'official' language and conventions, useful for reading short formulas and sanctioned by tradition, will be used.

(i) The associative and commutative laws satisfied by both conjunction and disjunction (Example 2.1 (viii)) justify obvious conventions for omitting brackets in conjunctions and disjunctions. Thus we shall write $A \vee B \vee C$ instead of any of the logically equivalent formulas $[[A \vee B] \vee C]$, $[A \vee [B \vee C]]$, $[[C \vee A] \vee B]$, $[[B \vee C] \vee A]$. Further $[A \wedge B]$ will be abbreviated to AB and we shall use the convention that conjunction binds tighter than disjunction: that $AB \vee CD$ is an abbreviation of any one of the logically equivalent formulas $[[A \wedge B] \vee [C \wedge D]]$, $[[B \wedge A] \vee [D \wedge C]]$, $[[C \wedge D] \vee [B \wedge A]]$, ... It is also convenient to treat negation as an operator which acts on the left: thus A' is an abbreviation of $\neg A$, so that $AB' \vee C'$ is an abbreviation of $[[A \wedge \neg B] \vee \neg C]$.

Propositional logic

(ii) There will be occasions later when arbitrary conjunctions and disjunctions of formulas will be used, though we shall not pursue a formal development of logics with infinite formulas (e.g. Kreisel and Krivine 1967, Chapter 7). If Δ is a set of formulas (of the official language—in this case $\Delta \subseteq \mathbb{F}(J, P)$), then $\bigvee \Delta$ and $\bigwedge \Delta$ denote the disjunction and conjunction respectively of the formulas in Δ. Neither $\bigvee \Delta$ nor $\bigwedge \Delta$ are official formulas.

(iii) A useful notation due to Schütte (1961) is as follows: $F[A]$ denotes a formula F with a distinguished occurrence of a subformula A. Similarly, $F[A, B]$ denotes a formula F with distinguished occurrences of subformulas A and B. Given a formula $F[A]$ and a formula B then $F[B]$ denotes the formula which arises from $F[A]$ by replacing that occurrence of A, and that occurrence only, by B.

EXAMPLE 2.2.

(i) Let $v \in \mathbb{V}$. Then
$$v(A) = v(B) \Rightarrow v(F[A]) = v(F[B]).$$
Hence if $A \leftrightarrow B$ then $F[A] \leftrightarrow F[B]$.

(ii) Let $F \in \mathbb{F}(J, P)$. Let Y be a distinguished occurrence in F of a subformula. Then for any subformula Z of F (thus Z is a subword of F) (a) Z is a subword of Y

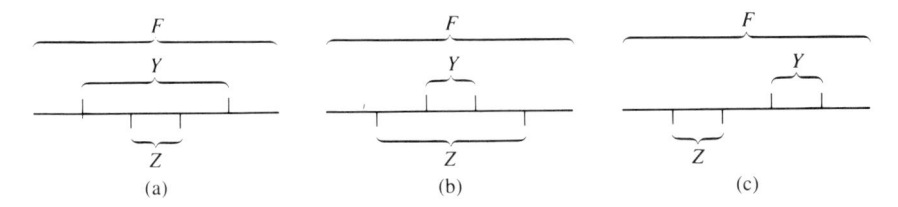

or (b) Y is a subword of Z or (c) Z or Y do not overlap. This fact can be established by examining the role of brackets in the construction of formulas. It does not necessarily hold with the informal abbreviations. Thus the informal expression $A \vee B \vee C$ has two properly overlapping subformulas. $A \vee B$ and $B \vee C$; this overlap does not occur in the official versions $[A \vee [B \vee C]]$ or $[[A \vee B] \vee C]$. ◆

2.2.3 Normal forms

(i) *Disjunctive normal form*: each formula is logically equivalent to a disjunction of formulas where each disjunct is a conjunction of literals.
(ii) *Conjunctive normal form*: each formula is logically equivalent to a conjunction of formulas, each of which is a disjunction of literals.

These results are attributed to Peirce. They are easily proved by induction on the number of proposition letters occurring in the formula (Church 1956, Chapter II).

In automated theorem-proving it is convenient to represent a disjunction of literals $D = a_1 \vee \cdots \vee a_n$ as a *clause*, i.e. the set $\hat{D} = \{a_1, \ldots, a_n\}$. The conjunctive normal form $C_1 \wedge \cdots \wedge C_k$ can then be represented as a list $\hat{C}_1, \ldots, \hat{C}_k$ of clauses.

2.2.4 Minimal models

The concept of minimum model is of some importance in logic programming where the predicate logic is used. The essence of the construction of minimum models, however, already appears in the propositional logic with an ordering of the positive diagrams of valuations as follows.

DEFINITION 2.5.

(i) For $u \in \mathbb{V}$, $pd(u) =_{\text{def}} \{p \in P; u(p) = 1\}$. This is the *positive diagram* of u. Thus $pd: \mathbb{V} \to \mathbb{P}(P)$. In fact, the mapping is surjective. Thus, for a set Q of proposition letters define a corresponding valuation v by

$$v() = 1 \Leftrightarrow p \in Q \quad (p \in P).$$

Then $pd(v) = Q$.

(ii) Let Δ be a set of formulas. The positive diagrams of models in Δ^* are sets of proposition letters and so are ordered by inclusion. A *minimal model* of Δ is a model u of Δ such that there is no model v of Δ other than u with $pd(v) \subseteq pd(u)$. A *mimimum model* of Δ is a model u of Δ such that $pd(v) \subseteq pd(u)$ for all $v \in \Delta^*$.

(iii) Let $U \subseteq \mathbb{V}$. Then $\wedge U$ is the valuation defined by

$$\wedge U(p) =_{\text{def}} glb(\{u(p); u \in U\}) \quad (p \in P)$$

(see Definition 2.2(b) for glb). A set Δ of formulas is $\wedge$-*closed* if $\wedge \Gamma^* \in \Delta^*$ for every subset Γ of Δ.

EXAMPLE 2.3.

(i) Every set of formulas has minimal models, but not every set has a minimum model.
(ii) The minimum model of a subset Y of P is the valuation v where $v(p) =_{\text{def}} 1$ iff $p \in P$.
(iii) The minimum model of $\{p_1, \ldots, p_n, (p_1 \wedge \cdots \wedge p_n) \supset q\}$, where $p_1, \ldots, p_n, q \in P$, is the valuation which has the value 1 only on this set of proposition letters.
(iv) The single formula $p_1 \vee p_2$ has no mimimum model. It has a minimal model u with $u(p) = 1$ iff $p = p_1$ and a minimal model v with $v(p) = 1$ iff $p = p_2$. The only valuation below both u and v is the one that puts every proposition letter false, and this valuation is not a model of $p_1 \vee p_2$.
(v) If Δ is a $\wedge$-closed set of formulas, then $\wedge \Delta^*$ is the minimum model of Δ. ◆

Propositional logic

The following theorem defines some classes of formulas with minimum models.

THEOREM 2.1.

(i) If A is either a positive literal or a disjunction of literals, at most one of which is positive, then A has a minimum model.
(ii) Let K be a collection of $\wedge$-closed classes of formulas. Then $\sum K$ is also $\wedge$-closed.

Proof.

(i) Exercise.
(ii) Let $U \subseteq (\sum K)^*$. Then $U \subseteq \prod \{\Delta^*; \Delta \in K\}$ (Example 2.1(i)). But each $\Delta \in K$ is $\wedge$-closed. Hence $\wedge U \in \Delta$ for all $\Delta \in K$, i.e. $\wedge U \in \prod \{\Delta^*; \Delta \in K\}$ $= (\sum K)^*$. ∎

COROLLARY 2.2. Every set of formulas of the above forms has a minimum model.

2.2.5 Inductive definitions and logic programming

As with the concept of the minimum model, another important concept of logic programming, the *procedural interpretation* of logic programs already occurs in the propositional logic.

DEFINITION 2.6. A *definite program clause* is a disjunction of literals just one of which is positive. A *definite program* is a set of definite program clauses. This includes the case of a clause consisting of just one (positive) literal only—such clauses are *unit clauses*. In a definite program clause the positive literal is known as the *head* of the clause and the set of literals which occur negatively in it is known as the *body* of the clause. Thus a unit clause is a clause with an empty body.

EXAMPLE 2.4.

(i) Every definite program has a minimum model (Corollary 2.2).
(ii) Let Δ be a definite program and $p, q, \ldots, r$ be a finite set of proposition letters. $\Delta + \{\neg p \vee \neg q \vee \cdots \vee \neg r\}$ is inconsistent iff the minimum model of Δ satisfies $(p \wedge q \wedge \cdots \wedge r)$.
(iii) Every definite program clause which is not a unit clause is logically equivalent to a formula of the form $A \supset B$ where B is the positive literal and A is the conjunction of the (positive) literals in the body of the clause (Example 2.1(ix)). ◆

There arises the problem of constructing such a model when the program is given effectively. The key step is to associate with each non-empty definite program Δ an inductive definition Δ_{ind} as follows.

DEFINITION 2.7. Axioms of Δ_{ind}:

$$P \cdot \Delta.$$

Production rules of Δ_{ind}:

$$\frac{p_1, \ldots, p_n}{q}$$

(read 'from $p_1, \ldots, p_n$ derive q') for every definite program clause $\neg p_1 \vee \cdots \vee \neg p_n \vee q$ of Δ. The set of proposition letters generated by Δ_{ind} is denoted by $\mathfrak{H}(\Delta)$.

From the constructive point of view the production rules are procedural directions for effecting a construction. The 'implication' $\supset$ in the definite program clauses (see Example 2.2(iii)) is interpreted rather like IF ... THEN program instructions in BASIC or PASCAL. $\mathfrak{H}(\Delta)$ 'exists' only in the sense of this construction. To say $p \in \mathfrak{H}(\Delta)$ is *merely* another way of saying 'p is derived in Δ_{ind}'—the symbols $\in$ and $\mathfrak{H}(\Delta)$ have no other meaning than that provided by this context (see Lorenz 1961a). From the classical point of view the axioms and rules are replaced by statements of *conditions* which must be satisfied by the set $\mathfrak{H}(\Delta)$, i.e. $\mathfrak{H}(\Delta)$ is regarded as a pre-existing entity which satisfies the following conditions:

(i) $p \in \mathfrak{H}(\Delta)$ for all $p \in P \cdot \Delta$;
(ii) If $p_1 \in \mathfrak{H}(\Delta), \ldots, p_n \in \mathfrak{H}(\Delta)$, then $q \in \mathfrak{H}(\Delta)$ for every definite program clause $\neg p_1 \vee \cdots \vee \neg p_n \vee q$ of Δ.

To make this into a classical *definition* of $\mathfrak{H}(\Delta)$ one most add the *extremal clause*:

(iii) $\mathfrak{H}(\Delta)$ is the smallest class satisfying (1) and (2) or, equivalently,
(iii') a proposition letter is in $\mathfrak{H}(\Delta)$ if it is so only by (i) and (ii) (cf. (F2) of Definition 2.1(i)).

Extremal clauses in inductive definitions are constructively redundant as they do not affect the process of construction—it is not necessary to write them after IF ... THEN instructions, though there might be some virtue in putting one in a COMMENT. For the remainder of this book extremal clauses will be omitted from inductive definitions: the classically minded reader can systematically insert them.

$\mathfrak{H}(\varDelta)$ is an example of an *inductive class*, to use Curry's (1963, pp. 38–9) terminology. In general there are two components to the definition of an inductive class $\varGamma$—the *initial specifications* and the *generating specifications*.

The initial specifications define the class I of *initial elements*. The generating specifications define a class $\mathfrak{G}$—usually, but not necessarily finite—of generating rules. It is understood that the application of each such rule to a sequence or set of arguments appropriate to that rule, which are either initial elements or have already been placed in $\varGamma$ by application of the rules, produces a (not necessarily unique) element. This requirement, frequently called the *closure specification*, is implicit in the definition of *inductive class*, in particular in the practice of generating rules.

The generating rules can be considered as a 'generating function' (also denoted by $\mathfrak{G}$ which, acting on a subset J of $\varGamma$, gives the set $\mathfrak{G}(J)$ of elements generated from J by application of the rules of $\mathfrak{G}$ to J. The generating function $\mathfrak{G}$ has the following properties:

$$J \subseteq \mathfrak{G}(J) \tag{2.2}$$

$$\mathfrak{G}(\mathfrak{G}(J)) \subseteq \mathfrak{G}(J) \tag{2.3}$$

$$J \subseteq K \;\Rightarrow\; \mathfrak{G}(J) \subseteq \mathfrak{G}(K). \tag{2.4}$$

EXAMPLE 2.5.

(i) $\mathbb{F}(J, P)$ is an inductive class with initial elements P ((F1) of Definition 2.1(i)). The generating specifications are the four modes of combination prescribed by (F2). In this context (2.3) means that every logical combination of formulas is a formula (i.e. $\mathfrak{G} = \varLambda$).
(ii) For the non-empty definite program $\varDelta$ the initial elements are the axioms of $\varDelta_{\text{ind}}$ and the generating specifications are the production rules. ◆

From the classical perspective it can be proved that the set of entities *generated* by an inductive definition interpreted as *construction rules* is the smallest class *satisfying* the inductive definition interpreted as *conditions*. This is the essence of the so-called *procedural interpretation* in logic programming. We then have the following theorem.

THEOREM 2.3. *Let $\varDelta$ be a non-empty definite program. Define $v \in \mathbb{V}$ by*

$$v(p) = 1 \;\Leftrightarrow\; p \in \mathfrak{H}(\varDelta) \qquad (p \in P).$$

Then v is the minimum model of $\varDelta$.

COROLLARY 2.4. $\mathfrak{H}(\varDelta) = \{p \in P; \varDelta \to p\}$.

Proof. Let $p \in \mathfrak{H}(\Delta)$. Then $v(p) = 1$. By Theorem 2.3 v is the smallest model of Δ. Hence $u(p) = 1$ for every $u \in \Delta^*$. Thus $\Delta \to p$. Conversely, suppose that $\Delta \to p$. As $v \in \Delta^*$, $v(p) = 1$ so that $p \in \mathfrak{H}(\Delta)$. ∎

2.2.6 Compactness

The power of the classical propositional logic is derived from the *compactness theorem* or *finiteness theorem* from which can be derived the compactness theorem of classical predicate logic on which the early productivity of model theory rested (see Robinson 1951, Chapter 2, Chang 1974, Vaught 1974).

The compactness theorem of the classical predicate logic originated from Gödel's completeness theorem (1930)—if a sentence of a first-order predicate logic holds in every structure in which it is defined, then it is a theorem. Closely related to this is the 'extended completeness theorem' (Henkin 1949b, Robinson 1951)—every consistent set of sentences of a first-order predicate logic possesses a model. Mal'cev (1941) stated a similar 'principle of localization'—if every finite subset of a set Δ of first-order sentences has a model, then Y itself has a model. The compactness theorem for first-order predicate logic follows from the principle of localization: let Δ be a set of first-order sentences and let S also be such a sentence. If every model of Δ is also a model of S then there exists a finite subset Δ_0 of Δ such that every model of Δ_0 is also a model of S.

Observe that, unlike the completeness theorem, only semantic notions occur in the principle of localization—the principle depends only on the notion of *model*, and not on the deductive apparatus of the predicate calculus. In fact, it can be established entirely from semantic notions (e.g. Frayne *et al.* 1962, Kreisel and Krivine 1967).

The term *compactness* was used by Tarski (1952) because the localization principle has a topological interpretation (see Appendix, §T). Thus, let $\mathfrak{S}(L)$ be the set of sentences of a first-order language L and let M be the set of structures whose domains are subsets of a fixed infinite set and whose type is appropriate for L. For each subset K of $\mathfrak{H}(L)$ let K^* be the set of structures in M which satisfy all members of K. Then the collection $\{\{S\}^*: S \in \mathfrak{S}(L)\}$ of subsets of M is a basis for a topology on M in which the closed sets are the sets K^* where $K \subseteq \mathfrak{S}(L)$. With this topology, M has the *finite intersection property*, i.e. if $\{C_i; i \in I\}$ is a collection of closed sets such that every finite number of them has non-empty intersection, then the total intersection $\prod \{C_i; i \in I\}$ is also non-empty. This follows trivially from the principle of localization.

Like Kreisel and Krivine (1967), we begin with the compactness theorem of the propositional calculus and later use it for the compactness of the predicate calculus. There are several versions of the compactness theorem of the propositional calculus which are all equivalent in classical logic.

Propositional logic 31

Part (ii) of the following theorem is more general than part (i) and is sometimes more useful.

THEOREM 2.5.

(i) If every valuation is a model of some member of the non-empty set Δ of formulas, then there exists a non-empty finite subset Γ of Δ such that Γ is a tautology.
(ii) If every model of a set Θ of formulas is also a model of some member of the set Δ of formulas then there is a finite subset Γ of Δ such that $\Theta \to \Gamma$.

This theorem is reminiscent of the Heine–Borel theorem of elementary analysis. Both can be proved using König's (1926) *infinity lemma*.

THEOREM 2.6. A set Δ of formulas has a model iff every finite subset of Δ has a model.

Proof. It follows from Theorem 2.5 that Δ has a model iff there is a finite subset of Δ without a model. ∎

THEOREM 2.7. Let Δ be a set of formulas and A a formula. Then $\Delta \to A$ iff there exists a finite subset Γ of Δ such that $\Gamma \to A$.

Proof. $\Delta \to A$ iff $\Delta + \{\neg A\}$ is inconsistent (Example 2.1(v)). The theorem then follows from Theorem 2.6. ∎

For the many-valued logics considered later, the notion of *model* does not occupy the central place that it has in classical logic; the proof of Theorem 2.7 given below can be used with little change in applications to lattice-valued logics (Chapter 6, §6.6).

Preliminary Definitions and Results

(i) Let $\Gamma, \Theta \subseteq \Delta$ and A be a fixed formula:

$$cx(\Gamma) =_{\text{def}} \{v \in \mathbb{V}; glb(v(\Gamma)) \not\leq v(A)\}.$$

By (2.1) each $v \in cx(\Gamma)$ is a Γ-*counterexample* to the claim that $\Gamma \to A$:

$$\Gamma \to A \quad \text{iff} \quad cx(\Gamma) = \emptyset \tag{2.5}$$

$$\text{if} \quad \Theta \supseteq \Gamma \quad \text{then} \quad cx(\Theta) \subseteq cx(\Gamma) \tag{2.6}$$

$$\text{if} \quad v \in cx(\Gamma) \quad \text{and} \quad v(\Gamma) = v(\Delta) \quad \text{then} \quad v \in cx(\Theta). \tag{2.7}$$

(ii) Let $\tau = (\tau_1, \ldots, \tau_n)$ be a finite sequence of truth-values in $\mathbb{T}v$ and $v \in \mathbb{V}$. Define

$$v \supseteq \tau \Leftrightarrow_{\text{def}} \tau_1 = v(p_i) \qquad (1 \leqslant i \leqslant n)$$

$$val(\tau) =_{\text{def}} \{v \in \mathbb{V}; v \supseteq \tau\}.$$

Let $\tau t =_{\text{def}} \tau^\frown \langle t \rangle$. Then

$$val(\tau) = \sum \{val(\tau t); t \in \mathbb{T}v\}. \tag{2.8}$$

The proof of Theorem 2.7 depends on a step-by-step construction of a Δ-counterexample, the possibility of which is guaranteed by the following lemma.

LEMMA 2.8. Let $P(\tau)$ be the statement: 'for every finite subset Γ of Δ there exists a $v \in cx(\Gamma)$ such that $v \supseteq \tau$'. Then

$$P(\tau) \Rightarrow (Et \in \mathbb{T}v)P(\tau t).$$

Proof. Suppose $P(\tau)$. We derive a contradiction from the assumption that not-$P(\tau t)$ for all $t \in \mathbb{T}$. Now

$$P(\tau) \Leftrightarrow (\Gamma \in \mathbb{P}_\omega(\Delta))(cx(\Gamma) \cdot val(\tau) \neq \emptyset).$$

Thus for each $t \in \mathbb{T}v$ there exists $\Gamma_t \in \mathbb{P}_\omega(\Delta)$ such that

$$cx(\Gamma_t) \cdot val(\tau t) = \emptyset. \tag{2.9}$$

Let $\Gamma = \sum \{\Gamma_t; t \in \mathbb{T}v\}$. Then $\Gamma \in \mathbb{P}_\omega(\Delta)$ and, by (2.6). $cx(\Gamma) \subseteq cx(\Gamma_t)$ $(t \in \mathbb{T}v)$. Thus $cx(\Gamma) \cdot val(\tau t) = \emptyset$ $(t \in \mathbb{T}v)$. Then, by (2.8), $cx(\Gamma) \cdot val(\tau) = \emptyset$. This contradicts $P(\tau)$. Thus $P(\tau)$ implies that $P(\tau t)$ for some $t \in \mathbb{T}v$. ∎

Proof of Theorem 2.7. Suppose $\Delta \to A$. Suppose also, contrary to the statement of the theorem, that A is not a logical consequence of any $\Gamma \in \mathbb{P}_\omega(\Delta)$. Then for all $\Gamma \in \mathbb{P}_\omega(\Delta)$, $cx(\Gamma) = \emptyset$. Thus $P(\lambda)$, where λ is the null sequence of truth-values. By Lemma 2.8 there exists an infinite sequence τ of truth-values, i.e. $\tau \in \mathbb{V}$, which is a counterexample to every $\Gamma \in \mathbb{P}_\omega(\Delta)$, i.e. $\tau \in cx(\Gamma)$ for all $\Gamma \in \mathbb{P}_\omega(\Delta)$. As $\tau(\Delta) \subseteq \mathbb{T}v$ and $\mathbb{T}v$ is finite there is a $\Gamma \in \mathbb{P}_\omega(\Delta)$ such that $\tau(X) = \tau(\Gamma)$. By (2.7), $\tau \in cx(\Gamma)$. By (2.5) this contradicts the supposition that $\Delta \to A$. Hence there exists $\Gamma \in \mathbb{P}_\omega(\Delta)$ such that $\Gamma \to A$.

The converse is trivial ∎

2.2.7 Syntactical concepts

The two principle methods of axiomatizing logics are the Hilbert and the Gentzen types of system. Hilbert spaces are theorem-oriented. They generally give easy proofs of the completeness theorem but are not so easy to use.

Propositional logic

Gentzen systems, on the other hand, are inference-oriented. They diminish the role of logical axioms and emphasize the rules of inference.

Hilbert systems There are many ways of axiomatizing the classical propositional logic (see Church 1956). The standard procedure is to choose a simple set of tautologies as *axioms* and then define a concept of *proof* or *derivation* such that a formula is provable if and only if it is a tautology. The Hilbert type of axiomatization is a system for generating theorems. As an example we choose a system H defined below.

The formulas of H are $\mathbb{F}_{\text{prop}} = \mathbb{F}(J, P)$ where $J = \{t, f, \wedge, \vee, \supset, \neg\}$.

Axioms of H. Any formula of one of the forms **A1**–**A10** below is an *axiom* of H. Ax_H is the set of such axioms:

A0: $t \qquad f \supset A$

A1: $A \supset [B \supset A]$

A2: $[A \supset B] \supset [[[A \supset [B \supset C]] \supset [A \supset C]]$

A3: $A \supset [B \supset [A \wedge B]]$

A4: $[A \wedge B] \supset B$

A5: $[A \wedge B] \supset A$

A6: $A \supset [A \vee B]$

A7: $B \supset [A \vee B]$

A8: $[A \supset C] \supset [[B \supset C] \supset [[A \vee B] \supset C]]$

A9: $[A \supset B] \supset [[A \supset \neg B] \supset \neg A]$

A10: $\neg\neg A \supset A$.

The single *rule of inference* is modus ponens, i.e.

MP: $\dfrac{A, A \supset B}{B}$

(read 'from A and $A \supset B$ derive B').

DEFINITION 2.8. The above definition of H is, in fact, an inductive definition of the *theorems* of H: the axioms are the initial elements and **MP** is the single generating rule.

(i) B is said to be *directly derivable* from A and $A \supset B$.
(ii) Let $\Delta \subseteq \mathbb{F}_{\text{prop}}$. A sequence of formulas $F_1, \ldots, F_m$ is a *derivation* (or *proof*) of F_m from hypotheses Δ in H if, for each $k = 1, 2, \ldots, m$, either $F_k \in \Delta + Ax_H$ or, for some $i, j < k$, F_k is directly derivable from F_i and F_j.

(iii) Write $\Delta \vdash_H A$ when there is a derivation of A from Δ in H. We write $\vdash_H A$ instead of $\emptyset \vdash_H A$. (It is easy to see that A is a *theorem* of H iff $\vdash_H A$.)

The connection between *truth* and *proof* had already been discussed by Bolzano before the notions had been formalized in their modern form—a correct proof must have the property that any example which satisfies the premiss must also satisfy the conclusion. Gödel's completeness theorem establishes the link between Frege's rules and Tarski's truth-definition—if $F \supset G$ cannot be proved (by Frege's rules), then there exists a model of F which does not satisfy G. The following theorem is proved in the standard treatises.

THEOREM 2.9.

(i) (Completeness) If B is a tautology then $\vdash_H B$.
(ii) (Soundness) If $\vdash_H B$ then B is a tautology.

There is a certain amount of redundancy in system H as t is deductively equivalent to $f \supset f$, i.e. $t \supset [f \supset f]$ and $[f \supset f] \supset t$ are both theorems. Also $\neg A$ is deductively equivalent to $A \supset f$, t to $A \supset A$, and f to $\neg[A \supset A]$ for every $A \in \mathbb{F}_{\text{prop}}$. Thus there are several systems which are equivalent to H but differ from it by the axioms and by the logical symbols which are taken as primitive and those that are defined.

(i) $J = \{\wedge, \vee, \supset, \neg\}$. The axioms are all formulas of $F(J, P)$ which are instances of the schemas **A1–A10**. This is the system $\hat{H}$.
(ii) $J = \{f, \vee, \wedge, \supset\}$, $t =_{\text{def}} [f \supset f]$, $\neg A =_{\text{def}} [A \supset f]$. The axioms are all formulas of $F(J, P)$ which are instances of the schemas **A0–A8, A10**. Omission of the axiom scheme **A10** from this system gives the system I whose theorems are the *theses of intuitionistic logic* as formulated by Heyting, though this formal system does not coincide with that of Heyting (see Heyting 1930, Chapter 7). It is well known that in the classical system some of the connectives can be defined in terms of the others, whereas in the intuitionistic system they are independent—none can be expressed in terms of the others (Wajsberg 1938, McKinsey 1939).

EXAMPLE 2.6. In I every formula of the form

$$\{\{D \supset B\} \supset [[[B \wedge C] \supset A] \supset [[D \wedge C] \supset A]]\}$$

is a theorem. ◆

Resolution The most simple-minded method of proving that a formula A is a tautology is to check through all possible valuations of the proposition letters in A. If there are n such letters, there are 2^n cases. In automated theorem-proving the complexity of this procedure is avoided by choosing a *refutation procedure* rather than a *proof procedure*. Thus, to prove A, we *refute* ¬A, i.e. we show that ¬A is inconsistent (Example 2.1(vi)). Refutation procedures are not directly applicable to logics which lack the connection between provability and inconsistency. In particular, it is not obvious how to apply them to the lattice-valued logics defined in Chapter 6 or the three- or four-valued logics of Chapter 8.

The modern refutation technique is based on the *resolution principle* of Robinson (1965). (For a brief history of automated theorem-proving see Lloyd 1984.) First express ¬A in conjunctive normal form (§2.2.3), i.e. as a list of clauses. Then prove them to be jointly contradictory by deriving from them the empty clause $\square$ by using the single *resolution rule*: for clauses C, D and $p \in P$

res: from $C + \{p\}, D + \{\neg p\}$ derive $C + D$.

The result $C + D$ of the application of *res* is called the *resolvant* and we say that it is *resolved* by p. The extension of this procedure to the predicate logic uses a process of *unification* to handle the proliferation of terms (Chang and Lee 1973, Lloyd 1984, Börger 1989).

DEFINITION 2.9.

(i) For a set Δ of clauses and clause A write $\Delta \vdash_{res} A$ when A is derivable from Δ by applications of *res*:

$$C_{res}(\Delta) =_{def} \{B; B \text{ a clause and } \Delta \vdash_{res} B\}.$$

(ii) For all valuations v, $v(\square) =_{def} 0$.

It is simple to verify that if C, D are clauses and $p \in P$ then

$$C + \{p\}, D + \{\neg p\} \quad \rightarrow \quad C + D.$$

Thus if $\Delta \vdash_{res} A$, where A is a non-empty set, then $\Delta \rightarrow A$. In particular

if $\Delta \vdash_{res} \square$ then Δ is contradictory.

The converse of this is not so easy.

THEOREM 2.10. *If Δ is a contradictory finite set of clauses then $\Delta \vdash_{res} \square$.*

Proof. By induction on the number of propositional letters occurring in clauses. Let

$$C_k =_{def} \mathbb{P}_\omega(\{p_1, \ldots, p_k, \neg p_1, \ldots, \neg p_k\}).$$

The members of C_k are clauses whose propositional letters are amongst $\{p_1, \ldots, p_k\}$. We prove that $\Delta \vdash_{res} \square$ indirectly. Suppose that

$$\Delta \text{ is contradictory.} \quad (2.10)$$

Suppose that $\Delta \not\vdash_{res} \square$ and let $S(k)$ be the statement '$C_k \cdot C_{res}(\Delta)$ is consistent'. It will be proved by induction on k that

$$(k)S(k). \tag{2.11}$$

Basis: $\qquad\qquad\qquad\qquad S(0). \tag{2.12}$

For $C_0 = \{\square\}$. From $\Delta \not\vdash_{res} \square$ it follows that $\Delta \vdash_{res} \square$ implies $v(\square) = 1$ for all $v \in \mathbb{V}$. Hence (2.12) holds.

Inductive step: $\qquad\qquad S(k) \Rightarrow S(k+1). \tag{2.13}$

For suppose $S(k)$. Then there exists $v \in \mathbb{V}$ such that

$$v(B) = 1 \qquad (B \in C_k \cdot C_{res}(\Delta)). \tag{2.14}$$

Fix such a v and for $t \in \{0, 1\}$ define $v_t \in \mathbb{V}$ by

$$v_t(p_i) = \begin{cases} v(p_i) & \text{if } i \leq k \\ t & \text{if } i > k. \end{cases}$$

Then, since the proposition letters of clauses in C_k are amongst $\{p_1, \ldots, p_k\}$,

$$v_1(A) = v_0(A) = v(A) \qquad (A \in C_k). \tag{2.15}$$

Now suppose that, contrary to (2.13), not-$S(k+1)$. Then

$$(u \in \mathbb{V})(EA \in C_{k+1})(\Delta \vdash_{res} A \;\&\; u(A) = 0).$$

Hence there exist $A_0, A_1 \in C_{k+1} \cdot C_{res}(\Delta)$ such that

$$v_0(A_0) = v_1(A_1) = 0. \tag{2.16}$$

By (2.14) and (2.15), $A_0, A_1 \notin C_k$. Hence they contain p_{k+1} or $\neg p_{k+1}$. So there exist $X, Y \in C_k$ such that

$$A_0 = X + \{p_{k+1}\} \qquad A_1 = Y + \{\neg p_{k+1}\}. \tag{2.17}$$

For suppose to the contrary that $A_0 = X + \{\neg p_{k+1}\}$ for some $X \in C_k$. Then $v_0(A_0) = v_0(X) \cup v_0(\neg p_{k+1}) = v(X) \cup 0' = 1$ by (2.14) (see Definition 2.2(ii) for $\cup$). This contradicts (2.15). Hence $A_0 = X + \{p_{k+1}\}$. Similarly, $A_1 = Y + \{p_{k+1}\}$. From (2.15) and (2.16) we then have $v(X) = v_0(A_0)$ and $v(Y) = v_1(A_1)$, and hence

$$v(X + Y) = v(X) \cup v(Y) = v_0(A_0) \cup v_1(A_1) = 0$$

by (2.16). Resolution by p_{k+1} then gives $X + Y \in C_k \cdot C_{res}(\Delta)$. This contradicts (2.14). Thus $S(k)$ contradicts not-$S(k+1)$, thus proving (2.13).

(S.11) follows from (2.12) and (2.13) by induction on k.

Next, as Δ is finite there is a k such that $\Delta \subseteq C_k$ so that $\Delta \subseteq C_k \cdot C_{res}(\Delta)$. By $S(k)$, Δ is consistent, contrary to (2.10). Hence $\Delta \vdash_{res} \square$. ∎

COROLLARY 2.11. Let X be a set of definite program clauses and $p \in P$. Then $X \to p$ iff $X + \{\neg p\} \vdash_{res} \square$.

Theorem 2.7 is needed for this result.

Gentzen Systems The sequent calculus and natural deduction systems for predicate calculus were developed by Gentzen (Szabo 1969), who also described a transformation of derivation from each system to the other. A normal form theorem for proofs in the predicate calculus, which has profound consequences in proof theory, was obtained in 1934–5. Gentzen's classification of deductive operations is expressed in a formal system in which the rules of inference apply to *sequents*, i.e. ordered pairs (Γ, Δ) of finite sequences Γ, Δ of formulas. Γ is called the *antecedent* and Δ the *succedent* of (Γ, Δ).

If $\Gamma = \langle A_1, \ldots, A_m \rangle$ and $\Delta = \langle B_1, \ldots, B_k \rangle$, then the ordered pair (Γ, Δ) can be understood as the formula $(A_1 \wedge \cdots \wedge A_m) \supset (B_1 \vee \cdots \vee B_k)$ or as the relation of logical consequence $A_1 \wedge \cdots \wedge A_m \to B_1 \vee \cdots \vee B_k$. The interpretation can be extended to the cases where Γ or Δ is λ, the empty sequence, by regarding Γ as true when $\Gamma = \lambda$ and Δ as false when $\Delta = \lambda$. For a formula A we write Γ, A, Δ instead of $\Gamma \frown \langle A \rangle \frown \Delta$, and $(\Gamma : \Delta)$ instead of (Γ, Δ).

There are numerous calculi of sequents available (Curry 1963). Schütte (1961) gives an efficient system related to Gentzen systems which we adapt to a completeness proof for a three-valued logic in Chapter 8. It is sometimes convenient to define sequents as ordered pairs of finite *sets* rather than sequences (e.g. Dummett (1977); we shall use this device in Chapter 10.

Gentzen systems are all characterized by having rules of inference associated with each junctor and quantifier which *introduce* the junctor or quantifier into the antecedent and the succedent. Thus, for the system G1 described in full by Kleene (1962, Chapter 15), the formulas are $F(\{\wedge, \vee, \supset, \neg\}, P)$ and the axioms (*initial sentences*) are all sequents of the form $(C : C)$ for formulas C. For the junctor $\supset$, for instance, there are the rules

$R(\supset)$: $\dfrac{(\Gamma : \Theta, A) \quad (\Gamma : \Theta, B)}{(\Gamma : \Theta, A \supset B)}$

$L(\supset)$: $\dfrac{(\Delta : \Psi, A) \quad (B, \Gamma : \Theta)}{(A \supset B, \Delta, \Gamma : \Psi, \Theta)}$

For negation there are the rules

$R(\neg)$: $\dfrac{(A, \Gamma : \Theta)}{(\Gamma : \Theta, \neg A)}$

$L(\neg)$: $\dfrac{(\Gamma : \Theta, A)}{(\neg A, \Gamma : \Theta)}$.

In addition to the logical rules there are *structural rules* which, with one exception, do not introduce new elements or delete essential information. For instance, there is a *thinning rule*:

$$R(\text{thinning}): \quad \frac{(\Gamma : \Theta)}{(\Gamma : \Theta, C)}$$

$$L(\text{thinning}): \quad \frac{(\Gamma : \Theta)}{(C, \Gamma : \Theta)}$$

The exception is the important cut rule:

$$(\text{cut}): \quad \frac{(A : \Psi, C) \quad (C, \Gamma : \Theta)}{(\Delta, \Gamma : \Psi, \Theta)}$$

G1 becomes an intuitionistic system when $\Theta = \lambda$ in $R(\neg)$ and thinning can only be applied to the antecedent, i.e. L(thinning) only can be used.

G1 is equivalent to the previous Hilbert-type formulation $\check{H}$ of the propositional calculus in the following sense (Kleene 1962, Theorems 46, 47).

THEOREM 2.12. $(\Gamma : A)$ is provable in G1 iff $\Gamma \vdash_{\check{H}} A$.

The elimination of (cut) from proofs has deep implications for proof theory by facilitating normal forms.

THEOREM 2.13. (Gentzen's *Haupsatz*, or normal form theorem). Given a proof in G1 of a sequent there can be constructed a cut-free proof of the same sequent.

2.3 Predicate logic

The predicate logic gives content to the proposition letters of the propositional calculus and by introducing variables enables quantification to be effected. Many of the important elementary facts about the predicate logic, such as compactness, are established by reducing it, in a certain sense, to the propositional logic.

2.3.1 Formation rules

The alphabet M of the predicate language $L(\tau)$ (of *type* τ) consists of the pairwise disjoint subalphabets Fv, Bv, R, Fn, J, Q, G, where the following hold.

(i)(a) *Fv* is an infinite set of symbols

$$x_1, x_2, x_3, \ldots$$

called *free variables*.

(i)(b) *Bv* is an infinite set of symbols

$$a_1, a_2, a_3, \ldots$$

called *bound variables*.

(ii) *R* is a (finite or infinite) set

$$P_1, P_2, P_3, \ldots$$

of *predicate letters*.

Fn is a set (finite or infinite)

$$f_1, f_2, f_3, \ldots$$

of *function letters*. ($R + Fn$ is the set of *non-logical constants*.) Associated with R and Fn is a function $\tau: R + Fn \to \mathbb{N}$, the *type* of the language, such that $\tau(P) > 0$ for $P \in R$. τ defines the number of argument places of both predicate and function letters. We shall assume that Fn contains at least one letter f such that $\tau(f) = 0$. Such function letters are called *individual constants*.

(iii) *J* is the set of junctors as in §2.2.1.

(iv) *Q* is the list

$$\forall, \exists$$

of *quantifier symbols*. $\forall$ is the *universal quantifier* and $\exists$ is the *existential quantifier*. ($J + Q$ is the set of *logical constants*.)

(v) *G* is the set of separators as in §2.2.1.

For any word W on the alphabet M (i.e. $W \in M^*$, see Appendix, §W), $Fv(W)$ and $Bv(W)$ respectively denote the sets of free variables and bound variables which occur in W.

The predicate language $L(\tau)_=$ is an extension of $L(\tau)$ obtained by adjoining a binary predicate symbol which is interpreted as equality.

DEFINITION 2.10.

(i) The class $\mathbb{T}$ of *terms* is defined inductively as follows.
   (T1) Every individual constant is a term. If $f \in Fn$ with $n = \tau(f) > 0$ and $y_1, \ldots, y_n \in Fv$ then $f[y_1, \ldots, y_n] \in \mathbb{T}$.
   (T2) If $P \in R$ with $n = \tau(P)$ and $t_1, \ldots, t_n \in \mathbb{T}$ then $P[t_1, \ldots, t_n] \in \mathbb{T}$.

(ii) For $s \in \mathbb{T}$, $Fv(s)$ is the set of free variables which occur in s. If $X \subseteq Fv$ then $\mathbb{T}(X) =_{\text{def}} \{s \in \mathbb{T}; Fv(t) \subseteq X\}$. $\mathbb{T}(\emptyset)$ is the set of *constant terms*. As Fn contains at least one individual constant, $\mathbb{T}(\emptyset) \neq \emptyset$.

Definition 2.11.

(i) Let $Y \subseteq Fv$. $At(Y)$ is the set of *atoms* (also *atomic formulas, prime formulas,*...) of M, whose free variables are amongst Y, defined as follows.
  (A) If $P \in R$, $n = \tau(P)$, and $t_i \in \mathbb{T}(Y)$ for $1 \leq i \leq n$ then $P[t_1, \ldots, t_n] \in At(Y)$:
  $$At =_{\text{def}} At(Fv).$$

(ii) $At(\emptyset)$ is the set of *ground atoms*, i.e. atoms in which no free variables occur.

(iii) A *positive literal* is an atom. A *negative literal* is the negation of an atom. A *literal* is a positive or negative literal.

(iv) $J + Q$ is counted as the set of logical symbols.

Definition 2.12.

(i) The class $\mathbb{F}_{\text{pred}}(\tau)$ of *formulas* is defined by induction.
  (F1) $At \subseteq \mathbb{F}_{\text{pred}}(\tau)$.
  (F2) If $A, B \in \mathbb{F}_{\text{pred}}(\tau)$ then $\neg A$, $[A \vee B]$, $[A \wedge B]$, $[A \supset B] \in \mathbb{F}_{\text{pred}}(\tau)$.
  (F3) If $A \in \mathbb{F}_{\text{pred}}(\tau)$, $x \in Fv$, $a \in Bv - Bv(A)$ then $\forall a A(a/x) \in \mathbb{F}_{\text{pred}}(\tau)$ and $\exists a A(a/x) \in \mathbb{F}_{\text{pred}}(\tau)$ (see Appendix, §W for the substitution operator (a/x)). Note that
  $$Fv(\forall a A(a/x)) = Fv(\exists a A(a/x)) = Fv(A) - \{x\}$$
  and
  $$Bv(\forall a A(a/x)) = Bv(\exists a A(a/x)) = Bv(A) + \{a\}.$$

(ii) The notion of *logical combination* is constituted by the formation rules as in Definition 2.1, but account must now be taken of the rules concerning quantifier symbols.

Example 2.7. The set of terms is closed under substitution of terms for free variables. Therefore so too are the sets of atomic formulas and formulas. ◆

We have followed the practice of Schütte (1961) in making a typographic distinction between free and bound variables—there is a minor systematic advantage in doing this when framing rules of inference. Note also that the substitution operators (a/x) used in the formation of quantified formula 'pass through' quantifiers, i.e. if a, b are distinct free variables and x, y are distinct free variables then

$$Q_1 a(Q_2 b A(b/y))(a/x) = Q_1 a Q_2 b A(a/x)(b/y)$$

where Q_1, Q_2 are $\exists$ or $\forall$ (see Appendix, §W).

$\mathbb{F}_{\text{pred}}(\tau)$ clearly depends on both the type of the language and the set J of junctors. As before we shall omit reference to J where it has been fixed.

EXAMPLE 2.8. Let $L(\tau)$ be a language with one three-place predicate letter P (i.e. $\tau(P) = 3$), one constant c and one one-place function symbol f (i.e. $\tau(iii) = 0$ and $\tau(f) = 1$). Then $\mathbb{T}(\emptyset)$, the set of constant terms, is $\{c, f[c], f[f[c]], \ldots\}$. $P[f[c], c, f[f[c]]]$ is a ground atom and

$$\forall a_1 P[x_1, x_2, x_3](a_1/x_2) = \forall a_1 P[x_1, a_1, x_3] \in \mathbb{F}_{\text{pred}}$$

$$\forall a_1 P[x_1, x_2, x_3](a_1/x_4) = \forall a_1 P[x_1, x_2, x_3] \in \mathbb{F}_{\text{pred}}.$$

Both $\forall a_1 P[x_1, a_1, x_3]$ and $\forall a_1 P[x_1, x_2, x_3]$ have unit logical length and are logical combinations of $\{P[x_1, x_2, x_3]\}$. ◆

As in §2.1.1 the set of formulas can be considered as an algebra. In the same way the terms can be considered as an algebra whose functions are the function letters. We shall use the symbol $\mathbb{T}$ to denote both the set of terms and the algebra. For any set $X \subseteq Fv$, $\mathbb{T}(X)$ is a subalgebra of $\mathbb{T}$.

2.3.2 Semantical concepts

The semantics of the propositional logic was based on assigning truth-values to the proposition letters in P. The semantics of predicate logic is likewise based on assigning truth-values to atoms in At, but now the assignments are mediated by interpreting the function and predicate letters as suitable functions and relations respectively on relational systems of the appropriate type. The following definition of *truth* shows how the truth of a formula is systematically determined by the truth of its subformulas.

DEFINITION 2.13.

(i) A *realization* of $L(\tau)$ is a relational system (structure) of type τ (Appendix, §R). For a non-empty set A a *realization on A* is a realization $\mathfrak{A}$ such that $|\mathfrak{A}| = A$.

(ii) An *assignment* to a structure $\mathfrak{A}$ of type τ is a mapping $\alpha: Fv \to |\mathfrak{A}|$. This function can be extended to a mapping $\alpha: \mathbb{T} \to |\mathfrak{A}|$ by the inductive definition:
 (a) if $t \in Fv$ then α has already been defined;
 (b) if $t_1, \ldots, t_n \in \mathbb{T}$ where $n = \tau(f_i)$ then

$$\alpha(f_i[t_1, \ldots, t_n]) = g_i(\alpha(t_1), \ldots, \alpha(t_n)).$$

(Assignments interpret terms as elements of $\mathfrak{A}$.)

(iii) Given a realization $\mathfrak{A}$ and an assignment α to $\mathfrak{A}$ there is a valuation $[\mathfrak{A}, \alpha]$ which assigns truth-values to the atoms in At defined by

$$[\mathfrak{A}, \alpha](P_i[t_1, \ldots, t_n]) =_{\text{def}} R_i(\alpha(t_1), \ldots, \alpha(t_n))$$

$$[\mathfrak{A}, \alpha](t) =_{\text{def}} 1 \qquad [\mathfrak{A}, \alpha](f) =_{\text{def}} 0.$$

The valuation $[\mathfrak{A}, \alpha]$ extends to the determination of truth-values of all formulas in $\mathbb{F}_{pred}$ by the following inductive definition which follows the definition of *formula*:
(a) If $A \in At$ then $[\mathfrak{A}, \alpha](A)$ has already been defined.
(b) If $A, B \in \mathbb{F}_{pred}$ then

$$[\mathfrak{A}, \alpha](\neg A) = [\mathfrak{A}, \alpha](A)'$$

$$[\mathfrak{A}, \alpha]([A \vee B]) = [\mathfrak{A}, \alpha](A) \cup [\mathfrak{A}, \alpha](B)$$

$$[\mathfrak{A}, \alpha]([A \wedge B]) = [\mathfrak{A}, \alpha](A) \cap [\mathfrak{A}, \alpha](B)$$

$$[\mathfrak{A}, \alpha]([A \supset B]) = [\mathfrak{A}, \alpha](A)' \cup [\mathfrak{A}, \alpha](B).$$

(c) If $A \in \mathbb{F}_{pred}$, $x \in Fv$, $a \in Bv - Bv(A)$, then

$$[\mathfrak{A}, \alpha](\forall a A(a/x)) = \text{glb}\{[[\mathfrak{A}, \alpha[s/x]](A); s \in |\mathfrak{A}|\}$$

$$[\mathfrak{A}, \alpha](\exists a A(a/x)) = \text{lub}\{[[\mathfrak{A}, \alpha[s/x]](A); s \in |\mathfrak{A}|\}$$

where $\alpha[s/x]$ is the assignment defined by

$$\alpha[s/x](x_i) = \begin{cases} s & \text{if } x_i = x \\ \alpha(x_i) & \text{otherwise.} \end{cases}$$

$\mathbb{V}_{pred}$ is the class of *valuations*. Where it is necessary to specify the dependence of $\mathbb{V}_{pred}$ on the language $L(\tau)$ we shall write $\mathbb{V}_{pred}(\tau)$. If $[\mathfrak{A}, \alpha](A) = 1$ we write $(\mathfrak{A}, \alpha) \vDash A$ and say that α *satisfies the formula A in the realization* $\mathfrak{A}$. For a set Δ of formulas we write $(\mathfrak{A}, \alpha) \vDash \Delta$ when $(\mathfrak{A}, \alpha) \vDash A$ for all $A \in \Delta$ and say that $(\mathfrak{A}, \alpha)$ is a *model* of Δ. Thus $(\mathfrak{A}, \alpha) \vDash \forall a A(a/x)$ iff $(\mathfrak{A}, \alpha[s/x]) \vDash A$ for all $s \in |\mathfrak{A}|$, and $(\mathfrak{A}, \alpha) \vDash \exists a A(a/x)$ iff $(\mathfrak{A}, \alpha[s/x]) \vDash A$ for some $s \in |\mathfrak{A}|$. A formula A is *true in* $\mathfrak{A}$ if $(\mathfrak{A}, \alpha) \vDash A$ for all assignments $\alpha: Fv \to |\mathfrak{A}|$ (see (vii) below.)
(iv) A formula is said to be a *tautology* or to be *valid* if every valuation is a model of it. A set of formulas is *satisfiable* or *consistent* if it has a model and *inconsistent* if it has no model.
(v) A formula A is a *logical consequence* of the set Δ of formulas if every model of Δ is a model of A. This is equivalent to saying

$$\text{glb}(v(\Delta)) \leq v(A) \qquad (v \in \mathbb{V}_{pred}). \qquad (2.17)$$

(vi) Two formulas A, B are *logically equivalent* when each is a logical consequence of the other. We shall use the same symbols as in Definitions 2.4(iii) and 2.4(iv) for logical consequence and logical equivalence. For any set Δ of formulas we write $A \leftrightarrow_\Delta B$ when $\Delta, A \to B$ and $\Delta, B \to A$ and say that A is equivalent to B relative to Δ.
(vii) There is a subsidiary notion of logical consequence which is of special concern to the theorem-oriented rules of inference of Hilbert systems: a formula A is said to be a *th-consequence* of the set Δ (written $\Delta \to^{th} A$)

Predicate logic   43

provided that for every realization $\mathfrak{A}$, if B is true in $\mathfrak{A}$ for every $B \in \Delta$ then A is also true in $\mathfrak{A}$ (see Rasiowa and Sikorski 1963, p. 174).

Note that the value of $[\mathfrak{A}, \alpha]$ on the formula A depends only on the free variables which occur in A, i.e. for valuations α, β if $\alpha(x) = \beta(x)$ for all $x \in Fv(A)$ then $[\mathfrak{A}, \alpha](A) = [\mathfrak{A}, \beta](A)$. Further,

$$[\mathfrak{A}, \alpha[a(t)/x]](A) = [\mathfrak{A}, \alpha](A(t/x)) \qquad (t \in \mathbb{T}, x \in Fv). \qquad (2.18)$$

EXAMPLE 2.9.

(i) If $\Delta \to A$ then $\Delta \to^{\text{th}} A$ but the converse does not always hold. Thus for all formulas A, $A \to^{\text{th}} \forall a A(x/a)$ but formulas such that $A \not\to \forall a A(x/a)$ can easily be found.
(ii) In addition to the equivalences based on the proposition connectives (§2.2.2) there are the following well-known equivalences concerned with quantifiers:

   (a) Let $x \in Fv$. If $x \notin Fv(A)$ then $Qa A(a/x) \leftrightarrow A$ for $Q = \forall$ or $\exists$.
   (b) $Q_1 a A(a/x) \leftrightarrow \neg Q_2 a \neg A(a/x)$ and $\neg Q_1 a A(a/x) \leftrightarrow Q_2 a \neg A(a/x)$ for $Q_1 = \forall$ and $Q_2 = \exists$, or $Q_1 = \exists$ and $Q_2 = \forall$. etc. (e.g. Kleene 1962, §35). ◆

There are occasions (e.g. Chapter 6, §6.6) where it is expedient systematically to adjoin an additional one-place predicate symbol S to the list of predicate symbols of $L(\tau)$. This changes the type of the language to τ', say, where τ' is τ extended by a final one-place relation, i.e. $\tau'(S) = 1$. Thus, if $\mathfrak{A}$ is a structure of type τ and $X \subseteq |\mathfrak{A}|$ (i.e. X is a one-place relation) then $(\mathfrak{A}, X)$ (i.e. $\mathfrak{A}$ expanded by X) has type τ'. Let $A \in \mathbb{F}_{\text{pred}}(\tau)$ and let Φ be the $\mathbb{F}_{\text{pred}}(\tau')$ formula $\exists a [S(i) \wedge A(a/x)]$. Then, for any valuation α,

$$(\mathfrak{A}, X, \alpha) \vDash \Phi \qquad \text{iff} \qquad (\mathfrak{A}, x, \alpha[y/x]) \vDash A \text{ for some } y \in X.$$

In particular, if $X = \alpha(\Gamma)$, where Γ is the finite set $\{t_1, \ldots, t_n\}$ of terms, then in the expanded structure $(\mathfrak{A}, \alpha(\Gamma), \alpha)$, S is interpreted as the finite set $\{\alpha(\tau_1), \ldots, \alpha(\tau_n)\}$ and so

$$(\mathfrak{A}, \alpha(\Gamma), \alpha) \vDash \Phi \quad \Leftrightarrow \quad (\mathfrak{A}, \alpha) \vDash A(t_1/x) \vee \cdots \vee A(t_n/x).$$

Thus Φ is equivalent to a finite conjunction of formulas of $L(\tau)$. If, however, Γ is an arbitrary set of terms then Φ is equivalent to an *arbitrary disjunction*—naturally written as $\bigvee \{A(t/x); t \in \Gamma\}$ (see §2.2.2).

THEOREM 2.14. Let F be a formula of $L(\tau)$ and $a \in Bv - Bv(F)$.

(i) Let Δ be a set of formulas and let $x \in Fv$ such that $x \notin Fv(\Delta)$. If $\Delta \to F$ then $\Delta \to \forall a F(a/x)$.
(ii) $\forall a F(a/x) \to F(t/x)$ for all $t \in \mathbb{T}$.
(iii) $F(t/x) \to \exists a F(a/x)$ for all $t \in \mathbb{T}$.

(iv) Let Δ be a set of formulas and let $x \in Fv$ such that $x \notin Fv(\Delta, G)$. For any formulas F, G, if $\Delta, F \to G$ then
$$\Delta, \exists a F(a/x) \to G.$$

Proof. The proof uses Definition 2.13(iii)(c).

(i) Suppose that $\Delta \to F$ and $(\mathfrak{A}, \alpha) \vDash \Delta$. As $x \notin Fv(\Delta)$, $(\mathfrak{A}, \alpha[y/x]) \vDash \Delta$ for all $y \in |\mathfrak{A}|$. As $\Delta \to F$, $(\mathfrak{A}, \alpha[y/x]) \vDash F$ for all $y \in |\mathfrak{A}|$. Hence $(\mathfrak{A}, \alpha) \vDash \forall a F(a/x)$. Thus $\Delta \to \forall a F(a/x)$.
(ii) Suppose that $(\mathfrak{A}, \alpha) \vDash \forall a F(a/x)$. Then $(\mathfrak{A}, \alpha[y/x]) \vDash F$ for all $y \in |\mathfrak{A}|$. Hence $(\mathfrak{A}, \alpha[\alpha(t)/x]) \vDash F$ for all $t \in \mathbb{T}$. Thus $(\mathfrak{A}, \alpha) \vDash F(t/x)$.
(iii) Suppose that $(\mathfrak{A}, \alpha) \vDash F(t/x)$ where $t \in \mathbb{T}$. Then $(\mathfrak{A}, \alpha[\alpha(t)/x]) \vDash F$. Hence $(\mathfrak{A}, \alpha) \vDash \exists a F(a/x)$.
(iv) Suppose that $\Delta, F \to G$ and that $(\mathfrak{A}, \alpha) \vDash H$ for all $H \in \Delta, \exists a F(a/x)$. Then $(\mathfrak{A}, \alpha[y/x]) \vDash F$ for some $y \in |\mathfrak{A}|$. But $x \notin Fv(\Delta)$. Hence $(\mathfrak{A}, \alpha[y/x]) \vDash H$ for all $H \in \Delta, F$. Thus $(\mathfrak{A}, \alpha[y/x]) \vDash G$. Since $x \notin Fv(G)$, by (2.16), $(\mathfrak{A}, \alpha) \vDash G$. ∎

COROLLARY 2.15. *Let Δ be a set of formulas. For any $x \in Fv$ such that $x \notin Fv(\Delta)$ and any formulas F, G and bound variables $a, b \notin Fv(F, G)$,*
$$F \leftrightarrow_\Delta G \implies Qa F(a/x) \leftrightarrow_\Delta Qb G(b/x\}$$
where Q is $\forall$ or $\exists$.

2.3.3 Normal forms

In addition to the normal forms provided by propositional logic, predicate logic has normal forms based on the pattern of occurrence of quantifier symbols.

DEFINITION 2.14.

(i) A *sentence* or *closed formula* is a formula A such that $Fv(A) = \emptyset$.
(ii) A word on the alphabet M is *quantifer-free* if there are no occurrences of quantifier symbols in it.
(iii) A *prenex formula* is a formula WA where A is quantifier-free and W is a sequence of expressions $\forall a$ or $\exists b$ where $a, b \in Fv$. For such a formula, A is its *matrix* and W is the prenex. A *universal (existential) formula* is a prenex formula WF with matrix F and prenex W in which the only quantifiers that occur are universal (existential). Note that a matrix is not necessarily a formula according to the criteria of Definition 2.12 because bound variables could occur in it. A *variant* of the matrix A is a formula obtained from A by substituting distinct free variables from $Fv - Fv(A)$ for the bound variables in A.

Predicate logic

(iv) Let A be a formula. Suppose that $Fv(A) \neq \emptyset$. Let x be the first free variable in the list Fv which occurs in $Fv(A)$ and let a be the first bound variable which does not occur in A. Define

$$\alpha(A) =_{\text{def}} \forall a A(a/x) \qquad \varepsilon(A) =_{\text{def}} \exists a A(a/x).$$

Then $Fv(\alpha(A)) = Fv(\varepsilon(A)) = Fv(A) - \{x\}$. If $Fv(A)$ has $n > 0$ members

$$\forall A =_{\text{def}} \alpha^n(A) \qquad \exists A =_{\text{def}} \varepsilon^n(A).$$

The sentences $\forall A$ and $\exists A$ are the *universal closure* and *existential closure* of A respectively.

As Fv is infinite, the following theorem can be proved by induction on the construction of formulas (Church 1956, §39, Kreisel and Krivine 1967, Chapter 2, etc.).

THEOREM 2.16. *For every formula A there can be constructed a prenex formula B such that $A \leftrightarrow B$.*

EXAMPLE 2.10.

(i) For all formulas A, B if A is closed then by Example 2.9:
  (a) $A \leftrightarrow \forall A$
  (b) $A \supset \forall B \leftrightarrow \forall [A \supset B]$.
(ii) $\Delta \to^{\text{th}} A$ iff $\forall \Delta \to \forall A$ where $\forall \Delta =_{\text{def}} \{\forall B; B \in \Delta\}$.
(iii) $\Delta, A \to B$ iff $\Delta \to [A \supset B]$ (cf. Example 2.1(iii)). Further, if A is closed then $\Delta, A \to^{\text{th}} B$ iff $\Delta \to^{\text{th}} [A \supset B]$. Theorems of this type are known as 'deduction theorems' (Rasiowa and Sikorski 1963, p. 313). The rules of inference and the concept of 'proof from hypotheses' in a Hilbert system of classical predicate calculus can be so formulated that the restriction 'A is closed' can be avoided (e.g. Church 1956, §36, Barnes and Mackey 1975, Chapter 4). ◆

2.3.4 Canonical realizations

Many elementary results in the theory of models are based on the close connection between valuations of formulas of propositional logic and realizations in the algebra of terms.

DEFINITION 2.15.

(i) Let $X \subseteq Fv$. $Can(X)$ is the set of realizations on $\mathbb{T}(X)$. These are *canonical realizations* (Definition 2.13(i)). $Can(\emptyset)$ is the set of *Herbrand realizations* (the domain of a Herbrand realization is $\mathbb{T}(\emptyset)$, the set of constant terms). For $B \in Can(X)$, the *positive diagram* $pd(B)$ is defined by

$$pd(B) = \{p \in At(X); B \vDash p\}.$$

(ii) $\mathbb{V}_{Can(X)}$ is the set of valuations $[\mathfrak{A}, \alpha] \in \mathbb{V}_{pred}$ where $\mathfrak{A} \in Can(X)$. These valuations are *canonical valuations*; if $X = \emptyset$ they are called *Herbrand valuations*. The assignments in a canonical valuation are mappings from Fv to $\mathbb{T}(X)$—we call these *canonical assignments*. $\mathbb{V}_{Can(\emptyset)}$ is the set of Herbrand valuations. A *normal canonical (Herbrand) valuation* is canonical (Herbrand) valuation $[\mathfrak{A}, \iota]$ is the identity on the set of terms. A *canonical (Herbrand) model* of a formula A is a canonical (Herbrand) valuation v such that $v(A) = 1$.

EXAMPLE 2.11.

(i) Canonical assignments can be considered as substitution operators. Thus a canonical assignment σ can be extended to an operation on formulas A by defining $A\sigma$ to be the result of simultaneously substituting $\sigma(x_1)$ for x_1, $\sigma(x_2)$ for $x_2, \ldots$. Of course, only the restriction of σ to $Fv(A)$ matters here.
(ii) Let $S \subseteq At(X)$. There exists a unique $S^+ \in Can(X)$ whose positive diagram is in S, i.e. each set of atoms determines a canonical realization in $Can(X)$. Compare this with the propositional case (Definition 2.5). ◆

The connection between propositional logic and predicate logic is based on the following simple observation: each subset of atomic formulas in $At(X)$ corresponds uniquely to a propositional valuation on $At(X)$ and simultaneously to the positive diagram of a canonical realization in $Can(X)$. As far as quantifier-free formulas are concerned, predicate logic valuations are equivalent to propositional logic valuations on atoms.

LEMMA 2.17. *Let $[\mathfrak{A}, \alpha] \in \mathbb{V}_{pred}$. Then there exists $v \in \mathbb{V}_{prop}(At)$ such that $[\mathfrak{A}, \alpha](A) = v(A)$ for all quantifier-free formulas A.*

Proof. $v(Q) =_{def} [\mathfrak{A}, \alpha](Q)$, $(Q \in At)$. Thus $v \in \mathbb{V}_{prop}(At)$. By Definition 2.3 and Definition 2.13(iii)(c) the identity of the $[\mathfrak{A}, \alpha]$ and v on At extends to all quantifier-free formulas. ■

It easily follows that a normal canonical realization is equivalent to a propositional logic valuation on At in the following sense.

THEOREM 2.18. *Let $Y \subseteq Fv$. There is a bijection $\Theta: Can(Y) \to \mathbb{V}_{prop}(At(Y))$ such that $[\mathfrak{B}, \iota](A) = \Theta(\mathfrak{B})(A)$ for all quantifier-free formulas A with $Fv(A) \subseteq Y$ and all $\mathfrak{B} \in Can(Y)$.*

COROLLARY 2.19. *Let Δ be a set of quantifier-free formulas and Y a set of free variables such that $Fv(\Delta) \subseteq Y$. If Δ is satisfiable then it has a normal canonical model in $Can(Y)$.*

2.3.5 *The uniformity theorem and Herbrand's theorem*

To apply the preceding results to universal formulas we need the notion of *substitution instance*.

DEFINITION 2.16. Let A be a universal or existential formula and $T \subseteq Fv$.

(i) Let $J(Y)$ be the set of canonical assignments, i.e. mappings of free variables to $\mathbb{T}(Y)$ (substitutions of terms of $\mathbb{T}(Y)$). For $Z \subseteq Fv$, $J(Y, Z)$ denotes the set of assignments which leave Z unchanged. Thus, for any formula D, if $\sigma \in J(Y, Fv(D))$ then $D\sigma = D$ and $Fv(D\sigma) \subseteq Fv(D) + Y$.

(ii) A *Y-instance* of A is a formula resulting from the substitution of terms in $\mathbb{T}(Y)$ for the bound variables in the matrix of A. $Inst(Y, A)$ is the set of all such instances. Thus $Inst(Y, A) = \{A'\sigma; \sigma \in J(Y)\}$ and A' is a variant (Definition 2.14(iii)) of the matrix of A (any variant will do here). A $\emptyset$-instance of a universal or existential sentence is also called a *ground instance* (no free or bound variables occur in ground instances). $gd(A) =_{\text{def}} Inst(\emptyset, A)$ is the set of all ground instances of A. For a set U of universal sentences, $gd(U) =_{\text{def}} \sum \{gd(A); A \in U\}$.

EXAMPLE 2.12.

(i) Let $L(\tau)$ be a language with one constant c, one functional symbol f with $\tau(f) = 1$, and one predicate symbol P such that $\tau(P) = 2$. Then $\mathbb{T}(\emptyset) = \{c, f[c], f[f[c]], \ldots\}$. Let $A = \exists a P[a, x_1]$. Then

$$Inst(\emptyset, A) = \{P[c, x_1], P[f[c], x_1], P[f[f[c]], x_1], \ldots\}.$$

(ii) By Definition 2.13(iii)(c), for a universal formula A, set Y of free-variables such that $Fv(A) \subseteq Y$, and $\mathfrak{B} \in Can(Y)$,

$$(\mathfrak{B}, \iota) \vDash Inst(Y, A) \quad \Rightarrow \quad (\mathfrak{B}, \iota) \vDash A. \tag{2.19}$$

(iii) For any quantifier-free formula A, valuation $[\mathfrak{A}, \alpha]$, and canonical assignment σ,

$$[\mathfrak{A}, \alpha](A\sigma) = [\mathfrak{A}, \alpha \circ \sigma](A). \tag{2.20}$$

(iv) For any valuation $[\mathfrak{A}, \alpha]$, universal formula A, and $Y \subseteq Fv$,

$$(\mathfrak{A}, \alpha) \vDash A \quad \Rightarrow \quad (\mathfrak{A}, \alpha) \vDash Inst(Y, A). \tag{2.21}$$

♦

THEOREM 2.20. Let Δ be a set of universal formulas and Y a set of free variables such that $Fv(\Delta) \subseteq Y$. If Δ is satisfiable then it has a normal canonical model in $Can(Y)$.

Proof. As the formulas of Δ are universal, by (2.21), $\sum \{Inst(Y, A); A \in \Delta\}$ is also satisfiable. This is a set of quantifier-free formulas whose free variables are in Y. By Corollary 2.19 it has a normal canonical model in $Can(Y)$. By (2.19), this model is also a model of Δ. ∎

COROLLARY 2.21. *If a set of universal sentences has a model then it has a Herbrand model.*

EXAMPLE 2.13. The term 'universal' is essential in Corollary 2.21. Let $L(\tau)$ be a language with no function symbols except for one constant c and one unary predicate symbol P. Let $\Delta = \{P[c], \exists a \neg P[a]\}$. Then $\mathbb{T}(\emptyset) = \{c\}$ and so there are just two Herbrand realizations, $\mathfrak{H}_1 = \langle \{c\}, c, \{c\} \rangle$ and $\mathfrak{H}_2 = \langle \{c\}, c, \emptyset \rangle$, where the symbol c of $L(\tau)$ is interpreted as the constant c in the structures $\mathfrak{H}_1$ and $\mathfrak{H}_2$, and $\{c\}$ and $\emptyset$ are the interpretations of P in $\mathfrak{H}_1$ and $\mathfrak{H}_2$ respectively. Clearly, $\mathfrak{H}_1 \not\models \exists \neg P[a]$ and $\mathfrak{H}_2 \not\models P[c]$. Thus Δ does not have a Herbrand model. On the other hand, $\langle \{0, 1\}, 1, \{0\} \rangle$ is a model of Δ.

COROLLARY 2.22. *An existential sentence is valid iff all canonical realizations satisfy it.*

THEOREM 2.23. *Let Δ be a set of universal formulas. Then Δ has a model iff every finite subset of Δ has a model.*

Proof. Clearly, if Δ has a model then every finite subset of Δ has a model. Conversely, suppose that every finite subset of Δ has a model. Put $I(\Delta) =_{\text{def}} \sum \{I(Fv(\Delta), A); A \in \Delta\}$. By (2.21), every finite subset of $I(\Delta)$ has a model. By Lemma 2.17, every finite subset Y of $I(\Delta)$ has a propositional model. By the compactness theorem of propositional logic (Theorem 2.6) $I(\Delta)$ also has a propositional model. By Corollary 2.19, it has a canonical model in $Can(Fv(\Delta))$. By (2.19) Δ is satisfiable. ∎

THEOREM 2.24 (uniformity theorem). *An existential sentence E is valid iff there exists a finite set of ground instances of E whose disjunction is a tautology of propositional logic.*

Proof. Let C be a finite disjunction of ground instances of E. If C is a tautology of propositional logic, then by Theorem 2.18 it holds in all Herbrand realizations. By Definition 2.12(iii), E also holds in all Herbrand realizations. By Corollary 2.22, E is valid.

Conversely, suppose that E is valid. Then it certainly holds in every Herbrand realization. Thus, by Definition 2.13(iii), for every Herbrand realization there is a ground instance of E which holds in it. By Theorem 2.18, for each $v \in \mathbb{V}_{\text{prop}}(At(\emptyset))$ there is a ground instance F such that $v(F) = 1$. By Theorem 2.5 there exists a finite disjunction of ground instances of E which is a tautology of propositional logic. ∎

Predicate logic

THEOREM 2.25 (consistency theorem (Hilbert, Ackermann)). *The set U of universal sentences is not satisfiable iff there is a finite subset of $gd(U)$ which is inconsistent in propositional logic.*

EXAMPLE 2.14. Let U be a set of universal sentences and E a ground atom (or an existential sentence). Then $U \to E$ iff E (the disjunction of a finite number of instances of E) is a logical consequence in propositional logic of a conjunction of a finite number of ground instances of U. Thus, by Theorem 2.10, $U \to E$ iff $gd(U, \neg E) \vdash_{res} \square$. ♦

Let B be a quantifier-free formula with $Fv(B) \subseteq \{x_1, \ldots, x_n, y_1, \ldots, y_m\}$. Let F be the sentence

$$\forall a_1 \cdots \forall a_n \exists b_1 \cdots \exists b_m B(a_1, \ldots, a_n, b_1, \ldots, b_m / x_1, \ldots, x_n, y_1, \ldots, y_m)$$

where the indicated free variables are pairwise distinct. Let Γ be the set of canonical assignments to $T(\{x_1, \ldots, x_m\})$ which leave $x_1, \ldots, x_n$ fixed. For $\sigma_1, \ldots, \sigma_k \in \Gamma$ let

$$S(\sigma_1, \ldots, \sigma_k) =_{\text{def}} \forall a_1 \cdots \forall a_n [B\sigma_1 \vee \cdots \vee B\sigma_k](a_1, \ldots, a_n / x_1, \ldots, x_n)).$$

THEOREM 2.26 (Herbrand). *If F is valid there exist canonical assignments $\sigma_1, \ldots, \sigma_k$ such that $S(\sigma_1, \ldots, \sigma_k)$ is valid.*

Proof. Suppose, contrary to the above statement, that for every sequence $\sigma_1, \ldots, \sigma_k \in \Gamma$, $S(\sigma_1, \ldots, \sigma_k)$ is not valid. Then for each such sequence there is a structure $\mathfrak{A}$ satisfying $\neg S(\sigma_1, \ldots, \sigma_k)$ and so there is an assignment α to $\mathfrak{A}$ such that $(\mathfrak{A}, \alpha) \vDash \neg B\sigma_1 \wedge \cdots \neg B\sigma_k$. As $Fv(\neg B\sigma_1 \wedge \cdots \wedge \neg B\sigma_k) \subseteq \{x_1, \ldots, x_n\}$, by Theorem 2.17 there is a canonical model in $Can(\{x_1, \ldots, x_n\})$. By Corollary 2.19 there is a model in $\mathbb{V}_{\text{prop}}(At(\{x_1, \ldots, x_n\})$. Thus every finite subset of $\{\neg B\sigma, \sigma \in \Gamma\}$ has a model. By Theorem 2.6 the whole set has a propositional logic model, and so by Theorem 2.18 it has a model $\mathfrak{B} \in Can(\{x_1, \ldots, x_n\})$. But $\mathfrak{B} \vDash F$, so that $\mathfrak{B} \vDash B\sigma$ for some $\sigma \in \Gamma$. This is a contradiction. ∎

2.3.6 Skølem functions

The compactness of predicate logic follows from the previous section by augmenting the language M with additional appropriate function letters and finding for each formula an equivalent universal formula in the extended language.

DEFINITION 2.17. A *Skølem extension* of the predicate language $L(\tau)$ is a language which differs from $L(\tau)$ *only by additional function letters*. The type of this language is $\tau^{\clubsuit}$, say; $\tau^{\clubsuit}$ extends τ. The syntactical and semantical concepts are defined as for $L(\tau)$. Thus a realization of $L(\tau^{\clubsuit})$ differs from a realization of $L(\tau)$ only by having extra functions on its domain corresponding to the additional functions.

A *Skølem extension* of the $L(\tau)$ realization $\mathfrak{A}$ is obtained by adding appropriate functions (*Skølem functions*) to it.

By adjoining suitable functions letters to $L(\tau)$ arbitrary prenex formulas of $L(\tau^{\clubsuit})$ can be replaced by universal formulas which are equivalent in the sense of the following proposition.

PROPOSITION. Let $B \in \mathbb{F}_{\text{pred}}(\tau)$, $Fv(B) = \{x_0, \ldots, x_n\}$ and let $A = \exists a B(a/x_0)$. Let $L(\tau^{\clubsuit})$ be the Skølem extension of $L(\tau)$ formed by adjoining the new n-place function letter f to M. Then $A^{\clubsuit}$, the formula of $\mathbb{F}_{\text{pred}}(\tau^{\clubsuit})$ which is obtained from B by substituting the term $f[x_1, \ldots, x_n]$ for every occurrence of x_0 in B, has the following properties:

(i) $Fv(A) = Fv(A^{\clubsuit})$, $Bv(A) \supset Bv(A^{\clubsuit})$;
(ii) $A^{\clubsuit} \to A$;
(iii) for each realization $\mathfrak{A}$ and assignment α to $\mathfrak{A}$ there is a Skølem extension $\mathfrak{A}^{\clubsuit}$ of $\mathfrak{A}$ such that $[\mathfrak{A}, \alpha](A) = [\mathfrak{A}^{\clubsuit}, \alpha](A^{\clubsuit})$.

Assuming the axiom of choice, there exists a suitable function for adjoining to $\mathfrak{A}$ to make $\mathfrak{A}^{\clubsuit}$. Then Theorem 2.27 below follows from the above proposition by induction on length of prenex formula.

THEOREM 2.27. There is a Skølem extension $L(\tau^{\clubsuit})$ of $L(\tau)$ such that for each prenex formula $A \in \mathbb{F}_{\text{pred}}(\tau)$ there is a universal formula $A^{\clubsuit} \in \mathbb{F}_{\text{pred}}(\tau^{\clubsuit})$ such that:

(i) $A^{\clubsuit} \to A$;
(ii) for each realization $\mathfrak{A}$ and assignment α to $\mathfrak{A}$ there is a Skølem extension $\mathfrak{A}^{\clubsuit}$ of $\mathfrak{A}$ such that $[\mathfrak{A}, \alpha](A) = [\mathfrak{A}^{\clubsuit}, \alpha](A^{\clubsuit})$.

The following corollary shows that if a set of formulas has a model then it has a denumerable model. This is a well-known limitation on the expressive power of first-order predicate logic which is due to Skølem and Löwenheim.

COROLLARY 2.28. If a set of formulas is consistent then it has a Herbrand model (of $L(\tau^{\clubsuit})$).

Proof. Replace the set of formulas by an equivalent set of universal formulas in a suitable Skølem extension of $L(\tau)$ (Theorem 2.27). By Corollary 2.21 there exists a Herbrand model. ∎

THEOREM 2.29. Let $\Delta \subseteq \mathbb{F}_{\text{pred}}(\tau)$. Δ has a model iff every finite subset of Δ has a model.

Proof. By Theorem 2.27 we need only consider universal formulas. Theorem 2.23 then gives the stated compactness condition. ∎

2.3.7 Syntactical concepts

Hilbert systems Systems H and I of the classical and intuitionistic predicate calculus respectively are obtained by using the appropriate set of predicate formulas and axioms of the propositional calculus (cf. §2.2.7) and, in addition, the following schemas A11 and A12 for quantified formulas together with the two 'predicate rules' of inference (R∀) and (R∃):

- **A11:** $\forall a A(a/x) \supset A(t/x)$ for all $t \in \mathbb{T}$, where $a \in Bv - Bv(A)$
- **A12:** $A(t/x) \supset \exists a A(a/x)$ for all $t \in \mathbb{T}$, where $a \in Bv - Bv(A)$
- **R∀:** from $C \supset A$ derive $C \supset \forall a A(a/x)$ provided that $x \notin Fv(C)$
- **R∃:** from $A \supset C$ derive $\exists a A(a/x) \supset C$.

The resulting classical system is sound and complete.

Note that with R∀ and R∃

$$C \supset A \to^{\text{th}} C \supset \forall a A(a/x)$$
$$A \supset C \to^{\text{th}} \exists a A(a/x) \supset C$$

and so these rules preserve theoremhood (cf. Examples 2.9 and 2.10), but the conclusions in each case are not always logical consequences of their respective premisses.

Gentzen systems The system in §2.2.7 can similarly be extended to the predicate calculus by adding the following four quantifier rules:

R(∀): $\dfrac{(\Gamma : \Theta, A)}{(\Gamma : \Theta, \forall a A(a/x))}$ $\quad (a \notin Bv(A), x \notin Fv(\Gamma))$

L(∀): $\dfrac{(A(t/x), \Gamma : \Theta)}{(\forall a A(a/x), \Gamma : \Theta)}$ $\quad (a \notin Bv(A))$

R(∃): $\dfrac{(A, \Gamma : \Theta)}{(\exists a A(a/x)A, \Gamma : \Theta)}$ $\quad (a \notin Bv(A), x \notin Fv(\Gamma))$

L(∃): $\dfrac{(\Gamma : \Theta, A(t/x))}{(\Gamma : \Theta, \exists a A(a/x))}$ $\quad (a \notin Bv(a))$.

Theorem 2.12 also holds for this predicate version of G1. A version of Theorem 2.13, the *extended Hauptsatz*, gives a normal form of proofs in which there is a clear separation of propositional inferences from this involving quantifiers.

THEOREM 2.30. From a proof in the classical system G1 of a sequent containing only prenex formulas there can be constructed another proof of the same sequent which is cut-free and in which there occurs a sequent S, the mid-sequent, such that:

(i) no quantifers occur in S;
(ii) the part of the proof from S to the end-sequent consists solely of applications of quantifier and structural rules.

Herbrand's theorem (Theorem 2.26) was originally proved constructively. By Theorem 2.28, the normal form can be found from a proof of the sequent $(\lambda: F)$. Then from the propositional part of the proof (above the mid-sequent) a propositional proof of $(\lambda: S(\sigma_1, \ldots, \sigma_1))$ can be extracted.

Gentzen-type systems yield a 'goal-directed' strategy for establishing completeness theorems for classical systems (e.g. Schütte 1961).

Attempt to prove a sequent $(\lambda: F)$ by constructing possible proof-trees; from an infinite path in the proof-tree generated by failure to find a proof a realization in which F is false can be read off. Such a method will be used in Chapter 8 for three- and four-valued logics.

Elementary formal systems Definition 2.6 of the propositional calculus can be extended to the quantifier-free formulas of the predicate calculus.

DEFINITION 2.18

(i) The class FVH of *free-variable Horn clauses* is defined inductively as follows:
   (h1) $At \subseteq FVH$;
   (h2) if $B \in FVH$ and $A \in At$ then $(A \supset B) \in FVH$.
(ii) For a set $\Delta \subseteq FVH$, $gd(\Delta)$ is the set of all constant formulas obtained from formulas of Δ by substituting constant terms for free variables. Thus $gd(\Delta)$ is the set of ground instances (Definition 2.16) of formulas $\forall A$ where $A \in \Delta$.

Clearly $FVH \subseteq \mathbb{F}_{pred}(\tau)$. From Example 2.1(i), every free-variable Horn clause is logically equivalent to a quantifier-free definite program clause. Although Horn formulas are of the greatest importance in logic programming, they were first investigated in connection with direct products of relational systems (Horn 1951). A direct connection between direct products and logic programming has yet to be established.

DEFINITION 2.19. The *elementary system* $E(\tau)$ is the subsystem of H with no axioms, *modus ponens* as the sole rule of inference, and with formulas restricted to FVH. (Although there are no axioms the concept of *derivation from hypotheses* (Definition 2.8(b)) still applies.)

EXAMPLE 2.15.
(i) Every formula derivable in $E(\tau)$ from free-variable Horn clauses is also an FVH.
(ii) For any $\Delta \subseteq FVH$ and $A \in At$

$$\Delta \vdash_H A \quad \Leftrightarrow \quad \Gamma \vdash_{E(\tau)} A.$$ ◆

If the language $L(\tau)$ contains an individual constant z, say, for at least one function letter f with $n = \tau(f) > 0$ then $\mathbb{T}$ contains the constant terms

$$z, f[z, z, \ldots, z], f[f[z, z, \ldots, z], z, \ldots, z],$$
$$f[f[f[z, z, \ldots, z], z, \ldots, z], z, \ldots, z] \ldots$$

which can be considered as representatives in $E(\tau)$ of the numbers 0, 1, 2, 3, ... In fact, put $v(0) = z$ and $v(n + 1) = f[v(n), z, \ldots, z]$. Then $v(n)$ is the numeral representing n. If $L(\tau)$ has a one-place predicate letter, N say, then these numerals can be 'generated' in $E(\tau)$ by a certain set of free-variable Horn clauses as follows. Let $\Delta = \{N[z], N[x] \supset N[f[x, z, \ldots, z]]\}$ where $x \in Fv$. Then, for $t \in T$,

$$gd(\Delta) \vdash_{E(\tau)} N[t] \quad \Leftrightarrow \quad (En \in \mathbb{N})(t = v(n)).$$

The possibility of numerical computations in $E(\tau)$ springs from this sort of representation.

DEFINITION 2.20. A relation $R \subseteq \mathbb{N}^n$ is *representable* in $E(\tau)$ if there exists an n-place predicate letter Q of $L(\tau)$ and a finite set $\Delta \subseteq FVH$ such that, for all $r_1, \ldots, r_n \in \mathbb{N}$,

$$gd(\Delta) \vdash_{E(M)} Q[v(r_1), \ldots, v(r_n)] \quad \Leftrightarrow \quad (r_1, \ldots, r_n) \in R.$$

Every relation which is representable in $E(\tau)$ is recursively enumerable. Conversely, by a trivial modification of Smullyan (1961), we have the following theorem.

THEOREM 2.31. If $L(\tau)$ has a sufficient supply of predicate and function letters, then every recursively enumerable relation is representable in $E(\tau)$.

Thus $E(\tau)$ has the maximum possible computing power.

The devices described here for generating recursively enumerable relations can easily be transformed to Smullyan's *elementary formal systems* by reduction to finite alphabets using the well-known method of Post.

The elementary system $E(\tau)$ is connected with predicate logic via the expression of free-variable Horn clauses as definite program clauses and by Definition 2.7.

THEOREM 2.32. Let $H_1, \ldots, H_m \in FVH$. Then for each ground atom G

$$\forall H_1, \ldots, \forall H_m \to G \quad \Leftrightarrow \quad G \in \mathfrak{H}(gd(\forall H_1) + \cdots + gd(\forall H_m))$$
$$\Leftrightarrow \quad gd(\forall H_1, \ldots, \forall H_m) + \{\neg G\} \vdash_{res} \square.$$

Proof. By Example 2.14, $\forall H_1, \ldots, \forall H_m \to G$ is equivalent to $gd(\forall H_1) + \cdots + gd(\forall H_m) \to_{\text{prop}} G$. By Corollary 2.4, this is equivalent to $G \in \mathfrak{H}(gd(\forall H_1) + \cdots + gd(\forall H_m))$. The connection with resolution follows from Example 2.14. ∎

COROLLARY 2.23 (see Kowalski 1974). *If $L(\tau)$ has a sufficient supply of predicate and function letters, then every recursively enumerable relation is representable, i.e. for every recursively enumerable relation R there exists a predicate letter of $L(\tau)$ and a conjunction H of free-variable Horn clauses such that for all $r_1, \ldots, r_n \in \mathbb{N}$*

$$H \to Q(v(r_1), \ldots, v(r_n)) \quad \Leftrightarrow \quad (r_1, \ldots, r_n) \in R.$$

Note that negation does not occur explicitly in the definitions of FVH and $E(\tau)$. However, negative information can be dealt with positively by the notion of finite refutability (see Chapter 4, §4.2.2).

DEFINITION 2.21. Let Δ be a set of formulas. *Ref*(Δ) is defined thus: for any formula F, $F \in Ref(\Delta)$ (read 'F is *finitely refutable*') if there exists a finite model of Δ in which F is false. (Thus, if $F \in Ref(\Delta)$ then there is a finite realization which is a counterexample to the claim that $\Delta \to F$.)

It follows from Corollary 2.33 that there are formulas $H_1, \ldots, H_n \in FVH$ such that the set of ground atoms which are not consequences of $\Delta =_{\text{def}} \{\forall H_1, \ldots, \forall H_n\}$ is not recursively enumerable. However, the set of ground atoms in $Ref(\Delta)$ is recursively enumerable and none of them is a consequence of Δ. $Ref(\Delta)$ is about the best that can be done to obtain a recursively enumerable set containing 'negative' information.

EXAMPLE 2.16. Consider a language with one one-place relation symbol E, one individual constant z, and one one-place function symbol S, so that the numerals $v(n)$ are defined by

$$v(0) = z$$

$$v(n+1) = S[v(n)].$$

Let H_1, H_2 be the formulas $E[z]$ and $E[x_1] \supset E[S[S[x_1]]]$ respectively. Then $\mathfrak{H}(gd(\forall H_1) + gd(\forall H_2))$ is the set of ground atoms $E[v(2n)]$ representing the even numerals. Now $E[v(2n+1)] \notin \mathfrak{H}(gd(\forall H_1) + gd(\forall H_2))$. By Theorem 2.32 $E[v(2n+1)]$ is not a consequence of $\{\forall H_1. \forall H_2\}$. Consider the finite realization $\mathfrak{M} = \langle \{0, 1, 2, \ldots, 2n+2\}; 0, f, \bar{E} \rangle$, where

$$f(r) = \begin{cases} r+1 & \text{if } r < 2n+2 \\ 2n+2 & \text{if } r = 2n+2 \end{cases}$$

and $\bar{E} = \{0, 2, 4, \ldots, 2n+2\}$. Then $\mathfrak{M}$ is a model of $\{\forall H_1, \forall H_2\}$ when z, S, E are interpreted as $0, f, \bar{E}$ respectively. But $\mathfrak{M} \nvDash E[v(2v+1)]$. Thus, $E[v(2n+1)] \in Ref(\{\forall H_1, \forall H_2\})$. ◆

Predicate logic

There are, however, some pathological cases which can arise with finite refutability. There exist finite sets Δ of free-variable Horn clauses (namely those which generate *simple sets* (see Rogers 1967)) such that the set of ground atoms in $Ref(\Delta)$ is finite, but the set of ground atoms which are not consequences of Δ is infinite.

2.3.8 Logic programming

The definitions given in §2.2.5 can be extended to formulas of the predicate calculus.

DEFINITION 2.22.

(i) A *clause* is a sentence of the form $\forall[L_1 \vee \cdots \vee L_m]$ where $L_1, \ldots, L_m$ are literals. A *definite program clause* is a clause in which just one of the conjuncts is a positive literal. A definite program is therefore logically equivalent to a formula $\forall H$ where $H \in FVH$. A definite program is a set of definite program clauses.
(ii) For positive literals $A, B_1, \ldots, B_n$, the expression $A \leftarrow B_1, \ldots, B_n$ denotes the clause $\forall[\neg B_1 \vee \cdots \vee \neg B_n \vee A]$ which is logically equivalent to $\forall[B_1 \supset \cdots [B_n \supset A] \cdots]$. A *unit clause* is a clause of the form $A \leftarrow$, i.e. $\forall A$.
(iii) A *definite goal* is a clause of the form $\leftarrow B_1, \ldots, B_k$ where $B_1, \ldots, B_k$ are positive literals; this is logically equivalent to $\neg \exists [B_1 \wedge \cdots \wedge B_k]$. The B_k are *subgoals*.

A problem in the elementary phase (i.e. the Horn clause phase) of logic programming is to prove that, for a given definite goal $\leftarrow B_1, \ldots, B_k$ and definite program $\{H_1, \ldots, H_m\}$, $\exists[B_1 \wedge \cdots \wedge B_k]$ is a logical consequence of $\{H_1, \ldots, H_m\}$, i.e. that the set of universal sentences $\{H_1, \ldots, H_m, \forall[\neg B_1 \vee \cdots \vee \neg B_k]\}$ is inconsistent. By Corollary 2.22 there is a finite conjunction of instances of these sentences which is inconsistent in the propositional logic. The procedure for performing the proof is therefore to generate the instances methodically using resolution (§2.2.7(ii)) to derive the empty clause.

Definite programs do not admit negation in the bodies of clauses. For the extensive theoretical and practical problems concerning implementation of negation in logic programming see Shepherdson (1988).

As the elementary phase of logic programming concerns Horn clauses and universal sentences, for semantical considerations it suffices to deal with Herbrand models. By analogy with the propositional case (§2.2.4) we have the following definition.

DEFINITION 2.23.

(i) For Herbrand valuation u, $pd(u)$, the *positive diagram* of u, is the set of ground atoms which hold in u, i.e. $pd(u) =_{\text{def}} \{p \in At(\emptyset); u(p) = 1\}$ (Definition 2.5). For $Q \subseteq At(\emptyset)$, Q^+ is the (unique) valuation such that $pd(Q^+) = Q$.

(ii) A *minimal (minimum) Herbrand model* of a set Δ of formulas is a model $\mathfrak{B}$ such that $[\mathfrak{B}, \iota]$ is minimal (minimum) with respect to the inclusion relation between the positive diagrams of models of Δ.

(iii) Let $\Gamma \subseteq Can(\emptyset)$. The *intersection* of Γ is defined by

$$\wedge \Gamma =_{\text{def}} (\bigcap \{pd(\mathfrak{A}); \mathfrak{A} \in \Gamma\})^+$$

As in Corollary 2.2 the set of Herbrand models of a definite program is closed under arbitrary intersection. Hence there exists a minimum Herbrand model associated with the procedural interpretation of the program.

Let Δ be the set of definite programs. The inductive definition $I(\Delta)_{\text{ind}}$ constructed from the set $I(\Delta)$ of substitution instances of Δ (cf. §2.2.5) as in the propositional case then gives the following theorem (see van Emden and Kowalski 1976).

THEOREM 2.34. Let Δ be a definite program. Let Y be the set of ground atoms generated by $I(\Delta)_{\text{ind}}$. Then Y^+ is the minimum model of Δ.

2.3.9 Variants and applications of predicate logic

There are several useful extensions of classical first-order predicate logic— second- and higher-order logics, weak second-order logic (with quantification over finite sets), first-order logic extended by the quantifiers 'there exist infinitely many' or 'there exist uncountably many', and logic with infinite conjunctions and disjunctions, etc. An abstract model theory, formulated by Barwise (1974), provides a conceptual apparatus for the contrast and comparison of logics, for generalization of such facts of classical first-order logic as the theorems of Löwenheim and Skølem, Hanf, etc., and for co-ordinating syntactical transformations with corresponding semantical operations with corresponding semantical operations. This area of research has since attracted much attention (see Barwise and Feferman 1985).

A large number of logical systems based on classical logic have found application in computer science—equational logic as in the theory of abstract data types, Horn clause logic as in logic programming (§2.3.8), modal logic (Chapter 11), temporal logic (Chapter 6, §6.10), etc. An important application of first-order predicate logic is in the construction of a logic of computer programs. Its aim is to provide a set of axioms and rules of inference for proving things about programs such as 'termination', 'equivalence' and 'correctness'. An influential initiative in this direction was the axiomatic method of Hoare (1969, 1971) whose approach was essentially due to Floyd (1967) but applied to program texts rather than flow diagrams. A serious mathematical examination of Hoare's ideas was undertaken in the seminal paper by Cook (1978).

The formulas of Hoare's system are triples of the form $\{A\}\ \pi\ \{B\}$, where π is a statement in the programming language and A and B are formulas of classical predicate logic (the *assertion language*). The *partial correctness formula* $\{A\}\ \pi\ \{B\}$ is *true* if, and only if, whenever A holds for the initial values of the program variables and π is executed, then either π will fail to terminate or B will be satisfied by the final values of the program variables. A typical rule of inference is

$$\frac{\{A\ \&\ B\}\ \pi\ \{A\}}{\{A\}\ \text{while}\ B\ \text{do}\ \pi\ \{A\ \&\ \neg B\}}.$$

The axioms and rules of inference embody the meanings of the individual statements of the programming language. Proofs of correctness are constructed by using these axioms together with a proof system for the assertion language. Clarke (1978) showed that for sufficiently complex ALGOL-like languages there cannot be a Hoare axiom system which is both sound and relatively complete in the sense of Cook (1978). The delicate issues of completeness, decidability, etc. are discussed by Apt (1981).

An abstract model theory of such logics arising in computer science, which resembles the previously mentioned abstract model theory in its aim of generalizing some basic results in model theory, but differs in the direction of development and in the results generalized, has been erected by Burstal and Goquen (1980, 1984). Mesequer (1989) added an abstract notion of the proof theory of a logic to abstract model theory to give a general concept of a 'logic' which integrates the semantics with the proof theory. As with abstract model theory, commitment to particular constructions—'language', 'proof', 'structure', 'model', etc.—is avoided by using category theory. An abstract notion of 'logic programming' results from this work.

3
Abstract logics

3.1 Introduction

There are many formal logics—classical logic (already encountered in Chapter 2), intuitionistic logic, various modal and temporal logics, and numerous logics arising in various ways from applications such as the logic of automata and the logic of quantum mechanics. Moreover, there are many structures—such as context-free languages, combinatorial systems, etc.—which are sufficiently logic-like to deserve the name *logic*. Many logics are primarily the embodiment of some notion of *truth*. They are thereby mainly concerned with the generating *theories*—a corpus of theorems or 'universal truths'—thus tending to obscure the view of formal logic as the science of valid inference. In Chapter 1, §1.3.2, there arose the problem of treating *neutrality* as if it were a truth-value. What has to be done to it to make a *logic* out of *truth*, *falsehood* and *neutrality*? Truth-values alone do not constitute a logic. Our strategy must clearly be guided by the various extant logics. In the brief treatment of classical logic in Chapter 2 the notion of *logical consequence* played some part, though it was subordinate to the idea of *theoremhood*. But in the discussion of scientific theories in Chapter 1 special significance was given to the deductive postulate. On grounds of generality, it can be claimed that logical consequence is more fundamental than theoremhood, as all that needs to be said about theoremhood can also be stated in terms of logical consequence. This view has been well expressed by Smiley (1981):

> We ought to distinguish between a *logic* and a *theory*. The principle constituent of a system of logic is a relation, the relation of consequence, whereas a theory is constituted by a property, the property of theoremhood. Moreover the definition of a theory presupposes an underlying logic, or emerges in the requirement that a set of theorems shall be closed under consequence. ... On the present account no logic is a theory, nor is it possible for a theory to be reducible to a logic. On the other hand this sharp distinction between logics and theories is a purely formal one which ignores the question of content and leaves open the possibility of one man's logic being as strong as another man's theory (e.g. the underlying logic for a theorem of mechanics might well include geometrical and numerical principles).

Accordingly, in this chapter (§3.5) a definition of *logics* will be given based on Tarski's well-known axiomatization of logical consequences (Tarski 1956, Chapter 3) and some general properties of logics derived from it. From this point of view, a logic is then a particular formalized science.

Prelogics 59

> Formalized deductive disciplines form the field of research of metamathematics roughly in the same sense in which spatial entities form the field of research of geometry. These disciplines are regarded, from the standpoint of metamathematics as sets of *sentences*. Those sentences ... are themselves regarded as inscriptions of a well-defined form. ... From the sentences of any set X certain other sentences can be obtained by certain operations called *rules of inference*. These sentences are called *consequences of the set X*. ... An exact definition of the two concepts of sentence and of consequence can only be given in those branches of metamathematics in which the field of investigation is a concrete formalized discipline.
>
> Tarski 1956, Chapter 3

In Chapter 6 we shall be concerned with constructing logics from sets of truth-values themselves organized as logics. The principles of construction involve such 'algebraic' operations on logics as homomorphism and restriction to subsystems. Accordingly we define a formal language with sufficient expressive power to define properties which are transferred by these operations (§§3.3–3.5) and, indeed, express the property of being a logic.

Following the definition of logics, various broad classes of logics are discussed under the following headings: compact logics (§3.8), formal systems (§3.9), finite logics (§3.10), and Gentzen systems (§3.11).

3.2 Prelogics

All extant logics consist at least of a set of sentences or propositions which combine by means of connectives and which are related by some concept of logical consequence. Our starting point, then, is a set Σ of objects which we can call *sentences*, *propositions*, or *formulas*. The sentences in Σ can be joined together in various ways to form new sentences. These modes of combination, we suppose, are effected by means of functions which, under certain conditions, will be identified with connectives. Thus sentences form an algebra, e.g. $\mathbb{F}_{prop}$ of Chapter 2, §2.1. But what distinguishes a logic from an algebra is a concept of *logical consequence*. With each set Δ of sentences there is associated the set $Cn(\Delta)$ of logical consequences of Δ, the set of sentences which can be inferred from Δ. Cn is a set-to-set function mapping $\mathbb{P}(\Sigma)$ to $\mathbb{P}(\Sigma)$.

In many logical systems, the function Cn is constructed, in the first instance by finitist motives, from a notion of *direct derivation* in which a sentence is derived from a finite number of hypotheses. In others, particularly those which are primarily concerned with applications springing from a semantic notion of 'satisfaction', Cn rests on non-finitist concepts. Yet there are logics, such as classical logic, in which (by non-finitist arguments) a 'completeness theorem' establishes the coincidence of the deductive notion with the semantic notion. Thus it cannot be presumed, without violation of the philosophical starting point, that all logics have a semantic basis or that all logics are intrinsically deductive. In concrete examples a logic might be armed with a suitable syntactical system or an appropriate semantic structure or both.

Before entering further into the definition of logics it is necessary to introduce some terminology for handling set-to-set functions defined on sets of sentences or on finite sets of sentences.

DEFINITION 3.1. Operations and relations on sets induce corresponding operations on set-to-set functions. Let C, D be set-to-set functions on a set Σ

(i) The *union* and *intersection*, $C + D$ and $C \cdot D$ respectively, of C and D are defined by

$$(C + D)(\Delta) =_{\text{def}} C(\Delta) + D(\Delta) \qquad (\Delta \subseteq \Sigma)$$

$$(C \cdot D)(\Delta) =_{\text{def}} C(\Delta) \cdot D(\Delta) \qquad (\Delta \subseteq \Sigma).$$

Likewise, if K is a family of set-to-set functions, the $\sum K$ and $\prod K$ are the set-to-set functions defined by

$$(\sum K)(\Delta) =_{\text{def}} \sum \{D(\Delta); D \in K\}$$

$$(\prod K)(\Delta) =_{\text{def}} \prod \{D(\Delta); D \in K\}.$$

The inclusion relation between sets induces such a relation between set-to-set functions:

$$C \subseteq D \quad \Leftrightarrow \quad C(\Delta) \subseteq D(\Delta) \qquad \text{for all } \Delta \subseteq D$$

$$D \supseteq C \quad \Leftrightarrow \quad C \subseteq D.$$

(ii) C is said to be *monotone increasing* if

$$C(\Delta) \subseteq C(\Delta + \Gamma) \qquad (\Delta, \Gamma) \subseteq \Sigma)$$

and *monotone decreasing* if

$$C(\Delta) \supseteq C(\Delta + \Gamma) \qquad (\Delta, \Gamma \subseteq \Sigma).$$

(iii) A *finitary* set-to-set function C is a set-to-set function defined only on finite sets: $C \colon \mathbb{P}_\omega(\Sigma) \to \mathbb{P}(\Sigma)$. The notions of *union* and *intersection* of (i) above and of *monotone increasing* and *monotone decreasing* apply also to finitary set-to-set functions.

(iv) Associated with each set-to-set function C is its restriction $C^\heartsuit$ to finite sets, so that $C^\heartsuit$ is a finitary set-to-set function. Conversely, associated with each finitary set-to-set function D is the set-to-set function $D^\heartsuit$ where

$$D^\heartsuit(\Delta) =_{\text{def}} \sum \{D(\Gamma); \Gamma \in \mathbb{P}_\omega(\Delta)\} \qquad (\Delta \in \mathbb{P}(\Sigma)).$$

For a set-to-set mapping C, $C^{\heartsuit\heartsuit}$ is the *compact kernel* of C: C is said

Prelogics

to be compact if $C = C^{♥♥}$. Thus C is compact iff for all $\Delta \subseteq \Sigma$

$$C(\Delta) = \sum \{C(\Gamma); \Gamma \in \mathbb{P}_\omega(\Delta)\}.$$

EXAMPLE 3.1.

(i) If C is compact it is monotone increasing. For if $\Gamma, \Delta \subseteq \Sigma$ then $\mathbb{P}_\omega(\Gamma) \subseteq \mathbb{P}_\omega(\Gamma + \Delta)$. Hence

$$C(\Gamma) = \sum \{C(\Psi); \Psi \in \mathbb{P}_\omega(\Gamma)\}$$
$$\subseteq \sum \{C(\Psi); \Psi \in \mathbb{P}_\omega(\Gamma + \Delta)\} = C(\Gamma + \Delta).$$

(ii) Let C be a finitary set-to-set function of Σ. Then $C^♥ = C^{♥♥♥}$ is compact. $C^{♥♥}$ is compact. ♦

For the moment, then, no relation between the algebra and the consequence function will be assumed. For want of a better term, an algebra equipped with a set-to-set function will be called a *prelogic* and, in anticipation of the later definition of *logics*, the set-to-set function will be called the *consequence function* of the prelogic. The definition is sufficiently broad to cover some of the strange mutants appearing in artificial intelligence. Logics are a special kind of prelogics which will be defined in §3.5. From the logical consequence function of a prelogic there is derived the notion of *logical equivalence*: two sentences (elements of the prelogic) or two sets of sentences are *logically equivalent* if they have the same set of logical consequences (Tarski 1956, Chapter 3).

DEFINITION 3.2.

(i) Let C be a set-to-set function on a set Σ. C determines an equivalence relation C^e on the subsets of Σ by

$$\Delta = \Gamma \, mod(C^e) \quad \Leftrightarrow_{\text{def}} \quad C(\Delta) = C(\Gamma) \qquad (\Gamma, \Delta \subseteq \Sigma).$$

This induces an equivalence relation on Σ, again denoted by C^e:

$$A = B \, mod(C^e) \quad \Leftrightarrow_{\text{def}} \quad \{A\} = \{B\} \, mod(C^e).$$

In a context where C is fixed, $A \leftrightarrow B$ is an abbreviation of $A = B \, mod(C^e)$.

(ii) A *prelogic* $\mathfrak{L}$ *of type* τ is an algebra $\mathfrak{A}$ of type τ together with a set-to-set function $C: \mathbb{P}(|\mathfrak{A}|) \to \mathbb{P}(|\mathfrak{A}|)$; we write $\mathfrak{A} = \langle \mathfrak{A}, C \rangle$. $\mathfrak{A}$ is the *algebra of formulas* of $\mathfrak{L}$, C is the *consequence function* of $\mathfrak{L}$, and C^e is the *logical equivalence* on $\mathfrak{L}$.

(iii) A prelogic $\langle \mathfrak{A}, C \rangle$ is *monotone increasing*, *monotone decreasing*, or *compact* if C is monotone increasing, monotone decreasing, or compact respectively.

EXAMPLE 3.2.

(i) Any algebra $\mathfrak{A}$ can be made into a prelogic trivially by adjoining the identity function on $\mathbb{P}(|\mathfrak{A}|)$. The prelogic so formed is compact and, according to Definition 3.1(ii), both monotone increasing and monotone decreasing.

(ii) Consider the algebra of type $\langle 1, 2, 2, 2 \rangle$ (Appendix, §A) of formulas $\mathbb{F}_{\text{prop}}$ of Chapter 2, §2.1. For any set X of formulas let $C(X) =_{\text{def}} \{A; X \to_{\text{prop}} A\}$ (see Definition 2.4(iii)). Then $\mathfrak{L} = \langle \mathbb{F}_{\text{prop}}, C \rangle$ is a monotone prelogic of type $(1, 2, 2, 2)$.

(iii) Consider the prelogic $\mathfrak{M} = \langle \mathbb{F}_{\text{prop}}, D \rangle$ where

$$D(\Delta) = \{A; \Delta + \{A\} \text{ is consistent}\}.$$

Then for all $A, B \in \mathbb{F}_{\text{prop}}$,

$$A = B \, mod(C^e) \quad \Leftrightarrow \quad A = B \, mod(D^e). \tag{3.1}$$

It is trivial to prove that $A = B \, mod(C^e)$ implies $A = B \, mod(D^e)$. For the converse, suppose that $A = B \, mod(D^e)$. Then $A \to B$. For suppose to the contrary that there exists $v \in \mathbb{V}$ such that $v(A) = 1$ but $v(B) \neq 1$. Then $v(\neg B) = 1$. Thus $\{A, \neg B\}$ is consistent so that $\neg B \in D(A)$. Hence $\neg B \in D(B)$, i.e. $\{\neg B, B\}$ is consistent. This is a contradiction. Hence $A \to B$. Likewise $B \to A$. Thus $A = B \, mod(D^e)$. From (3.1) it follows that the logical equivalence D^e of $\mathfrak{M}$ coincides with the logical equivalence C^e of $\mathfrak{L}$. $\mathfrak{M}$ is a prelogic but it is clearly not monotone increasing. In fact it is monotone decreasing:

$$D(\Delta + \Gamma) \subseteq D(\Delta) \quad (\Delta, \Gamma \subseteq \mathbb{F}_{\text{prop}}).$$

$\mathfrak{M}$ is a simple example of a 'non-monotonic logical system' in which the inference of A from hypotheses (i.e. Δ) could be invalidated when the hypotheses are augmented by further information Γ. It is alleged that this phenomenon is implicit in artificial intelligence (see Doyle and McDermott 1980, 1982).

(iv) The notion of a prelogic covers many kinds of structure which are not considered to be within the domain of logic. For instance, let $\mathfrak{L} = \langle \mathfrak{A}, C \rangle$ be a prelogic. Suppose that C satisfies the following *closure conditions* due to Kuratowski (1922):

$$C(\Delta + \Gamma) = C(\Delta) + C(\Gamma) \quad \Delta \subseteq C(\Delta)$$

$$C(C(\Delta)) = C(\Delta) \quad C(\emptyset) = \emptyset$$

for all $\Delta, \Gamma \subseteq |\mathfrak{A}|$. Then the sets $C(\Delta)$ are the closed sets of a topology on $|\mathfrak{A}|$ (Appendix, §T). Also, in the study of combinatorial geometries, a *geometry* is defined as a set Σ on which there is defined a set-to-set function C such that the following conditions are satisfied:

(a) for all $\Delta \subseteq \Sigma$, $A \in \Sigma$,

$$\Delta \subseteq C(\Delta), \quad C(C(\Delta)) \subseteq C(\Delta), \quad C(\emptyset) = \emptyset, \quad C(\{A\}) = \{A\};$$

(b) the 'localization condition' (this is the same as compactness, Definition 3.1(iv))

$$C(\Delta) = \sum \{C(\Gamma); \Gamma \in \mathbb{P}_\omega(\Delta)\};$$

Prelogics 63

(c) the 'exchange condition'

$$C(\Delta) = \Delta \quad \& \quad B \in \Sigma - \Delta \quad \& \quad A \in C(\Delta + \{B\}) - \Delta \quad \Rightarrow \quad B \in C(\Delta + \{A\}).$$

This does not mean that the subject of geometry is included in logic. Subjects are identified by their concepts and applications: two *different* subjects may well contain isomorphic structures. ◆

The customary notions of general algebra—*subsystem, homomorphism, derived system,* ... (Appendix, §A)—can be extended to pre-logics.

DEFINITION 3.3. Let $\mathfrak{L}_1 = \langle \mathfrak{A}_1, C_1 \rangle$, $\mathfrak{L}_2 = \langle \mathfrak{A}_2, C_2 \rangle$ be pre-logics of the same type.

(i) A homomorphism θ of $\mathfrak{A}_1$ into $\mathfrak{A}_2$ is a *weak homomorphism* of $\mathfrak{L}_1$ into $\mathfrak{L}_2$ if

$$C_1(\Delta) \subseteq \theta^{-1} C_2(\theta(\Delta)) \qquad (\Delta \subseteq |\mathfrak{A}_1|)$$

and a *strong homomorphism* if

$$C_1(\Delta) = \theta^{-1} C_2(\theta(\Delta)) \qquad (\Delta \subseteq |\mathfrak{A}_1|).$$

If θ maps $\mathfrak{L}_1$ onto $\mathfrak{L}_2$, then $\mathfrak{L}_2$ is a *weak (strong) homomorphic image* of $\mathfrak{L}_1$ if θ is a weak (strong) homomorphism. A strong homomorphism of $\mathfrak{L}_1$ into $\mathfrak{L}_2$ is an *isomorphism* if it is an (algebra) isomorphism (Appendix, §A) of $\mathfrak{A}_1$ into $\mathfrak{A}_2$.

(ii) If $\mathfrak{A}_1$ is a subalgebra of $\mathfrak{A}_2$ and the identity on $|\mathfrak{A}_1|$ is a weak (strong) homomorphism of $\mathfrak{L}_1$ into $\mathfrak{L}_2$, i.e. $C_1(\Delta) \subseteq |\mathfrak{A}_1| \cdot C_2(\Delta)$ (or $C_1(\Delta) = |\mathfrak{A}_1| \cdot C_2(\Delta)$) for all $\Delta \subseteq |\mathfrak{A}_1|$, then $\mathfrak{L}_1$ is a weak (strong) *subprelogic* of $\mathfrak{L}_2$.

(iii) If $\mathfrak{B}$ is a derived algebra of $\mathfrak{A}$ (Appendix, §A) (so that $|\mathfrak{B}| = |\mathfrak{A}|$) then the pre-logic $\langle \mathfrak{B}, C \rangle$ is said to be a *derived prelogic* of $\langle \mathfrak{A}, C \rangle$. A weak (strong) subprelogic of a derived prelogic is said to be a *weak (strong) subderived prelogic*.

The concept of homomorphism for prelogics (which by Definition 3.3(ii) includes the concept of *subprelogic*) is an important tool for constructing prelogics, particularly for semantically based logics (Chapter 5). Thus, let $\mathfrak{L}_2 = \langle \mathfrak{A}_2, C_2 \rangle$ be a prelogic, $\mathfrak{A}_1$ an algebra of the same type as $\mathfrak{A}_2$, and Θ a homomorphism of $\mathfrak{A}_1$ into $\mathfrak{A}_2$. Define C_1 to be the set-to-set function on $|\mathfrak{A}_1|$ by

$$C_1(\Delta) =_{\text{def}} \{A \in |\mathfrak{A}_1|; \Theta(A) \in C_2(\Theta(\Delta))\} \qquad (\Delta \subseteq |\mathfrak{A}_1|). \qquad (3.2)$$

Then $\mathfrak{L}_1 = \langle \mathfrak{A}_1, C_1 \rangle$ is a prelogic which is monotone increasing or decreasing

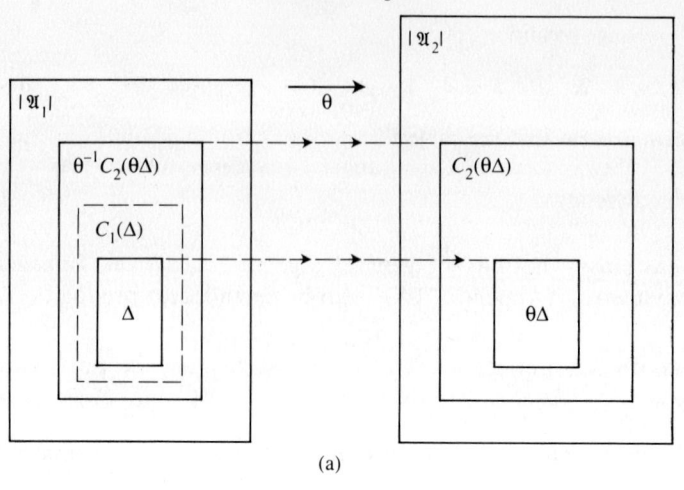

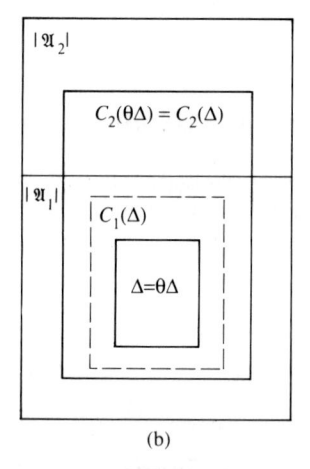

Fig. 3.1. (a) Strong/weak homomorphism; (b) strong/weak subsystems (θ = identity).

if $\mathfrak{L}_2$ is. Further, θ is a strong homomorphism of $\mathfrak{L}_1$ to $\mathfrak{L}_2$. In the application to the construction of logics (Chapter 5) $\mathfrak{L}_2$ is the structure of truth-values organized as a logic whilst $\mathfrak{A}_1$ is an algebra of formulas. (3.2) then shows how $\mathfrak{A}_1$ inherits a consequence function from $\mathfrak{L}_2$.

This raises the question of what properties $\mathfrak{L}_1$ inherits from $\mathfrak{L}_2$. Likewise we can ask what properties a subprelogic inherits from its parent. In order to give some answers to such questions we need a precise definition of conditions satisfied by a prelogic. This will enable us to define the conditions which a prelogic must satisfy in order to be a logic.

3.3 Conditions on prelogics

In this section a formal language $Qf_2(\tau)$ will be defined which is suitable for discussing prelogics of type τ. It is obvious that $Qf_2(\tau)$ must be second order because of the set-to-set function with which a prelogic is equipped. Moreover, it must have function symbols which denote the operations of prelogics of type τ. In this language it is possible to express the conditions relating the operations of the algebra to the attendant consequence and, in particular, the conditions which a prelogic must satisfy in order to be a logic. We shall mostly be concerned with 'algebraic' types of conditions so that it will suffice, as in many finitist investigations, to consider a free-variable (i.e. quantifier-free) language. $Qf_2(\lambda)$ is defined to be the language obtained by deleting all function symbols. Thus the formulas of $Qf_2(\lambda)$ can be interpreted in *all* prelogics.

The languages $Qf_2(\tau)$ need to be augmented in order to deal adequately with quantifiers in quantifier logics. This development will not be undertaken here.

VARIABLES

(i) $V^{(0)} = \{x_0, x_1, x_2, \ldots\}$ is the set of *individual variables*.
(ii) $V^{(1)} = \{X_0, X_1, X_2, \ldots\}$ is the set of *set variables*.

CONSTANTS

Logical constants

(i) The usual propositional junctors as in Chapter 2, §2.2.1.
(ii) *in* (membership), + (union of sets), $\varnothing$ (empty set), ι (unit operation—transforms an individual into the unit set whose only member is that individual).

Non-logical constants
A set *Fn* of function letters suitable for algebras of type τ, and a one-place function letter $\mathbb{C}$ (interpreted as the logical consequence function). There is also a function $\tau: Fn \to \mathbb{N}$ which determines the *type* of the language (see the formation rules of $L(\tau)$ (Chapter 2, §2.3.1)).

Brackets are used in the obvious way (cf. Chapter 2, §2.3.1).

The formulas of $Qf_2(\tau)$ are constructed by first defining terms and then atomic formulas. Here there are two types of terms—those that stand for individuals (type 0) and those that stand for sets (type 1) (cf. Definition 2.8).

TERMS

(i) The class $\mathbb{T}^{(0)}$ of *terms of type 0* is defined inductively:

(T0-1) $V^{(0)} \subseteq \mathbb{T}^{(0)}$;
(T0-2)   if $f \in Fv$, $\tau(f) = n$ and $t_1, \ldots, t_n \in \mathbb{T}^{(0)}$ then $f[t_1, \ldots, t_n] \in \mathbb{T}^{(0)}$.

(ii) The class $\mathbb{T}^{(1)}$ of *terms of type 1* is defined inductively:

(T1-1) $V^{(1)} \subseteq \mathbb{T}^{(1)}$;
(T1-2)   if $t \in \mathbb{T}^{(0)}$ then $\iota[t] \in \mathbb{T}^{(1)}$;
(T1-3) $\emptyset \in \mathbb{T}^{(1)}$;
(T1-4)   if $X, Y \in \mathbb{T}^{(1)}$ then $X + Y \in \mathbb{T}^{(1)}$;
(T1-5)   if $X \in \mathbb{T}^{(1)}$ then $\mathbb{C}[X] \in \mathbb{T}^{(1)}$.

As in Chapter 2, §§2.3.1 and 2.1.1, the set $\mathbb{T}^{(0)}$ of terms can be considered as an algebra generated by Fn on the generators $V^{(0)}$. Likewise $\mathbb{T}^{(1)}$ is an algebra formed by operations $+$ and $\mathbb{C}$ on the initial elements $V^{(1)} + \{\iota[t]; t \in \mathbb{T}^{(0)}\} + \{\emptyset\}$. But notice that $\mathbb{C}$ is not itself a set-to-set mapping although it is interpreted as such.

DEFINITION 3.4.

(i) *Atoms* are all expressions of the form t in T where t and T are terms of type 0 and 1 respectively. Following Definition 2.4, a *positive literal* is an atom, a *negative literal* is the negation of an atom, and a *literal* is a positive or a negative literal.

(ii) A term is said to be *C-free* if there is no occurrence of $\mathbb{C}$ in it. (Thus all terms in $\mathbb{T}^{(0)}$ are C-free). The set of C-free terms of type 1 is defined inductively by (T1-1)–(T1-5). A *C-free atom* is an atom t in T in which T is C-free. A *simple atom* is an atom of the form t in $\mathbb{C}[T]$ where T is a C-free term of type 1.

DEFINITION 3.5.

(i) Formulas are compounded from the junctors and atoms in the usual way as in Chapter 2, §2.1.1. The definition of *literal* and *clause* is as in Definition 2.4.

(ii) The conjunction of simple atoms, t_1 in $\mathbb{C}[\{t_2\}] \wedge t_2$ in $\mathbb{C}[\{t_1\}]$, is called a *logical identity*. Thus a logical identity is a conjunction of C-atoms. (Under certain circumstances—in a logic, for instance—logical identity defined in this way actually is an equivalence relation—see Chapter 5, §5.3.)

(iii) A *simple definite program clause* (SDPC) is a definite program clause (in $Qf_2(\tau)$) whose head is a simple atom. (This definition is given in anticipation of Example 3.2.)

Note that the set of formulas of $Qf_2(\lambda)$ consists of those formulas which are common to $Qf_2(\tau)$ for all types τ.

The concept of truth for $Qf_2(\tau)$ is defined in the usual way. First, by assigning values to the variables in a prelogic $\mathfrak{L}$ of type τ we obtain a propositional valuation on the atoms which then extends to a valuation of the formulas.

As there are two types of variable an *assignment* to the prelogic $\mathfrak{L} = \langle \mathfrak{U}, C \rangle$ is a function α on $V^{(0)} + V^{(1)}$ such that $\alpha(V^{(0)}) \subseteq |\mathfrak{U}|$ and $\alpha(V^{(1)}) \subseteq \mathbb{P}(|\mathfrak{U}|)$, i.e. α interprets variables of type 0 as individuals and variables of type 1 as sets. For later use in connection with compactness we define a *finite assignment* to be an assignment which assigns finite sets to variables of type 1. Assignments can then be extended canonically to all terms of the appropriate type. The extension of α to a mapping $\bar{\alpha}: \mathbb{T}^{(0)} \to |\mathfrak{U}|$ is straightforward enough. But whereas the values of α on the variables of type 0 and the values of the variables of type 1 can be chosen independently, the extension of α to terms of type 1 depends on the values of $\bar{\alpha}$ on the initial elements $\iota(t)$ of $\mathbb{T}^{(1)}$. This is because of the unit operation ((ii)(2) below) which transforms terms of type 0 into terms of type 1. Thus, corresponding to the clauses (T1-1)–(T1-5), we have:

(ii)(1) if $T \in V^{(1)}$ then $\bar{\alpha}(T) = \alpha(T)$;
(ii)(2) if $t \in T^{(0)}$ then $\bar{\alpha}(\iota[t]) = \{\bar{\alpha}(t)\}$ (unit set containing $\bar{\alpha}(t)$);
(ii)(3) $\bar{\alpha}(\emptyset) = \emptyset$ ($\emptyset$ is interpreted as the empty set);
(ii)(4) if $S, T \in \mathbb{T}^{(1)}$ then $\bar{\alpha}(S + T) = \bar{\alpha}(S) + \bar{\alpha}(T)$ (the logical constant $+$ of $Qf_2(\tau)$ is interpreted as the union of sets $\bar{\alpha}(S)$ and $\bar{\alpha}(T)$);
(ii)(5) if $T \in \mathbb{T}^{(1)}$ then $\bar{\alpha}(\mathbb{C}[T]) = C(\bar{\alpha}(T))$ (the non-logical constant $\mathbb{C}$ of $Qf_2(\tau)$ is interpreted as the consequence function of the given prelogic).

DEFINITION 3.6. Given a prelogic $\mathfrak{L} = \langle \mathfrak{U}, C \rangle$ of type τ and an assignment α to $\mathfrak{L}$, there is a function $[\mathfrak{L}, \alpha]$ which assigns truth-values to atomic formulas defined by

$$[\mathfrak{L}, \alpha](t \text{ in } T) =_{\text{def}} \begin{cases} 1 & \text{if } \alpha(t) \in \alpha(T) \\ 0 & \text{otherwise.} \end{cases}$$

$[\mathfrak{L}, \alpha]$ extends to all formulas of $Qf_2(\tau)$ as in Definition 2.13. For a formula F we write $(\mathfrak{L}, \alpha) \vDash F$ when $[\mathfrak{L}, \alpha](F) = 1$ and say that 'the assignment α satisfies the formula F in the structure $\mathfrak{L}$'. The effect of universal quantification is obtained by writing '$\mathfrak{L} \vDash F$' (i.e. F is true in $\mathfrak{L}$) when $(\mathfrak{L}, \alpha) \vDash F$ for all assignments α to $\mathfrak{L}$ and $(\mathfrak{L}, \alpha) \vDash_\omega F$ when $(\mathfrak{L}, \alpha) \vDash F$ for all finite assignments α.

Certain abbreviations in the arguments taken by the letter $\mathbb{C}$ make the language more usable informally: $\mathbb{CC}[T]$ will stand for $\mathbb{C}[\mathbb{C}[T]]$. Further, in accordance with our usual convention, for $T_1, \ldots, T_n \in \mathbb{T}^{(1)}$ and $t_1, \ldots, t_m \in \mathbb{T}^{(0)}$,

$\mathbb{C}[T_1, \ldots, T_n, t_1, \ldots, t_m]$ is an abbreviation of

$$\mathbb{C}[T_1 + \cdots + T_n + \{t_1\} + \cdots + \{t_m\}].$$

EXAMPLE 3.3.

(i) The logical consequence function C of the prelogic $\mathfrak{L} = \langle \mathfrak{A}, C \rangle$ is a closure operator (satisfies the Kuratowski closure axioms (Example 3.1)) iff $\mathfrak{L} \vDash F$ where F is the conjunction of the following formulas:

$$x_0 \text{ in } \mathbb{C}[X_0, X_1] \supset x_0 \text{ in } \mathbb{C}[X_0] + \mathbb{C}[X_1]$$

$$x_0 \text{ in } \mathbb{C}[X_0] + \mathbb{C}[X_1] \supset x_0 \text{ in } \mathbb{C}[X_0, X_1]$$

$$x_0 \text{ in } X_0 \supset x_0 \text{ in } \mathbb{C}[X_0]$$

$$x_0 \text{ in } \mathbb{C}\mathbb{C}[X_0] \supset x_0 \text{ in } \mathbb{C}[X_0]$$

$$\neg x_0 \text{ in } \mathbb{C}[\emptyset].$$

The compactness or localization condition of geometries (Example 3.1(vi)) cannot be expressed in this language.

(ii) Observe that the action of $\bar{\alpha}$ on terms of type 0 is independent of the logical consequence function C whereas C plays an essential part in the action on terms of type 1. Thus if $\mathbb{C}$ does not occur in T then $\langle \mathfrak{A}, C, \alpha \rangle \vDash t \text{ in } T$ iff for all set-to-set functions C, $\langle \mathfrak{A}, C, \alpha \rangle \vDash t \text{ in } T$. Further, if C, D are set-to-set functions on $|\mathfrak{A}|$ such that $C \subseteq D$ then

$$\langle \mathfrak{A}, C, \alpha \rangle \vDash t \text{ in } T \quad \Rightarrow \quad \langle \mathfrak{A}, D, \alpha \rangle \vDash t \text{ in } T.$$

(iii) Listed below are some formulas which are of importance for axiomatizing (traditional) connectives in logics. They are all SDPC's or conjunctions of them. D1, D2, and D3 constitute defining clauses in the definition of logics. These are formulas of $Qf_2(\langle 2, 2, 2, 1, \ldots \rangle)$ which has the two-place function letters f_1, f_2, f_3, and the one-place function letter f_4: in the application to logics these function symbols denote the connectives implication, conjunction, disjunction, and negation respectively.

D1: $\quad x_0 \text{ in } X_0 \supset x_0 \text{ in } \mathbb{C}[X_0]$

D2: $\quad x_0 \text{ in } \mathbb{C}\mathbb{C}[X_0] \supset x_0 \text{ in } \mathbb{C}[X_0]$

D3: $\quad x_0 \text{ in } \mathbb{C}[X_0] \supset x_0 \text{ in } \mathbb{C}[X_0, X_1]$

MP1: $\quad x_0 \text{ in } \mathbb{C}[x_1, f_1[x_1, x_0]]$

MP2: $\quad f_1[x_0, x_1] \text{ in } \mathbb{C}[X_0] \supset x_1 \text{ in } \mathbb{C}[X_0, x_0]$

MP3: $\quad x_1 \text{ in } \mathbb{C}[X_0, x_0] \wedge x_2 \text{ in } \mathbb{C}[X_1] \supset x_1 \text{ in } \mathbb{C}[X_0, X_1, f_1[x_2, x_1]]$

DT: $\quad x_1 \text{ in } \mathbb{C}[X_0, x_0] \supset f_1[x_0, x_1] \text{ in } \mathbb{C}[X_0]$

N1: $\quad x_0 \text{ in } \mathbb{C}[x_1, f_4[x_1]]$

N2: $\quad x_0 \text{ in } \mathbb{C}[x_1] \wedge x_0 \text{ in } \mathbb{C}[f_4[x_1]] \supset \mathbb{C}[\emptyset]$

N3: $\quad x_0 \text{ in } \mathbb{C}[x_1] \supset f_4[x_1] \text{ in } \mathbb{C}[f_4[x_0]]$

Cj1♣: x_0 in $\mathbb{C}[X_0, f_2[x_0, x_1]] \wedge x_1$ in $\mathbb{C}[X_0, f_2[x_0, x_1]]$

Cj2♣: x_0 in $\mathbb{C}[X_0, x_2] \wedge x_1$ in $\mathbb{C}[X_0, x_2] \supset f_2[x_0, x_1]$ in $\mathbb{C}[X_0, x_2]$

Dj1♣: $f_3[x_0, x_1]$ in $\mathbb{C}[X_0, x_0] \wedge f_3[x_0, x_1]$ in $\mathbb{C}[X_0, x_1]$

Dj2♣: x_2 in $\mathbb{C}[X_0, x_0] \wedge x_2$ in $\mathbb{C}[X_0, x_1] \supset x_2$ in $\mathbb{C}[X_0, f_3[x_0, x_1]]$. ♦

3.4 Transfer principles

In this section homomorphisms of prelogics are studied in order to determine some general classes of properties definable in $Qf_2(\tau)$ which are transferred by these operations. The most important case, studied in detail in Chapter 6, concerns a set of truth-values structured as a logic; an algebra of formulas maps homomorphically onto the logic of truth-values and thereby inherits many of its properties (e.g. (3.2)). We shall concentrate on SDPC's, not because they are optimal in any sense, but because they cover traditional axioms for connectives.

The following lemma establishes some transfer properties of atoms.

LEMMA 3.1. Let $\mathfrak{L} = \langle \mathfrak{A}, C \rangle$ and $\mathfrak{L}' = \langle \mathfrak{A}', C' \rangle$ be prelogics of type τ and θ a homomorphism of $\mathfrak{L}$ into $\mathfrak{L}'$. Let α be a valuation in $\mathfrak{A}$. Then $\theta \circ \alpha$ is a valuation in $\mathfrak{L}'$.

(i) If θ is a strong homomorphism of $\mathfrak{L}$ into $\mathfrak{L}'$ then for all $T \in \mathbb{T}^{(1)}$

$$\theta(\bar{\alpha}(T)) = (\theta \circ \alpha)(T) \tag{3.3}$$

$$(\mathfrak{L}, \alpha) \vDash t \text{ in } T \quad \Rightarrow \quad (\mathfrak{L}', \theta \circ \alpha) \vDash t \text{ in } T \tag{3.4}$$

$$(\mathfrak{L}', \theta \circ \alpha) \vDash t \text{ in } \mathbb{C}[T] \quad \Rightarrow \quad (\mathfrak{L}, \alpha) \vDash t \text{ in } \mathbb{C}[T]. \tag{3.5}$$

(ii) If θ is a weak homomorphism of $\mathfrak{L}$ into $\mathfrak{L}'$ and C' is monotone increasing, then for all C-free terms $T \in \mathbb{T}^{(1)}$

$$\theta(\bar{\alpha}(T)) = (\theta \circ \alpha)(T) \tag{3.6}$$

$$(\mathfrak{L}', \theta \circ \alpha) \vDash t \text{ in } \mathbb{C}[T] \quad \Rightarrow \quad (\mathfrak{L}, \alpha) \vDash t \text{ in } \mathbb{C}[T] \tag{3.7}$$

and for all $T \in \mathbb{T}^{(1)}$

$$\theta(\bar{\alpha}(T)) \subseteq (\theta \circ \alpha)(T) \tag{3.8}$$

$$(\mathfrak{L}, \alpha) \vDash t \text{ in } T \quad \Rightarrow \quad (\mathfrak{L}', \theta \circ \alpha) \vDash t \text{ in } T. \tag{3.9}$$

Proof.
ad (3.3) This is trivial for terms of type 0. For terms of type 1 we follow the inductive definition of $\mathbb{T}^{(1)}$ and the definition of valuation (ii)(1)–(ii)(5).

Basis:

(ii)(1) $T = X \in V^{(1)}$. Then $\alpha(X) \subseteq |\mathfrak{A}|$ as α is a valuation in $\mathfrak{L}$ and $(\theta \circ \alpha)(X) \subseteq |\mathfrak{A}'|$. So $\theta \circ \alpha$ is a valuation of $\mathfrak{L}'$ and (3.3) is trivial.

(ii)(2) $T = \iota(t)$ where $t \in \mathbb{T}^{(0)}$ then $\bar{\alpha}(\iota(t)) = \{\bar{\alpha}(t)\}$ and $\overline{\theta \circ \alpha}(\iota(t)) = \{\theta \circ \alpha(t)\}$ by (ii)(2). But $\theta(\{\bar{\alpha}(t)\}) = \{\theta(\bar{\alpha}(t))\} = \{\theta \circ \alpha(t)\}$. Hence (3.3) holds.

(ii)(3) $T = \emptyset$. Then both valuations α, $\theta \circ \alpha$ interpret the letter $\emptyset$ as the empty set by (ii)(3) and so (3.3) is again trivial.

Inductive step:

(ii)(4) Suppose (3.3) holds for $T = A, B \in \mathbb{T}^{(1)}$. Then

$$\theta(\bar{\alpha}(A + B)) = \theta(\bar{\alpha}(A) + \bar{\alpha}(B)) = \theta(\bar{\alpha}(A)) + \theta(\bar{\alpha}(B))$$
$$= \theta \circ \alpha(A) + \theta \circ \alpha(B) \quad \text{by hypothesis}$$
$$= \theta \circ \alpha(A + B) \quad \text{by (ii)(4)}.$$

(ii)(5) Suppose (3.3) holds for $T \in \mathbb{T}^{(1)}$. Then

$$\theta \circ \alpha(\mathbb{C}[T]) = C'(\theta \circ \alpha(T)) \quad \text{by (ii)(5)}$$
$$= C'(\theta(\bar{\alpha}(T))) \quad \text{by (3.3)}$$
$$= \theta C(\alpha(T)) \quad \text{as } \theta \text{ is a strong homomorphism}$$
$$= \theta(\bar{\alpha}(\mathbb{C}[T])) = \theta \circ \alpha(\mathbb{C}[T]) \quad \text{by (ii)(5)}.$$

It follows by induction that (3.3) holds for all terms of type 1.

ad (3.4) Suppose $(\mathfrak{L}, \alpha) \vDash t$ in T. Then $\bar{\alpha}(t) \in \bar{\alpha}(T)$. Hence $\theta(\bar{\alpha}(t)) \in \theta(\bar{\alpha}(T))$. By (3.3) $\theta \circ \alpha(t) \in \theta \circ \alpha(T)$. Hence $(\mathfrak{L}', \theta \circ \alpha) \vDash t$ in T.

ad (3.5) Suppose $(\mathfrak{L}', \theta \circ \alpha) \vDash t$ in $\mathbb{C}(T)$. Then $\theta \circ \alpha(t) \in C'(\theta \circ \alpha(T))$. By (3.4), $\theta(\bar{\alpha}(t)) \in C'(\theta(\bar{\alpha}(T)))$. As θ is a strong homomorphism (Definition 3.3), $\bar{\alpha}(t) \in C\langle \bar{\alpha}(T)\rangle$. Thus $(\mathfrak{L}, \alpha) \vDash t$ in $\mathbb{C}[T]$.

ad (3.6) By induction on the construction of C-free terms (cf. proof of (3.3)).

ad (3.7) As for (3.5) but using (3.6).

ad (3.8) By induction on the construction of terms as in the proof of (3.3), the only essential difference being in the inductive step where terms $\mathbb{C}[T]$ are considered, i.e. (ii)(5). Thus, suppose $\theta(\bar{\alpha}(T)) \subseteq \theta \circ \alpha(T)$. Then

$$\theta \circ \alpha(\mathbb{C}[T]) = C'(\theta \circ \alpha(T)) \quad \text{by (ii)(5)}$$
$$\supseteq C'(\theta(\bar{\alpha}(T))) \quad \text{as } C' \text{ is monotone increasing}$$
$$\supseteq \theta C(\bar{\alpha}(T)) \quad \text{as } \theta \text{ is a weak homomorphism (Definition 3.3)}$$
$$= \theta(\bar{\alpha}(\mathbb{C}[T])).$$

It follows by induction that (3.8) holds for all terms of type 1.

ad (3.9) As for (3.4) but using (3.8). ∎

Transfer principles

THEOREM 3.2. Let $\mathfrak{L} = \langle \mathfrak{A}, C \rangle$ and $\mathfrak{L}' = \langle \mathfrak{A}', C' \rangle$ be prelogics of type τ such that there is a strong homomorphism θ of $\mathfrak{L}$ into $\mathfrak{L}'$ (e.g. when $\mathfrak{L}$ is a strong subsystem of $\mathfrak{L}'$).

(i) Then for every clause F whose positive literals are of the form t in $\mathbb{C}[T]$

$$\mathfrak{L}' \vDash F \quad \Rightarrow \quad \mathfrak{L} \vDash F.$$

(ii) If θ is a homomorphism of $\mathfrak{L}$ onto $\mathfrak{L}'$ then for every clause F whose positive literals are of the form t in $\mathbb{C}[T]$

$$\mathfrak{L} \vDash F \quad \Rightarrow \quad \mathfrak{L}' \vDash F.$$

Proof.

(i) Suppose $\mathfrak{L}' \vDash F$ where $F = \neg A_1 \vee \cdots \vee \neg A_m \vee B_1 \vee \cdots \vee B_n$ and the A_i and B_j are atoms and the latter are of the form t in $\mathbb{C}[T]$. We show that $\mathfrak{L} \vDash F$, i.e. for all evaluations α in $\mathfrak{A}$, $(\mathfrak{L}, \alpha) \vDash F$. Suppose to the contrary that, for some evaluation α, $(\mathfrak{L}, \alpha) \vDash \neg F$. Then $(\mathfrak{L}, \alpha) \vDash A_i, \neg B_j$ for $1 \leq i \leq m$ and $1 \leq j \leq n$. By (3.4) $(\mathfrak{L}', \theta \circ \alpha) \vDash A_i$ and by (3.5) $(\mathfrak{L}', \theta \circ \alpha) \vDash \neg B_j$. Thus $(\mathfrak{L}', \theta \circ \alpha) \vDash \neg F$, contrary to the hypothesis.

(ii) Suppose that $\mathfrak{L} \vDash F$ where $F = \neg A_1 \vee \cdots \vee \neg A_m \vee B_1 \vee \cdots \vee B_n$ and the A_i and B_j are atoms and the former are of the form t in $\mathbb{C}[T]$. We show that $\mathfrak{L}' \vDash F$, i.e. for all evaluations α, $(\mathfrak{L}', \alpha) \vDash F$. Suppose to the contrary that for some evaluation α, $(\mathfrak{L}', \alpha) \vDash \neg F$. Since θ is onto, by the axiom of choice, there exists an assignment β to $\mathfrak{L}$ such that $\theta \circ \beta = \alpha$. Then by (3.4) and (3.5), as in (i), $(\mathfrak{L}, \beta) \vDash \neg F$. This contradicts the hypothesis. ∎

The principle on which the construction in Chapter 5 of semantic logics is based is as follows.

THEOREM 3.3. Let $\mathfrak{A}$ be an algebra of type τ and $\{\langle \mathfrak{A}_i, C_i \rangle; i \in I\}$ a set of monotone increasing prelogics of type τ. Suppose that for each $i \in I$, θ_i is a homomorphism (of algebras, see Appendix, §A) of $\mathfrak{A}$ into $\mathfrak{A}_i$. For each $i \in I$ define the set-to-set function C^i on $|\mathfrak{A}|$ by

$$C^i(\Delta) =_{\text{def}} \theta_i^{-1} C_i(\theta_i(\Delta)) \quad (\Delta \subseteq |\mathfrak{A}|).$$

Let F be an SDPC. Then $\langle \mathfrak{A}; C \rangle$ where $C =_{\text{def}} \prod \{C^i; i \in I\}$ is a prelogic such that

$$(i \in I)(\langle \mathfrak{A}_i, C_i \rangle \vDash F) \quad \Rightarrow \quad \langle \mathfrak{A}, C \rangle \vDash F.$$

Proof. Let $F = \neg A_1 \vee \cdots \vee \neg A_n \vee B$ where $A_1, \ldots, A_n, B$ are atoms and B is simple. Suppose that

$$\langle \mathfrak{A}_i, C_i \rangle \vDash F \quad (i \in I). \tag{3.10}$$

72   *Abstract logics*

We deduce that $\langle \mathfrak{A}, C \rangle \vDash F$. Suppose to the contrary that $\langle \mathfrak{A}, C \rangle \nvDash F$. Then there exists a valuation α such that $\langle \mathfrak{A}, C, \alpha \rangle \nvDash F$, i.e.

$$\langle \mathfrak{A}, C, \alpha \rangle \vDash A_m \quad \text{for } 1 \leq m \leq n \text{ and } \langle \mathfrak{A}, C, \alpha \rangle \vDash \neg B. \quad (3.11)$$

We deduce that, for some $j \in I$, $\langle \mathfrak{A}_j, C_j, \alpha \rangle \nvDash F$.

Now as B is simple it has the form t in $\mathbb{C}[T]$ where T is C-free. Suppose that $\langle \mathfrak{A}_i, C^i, \alpha \rangle \vDash t$ in $\mathbb{C}[T]$ for all $i \in I$. Then $\bar{\alpha}(t) \in C^i(\bar{\alpha}(T))$ for all $i \in I$. Since T is C-free (Example 3.3(ii)), $\alpha(T)$ is independent of the C^i so $\bar{\alpha}(t) \in \prod \{C^i(\bar{\alpha}(T))\} = C(\bar{\alpha}(T))$. Thus $\langle \mathfrak{A}, C, \alpha \rangle \vDash t$ in $\mathbb{C}[T]$, i.e. $\langle \mathfrak{A}, C^i, \alpha \rangle \vDash B$. But this contradicts (3.10). Hence for some $j \in I$, $\langle \mathfrak{A}, C^j, \alpha \rangle \vDash \neg B$. Now θ_j is a strong homomorphism of $\langle \mathfrak{A}, C^j \rangle$ to $\langle \mathfrak{A}_j, C_j \rangle$, so by (3.5)

$$\langle \mathfrak{A}_j, C_j, \theta_j \circ \alpha \rangle \vDash \neg B. \quad (3.12)$$

As C_j is monotone increasing, by (3.9) and (3.11)

$$\langle \mathfrak{A}_j, C_j, \theta_j \circ \alpha \rangle \vDash \neg A_m \quad (1 \leq i \leq m). \quad (3.13)$$

From (3.11) and (3.13) it follows that $\langle \mathfrak{A}_j, C_j, \theta_j \circ \alpha \rangle \nvDash F$. This contradicts (3.10). ∎

For transfer of properties from a prelogic to its compact kernel we need the following lemma.

LEMMA 3.4. *Let $\langle \mathfrak{A}, C \rangle$ be a monotone increasing prelogic which satisfies the condition*

$$\Delta \subseteq C(\Delta) \quad (\Delta \subseteq |\mathfrak{A}|)$$

and let α be an assignment.

(i) *Let β be an assignment such that $\alpha(x) = \beta(x)$ for all $x \in V^{(0)}$ and $\alpha(X) \subseteq \beta(X)$ for all $X \in V^{(1)}$. Then $\bar{\alpha}(T) \subseteq \bar{\beta}(T)$ for all $T \in \mathbb{T}^{(1)}$.*
(ii) *Suppose that C is compact. Let $T \in \mathbb{T}^{(1)}$. Then for any $\Delta \in \mathbb{P}_\omega(\bar{\alpha}(T))$ there exists a finite assignment β such that*

$$\beta(x) = \alpha(x) \quad (x \in V^{(0)}) \quad (3.14)$$

$$\Delta \subseteq \bar{\beta}(T) \subseteq \bar{\alpha}(T) \quad (3.15)$$

and if T is C-free then

$$\bar{\beta}(T) \in \mathbb{P}_\omega(\bar{\alpha}(T)). \quad (3.16)$$

Proof.

(i) This is proved by induction of the construction of terms of type 1.
(ii) By induction on the construction of terms of C-free terms.

Basis:
There are five cases corresponding to (ii)(1)–(ii)(5).

(ii)(1) $T = X_i$. Define β by $\beta(X) = \Delta$ if $X = X_i$, $\beta(X) = \varnothing$ if $X \in V^{(1)} - \{X_i\}$, and $\beta(x) = \alpha(x)$ if $x \in V^{(0)}$. Then (3.14)–(3.16) are satisfied.

(ii)(2) $T = \iota[t]$. Define β as for (ii)(1).

(ii)(3) $T = \varnothing$. Then $\bar{\alpha}(T) = \varnothing$. Hence $\Delta = \varnothing$ and so we can define β by $\beta(X) = \varnothing$ if $X \in V^{(1)}$ and $\beta(x) = \alpha(x)$ if $x \in V^{(0)}$. Then (3.14)–(3.16) are satisfied.

Inductive step:

(ii)(4) As an inductive hypothesis, assume that the lemma holds for the terms U_1, U_2. Let Δ be a finite subset of $\bar{\alpha}(U_1 + U_2)$. Put $\Delta_i = \Delta \cdot \bar{\alpha}(U_i)$. Then by the inductive hypothesis there exist assignments β_i such that (3.14) holds and $\Delta_i \subseteq \bar{\beta}_i(U_i) \subseteq \bar{\alpha}(U_i)$. Define β by $\beta(X) = \beta_1(X) + \beta_2(X)$ for $X \in V^{(1)}$ and $\beta(x) = \alpha(x)$ for $x \in V^{(0)}$. By (i), $\bar{\beta}(U_i) \supseteq \bar{\beta}_i(U_i) \supseteq \Delta_i$. Hence

$$\bar{\beta}(U_1 + U_2) = \bar{\beta}(U_1) + \bar{\beta}(U_2) \supseteq \Delta.$$

Further, since by hypothesis $\alpha(X) \supseteq \beta_i(X)$ we have $\alpha(X) \supseteq \beta(X)$ so by (i), $\bar{\alpha}(U_1 + U_2) \supseteq \bar{\beta}(U_1 + U_2)$. Thus the lemma also holds for $U_1 + U_2$.

(ii)(5) $T = \mathbb{C}[S]$ where (ii) holds for S. Let $\Delta \in \mathbb{P}_\omega(C(\bar{\alpha}(S)))$. As C is compact there exists $\Delta' \in \mathbb{P}_\omega(\bar{\alpha}(S))$ such that $\Delta \subseteq C(\Delta')$. By (ii) there exists a finite assignment β satisfying $\Delta' \subseteq \bar{\beta}(S) \subseteq \bar{\alpha}(S)$. As C is monotone increasing $C(\Delta') \subseteq C(\bar{\beta}(S)) \subseteq C\langle \bar{\alpha}(S))$. As $\Delta \subseteq C(\Delta') \subseteq C(\Delta)$, $\Delta \subseteq \bar{\beta}(\mathbb{C}[S]) \subseteq \bar{\alpha}(\mathbb{C}[S])$. (3.14)–(3.16) now follow by induction. ∎

THEOREM 3.5. *Let $\langle \mathfrak{A}, C \rangle$ be a monotone increasing prelogic which satisfies the condition*

$$\Delta \subseteq C(\Delta) \qquad (\Delta \subseteq |\mathfrak{A}|)$$

and let F be an SDPC. Then

$$\langle \mathfrak{A}, C \rangle \vDash_\omega F \;\Rightarrow\; \langle \mathfrak{A}, C^{\heartsuit\heartsuit} \rangle \vDash F.$$

Proof. $F = \neg A_1 \vee \cdots \vee \neg A_m \vee B$ where the A_i are atoms and B is a simple atom. Suppose that $\langle \mathfrak{A}, C \rangle \vDash_\omega F$. We deduce that $\langle \mathfrak{A}, C^{\heartsuit\heartsuit}, \delta \rangle \vDash F$ for all finite assignments δ.

Suppose, to the contrary, that for some finite assignment α, $\langle \mathfrak{A}, C^{\heartsuit\heartsuit}, \alpha \rangle \vDash \neg F$. Then $\langle \mathfrak{A}, C^{\heartsuit\heartsuit}, \alpha \rangle \vDash A_1 \wedge \cdots \wedge A_m \wedge \neg B$. $A_i = t_i$ in T_i, say, so that $\bar{\alpha}(t_i) \in \bar{\alpha}(T)$. Put $\Delta_i = \{t_i\}$. As $C^{\heartsuit\heartsuit}$ is compact (Example 3.1), by Lemma 3.3(ii) there exists a finite assignment β_i satisfying (3.14)–(3.16). Then $\bar{\beta}_i(t_i) \in \bar{\beta}_i(T_i) \subseteq \bar{\alpha}(T_i)$. Thus $\langle \mathfrak{A}, C^{\heartsuit\heartsuit}, \beta_i \rangle \vDash t_i$ in T_i. Define a new finite assignment γ by

$$\gamma(X) = \beta_1(X) + \cdots + \beta_m(X) \qquad \text{for } X \in V^{(1)}$$

$$\gamma(x) = \alpha(x) \qquad \text{for } x \in V^{(0)}.$$

Then $\gamma(X) \in \mathbb{P}_\omega(\alpha(X))$ and $\gamma(X) \supseteq \beta_i(X)$. By Lemma 3.3(i) and (3.15), $\bar{\alpha}(t_i) \in \bar{\beta}(T_i) \subseteq \bar{\gamma}(T_i)$. Thus $\langle \mathfrak{A}, C^{\heartsuit\heartsuit}, \gamma \rangle \vDash t_i$ in T_i. As the identity map on $|\mathfrak{A}|$ is a weak homomorphism of $\langle \mathfrak{A}, C^{\heartsuit\heartsuit} \rangle$ into $\langle \mathfrak{A}, C \rangle$, by (3.9), $\langle \mathfrak{A}, C, \gamma \rangle \vDash t_i$ in T_i. Thus

$$\langle \mathfrak{A}, C, \gamma \rangle \vDash A_1 \wedge \cdots \wedge A_m. \tag{3.17}$$

Next, $B = s \in \mathbb{C}[S]$ say, where S is C-free. As $\langle \mathfrak{A}, C^{\heartsuit\heartsuit}, \alpha \rangle \vDash \neg B$, $\bar{\alpha}(s) \notin \bar{\alpha}(\mathbb{C}[S]) = C^{\heartsuit\heartsuit}(\bar{\alpha}(S))$. Thus $\bar{\alpha}(s) \notin C(\Gamma)$ for all $\Gamma \in \mathbb{P}_\omega(\bar{\alpha}(S))$. By (3.16), $\bar{\gamma}(S) \in \mathbb{P}_\omega(\bar{\alpha}(S))$ and so $\bar{\gamma}(s) = \bar{\alpha}(s) \notin C(\bar{\gamma}(S))$. Thus $\langle \mathfrak{A}, C, \gamma \rangle \vDash \neg s$ in $\mathbb{C}[S]$. By (3.17), $\langle \mathfrak{A}, C, \gamma \rangle \vDash \neg F$. This contradicts the hypothesis, thus proving the theorem. ∎

Consider now derived prelogics (Definition 3.3(iii) and Appendix, §A). Let $\mathfrak{L} = \langle A, g_1, g_2, \ldots, C \rangle$ be a prelogic of type τ. Consider the derived prelogic $\mathfrak{L}' = \langle A, p[t_1], p[t_2], \ldots, C \rangle$, where $t_1, t_2, \ldots$ are terms of type τ and $p[t_1], p[t_2], \ldots$ are their associated polynomials. Let d be the corresponding term derivation.

Now the type of $\mathfrak{L}'$ is μ, say, which might differ from τ. The language $Qf_2(\tau)$ with which conditions in $\mathfrak{L}$ can be expressed differs from the language $Qf_2(\mu)$ which defines conditions in $\mathfrak{L}'$. But the term derivation d establishes a means of translating formulas of $Qf_2(\mu)$ into formulas of $Qf_2(\tau)$. For the set of terms of the algebra $\langle A, g_1, g_2, \ldots \rangle$ is already the set $\mathbb{T}^{(0)}(\tau)$ of terms of type 0 of $Qf_2(\tau)$. Likewise the set of terms of the algebra $\langle A, p[t_1], p[t_2], \ldots \rangle$ is the set $\mathbb{T}^{(0)}(\mu)$ of terms of type 0 of $Qf_2(\mu)$. So d maps $\mathbb{T}^{(0)}(\mu)$ to $\mathbb{T}^{(0)}(\tau)$: it can be extended to a mapping of the set $\mathbb{T}^{(1)}(\mu)$ of terms of type 1 of $Qf_2(\mu)$ to the set $\mathbb{T}^{(1)}(\tau)$ of terms of type 1 of $Qf_2(\tau)$ by following the inductive definition

(ii)(1) for $X \in V^{(1)}$, $d(X) = X$
(ii)(2) $d(\emptyset) = \emptyset$
(ii)(3) for $t \in \mathbb{T}^{(0)}(\mu)$, $d(\iota[t]) = \iota[d(t)]$
(ii)(4) for terms $X, Y \in \mathbb{T}^{(1)}(\mu)$

$$d(X + Y) = d(X) + d(Y)$$

(ii)(5) for $X, Y \in \mathbb{T}^{(1)}(\mu)$

$$d(\mathbb{C}[X]) = C[d(X)].$$

Next, d further extends to a mapping of atomic formulas of $Qf_2(\mu)$ to atomic formulas of $Qf_2(\tau)$. Let t and T be terms of types 0 and 1 respectively of

$Qf_2(\mu)$. Then

$$d(t \text{ in } T) =_{\text{def}} d(t) \text{ in } d(T).$$

Finally, by allowing d to preserve junctors d becomes a mapping of the formulas of $Qf_2(\mu)$ to formulas of $Qf_2(\tau)$. Then we have the following theorem.

THEOREM 3.6. Let $\mathfrak{L}$ be a logic of type τ, $\mathfrak{L}'$ a derived logic of type μ, say, and d the corresponding term derivation. Then for all formulas A of $Qf_2(\mu)$

$$\mathfrak{L}' \vDash A \quad \Leftrightarrow \quad \mathfrak{L} \vDash d(A).$$

Proof. Let the domain of $\mathfrak{L}$ (and therefore of $\mathfrak{L}'$) be A. Let α be an assignment to $\mathfrak{L}$. Then α maps variables of types 0 and 1 to A and to $\mathbb{P}(A)$ respectively (§3.3). Thus α is also an assignment to $\mathfrak{L}'$. α extends canonically to a mapping $\bar{\alpha}: \mathbb{T}^{(0)}(\tau) \to A$ and $\bar{\alpha}: \mathbb{T}^{(1)}(\tau) \to \mathbb{P}(A)$ and likewise to a mapping $\hat{\alpha}: \mathbb{T}^{(0)}(\mu) \to A$ and $\hat{\alpha}: \mathbb{T}^{(1)}(\mu) \to \mathbb{P}(A)$. Now by the Appendix, §A, for all $t \in \mathbb{T}^{(0)}(\mu)$,

$$\hat{\alpha}(t) = \bar{\alpha}(d(t)).$$

Then from the clauses (ii)(1)–(ii)(5) in the above definition of the extension of the term derivation d, it follows that for all $T \in \mathbb{T}^{(1)}(\mu)$

$$\hat{\alpha}(T) = \bar{\alpha}(d(T)).$$

Hence, by Definition 3.6, for all atomic formulas A of $Qf_2(\mu)$,

$$[\mathfrak{L}', \alpha](A) = [\mathfrak{L}, \alpha](d(A)).$$

Since d preserves junctors, this relationship also holds for all formulas A of $Qf_2(\mu)$. ∎

3.5 Logics

Tarski (1956, Chapter 5) formulated five axioms 'which express certain elementary properties of the primitive concepts (i.e. of *sentence* and *consequence*) and are satisfied in all known formalized disciplines'.

Let Σ be a set of sentences and let $Cn(A)$ denote the set of consequences of a set A of sentences: thus $Cn: \mathbb{P}(\Sigma) \to \mathbb{P}(\Sigma)$. Tarski's five axioms are as follows.

Ax1: $|\Sigma| \leq \aleph_0$

Ax2: $X \subseteq \Sigma \Rightarrow X \subseteq Cn(X) \subseteq \Sigma$

Ax3: $X \subseteq \Sigma \Rightarrow Cn(Cn(X)) = Cn(X)$

Ax4: $X \subseteq \Sigma \Rightarrow Cn(X) = \sum \{Cn(Y); Y \in \mathbb{P}_\omega(X)\}$

Ax5: there exists an f in Σ such that $Cn(\{f\}) = \Sigma$

Ax5 concerns a zero-place (i.e. constant) junctor related to negation. As junctors will be used to classify logics later, this axiom will not be used in our general definition of *logics*. Ax1 plays such a minor part in our dealings with logics that we shall drop it also, even though most examples we encounter satisfy it. Ax4, the *compactness* axiom, is not expressible in the formal language previously defined. Although it is an extremely important property of many logics, particularly classical logic (Theorem 2.7), there are significant 'logics' with an infinitary rule of inference which fail to satisfy Ax4. To include these systems in our definition of *logics* we therefore drop Ax4 from our definition but retain it as a principle of classification so that there are compact logics and non-compact logics.

DEFINITION 3.7.

(i) A prelogic $\mathfrak{L} = \langle \mathfrak{A}, Cn \rangle$ (of type τ) is a *logic* (of type τ) if Cn satisfies the conditions **D1, D2, D3**.

D1: $\Delta \subseteq Cn(\Delta)$ $\qquad (\Delta \subseteq |\mathfrak{A}|)$

D2: $Cn(Cn(\Delta)) \subseteq Cn(\Delta)$ $\qquad (\Delta \subseteq |\mathfrak{A}|)$

D3: $\Delta \subseteq \Gamma \Rightarrow (Cn(\Delta) \subseteq Cn(\Gamma))$ $\qquad (\Delta, \Gamma \subseteq |\mathfrak{A}|)$

(ii) $Th(\mathfrak{L}) =_{\text{def}} C(\emptyset)$ is the set of *theorems* of the logic $\mathfrak{L}$.
(iii) A logic is said to be *compact* if Ax4 is satisfied (cf. Definition 3.1).
(iv) For $A \in |\mathfrak{A}|$ and $\Delta \subseteq |\mathfrak{A}|$, A is said to be a *logical consequence* of Δ if $A \in Cn(\Delta)$ (also written as $\Delta \to A$).

D1 is Tarski's Ax2. **D2** is a weaker version of Ax3. In fact, Ax3 follows from **D1** and **D2**. The monotonicity property **D3**, which can also be expressed as

$$Cn(\Delta) \subseteq Cn(\Delta + \Gamma) \qquad (\Delta, \Gamma \subseteq |\mathfrak{A}|),$$

follows from the compactness axiom Ax4 (Tarski 1956, Chapter 5, Theorem 1). Thus **D1, D2, D3**, Ax4 are together equivalent to Ax1, Ax2, Ax3, Ax4. **D1, D2, D3** are expressible in Qf_2 by SDPC's.

The above definition of a *logic* covers those structures which we have hitherto called 'deductive systems', 'formal systems', 'logistical systems', etc. There are two kinds of logics—logics with a strictly finitist ancestry (e.g. Post canonical systems, formal grammars, etc.) and semantic logics which rest on the infinitistic notions of 'truth', 'satisfaction', etc. It is quite clear that most logics which occur in applications are founded on a classical notion of *satisfaction*, the *point* of such logics being directed towards models and theoremhood. For that reason it has rightly been observed that what we here call a *logic* is only the skeleton of a logic in a wider sense: the vital organs lie elsewhere in the models. Indeed, abstract model theories (Chapter 2, §2.3.9) regard a 'logic' as a symbiosis of a syntactical structure and a model theory which in some sense is independent of the particular formulation of the logic.

However, we shall not allow a preoccupation with semantics to preclude us from applying the term 'logic' to structures which are defined in purely formal terms. Moreover, too great a generality in the notion of semantics should not be admitted since it is trivially possible to associate a (loosely defined!) semantics with *every* logic (Example 3.8).

There is another usage of the term 'logic' where, for instance, we speak of *classical logic* as being independent of particular formulations. This sounds as though a logic is something over and above its instances. The framework of our definitions can accommodate this usage by defining various notions of *equivalence of logics* (§3.12). Then a *logic* in the wider sense can be considered as an equivalence class of logics. Thus, such a logic really is superior (in logical type) to its particular formulations. To the sceptic, however, this type of escalation is a superfluous and extravagant linguistic trick. The reality lies with the relation of equivalence.

Definition 3.7 acquires more substance by defining internal structure—logical operations (Definition 3.11 and Chapter 4) and quantifiers (Definition 3.14)—and external structure—criteria of comparison of logics (§3.12).

EXAMPLE 3.4.

(i) By Theorem 3.2 every strong subsystem and every strong homomorphic image of a logic is also a logic. Every subderived prelogic of a logic is also a logic—though possibly of a different type.
(ii) Let $\mathfrak{L}$ be a logic and A a formula of $Qf_2(\lambda)$. Suppose $(\mathfrak{L}, \alpha) \vDash A$. The atomic formulas of A are t_i in T_i for $1 \leqslant i \leqslant n$, say. Define an equivalence relation $\equiv$ on $|\mathfrak{L}|$ by

$$a \equiv b \quad \Leftrightarrow \quad \bigwedge_{1 \leqslant i \leqslant n} (a \in \bar{\alpha}(T_i) \Leftrightarrow b \in \bar{\alpha}(T_i)).$$

The canonical projection $a \to [a]$ is a strong homomorphism and, by (i), the strong homomorphic image $\hat{\mathfrak{L}}$ is a logic such that $(\hat{\mathfrak{L}}, \hat{\alpha}) \vDash A$. Moreover, $\hat{\mathfrak{L}}$ has $\leqslant 2^n$ elements. A is true in all logics iff it is true in all logics with $\leqslant 2^n$ elements.
(iii) The set of $Qf_2(\lambda)$ sentences which are true in all logics could be called the *logic of logics* if one is addicted to the theoremhood view of logics. By (ii) above, the logic of logics is a decidable set. It would be of some interest to form a syntactical characterization of this set. A 'logic of logics' is treated from a different point of view in Chapter 10. ◆

EXAMPLE 3.5. Consider an inductive definition with generating function $\mathfrak{C}$ which generates a set Γ of elements from initial elements I. Then $\langle \Gamma, \mathfrak{C} \rangle$ is a logic by (2.2)–(2.4). If the generating rules each have finitely many premisses, then $\langle \Gamma, \mathfrak{C} \rangle$ is compact (Examples 2.4(i) and 3.17(i)). ◆

Some properties of the logical consequence function (cf. Tarski 1956, Chapter V, §1) are recorded in the following theorem.

THEOREM 3.7. Let $\langle \mathfrak{A}, Cn \rangle$ be a logic.

(i) If $\Delta, \Gamma \subseteq |\mathfrak{A}|$ then
$$Cn(\Delta, \Gamma) = Cn(\Delta, Cn(\Gamma)) = Cn(Cn(\Delta), Cn(\Gamma)).$$

(ii) If K is an arbitrary class of subsets of $|\mathfrak{A}|$ then
$$Cn(\sum K) = Cn(\sum \{Cn(\Delta); \Delta \in K\}).$$

Proof.

(i) By **D1** and **D3**
$$\Delta + \Gamma \subseteq \Delta + Cn(\Gamma) \subseteq Cn(\Delta) + Cn(\Gamma) \tag{3.18}$$
$$Cn(\Delta) + Cn(\Gamma) \subseteq Cn(\Delta, \Gamma). \tag{3.19}$$

By (3.18), (3.19), and **D3**
$$Cn(\Delta, \Gamma) \subseteq Cn(\Delta, Cn(\Gamma)) \subseteq Cn(Cn(\Delta), Cn(\Gamma)) \subseteq Cn(Cn(\Delta, \Gamma)). \tag{3.20}$$

But by **D2**, $Cn(Cn(\Delta, \Gamma)) \subseteq Cn(\Delta, \Gamma)$. So by (3.20),
$$Cn(\Delta, \Gamma) = Cn(\Delta, Cn(\Gamma)) = Cn(Cn(\Delta), Cn(\Gamma)) = Cn(Cn(\Delta, \Gamma)).$$

(ii) By **D1**,
$$\sum K \subseteq Cn(\sum \{Cn(\Delta); \Delta \in K\}). \tag{3.21}$$

Since $\Delta \subseteq \sum K$ for every $\Delta \in K$, by **D3** we have $Cn(\Delta) \subseteq Cn(\sum K)$. So by (3.21)
$$\sum K \subseteq \sum \{Cn(\Delta); \Delta \in K\} \subseteq Cn(\sum K). \tag{3.22}$$

By **D3** and (3.22)
$$Cn(\sum K) \subseteq (\sum \{Cn(\Delta); \Delta \in K\} \subseteq Cn(Cn(\sum K)) \subseteq Cn(\sum K).$$

Thus
$$Cn(\sum K) = Cn(\sum \{Cn(\Delta); \Delta \in K\}). \qquad \blacksquare$$

Frequently in a logic $\mathfrak{L} = \langle \mathfrak{A}, Cn \rangle$, a certain subset Ax of $|\mathfrak{A}|$ has a privileged status because of mathematical exigency or philosophical prejudice. The set Ax is then called a set of *axioms* and the set $Cn(Ax)$ is called a *theory*. The axioms can be absorbed into the definition of a new logic by defining the new relativized consequence function as in the following definition.

Closed sets

DEFINITION 3.8. Let $\mathfrak{L} = \langle \mathfrak{A}, Cn \rangle$ be a logic and $\Gamma \subseteq |\mathfrak{A}|$. Define $Cn_\Gamma: \mathbb{P}(|\mathfrak{A}|) \to \mathbb{P}(|\mathfrak{A}|)$:

$$(\Delta) =_{\text{def}} Cn(\Gamma + \Delta).$$

Thus Cn determines a family $\{Cn_\Gamma; \Gamma \subseteq |\mathfrak{A}|\}$ of set-to-set functions. Each such function determines a binary relation imp_Γ on $|\mathfrak{A}|$ defined by

$$A\ imp_\Gamma B \Leftrightarrow_{\text{def}} B \in Cn_\Gamma(A)$$

(read 'A implies B by hypotheses Γ' or 'A implies B' when $\Gamma = \varnothing$). We write $\Delta \to_\Gamma A$ when $A \in Cn_\Gamma(\Delta)$. Thus

$$A\ imp_\Gamma B \Leftrightarrow A \to_\Gamma B.$$

The reader should note that the sentence 'A implies B' affirms a *relation* between propositions. It should not be confused with the *connective* (function on propositions) called 'implication' which will be discussed in Chapter 4, §4.2.

EXAMPLE 3.6.

(i) Let $\mathfrak{L} = \langle \mathfrak{A}, Cn \rangle$ be a logic and $\Gamma \subseteq |\mathfrak{A}|$. Then $\mathfrak{L}' = \langle \mathfrak{A}, Cn_\Gamma \rangle$ is also a logic and $Th(\mathfrak{L}') = Cn(\Gamma)$. By Definition 3.3(ii), $\mathfrak{L}$ is a weak sublogic of $\mathfrak{L}'$. If $\mathfrak{L}$ is compact so also is $\mathfrak{L}'$. Thus the intuitionistic propositional logic I of Chapter 2, §2.2.7, is a weak sublogic of the classical logic formulated with the junctors f, $\vee, \wedge, \supset$.

(ii) imp_Γ is reflexive and transitive:

$$A\ imp_\Gamma A$$

$$A\ imp_\Gamma B\ \&\ B\ imp_\Gamma C \Rightarrow A\ imp_\Gamma C.$$

(iii) For a fixed algebra $\mathfrak{A}$, the logics $\langle \mathfrak{A}, Cn \rangle$ are ordered by the inclusion relation on the set of consequence functions Cn. The maximum and minimum logics in this ordering are trivial:
  (a) the *maximum* logic has $Cn(\Delta) = |\mathfrak{A}|$ for all $\Delta \subseteq |\mathfrak{A}|$;
  (b) the *minimum* logic has $Cn(\Delta) = \Delta$ for all $\Delta \subseteq |\mathscr{A}|$.
Thus Cn is the identity function on $\mathbb{P}(|\mathfrak{A}|)$. ◆

3.6 Closed sets

As in various branches of algebra, the collection of certain kinds of subsystems of a given algebra (e.g. normal subgroups of a group) is ordered in a special way. Such orderings are studied in Chapter 5. The special kinds of subsystems of logics which concern us now are *closed sets*.

DEFINITION 3.9. Let $\langle \mathfrak{A}, Cn \rangle$ be a logic.

(i) A set $\Delta \subseteq |\mathfrak{A}|$ is said to be a *closed set* (also *fixed point* or *theory*) if $Cn(\Delta) = \Delta$. The set of closed sets of a logic $\mathfrak{L}$ is denoted by $cl(\mathfrak{L})$.
(ii) A set $\Delta \subseteq |\mathfrak{A}|$ is said to be *consistent* if $Cn(\Delta) \neq |\mathfrak{A}|$.

EXAMPLE 3.7. By **D1, D2, D3** the closed sets of the logic $\langle \mathfrak{A}, Cn \rangle$ are exactly the sets $Cn(\Delta)$ where $\Delta \subseteq |\mathfrak{A}|$. In the minimum logic on Σ every subset of Σ is closed. The set $cl(\mathfrak{L})$ of closed sets of a logic $\mathfrak{L}$ forms a lattice (Example 5.12(i) and Theorem 3.8 below). ◆

THEOREM 3.8. Let $\mathfrak{L} = \langle \mathfrak{A}, Cn \rangle$ be a logic.

(i) $cl(\mathfrak{L})$ is closed under arbitrary intersection.
(ii) Let $\Delta \subseteq |\mathfrak{A}|$. $Cn(\Delta)$ is the intersection of all closed sets which include Δ; $Cn(\varnothing) = \prod cl(\mathfrak{L})$.

Proof.

(i) Let K be a non-empty collection of closed sets of $\mathfrak{L}$. Let $\Delta \in K$. Then $\prod K \subseteq \Delta$. By **D3** $Cn(\prod K) \subseteq Cn(\Delta)$. As Δ is closed, $Cn(\prod K) \subseteq \Delta$. But this holds for all $\Delta \in K$. Hence $Cn(\prod K) \subseteq \prod K$. By **D1**, $Cn(\prod K) = \prod K$. Thus $\prod K$ is closed.
(ii) Let '$Q(\Delta)$' denote the collection of all closed sets which include Δ. Clearly $Cn(\Delta) \in Q(\Delta)$. Let $\Gamma \in Q(\Delta)$. Then $Cn(\Gamma) = \Gamma \supseteq \Delta$. By **D3**, $\Gamma \supseteq Cn(\Delta)$. Hence $\prod Q(\Delta) = Cn(\Delta)$. Putting $\Delta = \varnothing$ gives $Cn(\varnothing)$ as the intersection of all closed sets. ∎

Theorem 3.8(ii) shows that the theorems of a logic are exactly those sentences which are logical consequences of every set of sentences. But they do not necessarily coincide with the sentences which are the logical consequences of every sentence.

THEOREM 3.9. Let $\mathfrak{L} = \langle \mathfrak{A}, Cn \rangle$ be a logic. Let $T_{\mathfrak{L}} = \prod \{Cn(A); A \in |\mathfrak{A}|\}$. Then the following hold.

(i) $Th(\mathfrak{L}) \subseteq T_{\mathfrak{L}}$.
(ii) $T_{\mathfrak{L}}$ is the intersection of all non-empty closed sets of $\mathfrak{L}$.
(iii) If $Th(\mathfrak{L}) \neq \varnothing$ then $Th(\mathfrak{L}) = T_{\mathfrak{L}}$.
(iv) There exists a logic in which $Th(\mathfrak{L}) = \varnothing$ but $T_{\mathfrak{L}} \neq \varnothing$.

Proof.

(i) $Cn(\varnothing) \subseteq Cn(A)$ by **D3**. Hence $Th(\mathfrak{L}) = Cn(\varnothing) \subseteq T_{\mathfrak{L}}$.
(ii) Let Δ be the intersection of all non-empty closed sets of $\mathfrak{L}$. Now for all $A \in |\mathfrak{A}|$, A is a member of the closed set $Cn(A)$. Thus $Cn(A)$ is a

non-empty closed set. Hence $\Delta \subseteq T_\varrho$. Next, if Θ is a non-empty closed set then $A \in \Theta$ for some $A \in |\mathfrak{A}|$. Hence $T_\varrho \subseteq Cn(A) \subseteq Cn(\Theta) = \Theta$. So $T_\varrho \subseteq \Delta$. Thus, $T_\varrho = \Delta$.

(iii) Suppose $t \in Cn(\varnothing)$. Then $Cn(\varnothing) = Cn(t)$ and for every closed set Δ, $t \in Cn(\varnothing) \subseteq Cn(\Delta) = \Delta$. Thus every closed set is non-empty and $Cn(\varnothing)$ is the minimum closed set. By (ii), $Cn(\varnothing) = T_\varrho$.

(iv) Let $\Sigma = \{A, B\}$. Define the set-to-set function Cn by

$$Cn(\varnothing) = \varnothing, \quad Cn(A) = \{A\}, \quad Cn(B) = Cn(A, B) = \{A, B\}.$$

Then $\mathfrak{L} = \langle \Sigma, Cn \rangle$ is a logic and $T_\varrho = Cn(A) \cdot Cn(B) = \{A\}$. ∎

The existence of logical operations imposes conditions on the structure of the closed sets. Thus, if the logic satisfies Cn then the *weak conjunction condition* is satisfied:

WC: $(x, y)(Ez)(Cn(x, y) = Cn(z))$.

If the logic satisfies D then the *weak disjunction condition* is satisfied:

WD: $(x, y)(Ez)(z \in Cn(x) \cdot Cn(y))$.

THEOREM 3.10 (Lindenbaum). Let $\langle \Sigma, Cn \rangle$ be a logic satisfying **WC**. Let $\Delta_1, \ldots, \Delta_n$ and $\Gamma = \sum \{\Delta_i; 1 \leqslant i \leqslant n\}$ be closed sets and suppose that Ψ is a non-empty closed subset of Γ. Then $\Psi \subseteq \Delta_r$ for some r, $1 \leqslant r \leqslant n$.

Proof. Suppose not. Then $\Psi \not\subseteq \Delta_i$ for $1 \leqslant i \leqslant n$. Then for each i there exists $A_i \in \Psi - \Delta_i$. Thus $A_1, \ldots, A_n \in \Psi \subseteq \Gamma$. By **WC** there exists $t \in \Sigma$ such that $Cn(t) = Cn(A_1, \ldots, A_n) \subseteq \Psi \subseteq \Gamma$. As Γ is closed, $t \in \Delta_i$, say. Hence $A_i \in Cn(A_1, \ldots, A_n) = Cn(t) \subseteq Cn(\Delta_i) = \Delta_i$. This is a contradiction. ∎

EXAMPLE 3.8. Let $\mathfrak{L} = \langle \mathfrak{A}, Cn \rangle$ be a logic and let $\mathbb{R}$ be a set of objects ('realizations'). Further, let $\vDash$ be a subset of $\mathbb{R} \times |\mathfrak{A}|$. This can be considered as a 'satisfaction' relation between realizations and formulas of the logic. For $A \in |\mathfrak{A}|$ and $M \in \mathbb{R}$. M is a *model* of A if $M \vDash A$. Then for any $\Delta \subseteq |\mathfrak{A}|$, we say that $\Delta \to A$ if every realization which is a model of every member of Δ is also a model of A (cf. (2.1)). The satisfaction relation $\vDash$ is said to be *sound* if for all $\Delta \subseteq |\mathfrak{A}|$ and all $A \in |\mathfrak{A}|$

$$A \in Cn(\Delta) \quad \Rightarrow \quad \Delta \to A.$$

It is said to be *complete* if for all $\Delta \subseteq |\mathfrak{A}|$ and all $A \in |\mathfrak{A}|$

$$\Delta \to A \quad \Rightarrow \quad A \in Cn(\Delta).$$

Every logic can be trivially equipped with a sound and complete satisfaction relation: take $\mathbb{R} =_{\text{def}} cl(\mathfrak{L})$ and define the satisfaction relation $\vDash$ to be the membership relation. Thus

$$M \vDash A \quad \Leftrightarrow_{\text{def}} \quad A \in M.$$

◆

3.7 Logical operations

The typical examples of logical operations are connectives and quantifiers. In actual examples of logics the connectives and quantifiers are built into the logic before the notion of logical consequence is defined. In this section we reverse this development.

3.7.1 Connectives and logical equivalence

The definition of logic is very broad in that it does not require that there be any particular connection between the operations of the algebra $\mathfrak{A}$ and the logical consequence function Cn. But in the most important examples of logics there is a special relation mediated by the notion of logical equivalence (Tarski 1956, Chapter 5, §3).

The notion of logical equivalence given in Definition 3.2 can be relativized in an obvious way from Definition 3.9.

DEFINITION 3.10. Let $\mathfrak{L} = \langle \mathfrak{A}, Cn \rangle$ be a logic and $\Delta \subseteq |\mathfrak{A}|$. Let $\mathfrak{L}' = \langle \mathfrak{A}, Cn_\Delta \rangle$. Two sets $\Delta, \Gamma \subseteq |\mathfrak{A}|$ are said to be *logically equivalent relative to Δ* if $\Theta = \Gamma \, mod(Cn_\Delta^e)$ (see Definition 3.2(i)), i.e. $Cn(\Delta, \Theta) = Cn(\Delta, \Gamma)$. Elements $A, B \in |\mathfrak{A}|$ are *logically equivalent relative to Δ* if $A = B \, mod(Cn_\Delta^e)$ (also written $A \leftrightarrow_\Delta B$).

EXAMPLE 3.9.

(i) Logical equivalence (Definition 3.5) is exactly logical equivalence relative to the empty set (Definition 3.6).
(ii) Logical equivalence Cn^e and relative logical equivalence Cn_Δ^e are equivalence relations on the set of sentences of a logic, i.e. a reflexive, symmetric, and transitive relation (see Chapter 5, §5.4). This follows easily from **D1**, **D2**, and **D3**.
(iii) For compact logics the notion of relative logical equivalence can be reduced to logical equivalence relative to finite sets. Thus, if $\langle \Sigma, Cn \rangle$ is compact then, for all $A, B \in \Sigma$ and all $\Delta \subseteq \Sigma$, $A \leftrightarrow_\Delta B$ iff there exists a finite subset Δ' of Δ such that $A \leftrightarrow_{\Delta'} B$.
(iv) $A \leftrightarrow_\Delta B$ iff $A \, imp_\Delta B \,\&\, B \, imp_\Delta A$.
(v) For any non-empty set Σ the minimum logic on Σ is compact, it has no theorems, and logical equivalence is equality. ◆

Using the notion of logical equivalence we can formulate the concept of the *logical operation* or *connective* which is motivated by the following *substitution property*. In all the important examples of logics, if Δ is a set of sentences and UBW is a formula with a distinguished occurrence of a subformula B, and B' is logically equivalent to B relative to Δ, then the formula $UB'W$ which arises from UBW by substituting B' for that occurrence of B is logically equivalent to UBW (Example 2.2(i)).

Logical operations

DEFINITION 3.11. Let $\mathfrak{L} = \langle \mathfrak{A}, Cn \rangle$ be a logic and g an n-ary operation of the algebra $\mathfrak{A}$.

(i) g is a *logical operation* (in $\mathfrak{L}$) or a *connective* if it preserves logical equivalence in the sense

for all $A_1, \ldots, A_n, B_1, \ldots, B_n \in |\mathfrak{A}|$, if $A_i \leftrightarrow B_i$ for $1 \leqslant i \leqslant n$ then $g(A_1, \ldots, A_n) \leftrightarrow g(B_1, \ldots, B_n)$.

(ii) g is a *strongly logical operation* or a *strong connective* if g preserves logical equivalence relative to Δ (in the above sense) for all $\Delta \subseteq |\mathfrak{A}|$, i.e. if g is a logical operation in $\langle \mathfrak{A}, Cn_\Delta \rangle$ for all $\Delta \subseteq |\mathfrak{A}|$.

(iii) Let $\mathfrak{L} = \langle \Sigma, c_1, c_2, \ldots, Cn \rangle$ be a logic with connectives $c_1, c_2, \ldots$ The reduced logic $\bar{\mathfrak{L}}$ is $\langle \bar{\Sigma}, \bar{c}_1, \bar{c}_2, \ldots, \bar{C} \rangle$ (see Example 3.9(i) below) where the following hold.

 (a) The elements of $\bar{\Sigma}$ are the logical equivalence classes $[A]$ for $A \in \Sigma$. Let θ be the *canonical mapping* $A \to [A]$, i.e. $\theta(A) = [A]$.
 (b) The operations $\bar{c}_i$ are the operations induced on $\bar{\Sigma}$ by logical equivalence, i.e.

$$\bar{c}_i(\theta(A), \theta(B), \ldots) =_{\text{def}} \theta(c_i(A, B, \ldots)).$$

 (c) $\bar{C}$ is the set-to-set function on $\bar{\Sigma}$ defined by

$$\bar{C}(\Delta) =_{\text{def}} \theta Cn(\theta^{-1}\Delta) \qquad (\Delta \subseteq \bar{\Sigma}).$$

EXAMPLE 3.10.

(i) $Cn = \theta^{-1}\bar{C}\theta$ and so θ is a strong homomorphism of $\mathfrak{L}$ onto $\bar{\mathfrak{L}}$ (Definition 3.3(i)). By Theorem 3.2, $\bar{\mathfrak{L}}$ is a logic (Example 3.4).

(ii) $\bar{\mathfrak{L}}$ is a logic which shares the same logical identities with $\mathfrak{L}$ (Theorem 3.2). The reduced algebra (or quotient-algebra) $\mathfrak{L}\P =_{\text{def}} \langle \bar{\Sigma}, \bar{c}_1, \bar{c}_2, \ldots \rangle$ is usually called the *Lindenbaum algebra* of the logic $\mathfrak{L}$. The idea of constructing an algebra in this way was due to A. Tarski and A. Lindenbaum. Lindenbaum was a distinguished logician of the Polish school who was a victim of the Nazis (*Fundamenta Mathematicae*, **33**, 1945; see also Rasiowa and Sikorski 1963, p. 246n).

(iii) The connectives in classical logic are strong connectives (Example 2.1(vii)). The classical logic $\langle \mathbb{F}_{\text{pred}}, \vee, \wedge, t, f, Cn \rangle$, where Cn is the consequence function defined in Definition 2.13(v), is a logic according to Definition 3.5, and by Theorem 2.29 it satisfies Ax4 and so is compact. ♦

The substitution property mentioned above also justifies the definition of a certain class of logics whose elements are words on an alphabet.

DEFINITION 3.12. Let A be an alphabet and $\mathfrak{F}(A)$ be a non-empty subset of A^* which is *closed under substitution* of formulas for formulas in the following sense: for $UVW \in \mathfrak{F}(A)$, where also $V \in \mathfrak{F}(A)$, define the *substitution* operator $\Delta_{U,W}: \mathfrak{F}(A) \to A^*$ by

$$\Delta_{U,W}(P) = UP \qquad (P \in \mathfrak{F}(A)).$$

$\mathfrak{F}(A)$ is a *domain of formulas* if it is closed under all substitution operators $\Delta_{U,W}$. $\langle \mathfrak{F}(A), Cn \rangle$ is a *word logic* if each substitution operator is a logical operation (Definition 3.10) and a *strong word logic* if each substitution operator is a strongly logical operation.

EXAMPLE 3.11.

(i) The property of being a (strongly) logical operation is expressible in Qf_2 by SDPC's whose atoms are all of the form t in $\mathbb{C}[T]$. By Theorem 3.2 it follows that it is transmitted to homomorphic images.
(ii) If all the operations of the algebra $\mathfrak{A}$ of the logic $\langle \mathfrak{A}, Cn \rangle$ are strongly logical operations, then each relative logical equivalence relation Cn_A^e is a congruence on $\mathfrak{A}$ (Appendix, §A). ◆

Definition 3.11 and Example 3.11 raise the question of whether it is possible to equip an arbitrary algebra with a logical consequence function in such a way that the operations of the algebra become logical operations. The following easy example shows how this can be done by using congruences on the algebra.

EXAMPLE 3.12.

(i) Let $\mathfrak{A}$ be an algebra and θ a congruence on it. Let $[x]\theta$ denote the congruence class of the element x. Define the set-to-set function Cn on $\mathfrak{A}$ by

$$Cn(\Delta) =_{\text{def}} \sum \{[x]\theta; x \in \Delta\}.$$

Then $\langle \mathfrak{A}, Cn \rangle$ is a compact logic in which the operations of the algebra are connectives. Cn is in fact a closure operation (Example 3.1(v)). Observe that $A \in Cn(\Delta)$ iff there exists $B \in \Delta$ such that $A \in Cn(B)$. This is typical of a certain class of logics and so we make the definition: a logic $\langle \mathcal{A}, Cn \rangle$ is *strongly compact* if for all $\Delta \subseteq |\mathfrak{A}|$, $A \in Cn(\Delta)$ implies the existence of $B \in \Delta$ such that $A \in Cn(B)$.
(ii) Let $\langle \mathfrak{A}, Cn_1 \rangle$, $\langle \mathfrak{A}, Cn_2 \rangle$ be logics. Then $\langle \mathfrak{A}, Cn_1 \cdot Cn_2 \rangle$ is a logic. If Cn_1 and Cn_2 are both compact then $Cn_1 \cdot Cn_2$ is also compact. But it is not the case that if Cn_1 and Cn_2 are both strongly compact then $Cn_1 \cdot Cn_2$ is also strongly compact.
(iii) Let Σ be a non-empty set and $\Gamma \subseteq \Sigma$. Define the set-to-set function C_Γ on Σ by

$$A \in C_\Gamma \Leftrightarrow_{\text{def}} \Delta \cdot \Gamma^c = \emptyset \text{ or } A \in \Gamma.$$

The $\langle \Sigma, C_\Gamma \rangle$ is a strongly compact logic in which A is logically equivalent to B iff A and B are either both in Γ or both not in Γ. This result can be extended to the case where K is a collection of subsets of Σ: put $Cn =_{\text{def}} \prod \{C_\Gamma; \Gamma \in K\}$. But Cn is not necessarily strongly compact. ◆

Another way in which logics can be constructed from algebras is via the system of subalgebras (see Example 3.19(iii), below).

THEOREM 3.11. *Let $\mathfrak{A}$ be an algebra. For each subset Δ of $|\mathfrak{A}|$ define $Cn_{\mathfrak{A}}(\Delta)$ to be the subalgebra generated by Δ. Then $\langle \mathfrak{A}, Cn_{\mathfrak{A}} \rangle$ is a compact logic.*

The operations of an algebra are not necessarily logical operations of its subalgebra logic.

EXAMPLE 3.13. In any group the inverse operation is a logical operation, but this is not always the case with group multiplication. Consider an infinite cyclic group $\mathfrak{G}$, i.e. a group whose elements are the positive and negative powers of an element a. Then, using the notation of Theorem 3.11, the group multiplication is not a logical operation. For $a \notin Cn_{\mathfrak{G}}(a^3)$, i.e. a is not in the subgroup generated by a^3, and hence

$$a^3 \neq a \bmod(Cn_{\mathfrak{G}}^e). \tag{3.23}$$

But $a = a^{-1} \bmod(Cn_{\mathfrak{G}}^e)$. If group multiplication were a logical operation we would have $a^2 \cdot a = a^2 \cdot a^{-1} \bmod(Cn_{\mathfrak{G}}^e)$, contrary to (3.23).

This raises the question of necessary and sufficient conditions on a group for its multiplication to be a logical operation. We prove the following proposition.

PROPOSITION. *For any group the multiplication operation is a logical operation in the subgroup logic iff every element is of order 2.*

Proof. Let $\mathfrak{G}$ be a group and let $\leftrightarrow$ denote the logical equivalence $Cn_{\mathfrak{G}}^e$. First suppose that multiplication is a logical operation of $\langle \mathfrak{G}, Cn_{\mathfrak{G}} \rangle$. Let $x \in |\mathfrak{G}|$. Then x is in the subgroup generated by x^{-1} and x^{-1} is in the subgroups generated by x, i.e. $x^{-1} \in Cn_{\mathfrak{G}}(x)$ and $x \in Cn_{\mathfrak{G}}(x^{-1})$. Thus $x \leftrightarrow x^{-1}$. Since multiplication is a logical operation, $x^2 \leftrightarrow xx \leftrightarrow e$ where e is the unit element of $\mathfrak{G}$. Hence $x^2 \in Cn_{\mathfrak{G}}(e)$. But $Cn_{\mathfrak{G}}(e)$ is the subgroup generated by e, i.e. e itself. Thus $x^2 = e$, i.e. x has order 2. Conversely, suppose that every element of $\mathfrak{G}$ has order 2. To show that group multiplication is a logical operation it suffices to prove that

$$x \leftrightarrow y \quad \Rightarrow \quad xz \leftrightarrow yz \qquad (x, y, z \in |\mathfrak{G}|). \tag{3.24}$$

Suppose $x \leftrightarrow y$. Then each of x, y is in the subgroup generated by the other. Hence there exist integers n, m such that $x = y^n$ and $y = x^m$. Since x, y are of order 2 it can be assumed that $n, m \in \{0, 1\}$. If $n = 0$ then $x = e$ and hence $x = y$. If $n = 1$ then again $x = y$. Hence, in both cases, $xz = yz$ for all $z \in |\mathfrak{G}|$. Hence $xz \leftrightarrow yz$. Thus (3.24) holds. ∎◆

3.7.2 *Quantifier logics*

Predicate logics are certain logics whose sentences have a particular internal structure. We are therefore obliged now to consider the elements of our logics as words on certain alphabets in order to have a ready apparatus of substitution of expressions (Appendix, §W). Thus, quantifier logics are already word logics (Definition 3.8). An important function is assumed by operations of joining certain distinguished letters within a word together—*binding*. Bourbaki (1958, §1.1) develops a method of doing this which involves bars above the line (called 'links') joining the relevant symbols. However, this artifice 'would lead to insuperable difficulties for the printer and for the reader'. Resort is therefore made to a one-dimensional symbolism which is similar to the conventional use of 'free' or 'bound' variables which are linked by quantifiers. Many formulations employ a single list of variables, so that in a single expression one occurrence of a variable might be bound and another occurrence of the same variable might be free. This causes some minor technical complexity. Some of these irritations can be avoided by using typographically distinct lists of bound variables and free variables (e.g. Schütte 1961, Rasiowa and Sikorski 1963). This device has been employed in Chapter 2 and it will also be used later in the construction of various quantifier logics.

The modern development of quantification theory is crystallized in Frege's *Begriffschrift* (Frege 1879), where the Aristotelian quantifiers (the universal and existential quantifiers) are applied to individuals. Frege's ideas remained unknown for many years until Russell helped to establish them. The modern idea of quantification over individuals seems to have been developed independently by Mitchell (1883), Peirce (1885), and Peano (1889) (see Church 1956, §49). Frege's *Begriffschrift* introduces the concept of 'bound variable' without explicitly naming it. The terms 'real' and 'apparent' variable derive from Peano (1894) where the ideas are illustrated in the familiar expression of a definite integral. In the expression $\int_0^1 x^m \, dx$, x is an apparent variable, not a real variable—it is meaningful to substitute a particular number for m but not for x. The terms 'free' and 'bound' are the current usage.

There are many types of quantifier besides the traditional universal and existential quantifiers, particularly in logics with infinitely long formulas (Henkin 1961). Examples include:

'for infinitely many x, $A(x)$';
'generically A', i.e. the set of x's for which $A(x)$ holds is open and dense;
'for each x there exists a y and for each z there exists a w such that $A(x, y, z, w)$'.

A remarkable quantifier—'there are uncountably many $x \ldots$'—was first studied by Mostowski (1957) who posed the problem of finding an analogue of the Gödel completeness theorem for the logic obtained by adjoining this quantifier to the first-order predicate logic. Fuhrken (1964) proved the compactness of this logic. Vaught (1964) proved completeness using a complicated set of axioms. Keisler (1970) proved completeness with four simple axiom schemes. In this book we shall confine attention to the 'Aristotelian quantifiers'—'for all $x \ldots$' and 'there exists an $x \ldots$'—because

Logical operations

of their traditional importance and because of the completeness theorem of classical logic (Chapter 2, §2.3.7). The feature of actual quantifiers which we shall generalize below is the operation of binding of variables.

DEFINITION 3.13. Let M be an alphabet. M^* is the set of words on M (Appendix, §W). A *free-variable* logic on M is a word logic (Definition 3.8) $\langle \Sigma, Cn \rangle$ such that the following hold

(i) M includes an infinite subset Fv which are called free variables. For $\Gamma \subseteq M^*$, $Sb(\Gamma, Fv)$ is the set of all substitution operators on M^* which simultaneously substitute members of Γ for free variables. For $W \in M^*$, $Fv(W)$ denotes the set of free variables in W, and for $\Gamma \subseteq M^*$, $Fv(\Gamma) =_{\text{def}} \sum \{Fv(W); W \in \Gamma\}$. The variables occur in formulas in such a way that for each connective c of the logic

$$Fv(c(U, V, \ldots)) = Fv(U, V, \ldots) \qquad (U, V \in M^*).$$

(ii) M^* includes a set $\mathbb{T}$ of *terms* such that $\mathbb{T} \supseteq Fv$. Both $\mathbb{F}$ and $\mathbb{T}$ are closed under substitution of terms for free variables, i.e.

$$A\sigma \in \Sigma \quad (A \in \Sigma, \sigma \in Sb(\mathbb{T}, Fv))$$

$$t\sigma \in \mathbb{T} \quad (t \in \mathbb{T}, \sigma \in Sb(\mathbb{T}, Fv)),$$

and arbitrary substitutions for free variables commute with the connectives: for each connective c of the logic

$$c(U, V, \ldots)\sigma = c(U\sigma, V\sigma, \ldots) \qquad (U, V \in M^*, \sigma \in Sb(M^*, Fv)).$$

The terms t such that $Fv(t) = \emptyset$ are the *constant terms*.

(iii) Substitution of terms for free variables preserves logical consequence, i.e.

$$A \in Cn(\Delta) \quad \Rightarrow \quad A\sigma \in Cn(\Delta\sigma) \qquad (\Delta, A \subseteq \Sigma, \sigma \in Sb(\mathbb{T}, Fv)).$$

In a free-variable logic, substitution of terms for free variables is a logical operation (Definition 3.10), i.e.

$$A \leftrightarrow B \quad \Rightarrow \quad A\sigma \leftrightarrow B\sigma \qquad (A, B \in \Sigma, \sigma \in Sb(\mathbb{T}, Fv)).$$

EXAMPLE 3.14.

(i) Let Σ be the set of atomic formulas of the predicate language $L(\tau)$ (Chapter 2, §2.3.1) and let Cn be the identity function on $\mathbb{P}(\Sigma)$. Since the atomic formulas are closed under substitution of terms for free variables (Example 2.6), $\mathfrak{L}_0(\tau) =_{\text{def}} \langle \Sigma, Cn \rangle$ is a free-variable logic.

(ii) Consider the subset $\mathbb{F}$ of $\mathbb{F}_{\text{pred}}$ (the set of formulas of the classical predicate calculus (see Chapter 2, §2.3.1)) consisting of all formulas in which no quantifiers occur. Let Cn be the classical logical consequence function (Definition 2.13(v)) restricted to these quantifier-free formulas. Then $\langle \mathbb{F}, Cn \rangle$ is a compact free-variable logic. This logic is little more than the classical propositional logic (see Chapter 2, §2.3.4).

Quantifier logics are free-variable logics on which there are certain operations which *bind* free variables and preserve logical equivalence. ◆

Definition 3.14.

(i) Let $\langle \mathfrak{A}, Cn \rangle$ be a free-variable logic with the set Fv of free variables and set $\mathbb{T}$ of terms. (In this definition we consider Fv^n to be the set of words of length n on Fv. This allows us to apply substitution operations on Fv^n. Further, if $\alpha \in Fv^n$ then $|\alpha|$ denotes the set of free variables which occur in α.) A mapping $Q: Fv^n \times |\mathfrak{A}| \to |\mathfrak{A}|$ is an *n-ary quantifier* under the following conditions.

Q1: Q is a logical operation, i.e. logical equivalence is preserved:

$$F \leftrightarrow G \quad \Rightarrow \quad Q(\alpha, A) \leftrightarrow Q(\alpha, B) \qquad (A, B \in |\mathfrak{A}|, \alpha \in Fv^n).$$

Q2: Q annihilates ('binds') free variables, i.e.

$$Fv(Q(\alpha, A)) = Fv(A) - \{|\alpha|\}.$$

The free variables in $|\alpha|$ are said to be *bound* in $Q(\alpha, F)$.

Q3: Provided that no 'clashes' occur, the renaming of bound variables preserves logical equivalence, i.e.

$$Q(\alpha(x/y), A(x/y)) \quad \leftrightarrow \quad Q(\alpha, A)$$

for all $y \in Fv$ and all $x \in Fv - (|\alpha| + Fv(A))$.

Q4: Substitution of terms in which no bound variables occur preserves logical equivalence, i.e.

$$Q(\alpha, A)(t/y) \quad \leftrightarrow \quad Q(\alpha, A(t/y))$$

for any free variable y such that $y \notin |\alpha|$ and all $t \in \mathbb{T}$ with $|\alpha| \cdot Fv(t) = \varnothing$.

The operation Q is a *strong quantifier* if **Q1**–**Q4** hold together with

Q1*: $F \leftrightarrow_\Delta G \quad \Rightarrow \quad Q(\alpha, A) \leftrightarrow_\Delta Q(\alpha, B)$

for all sets Δ, A, B of formulas and all $\alpha \in Fv^n$ such that $|\alpha| \cdot Fv(\Delta) = \varnothing$.

(ii) A structure $\langle \mathfrak{A}, Q_1, \ldots, Cn \rangle$ is a *quantifier logic* if $\langle \mathfrak{A}, Cn \rangle$ is a free variable logic and $Q_1, \ldots$ are quantifiers on it.

Example 3.15.

(i) In the classical predicate logic (Chapter 2, §2.3) the mapping $Q_i: Fv \times \mathbb{F}_{pred} \to \mathbb{F}_{pred}$, $i = 1, 2$, can be defined by $Q_1(x, A) = \forall a A(a/x)$ and $Q_2(x, A) = \exists a A(a/x)$ where a is the first bound variable which does not occur in A (Definition 2.12). Thus $\forall, \exists$ are unary quantifiers. For, by Corollary 2.15, **Q1*** holds. **Q2**, **Q3**, and **Q4** follow from the properties of substitution and the formation rules in Definition 2.12. The structure $\langle \mathbb{F}_{pred}, Q_1, Q_2, Cn \rangle$, where Cn is the classical logical consequence function

(Definition 2.13(v)), is a quantifier logic. The traditional (Aristotelian) quantifiers will be discussed further in Chapter 4, §4.11.

(ii) The definition of *quantifier* is broad enough to allow objects which would not normally be considered as quantifiers. For instance, let t be a constant term of a free-variable logic $\langle \mathfrak{A}, Cn \rangle$. Define $Q_t: Fv \times |\mathfrak{A}| \to |\mathfrak{A}|$ by

$$Q_t(x, A) =_{\text{def}} A(t/x) \qquad (x \in Fv, A \in |\mathfrak{A}|).$$

Then it follows from the two preceding definitions that Q_t is a quantifier.

(iii) The quantifiers of a given quantifier logic $\mathfrak{L} = \langle \mathfrak{A}, Q_1, Q_2, \ldots, Cn \rangle$ can be composed amongst themselves and with the other logical operations in three ways to give further quantifiers.

(a) Suppose that Q_i and Q_j are n-ary and m-ary quantifiers respectively. Define $Q: Fv^{n+m} \times |\mathfrak{A}| \to |\mathfrak{A}|$ by

$$Q(\alpha \cap \beta, A) =_{\text{def}} Q_i(\alpha, Q_j(\beta, A))$$

for all $\alpha \in Fv^n$, $\beta \in Fv^m$, and $A \in |\mathfrak{A}|$. Then Q is an $n \geq m$-ary quantifier.

(b) Suppose now that the operations of the algebra $\mathfrak{A}$ are connectives. Then the polynomials of $\mathfrak{A}$ (Appendix, §A) are also connectives. Let p be an r-place polynomial of $\mathfrak{A}$. p is also a connective (Definition 3.11). Let $Q_i, \ldots, Q_j$ be r n-ary quantifiers. Define $p(Q_i, \ldots, Q_j): Fv^n \times |\mathfrak{A}| \to |\mathfrak{A}|$ by

$$p(Q_i, \ldots, Q_j)(\alpha, A) =_{\text{def}} p(Q_i(\alpha, A), \ldots, Q_j(\alpha, A))$$

for all $\alpha \in Fv^n$ and all $A \in |\mathfrak{A}|$. Then $p(Q_i, \ldots, Q_j)$ is also an n-ary quantifier.

(c) Let p be a unary polynomial. Let Q be an n-ary quantifier. Define $Q \circ p: Fv^n \times |\mathscr{A}| \to |\mathfrak{A}|$ by

$$Q \circ p(\alpha, A) =_{\text{def}} Q(\alpha, p(A)) \qquad (\alpha \in Fv^n, A \in |\mathfrak{A}|).$$

Then $Q \circ p$ is also an n-ary quantifier.

Quantifiers which are constructed from $Q_1, Q_2, \ldots$ by a finite number of operations of the above three forms are the *derived quantifiers* of $\mathfrak{L}$. ◆

A brief study of some axioms for quantifiers will be undertaken in Chapter 4, §4.4.

3.8 Compact logics

The notion of compactness of classical logic has already been discussed in Chapter 2, §2.2.6. The notion will now be applied to general logics. The most significant point about compactness is that the axioms for compact logics can be replaced by equivalent axioms concerning finitary consequence functions and finite sets. For this purpose, recall Definition 3.1(iv).

THEOREM 3.12. Let $\mathfrak{A}$ be an algebra and Cn be a finitary set-to-set function. Then we have the following.

(i) $\langle \mathfrak{A}, Cn^{\blacktriangledown} \rangle$ is a logic if Cn satisfies the following conditions $\mathbf{D1}^\omega$, $\mathbf{D2}^\omega$, $\mathbf{D3}^\omega$ for all $A, B \in |\mathfrak{A}|$, $\Delta, \Gamma \in \mathbb{P}_\omega(|\mathfrak{A}|)$.

  $\mathbf{D1}^\omega$: $A \in Cn(A)$

  $\mathbf{D2}^\omega$: $B \in Cn(\Delta, A)$ & $A \in Cn(\Gamma)$ $\Rightarrow$ $B \in Cn(\Delta, \Gamma)$

  $\mathbf{D3}^\omega$: $A \in Cn(\Gamma)$ $\Rightarrow$ $A \in Cn(\Gamma, \Delta)$.

(ii) If Cn is monotone increasing and $\langle \mathfrak{A}, Cn^{\blacktriangledown} \rangle$ is a logic then Cn satisfies $\mathbf{D1}^\omega$, $\mathbf{D2}^\omega$, and $\mathbf{D3}^\omega$.

Proof.

(i) Suppose that Cn satisfies $\mathbf{D1}^\omega$, $\mathbf{D2}^\omega$, and $\mathbf{D3}^\omega$. To prove that $\langle \mathfrak{A}, Cn^{\blacktriangledown} \rangle$ is a logic it is necessary to show that $\mathbf{D1}$, $\mathbf{D2}$, and $\mathbf{D3}$ hold. The proofs of $\mathbf{D1}$ and $\mathbf{D3}$ are straightforward deductions from Definition 3.1(v). To prove $\mathbf{D2}$, however, repeated application of $\mathbf{D2}^\omega$ is needed. Thus, let $A \in Cn^{\blacktriangledown} Cn^{\blacktriangledown}(\Delta)$: then there exists a finite set $\Gamma = \{B_1, \ldots, B_n\}$, say, such that $A \in Cn^{\blacktriangledown}(\Gamma)$ and $\Gamma \subseteq Cn^{\blacktriangledown}(\Delta)$. Again, there exist finite subsets $\Delta_1, \ldots, \Delta_n$ of Δ such that $B_i \in \Delta_i$. Put $\Psi = \Delta_1 + \cdots + \Delta_n$. Then $\Psi \in \mathbb{P}_\omega(\Delta)$ and by $\mathbf{D3}^\omega$, $A \in Cn(\Gamma)$ and $B_i \in Cn(\Psi)$ for $1 \leq i \leq n$. Application of $\mathbf{D2}^\omega$ n times gives $A \in Cn(\Psi)$. Hence $A \in Cn^{\blacktriangledown}(\Delta)$. Thus $Cn^{\blacktriangledown}(Cn^{\blacktriangledown}(\Delta)) \subseteq Cn^{\blacktriangledown}(\Delta)$.

(ii) Suppose that Cn is monotone increasing and $\langle \mathfrak{A}, Cn^{\blacktriangledown} \rangle$ is a logic. Then $Cn^{\blacktriangledown}$ satisfies $\mathbf{D1}$, $\mathbf{D2}$, and $\mathbf{D3}$.

  $\mathbf{D1}^\omega$: By $\mathbf{D1}$, $A \in Cn^{\blacktriangledown}(A)$. By Definition 3.1(iv) there exists a finite subset Γ of $\{A\}$ such that $A \in Cn(\Gamma)$. As Cn is monotone increasing, $A \in Cn(A)$.

  $\mathbf{D2}^\omega$: Let $\Delta, \Gamma \in \mathbb{P}_\omega(|\mathfrak{A}|)$. Suppose that $B \in Cn(\Delta, A)$ and $A \in Cn(\Gamma)$. By $\mathbf{D3}$, $B \in Cn^{\blacktriangledown}(\Delta, Cn^{\blacktriangledown}(\Gamma))$. By Theorem 3.6(i), $B \in Cn^{\blacktriangledown}(\Delta, \Gamma)$. Hence there exists $\Psi \in \mathbb{P}_\omega(\Delta + \Gamma)$ such that $B \in Cn(\Psi)$. As Cn is monotone increasing, $B \in Cn(\Delta, \Gamma)$.

  $\mathbf{D3}^\omega$: This follows directly from the hypothesis that Cn is monotone increasing. ∎

The property of compactness is not expressible in Qf_2 but it nevertheless has some important transfer properties:

EXAMPLE 3.16.

(i) Every strong sublogic and every strong homomorphic image of a compact logic is also compact.

(ii) Let $\mathfrak{L} = \langle \mathfrak{A}, Cn \rangle$. Then $Cn^{\blacktriangledown\blacktriangledown}$ is the compact kernel of Cn (Definition 3.1(iv)) and $\mathfrak{L}^{\blacktriangledown\blacktriangledown} = \langle \mathfrak{A}, Cn^{\blacktriangledown\blacktriangledown} \rangle$ is a compact logic. ◆

Compact logics

In many logics the notion of logical consequence is based on a definition of *direct derivation* which describes the one-step action of the rules of inference (cf. Definition 2.6). Iteration of this action yields the idea of a proof. In the next theorem we abstract from this well-known situation by considering a set-to-set function D which plays the role of direct derivation.

THEOREM 3.13. Let D be a set-to-set mapping on a set Σ satisfying the following conditions:

R1: $D(\Delta) \supseteq \Delta \qquad (\Delta \subseteq \Sigma)$

R2: $D(\Delta + \Gamma) \supseteq D(\Delta) \qquad (\Delta, \Gamma \subseteq \Sigma)$

Define $Cn(\Delta)$ by

$$Cn(\Delta) = \prod \{\Gamma; \Delta \subseteq \Gamma \subseteq \Pi \ \& \ D(\Gamma) \subseteq \Gamma\}.$$

Then Cn is a logical consequence function on Σ and $D(Cn(\Delta)) = Cn(\Delta)$.

Proof. Let $G[\Delta] = \{\Gamma \supseteq \Delta; D(\Gamma) \subseteq \Gamma\}$. It is trivial to verify that Cn satisfies **D1**. Further, $G[\Delta + \Gamma] \subseteq G[\Delta]$. Hence $\prod G[\Delta + \Gamma] \supseteq \prod G[\Delta]$. Thus $Cn(\Delta + \Gamma) \supseteq Cn(\Delta)$ and Cn satisfies **D2**.

Let $\Gamma \in G[\Delta]$. Then $Cn(\Delta) \subseteq \Gamma$. By **D2**, $D(Cn(\Delta)) \subseteq D(\Gamma) \subseteq \Gamma$. Hence $D(Cn(\Delta)) \subseteq \prod G[\Delta] = Cn(\Delta)$. By **D1**, $D(Cn(\Delta)) = Cn(\Delta)$. Since Cn satisfies **D1** it now follows that $Cn(\Delta) \in G[Cn(\Delta)]$. Hence $Cn(Cn(\Delta)) \subseteq Cn(\Delta)$. Thus Cn satisfies **D3**. ∎

If each rule of inference has finitely many premisses, then D is compact and Cn can be built up from 'below' by forming the increasing chain

$$\Delta \subseteq D(\Delta) \subseteq D^2(\Delta) \subseteq D^3(\Delta) \subseteq \cdots$$

to give a compact consequence function.

COROLLARY 3.14. Let $D^*(\Delta) = \sum \{D^n(\Delta); n \in \mathbb{N}\}$ where D satisfies **R1** and **R2**. If D is compact then $D^* = Cn$ and Cn is compact.

Proof. The proof is accomplished in four steps.

(i) First we show that $D(D^*(\Delta)) \subseteq D^*(\Delta)$. Let $A \in D(D^*(\Delta))$. By the compactness of D there exist $B_1, \ldots, B_n \in D^*(\Delta)$ such that $A \in D(B_1, \ldots, B_n)$. Hence there exists an m such that $B_1, \ldots, B_n \in D^m(\Delta)$. Thus, $A \in D^{m+1}(\Delta) \subseteq D^*(\Delta)$.

(ii) Then $\Delta \subseteq D^*$ and so, from (i), $D^*(\Delta) \in G[\Delta]$ so that $Cn(\Delta) \subseteq D^*(\Delta)$.

(iii) Conversely, $D^*(\Delta) \subseteq Cn(\Delta)$. For let $\Gamma \in G[\Delta]$. Then $\Gamma \supseteq \Delta$ and $D(\Gamma) \subseteq \Gamma$. By **R1** and **R2**, $\Gamma \supseteq D^n(\Delta)$ for all $n \in \mathbb{N}$ and so $\Gamma \supseteq D^*(\Delta)$. Hence $Cn(\Delta) = \prod G[\Delta] \supseteq D^*(\Delta)$.

(iv) For the compactness of Cn, let $A \in Cn(\Delta)$. Then $A \in D^m(\Delta)$ for some m. But, by induction on m, D^m is compact. Hence there exists a finite subset $Z \subseteq \Delta$ such that $A \in D^m(Z)$. Hence $A \in D^*(Z) = Cn(Z)$. ∎

The two following theorems on compact logics are given in Tarski (1965, Chapter 5, §1).

THEOREM 3.15. Let $\langle \mathfrak{A}, Cn \rangle$ be a compact logic. Let K be a collection of subsets of $|\mathfrak{A}|$ such that for every finite subset L of K there exists a member of K which includes $\sum L$. Then $Cn(\sum K) = \sum \{Cn(\Delta); \Delta \in K\}$.

Proof. Let $x \in Cn(\sum K)$. By compactness, there exists a finite subset Z of K such that $x \in Cn(\sum Z)$. But there exists a Γ in K such that $\Gamma \supseteq \sum Z$. Hence $Cn(\Gamma) \supseteq Cn(\sum Z)$. Thus $x \in Cn(\Gamma) \subseteq \sum \{Cn(\Delta); \Delta \in K\}$. So

$$Cn(\sum K) \subseteq \sum \{Cn(\Delta); \Delta \in K\}.$$

But for $\Delta \in K$, $\Delta \subseteq \sum K$. Hence

$$\sum \{Cn(\Delta); \Delta \in K\} \subseteq Cn(\sum K).$$

The stated equality follows from these two inclusions. ∎

COROLLARY 3.16. Let K be a non-empty collection of subsets of $|\mathfrak{A}|$ such that for all $\Delta, \Gamma \in K$, either $\Delta \subseteq \Gamma$ or $\Gamma \subseteq \Delta$. Then

$$Cn(\sum K) = \sum \{Cn(\Delta); \Delta \in K\}.$$

THEOREM 3.17. Let $\langle \mathfrak{A}, Cn \rangle$ be a compact logic. Let $\Delta \subseteq |\mathfrak{A}|$ and Γ be a finite subset of $Cn(\Delta)$. Then there exists a finite set Ψ such that $\Psi \subseteq \Delta$ and $\Gamma \subseteq Cn(\Psi)$.

Proof. As the logic is compact, for each $A \in \Gamma$ there exists a finite set $\Gamma(A)$ such that $A \in Cn(\Gamma(A))$ and $\Gamma(A) \subseteq \Delta$. Hence

$$\Gamma \subseteq \sum \{Cn(\Gamma(A)); A \in \Gamma\}. \tag{3.25}$$

Put $\Psi = \sum \{\Gamma(A); A \in \Gamma\}$. Then $\Psi \in \mathbb{P}_\omega(\Delta)$. Thus $Y(A) \subseteq \Psi \subseteq \Delta$. By D3, $Cn(Y(A)) \subseteq Cn(\Psi)$. This holds for every element A of Γ. Hence $\sum \{Cn(Y(A)); A \in Cn\} \subseteq Cn(\Psi)$. By (3.25), $\Gamma \subseteq Cn(\Psi)$. ∎

THEOREM 3.18. Let $\mathfrak{L} = \langle \mathfrak{A}, Cn \rangle$ be a logic and $\mathfrak{L}^{\blacktriangledown\blacktriangledown} = \langle \mathfrak{A}, Cn^{\blacktriangledown\blacktriangledown} \rangle$ its compact kernel. Then every logical identity which holds in $\mathfrak{L}$ also holds in $\mathfrak{L}^{\blacktriangledown\blacktriangledown}$. Moreover, if an operation of $\mathfrak{A}$ is (strongly) logical in $\mathfrak{L}$ then it is also (strongly) logical in $\mathfrak{L}^{\blacktriangledown\blacktriangledown}$.

Proof. By Example 3.9(i) and Theorem 3.5. ∎

3.9 Formal systems and combinatorial systems

Many examples of logics which spring to mind—propositional calculi of various sorts, predicate logics (cf. Chapter 2)—are presented as formal systems. Formal systems can be defined abstractly in the following way.

DEFINITION 3.15.

(i) A *formal system* is a structure $\mathfrak{F} = \langle \mathfrak{A}; R_1, R_2, \ldots \rangle$ where $\mathfrak{A}$ is a countable set of objects called *formulas* and $R_1, R_2, \ldots$ is a sequence (possibly infinite) of finitary (possibly unary) relations on $\mathfrak{A}$.
(ii) A formula A is *directly derivable* from a set Δ of formulas if either $A \in \Delta$ or there exists a sequence $B_1, \ldots, B_n$ of formulas in Δ and a relation R_k such that $(B_1, \ldots, B_n, A) \in R_k$. $D_{\mathfrak{F}}$ is defined to be the set-to-set function defined by

$$A \in D_{\mathfrak{F}}(\Delta) \quad \Leftrightarrow \quad A \text{ is directly derivable from } \Delta.$$

(iii) A *proof of A from Σ* (or a *derivation*) is a finite sequence $A_1, \ldots, A_m$ such that $A_m = A$ and each member of this sequence is either in Σ or is directly derivable from preceding members, i.e. $A \in \sum \{D_{\mathfrak{F}}^n(\Sigma); n \in \mathbb{N}\}$. An *admissible rule of inference* is a set $K \subseteq \mathbb{P}_\omega(|\mathfrak{A}|) \times |\mathfrak{A}|$ such that

$$(\Delta, A) \in K \ \& \ \Delta \subseteq \Gamma \quad \Rightarrow \quad A \in D_{\mathfrak{F}}(\Gamma) \qquad (\Gamma \subseteq |\mathfrak{A}|).$$

(iv) A is a *consequence* of Σ if there exists a proof of A from Σ (written $\Sigma \vdash A$). $Cn_{\mathfrak{F}}(\Delta)$ denotes the set of consequences of Δ. Clearly, $Cn_{\mathfrak{F}}(\Delta) = \sum \{D_{\mathfrak{F}}^n(\Sigma); n \in \mathbb{N}\}$.

Since all the relations $R_1, R_2, \ldots$ are finitary relations, $D_{\mathfrak{F}}$ is compact (Definition 3.1). Further, by construction, $D_{\mathfrak{F}}$ satisfies **R1** and **R2** of Theorem 3.13. By Corollary 3.14, $\langle \mathfrak{A}, Cn_{\mathfrak{F}} \rangle$ is a compact logic.

It is sometimes useful to give a graphical representation of a proof as follows: each element A_i is assigned to a node $n(i)$ of a tree. If in this proof A_i is directly derivable from $A_{i_1}, \ldots, A_{i_r}$ via rule R_S say, then the node $n(i)$ is below the nodes $n(i1), \ldots, n(ir)$ and there are edges of the graph joining each of the nodes $n(i1), \ldots, n(ir)$ to $n(i)$.

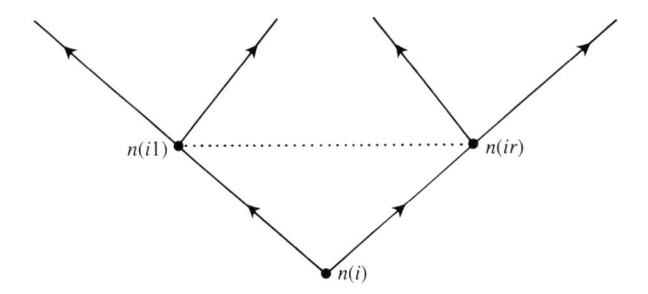

The association of formulas of a proof with nodes of a tree can be directly represented as

$$\frac{\frac{\cdots}{A_{il}} \qquad \frac{\cdots}{A_{ir}}}{A_i}$$

The root of the tree then represents the proven formula A and the leaves represent the members of the set of hypotheses from S which occur in the proof. This method of representing a sequence form of proof as a tree was introduced by Hilbert and Bernays (1934) and was called 'resolution into proof threads'.

EXAMPLE 3.17.

(i) Consider an inductive definition of a set Γ with generating function $\mathfrak{C}$. If $\mathfrak{C}$ has a set K of rules, each with finitely many premisses, then each such rule gives rise to a finitary relation R_r so that the structure $\mathfrak{F} = \langle \Gamma, \ldots, R_r, \ldots \rangle$ is a formal system and $D_{\mathfrak{F}}(\Delta) = \mathfrak{S}(\Delta)$. For example, consider an algebra $\mathfrak{A}$. For $\Delta \subseteq |\mathfrak{A}|$, $\mathfrak{S}(\Delta)$ denotes the set of elements generated by the algebra. $\mathfrak{S}(\Delta)$ is defined inductively.

(a) $\Delta \subseteq \mathfrak{S}(\Delta)$.
(b) For each operation F of $\mathfrak{A}$, if f is an n-ary operation and $x_1, \ldots, x_n \in \mathfrak{S}(\Delta)$ then $f(x_1, \ldots, x_n) \in \mathscr{S}(\Delta)$. Hence $\langle \mathfrak{A}, \mathfrak{S} \rangle$ is a logic.

(ii) The Hilbert system H of the classical propositional calculus (Chapter 2, §2.2.7) can be considered as a formal system $H = \langle \mathbb{F}_{prop}, \mathbf{MP} \rangle$ where $\mathbf{MP}$ is the *modus ponens* rule, i.e. $\mathbf{MP} = \{(A, [A \supset B]), A, B \in \mathbb{F}_{prop}\}$. Then $\langle \mathbb{F}_{prop}, Cn_H \rangle$ is a compact logic. Similarly, the system I of intuitionistic propositional calculus is also compact.

(iii) In the formal system H of (ii) let Ax consist of all formulas of the following forms.

A1: $[A \supset [B \supset A]]$

A2: $[[A \supset [B \supset C]] \supset [[A \supset B] \supset [A \supset C]]]$

A3: $[[\neg B \supset \neg A] \supset [[\neg B \supset A] \supset B]]$.

The logic $\langle \mathbb{F}_{prop}, Cn_{Ax} \rangle$ is the classical propositional calculus in the sense that $Cn_{Ax}(\varnothing)$ is precisely the set of classical tautologies (Mendelson 1964, p. 36). Let F_1, F_2, F_3, F_4, F_5 be the formulas

$$[A \supset [[A \supset A] \supset A]]$$

$$[[A \supset [[A \supset A] \supset A]] \supset [[A \supset [A \supset A]] \supset [A \supset A]]]$$

$$[[A \supset [A \supset A]] \supset [A \supset A]]$$

$$[A \supset [A \supset A]]$$

$$[A \supset A]$$

Formal systems and combinatorial systems 95

respectively. F_1, F_2, and F_4 are axioms of the forms **A1**, **A2** (Chapter 2, §2.2.7), and **A3** respectively. F_3 is directly derived from F_1 and F_2, and F_5 is directly derived from F_4 and F_3. Thus the sequence F_1, F_2, F_3, F_4, F_5 is a proof of $[A \supset A]$ from axioms. Its tree representation is

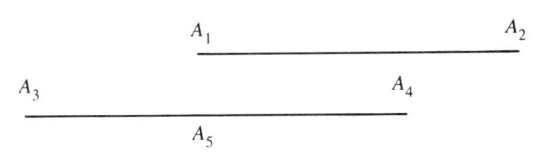

$$\frac{\dfrac{A_1 \qquad\qquad\qquad\qquad A_2}{A_3 \qquad\qquad\qquad A_4}}{A_5}$$

(iv) Note also that, although in these examples we have followed tradition in calling the signs ¬, ⊃, ∧, ∨ connectives, it has to be proved that they are actually connectives in the sense of Definition 3.10, i.e. strongly logical operations. For example, let $\mathbb{F}_2 =_{\text{def}} F(\{\wedge, \vee, \supset\}, P)$. *Modus ponens* can be used to give the formal system $\mathscr{F}_2 = \langle \mathbb{F}_2, \mathbf{MP} \rangle$. Let Ax consist of all formulas of the forms **A1** and **A2** above. The logic so obtained is a system of *pure implicational logic*. Beth (1962, p. 30) attributes the development of implicational logic to Tarski in about 1930. It was later discussed by Wajsberg (1935) and Henkin (1949a). *Classical implicational logic* is obtained by adding the 'Peirce law' (Peirce 1885) **A3'** to **A1** and **A2**.

A3': $[[[A \supset B] \supset A] \supset A]$.

A3', though valid for the classical propositional calculus, is not valid for the intuitionistic calculus. Other formulations of the implicational calculus can be found in Church (1956, §26). ◆

There are other types of structure which fall under the heading of 'formal system' but which are more remotely related to what is normally thought of as logic—for instance, *Post canonical languages* (e.g. Rosenbloom 1950, p. 162, Hermes 1965, p. 231). A special type of canonical language is known as a *combinatorial system* (Davis 1958). There is considerable interest in these structures as they lead to recursively unsolvable problems in mathematics such as group theory and geometric topology. For a discussion of the decision problems of combinatorial systems and of relations between them see Cudia and Singletary (1968a, b).

Our definition of 'combinatorial system' is more abstract than that given by Davis.

DEFINITION 3.16. A *combinatorial system* is a structure $\mathfrak{M} = \langle W, R_1, \ldots, R_n \rangle$ where W is a set of words on a finite alphabet and $R_1, \ldots, R_n$ are (computable) binary relations on W called *productions* with the following *finiteness property*: for $1 \leq i \leq n$ the sets $\{x; R_i(x, y)\}$ and $\{y; R_i(x, y)\}$ are finite for each $x, y \in W$ and can be computed (for a full discussion of the notion of *computability* see Davis (1958), Rogers (1967), etc.).

A useful device, due to W. E. Singletary, for investigating a combinatorial system $\mathfrak{M}$ (as in Definition 3.16) is its *system function*. This is the function $f_{\mathfrak{M}} \colon W \to \mathbb{P}_\omega(W)$ defined by

$$f_{\mathfrak{M}}(x) =_{\text{def}} \{x\} + \sum_{1 \leq i \leq n} \{y;\ R_i(x, y)\} \qquad (x \in W).$$

Every combinatorial system is a formal system (Definition 3.15). The productions are the rules of inference. The notion of *direct derivation* (Definition 3.15(i)) can now be given by means of the system functions: it is defined by

$$D_{\mathfrak{M}} =_{\text{def}} \sum \{f_{\mathfrak{M}}(x);\ x \in \Delta\}.$$

The concept of system function has been exploited by M. B. Thuraisingham (e.g. Thuraisingham 1987) and has been applied to problems of data-base security (Thuraisingham 1990).

In the actual cases defined by Davis (1958, p. 82), the finiteness property of the productions is easily verified. The productions are, in fact, one-premiss rules of inference. Thus the resulting logic is strongly compact (Example 3.12(i)). The interest in combinatorial systems is therefore focused on the consequences of unit sets.

Two classes of combinatorial systems can be distinguished by the following two conditions which can be expressed in Qf_2:

Det: $\quad A, B \in Cn(F) \quad \Rightarrow \quad A \in Cn(B) \lor B \in Cn(A) \qquad (A, B, F \in W)$

Sym: $\quad A \in Cn(B) \quad \Rightarrow \quad B \in Cn(A) \qquad (A, B \in W).$

A combinatorial system satisfying **Det** is said to be *deterministic*. Deterministic combinatorial systems include Turing machines, Markov algorithms, etc. Systems satisfying **Sym** are said to be *symmetric*—they include Thue systems, finitely presented groups, etc. (Davis 1958, Hermes 1965).

3.10 Finite logics

Where a logic is defined from semantical considerations, i.e. by truth-tables from a finite number of truth-values, the notion of logical consequence depends on the truth-values themselves being organized as a finite logic (cf. Chapter 5).

A convenient diagrammatic representation of finite logics is based on properties of its closed sets. To represent the finite logic $\langle \Sigma, Cn \rangle$ draw the lattice of subsets of Σ (Chapter 5, §5.7), marking the closed sets. Given any subset Δ of Σ, by Theorem 3.7(ii) $Cn(\Delta)$ is the least closed set above Δ in the lattice diagram. Thus, consider the logic $\langle \Sigma, Cn \rangle$ where $\Sigma = \{A, B, F\}$.

Finite logics

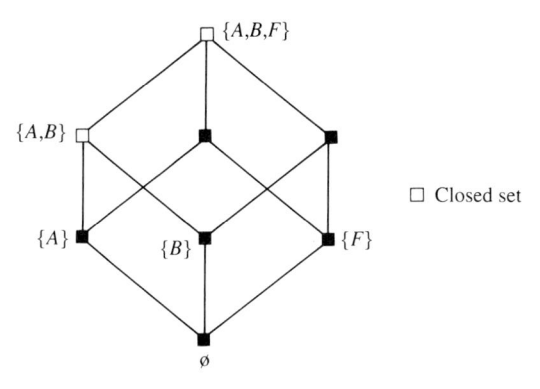

Fig. 3.2.

Figure 3.2 shows the lattice of subsets of Σ with the closed sets represented by □ and the remainder by ■. Note that Σ is closed.

To compute $Cn(A)$, find the closed sets which include $\{A\}$. These are $\{A, B\}$ and $\{A, B, F\}$. Then by Theorem 3.8(ii) $Cn(A) = \{A, B\} \cdot \{A, B, F\} = \{A, B\}$. In this logic A and B are logically equivalent because $Cn(A) = Cn(B) = \{A, B\}$.

DEFINITION 3.17. A logic is *simple* if logical equivalence is the identity.

EXAMPLE 3.18. The logic shown in Fig. 3.2 is not simple, but the logic shown in Fig. 3.3 is simple since $Cn(A) = \{A\}$, $Cn(B) = \{A, B\}$, and $Cn(F) = \{A, B, F\}$. ◆

There are three simple two-element logics (Fig. 3.4) and twelve three-element logics (Fig. 3.5) as classified up to isomorphism according to the structure of the closed sets. The only simple three-element logics satisfying **WC** are $6_3, 7_3, 11_3$; the only simple three-element logics satisfying **WD** are $2_3, 3_3, 6_3, 7_3$.

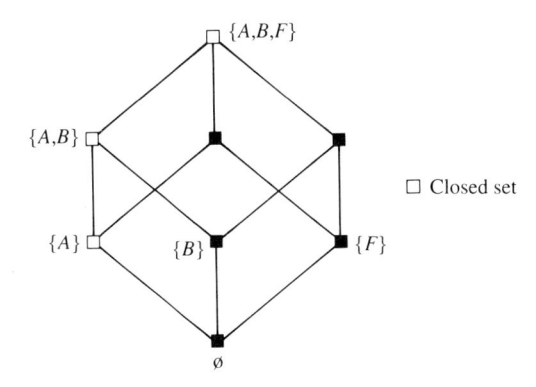

Fig. 3.3. A simple logic.

Abstract logics

Ø
1_2 2_2 3_2

☐ Closed set

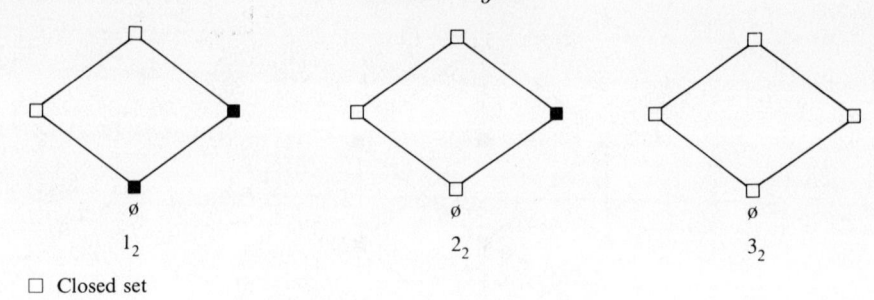

Fig. 3.4. Simple two-valued logics.

3.11 Gentzen systems

The development of logics into Gentzen type systems (cf. Chapter 2, §2.2.5) proceeds by a process of 'elementarization' (Bernays 1965), the first step of which consists of replacing the consequence function Cn of a compact logic $\mathfrak{L} = \langle \mathfrak{A}, Cn \rangle$ by its restriction $Cn^{\blacktriangledown}$ to finite sets of formulas (Definition 3.1(iv)). By Theorem 3.12(ii) the structure $\mathfrak{L}^{\blacktriangledown} =_{\text{def}} \langle \mathfrak{A}, Cn^{\blacktriangledown} \rangle$ satisfies $\mathbf{D1}^{\omega}$, $\mathbf{D2}^{\omega}$, and $\mathbf{D3}^{\omega}$. Conversely, by Theorem 3.12(i), from a structure $\langle \mathfrak{A}, Cn \rangle$, where $\mathfrak{A}$ is an algebra and Cn is a finitary set-to-set function which satisfies $\mathbf{D1}^{\omega}$, $\mathbf{D2}^{\omega}$, and $\mathbf{D3}^{\omega}$, there can be constructed a logic $\langle \mathfrak{A}, Cn^{\blacktriangledown} \rangle$. The second step of elementarization is the trivial one of replacing the finitary set-to-set function by its graph, i.e. the relation $R \subseteq \mathbb{P}_{\omega}(|\mathfrak{A}|) \times |\mathfrak{A}|$ given by

$$(\Delta, A) \in R \quad \Leftrightarrow_{\text{def}} \quad A \in Cn(\Delta) \qquad (A \in |\mathfrak{A}|, \Delta \in \mathbb{P}_{\omega}(|\mathcal{A}|)).$$

The axioms $\mathbf{D1}^{\omega}$, $\mathbf{D2}^{\omega}$, and $\mathbf{D3}^{\omega}$ can then be merely reworded in terms of this relation. It is expedient here to think of the pairs (Δ, A) as sequents and to write them as $(\Delta : A)$.

DEFINITION 3.18.

(i) A structure $\mathfrak{S} = \langle \mathfrak{A}, R \rangle$ where $\mathfrak{A}$ is an algebra and $R \subseteq \mathbb{P}_{\omega}(|\mathfrak{A}|) \times |\mathfrak{A}|$ is a *G-structure* if the following hold.

$\mathbf{D1}^{\omega}$: $(A : A) \in R \qquad (A \in |\mathfrak{A}|)$

$\mathbf{D2}^{\omega}$: for all $\Delta : \Gamma \in \mathbb{P}_{\omega}(|\mathfrak{A}|)$ and $A, B \in |\mathfrak{A}|$

$\qquad (\Delta : A) \in R \ \& \ (\Gamma, A : B) \in R \ \Rightarrow \ (\Delta, \Gamma : B) \in R$

$\mathbf{D3}^{\omega}$: $(\Delta : A) \in R \ \Rightarrow \ (\Delta, \Gamma : A) \in R \qquad (\Delta, \Gamma \subseteq |\mathfrak{A}|, A \in |\mathfrak{A}|)$

(ii) Let $\mathfrak{S} = \langle \mathfrak{A}, R \rangle$ be a G-structure. The relation R is the graph of a finitary set-to-set function C_R. By Theorem 3.12 the structure $\mathfrak{S}^{\blacktriangledown} =_{\text{def}} \langle \mathfrak{A}, C_R^{\blacktriangledown} \rangle$ is a compact logic.

1_3

2_3

3_3

4_3

5_3

6_3

7_3

8_3

9_3

10_3

11_3

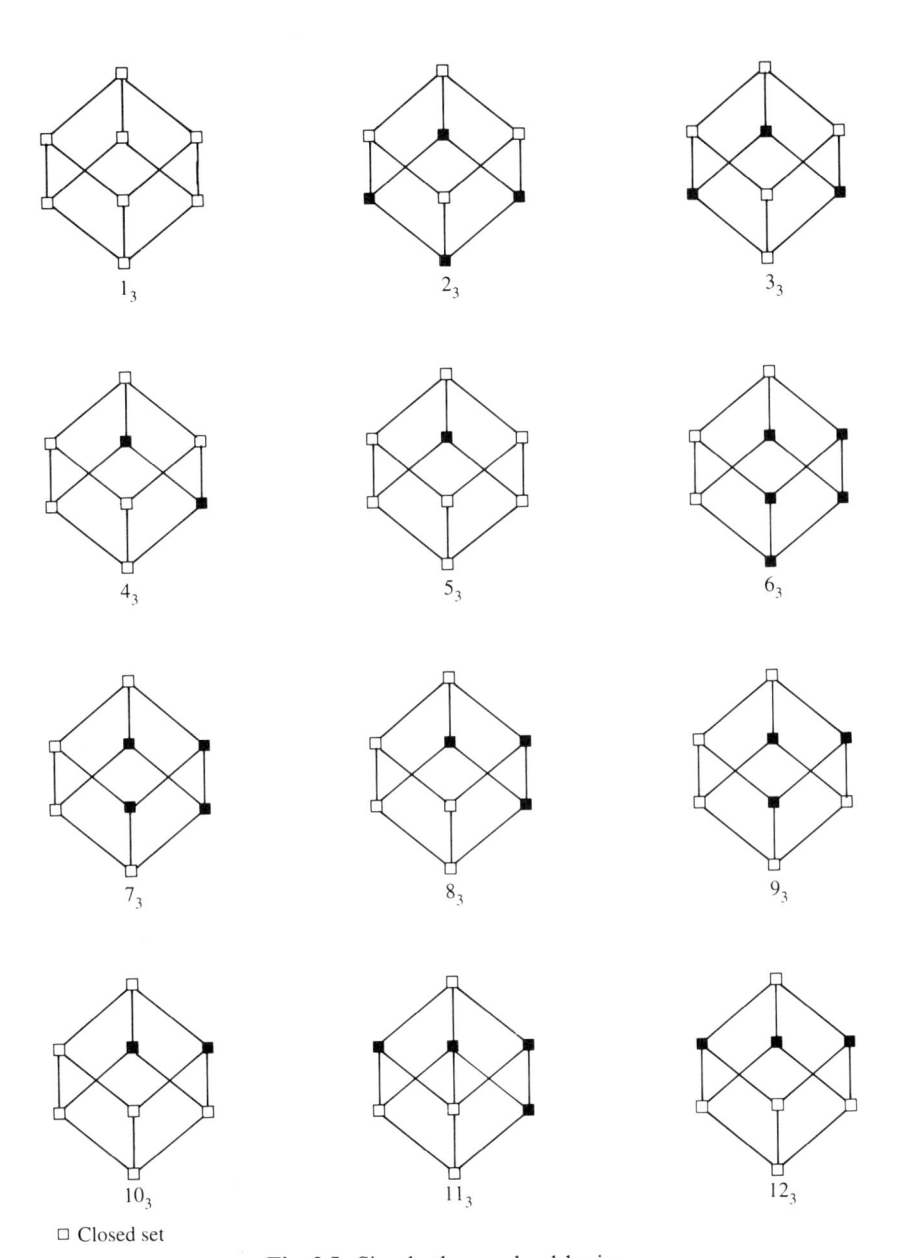

12_3

☐ Closed set

Fig. 3.5. Simple three-valued logics.

EXAMPLE 3.19.

(i) The axioms **D1**$^\omega$, **D2**$^\omega$, and **D3**$^\omega$ can conveniently be replaced by the following equivalent conditions.

H1: $(\Delta : A) \in R$ $\quad (A \in \Delta \in \mathbb{P}_\omega(|\mathfrak{A}|))$

H2: $(\Delta : A) \in R$ & $(B \in \Delta)((\Gamma : B) \in R)$ $\Rightarrow$ $(\Gamma : A) \in R$

for all $\Delta, \Gamma \in \mathbb{P}_\omega(|\mathfrak{A}|)$ and $A \in |\mathfrak{A}|$

(ii) The axioms **D2**$^\omega$ and **D3**$^\omega$ can then be expressed as 'cut' and 'thinning' rules:

$$\frac{(\Delta : A) \quad (\Gamma, A : B)}{(\Delta, \Gamma : B)} \quad \text{cut}$$

and

$$\frac{(\Delta : A)}{(\Delta, \Gamma : A)} \quad \text{thinning}$$

respectively. These conditions are clearly related to the cut and thinning rules of Gentzen systems (Chapter 2, §2.2.7(iii)).

(iii) Let Σ be a non-empty set. **D1**$^\omega$, **D2**$^\omega$, and **D3**$^\omega$ can be considered as unary, ternary, and binary relations on the set $\mathbb{P}_\omega(\Sigma) \times \Sigma$ of sequents. Hence the structure $\mathfrak{F} =_{\text{def}} \langle \mathbb{P}_\omega(\Sigma) \times \Sigma, \mathbf{D1}^\omega, \mathbf{D2}^\omega, \mathbf{D3}^\omega \rangle$ is a formal system (Definition 3.15). From this there can be constructed the associated logic $Sq(\Sigma) =_{\text{def}} \langle \mathbb{P}_\omega(\Sigma) \times \Sigma, Cn_{\mathfrak{F}} \rangle$: this is the *logic of sequents* on Σ.

Each subset Γ of the domain of $Sq(\Sigma)$ is a relation—a subset of $\mathbb{P}_\omega(\Sigma) \times \Sigma$. For any set $\Gamma \subseteq \Sigma$ the relation $Cn_{\mathfrak{F}}(\Gamma)$ is defined inductively from the initial elements $(A : A)$, where $A \in \Sigma$, together with the elements of Γ by the generating specifications **D2**$^\omega$, **D3**$^\omega$. The set $Cn_{\mathfrak{F}}(\emptyset)$ of theorems of this logic constitutes a trivial relation in the sense that

$$(\Delta : A) \in Cn_{\mathfrak{F}}(\emptyset) \Leftrightarrow A \in \Delta.$$

The structure $\mathfrak{S} = \langle \Sigma, Cn_{\mathfrak{F}}(\emptyset) \rangle$ is a G-structure such that $\mathfrak{S}^\blacktriangledown$ is the minimum logic on Σ (Example 3.6(iii)).

Note that the cut rule **D2**$^\omega$ cannot in general be eliminated in $Sq(\Sigma)$. For consider $\Sigma = \{1, 2, 3\}$ and let $\Gamma = \{(1 : 2), (2 : 3)\}$. Then $(1 : 3)$ cannot be generated from Γ by **D1**$^\omega$ and **D2**$^\omega$ only, but it can when **D3**$^\omega$ is used. ◆

The final step in the 'elementarization' consists of replacing finite sets by finite sequences. The entities $(\Delta : A)$ are thus replaced by *sequents* in the sense of Chapter 2, §2.2.5(iii). The definition of G-structure can be appropriately extended to embrace the sequent form.

G-structures in their sequent form are a form of an *abstract calculus* first considered by Hertz (1922, 1923), particularly in connection with the independence of axiom systems. In Hertz's treatment there is a domain Σ of elements from which one constructs 'sentences' of the form $A_1, \ldots, A_n \to B$

where $A_1, \ldots, A_n, B$ are in Σ and $\to$ is a new symbol. The sequence $A_1, \ldots, A_n$ is the *antecedent* and B is the *succedent*. A set Σ of sentences of the given form is called a *sentence system* and a sentence system is *closed* if it cannot be extended by (the sequent form of) **H1**, **H2** (Example 3.19(i)). A subset Π of Σ is called an *axiom system* for Σ if each sentence in Σ is obtained by application of **H1**, **H2** to Π.

H1 and **H2** can equivalently be replaced by other systems of rules (e.g. Popper 1947, Lorenzen 1955, §6, pp. 41–2). Hertz himself used the following two rules.

H*1: if $A_1, \ldots, A_n \to B$ and A_i occurs in $F_1, \ldots, F_k$ for $1 \leq i \leq n$ then $F_1, \ldots, F_k \to B$

H*2: if $A_{r1}, \ldots, A_{rn(r)} \to B_r$ for $1 \leq r \leq m$ and $B_1, \ldots, B_m, A_1, \ldots, A_k \to F$ then $A_{11}, \ldots, A_{1n(1)}, A_{21}, \ldots, A_{mn(m)} \to F$

(**H*2** was called a *syllogism*.) A sentence system is closed if and only if it contains all sentences of the form $A \to A$ and is closed under the operation of **H*1** and **H*2**.

Gentzen (1932) came to the sequent calculus partly through his early investigations into sentence systems and the problem of independent axioms. In this paper Gentzen noted that **H*2** can be replaced by the simplified rule with $m = 1$. This is the case he called the 'cut rule'.

The passage from sets to sequences entails the adjunction of two other conditions (rules) to **G1**, **G2**, **G3** (or **H1**, **H2**).

Contraction: $\dfrac{(A, A, \Delta : F)}{(A, \Delta : F)}$

Permutation: $\dfrac{(\Delta, A, B, \Gamma : F)}{(\Delta, B, A, \Gamma : F)}$

Calculi constructed from G-structures in this way can be characterized as 'pure' calculi in that they concern only the consequences from finitely many premisses. They can be extended to include logical operations on formulas.

The connection of the pure sequent calculus with arbitrary logics considered above will be considered further in Chapter 11. It formed only one part of Gentzen's motivation for the sequent calculus. The other part was the calculus of 'natural deduction' which is a form of hypothesis calculus, i.e. a calculus in which hypotheses are introduced and afterwards eliminated by certain rules. Gentzen (1934) first introduced sequent calculi in connection with natural deduction.

3.12 Comparison of logics

In this section we touch on certain relations between logics whereby one logic is 'embedded' or 'interpreted' in some way in another. Two logics can then be considered to be 'equivalent' if each is embedded in the other. But there is no unique notion of embedding which is suitable for all purposes. Moreover, for each general concept of embedding the condition of effectiveness can be imposed, and with the arithmetization of syntax, made possible in systems containing a sufficient subsystem of arithmetic, there is the possibility of *provable* representation as distinct from mere existence of an effective embedding (e.g. Kreisel 1955, Wang 1963, §4, Chapter 13). It is too ambitious to try to formulate a completely general concept of 'embedding' or 'equivalence'. A more feasible project is to classify a number of existing concepts that have proved their worth.

Consider two formulations of classical propositional logic. Let $\mathfrak{L}_1 = \langle \mathbb{F}(J_1, P), \wedge, \vee, \neg, C_1 \rangle$ where $\mathbb{F}(J_1, P)$ is the set of formulas constructed from the Aristotelian junctors $\wedge, \vee, \neg$ (Definition 2.1) and C_1 is the classical logical consequence function on such formulas (Definition 2.4(iii)). Let $\mathfrak{L}_2 = \langle \mathbb{F}(J_2, P), f, \supset, C_2 \rangle$ be the classical propositional logic whose set of formulas $\mathbb{F}(J_2, P)$ is constructed from falsehood f and implication $\supset$ (Chapter 2, §2.2.7), and whose consequence function C_2 is the classical consequence function on $\mathbb{F}(J_2, P)$. It is well known that the junctors f and $\supset$ can be defined in terms of $\wedge, \vee$, and $\neg$, and vice versa. Consider the mapping $\psi: \mathbb{F}(J_2, P) \to \mathbb{F}(J_1, P)$ defined inductively by

$$\psi(f) = [p_1 \wedge \neg p_1]$$
$$\psi(p) = p \qquad (p \in P)$$
$$\psi([A \supset B]) = [\neg \psi(A) \vee \psi(B)] \qquad (A, B \in \mathbb{F}(J_2, P)).$$

Let q be the polynomial of the algebra of formulas $\langle \mathbb{F}(J_1, P), \wedge, \vee, \neg \rangle$ (Appendix, §A) defined by

$$q(A, B) =_{\text{def}} [\neg A \vee B] \qquad (A, B \in \mathbb{F}(J_1, P)).$$

Let $\mathbb{F}'$ be the set of $\mathfrak{L}_2$ formulas generated by q from $P + \{[p_1 \wedge \neg p_1]\}$. Then ψ is an (algebra) isomorphism of the algebra $\langle \mathbb{F}(J_2, P), f, \supset \rangle$ onto $\langle \mathbb{F}', [p_1 \wedge \neg p_1], q \rangle$. Thus $\mathfrak{L}_2$ is a strong subderived logic of $\mathfrak{L}_1$ (Definition 3.3). In this case the embedding of $\mathfrak{L}_2$ in $\mathfrak{L}_1$ is obviously effective. Likewise, $\mathfrak{L}_1$ is a strong subderived logic of $\mathfrak{L}_2$. If the notion of equivalence of $\mathfrak{L}_1$ and $\mathfrak{L}_2$ means anything at all, it must at least imply that each is a strong subderived logic of the other. We therefore make the following definition.

DEFINITION 3.19.

(i) A logic $\mathfrak{L}_1$ is *reducible* to $\mathfrak{L}_2$—written $\mathfrak{L}_1 \leqslant_S \mathfrak{L}_2$—if $\mathfrak{L}_1$ is a strong subderived logic of $\mathfrak{L}_2$.
(ii) $\mathfrak{L}_1$ is *strongly equivalent* to $\mathfrak{L}_2$ if each is strongly reducible to the other.

The strong reducibility relation on logics is reflexive and transitive, and therefore logics under this relation are a quasi-ordering (Chapter 5, §5.5). If $\mathfrak{L}_1$ is strongly reducible to $\mathfrak{L}_2$ then $\mathfrak{L}_1$ inherits not only compactness from $\mathfrak{L}_2$ but certain properties expressible in Qf_2. This follows immediately from Theorem 3.2(ii) and 3.6.

THEOREM 3.19. If $\mathfrak{L}_1 \leqslant_S \mathfrak{L}_2$, where $\mathfrak{L}_1$ and $\mathfrak{L}_2$ are of types τ_1 and τ_2 respectively, then there is a term derivation d (Appendix, §A) of $Qf_2(\tau_2)$ to $Qf_2(\tau_1)$ such that, for every clause F of $Qf_2(\tau_1)$ whose positive literals are of the form $\mathbb{C}[T]$,

$$\mathfrak{L}_2 \vDash d(F) \quad \Rightarrow \quad \mathfrak{L}_1 \vDash F.$$

The notion of reducibility formulated in the above definition is closely related to the notion used by Kleene (1962, §81), where classical propositional logic is 'reduced' to intuitionistic logic by some results of Glivenko (1929) and Gödel (1932-3). In our terms, classical logic is isomorphic to a strong subderived system of intuitionistic propositional logic, i.e. $H \leqslant_S I$. However, I is a weak sublogic of H (Example 3.6(i)). To cover this contingency, Definition 3.19 can be modified to give a notion of *weak reducibility* by replacing 'strong' by 'weak'. We then have $I \leqslant_W H$.

Weak reducibility, like strong reducibility, is a quasi-ordering on logics, but it is a coarser relation in the sense that

$$\mathfrak{L}_1 \leqslant_S \mathfrak{L}_2 \quad \Rightarrow \quad \mathfrak{L}_1 \leqslant_W \mathfrak{L}_2$$

for all logics $\mathfrak{L}_1$ and $\mathfrak{L}_2$ but the converse does not hold. It should be remarked that I is strongly reducible to a certain modal extension H' of H (Gödel 1933; see also Chapter 11). Let N be a new one-place junctor and consider the enlarged set of formulas $\mathbb{F}(\{N, t, f, \wedge, \vee, \neg, \supset\}, P)$. The axiom schemas of H' are those of H together with the following.

A11: $NA \supset A$

A12: $NA \supset [N[A \supset B] \supset NB]$

A13: $NA \supset NNB$

The rules of inference are **MP** together with the rule

$$\frac{A}{NA}.$$

NA is to be read informally as 'A is provable'. Then I reduces to H by the transformation defined inductively by

$$\neg A \to \neg NA$$
$$[A \supset B] \to [NA \supset NB]$$
$$[A \vee B] \to [NA \vee NB]$$
$$[A \wedge B] \to [A \wedge B].$$

For quantifier logics the notions of reduction are much more delicate. The reductions now require the interpretation of the additional operations of quantifiers and substitution of terms as well as connectives. Moreover, the languages Qf_2 must be augmented to deal with quantifiers and free variables. This development requires further research beyond the scope of this book.

4

Logical operations

4.1 Connectives

In the preceding chapter the notion of a logic was defined. A language Qf_2, suitable for expressing conditions satisfied by logics, was defined. Various examples were given in Chapters 1 and 2. The present chapter is a taxonomic exercise—we define some important classes of logics axiomatically using logical operations as the principle of classification (for an algebraic treatment see Rasiowa (1974)). These axioms are all expressible as simple definite program clauses and are therefore subject to the transfer theorems of Chapter 3, §3.4.

What are the important logical operations? Consider the situation in algebra. From amongst the various mathematical structures which are recognized as algebraic, certain kinds are defined axiomatically. For example, a ring is a structure $\langle R, +, \cdot \rangle$ with two binary operations $+$ and $\cdot$ of addition and multiplication respectively, such that $\langle R, + \rangle$ is an Abelian group and multiplication is associative and distributes over addition. Observe that axioms for rings concern the behaviour of the two operations. Likewise the axioms for logics concern the behaviour of the connectives.

Why single out these particular axioms for rings? Why binary operations—why not ternary operations? Why are addition and multiplication so important? Discussion of these questions usually stops on a historical–cultural reference or an a priori invocation of symmetry, simplicity, etc., although preference for these principles is itself historical. Addition and multiplication entered into human affairs via the practice of accounting—counting, measurement of length and area, etc. The early development of algebra grew from this. It is sometimes argued that the axioms for rings are important because there are many examples of rings. This is correct, but when one asks why this is the case, one can only point to historical precedent.

The question concerning binary and ternary operations might be answered by pointing to the relative simplicity of the binary. This answer is not entirely satisfactory because unary operations are even simpler and it is a sociological fact that structures with only unary operations do not form an industry in any way comparable with the vast binary sector. As for functions of four variables, a fellow mathematician (a very human person) who knew a lot about them, once, when mellowed by Budweiser, confided to me that it was impossible for the human mind to comprehend them.

4.2 The historical connectives

Quantifiers and connectives were the stuff of logic in antiquity. The historically important connectives are undoubtedly implication, negation, conjunction, and disjunction. Various modern experiences in mathematical logic (Gödel's incompleteness theorem (Gödel 1931), intuitionism (Chapter 11), problems of negation in logic programming) indicate that these connectives cannot be taken naively—their properties must be studied in context.

Various axioms governing these logical operations will now be considered. Quantifiers will be considered later in the chapter.

In the remainder of this chapter various axioms for logics $\langle \mathfrak{A}, Cn \rangle$ will be formulated in the form:

$$\mathbf{Ax}: \quad \cdots Cn(A, \Delta) \cdots \quad \Rightarrow \quad \cdots$$

with elements of the $\mathfrak{A}$ being denoted by $A, B, C, \ldots$ and sets of elements by $\Delta, \Gamma, \ldots$. $\mathbf{Ax}^0$ denotes the axiom obtained from $\mathbf{Ax}$ by putting all set variables to $\emptyset$. We shall use the notation '$\rightarrow$' for logical consequence (Definition 3.7), '$\leftrightarrow$' for logical equivalence, and $\leftrightarrow_\Delta$ for relative logical equivalence (Definition 3.10) when convenient.

4.2.1 Implication

The theory of those arguments whose validity depends on the meaning of conditional sentences was, in antiquity, the work of the Stoics. Aristotle had no word for conditional statements (Kneale 1962, p. 99). The definition of implication was the subject of hot dispute amongst the Megarians and Stoics (Kneale 1962, Chapter III, §3):

> To the credit of the Megarian–Stoic school are some very subtle researches into the most important propositional functors. The thinkers of the school even succeeded in stating quite correct truth-matrices.
>
> Bocheński 1961, p. 115

In particular, Philo gave the truth-value table for what is now known as *material implication*. Thus, Philonian implication is a truth-value relation. Nevertheless, implication as a relation between propositions was also recognized:

> They [the Stoics] were well aware of the truth that the principle of a valid inference-schema can be represented in a conditional statement, such as 'If if-the-first-then-the-second, and the first, then the second.' Indeed they sometimes said that the validity of an inference scheme depended on the truth of the corresponding conditional.
>
> Kneale 1962, p. 159

The historical connectives 107

Peirce justified material implication on the grounds of its expressive power:

> ... both Professor Schröder and I prefer to build the algebra of relatives upon this conception of the conditional proposition. The inconvenience, after all, ceases to be important when we reflect that no matter what the conditional proposition be understood to mean, it can always be expressed by a complexus of Philonian conditionals and denials of conditionals.
>
> Bocheński 1961, p. 314

Material implication maintained its pre-eminence in mathematical logic until the advent of Lewis's modal logic and intuitionism. Intuitionism rejects the classical realist assumptions which enable implication to be defined in terms of truth-values. The concept of *construction* is the basic idea:

> The *implication* $p \to q$ can be asserted, if and only if we possess a construction r, which, joined to any construction proving p (supposing that the latter be effected), would automatically effect a construction proving q. In other words, a proof of p, together with r would form a proof of q.
>
> Heyting 1956, p. 99

An alternative, non-classical, approach is Lorenzen's dialogical conception of logic (Kamlah and Lorenzen 1967) which is not motivated by intuitionistic subjectivism:

> To assert a proposition makes sense only if there is someone on the other side, albeit fictitiously, who denies or at least doubts the asserted proposition. But it is not enough merely to argue about propositions, there must exist precise stipulations on the rules of argumentation, rules which, in a way, define the exact meaning of the proposition in question.
>
> Lorenz 1973, p. 359

The exact meaning of implication is defined by the argumentation rules for that logical particle.

Whatever conceptions of logic and conditional proposition are entertained, implication, whatever else it may mean, at least obeys the familiar rule, known to Megarian and Stoic logic (Bocheński 1961, p. 127), of *modus ponendo ponens*. This rule, by letting $A \supset B$ stand for 'A implies B', can be expressed as a condition on a logic $\langle \Sigma, \supset, Cn \rangle$ as

MP1: $C \in Cn(B, B \supset C)$.

(This corresponds to **MP** of Chapter 3.) There are certain technical advantages in considering conditions on such a logic which can also be taken as alternative characterizations of implication:

MP2: $B \supset C \in Cn(\Delta) \Rightarrow C \in Cn(\Delta, B)$ for $\Delta \subseteq \Sigma, B, C \in \Sigma$

MP3: $C \in Cn(\Delta \& B) \& A \in Cn(\Gamma) \Rightarrow C \in Cn(\Delta, \Gamma, A \supset B)$

for $\Delta, \Gamma, A, B, C \subseteq \Sigma$.

(These conditions can obviously be expressed in Qf_2.)

For a discussion of the fortunes of implication in quantum logics, see Greechie and Gudden (1973).

EXAMPLE 4.1.

(i) Every formal system $\mathfrak{F}$ which includes *modus ponens* amongst its rules of inference satisfies **MP2** (cf. Chapter 3, §3.9). For suppose $\Delta, \vdash_{\mathfrak{F}} A \supset B$. Then by the definition of *proof* (Definition 3.15(iii)), $\Delta, A \vdash_{\mathfrak{F}} A \supset B$. Hence, by *modus ponens*, $\Delta, A \vdash_{\mathfrak{F}} B$.

(ii) It is easy to prove, using Theorem 3.6, that in any logic $\mathfrak{L} = \langle \Sigma, \supset, Cn \rangle$, **MP1**, **MP2**, and **MP3** are equivalent. By this result we can refer to any of the equivalent versions **MP1**, **MP2**, or **MP3** simply as **MP**. ◆

Many important formal systems (rather, their corresponding logics $\langle \mathfrak{A}, Cn \rangle$) with the junctor $\supset$ and *modus ponens* as rule of inference also satisfy the further 'deduction theorem' conditions:

DT: $C \in Cn(\Delta, A) \;\Rightarrow\; A \supset C \in Cn(\Delta)$ for $A, C \in |\mathfrak{A}|, \Delta \subseteq |\mathfrak{A}|$.

This is the converse of **MP2**. There are restricted versions of this condition (Example 2.10(iii) dedicated to predicate logics). It is known as the 'deduction theorem'. In its application to *Principia Mathematica* it was first established by Tarski in 1921 (Tarski 1956, p. 32n). It was proved for classical logic independently by Herbrand (1930) and Tarski (1930) (see Curry 1963, p. 249). Church observed that it depended only on the axiom scheme **A1** and **A2**. A binary operation $\supset$ will therefore be called a *normal implication* if **MP** and **DT** (unrestricted) both hold. Thus $\supset$ is a normal implication iff

$$C \in Cn(\Delta, B) \;\Leftrightarrow\; B \supset C \in Cn(\Delta). \tag{4.1}$$

(Thus in a Hilbert system in which logical consequence is 'deduction from hypotheses' $\supset$ is not normal if **DT** holds only in the restricted form.)

We have not yet shown that implication in a logic is a connective in the terms of Definition 3.10. In fact, we have the following theorem.

THEOREM 4.1. *Normal implication is a strongly logical operation.*

Proof. Let $\mathfrak{L} = \langle \Sigma, \supset, Cn \rangle$ be a logic satisfying **MP** and **DT**. We show that

$$B \leftrightarrow_\Delta B' \;\Rightarrow\; B \supset C \leftrightarrow_\Delta B' \supset C \tag{4.2}$$

$$C \leftrightarrow_\Delta C' \;\Rightarrow\; B \supset C \leftrightarrow_\Delta B \supset C'. \tag{4.3}$$

ad (4.2) Suppose $B \leftrightarrow_\Delta B'$. Then $B \in Cn(\Delta, B')$. Hence, by **MP1**,

$$C \in Cn(\Delta, B, B \supset C) \subseteq Cn(\Delta, B', B \supset C).$$

By **DT**, $B' \supset C \in Cn(\Delta, B \supset C)$. Likewise $B \supset C \in Cn(\Delta, B' \supset C)$. Thus, $B' \supset C \leftrightarrow_\Delta B \supset C$. This completes the proof of (4.2).

ad (4.3) Suppose $C \leftrightarrow_\Delta C'$. Then $C \in Cn(\Delta, C')$. But by **MP1**

$$C \in Cn(B, B \supset C') \subseteq Cn(\Delta, B, B \supset C').$$

Hence $C \in Cn(\Delta, B, B \supset C')$. By **DT**, $B \supset C \in Cn(\Delta, B \supset C')$. Likewise $B \supset C' \in Cn(\Delta, B \supset C)$. Hence $B \supset C \leftrightarrow_\Delta B \supset C'$. This completes the proof of (4.3). ∎

A binary operation satisfying **MP** only is not necessarily a connective in the sense of Definition 3.7. A three-element logic can be constructed with an operation which satisfies **MP1** but is not a connective. Another axiom which ensures that an operation which satisfies **MP** is a connective is

I1: for all $A, B, C, D \in \Sigma$, $\Delta \subseteq \Sigma$
  $B \in Cn(A, \Delta)$ & $D \in Cn(C, \Delta)$ & $B \supset C \in Cn(\Delta)$ $\Rightarrow$ $A \supset D \in Cn(\Delta)$.

THEOREM 4.2. Let $\mathfrak{L} = \langle \Sigma, \supset, Cn \rangle$ be a logic which satisfies **MP** and **I1**. Then $\supset$ is a strongly logical operation.

Proof. We prove

$$B \leftrightarrow_\Delta A, D \leftrightarrow_\Delta C \Rightarrow A \supset D \leftrightarrow_\Delta B \supset C. \quad (4.4)$$

Suppose $B \leftrightarrow_\Delta A$ and $D \leftrightarrow_\Delta C$. Then $B \in Cn(\Delta, A)$ and $D \in Cn(\Delta, C)$. By **MP**, $B \in Cn(\Delta, A, B \supset C)$ and $D \in Cn(\Delta, C, B \supset C)$, and by **D1**, $B \supset C \in Cn(\Delta, B \supset C)$. By **I1**, $A \supset D \in Cn(\Delta, B \supset C)$. Similarly, $B \supset C \in Cn(\Delta, A \supset D)$. Thus, $A \supset D \leftrightarrow_\Delta B \supset C$. This proves (4.4). ∎

MP & **DT** imply **MP** & **I1** but not conversely (Exercise!).

DT and **MP** guarantee the uniqueness of normal implication in the following sense.

THEOREM 4.3. Let $\supset_1, \supset_2$ be normal implications on a logic $\langle \Sigma, Cn \rangle$. Then for all $A, B \in \Sigma$, $\Delta \subseteq \Sigma$, $A \supset_1 B \leftrightarrow_\Delta A \supset_2 B$.

Proof. $B \in Cn(\Delta, A, A \supset_1 B)$ by **MP** for $\supset_1$. By **DT** for $\supset_2$, $A \supset_2 B \in Cn(\Delta, A \supset_1 B)$. Similarly $A \supset_1 B \in Cn(\Delta, A \supset_2 B)$. ∎

In certain formal systems in which *modus ponens* is the only rule of inference implication is normal (see Example 3.17(ii)).

THEOREM 4.4. Let $\mathfrak{F} = \langle \mathfrak{A}, R \rangle$ be a formal system (Definition 3.8) with an operation $\supset$ where R is the relation $\{(A, A \supset B, B), A, B \in |\mathfrak{A}|\}$ and $Cn_\mathfrak{F}(\emptyset)$ includes all elements $A \supset (B \supset A)$ and $(A \supset B) \supset ((A \supset (B \supset C)) \supset (A \supset C))$ where $A, B, C \in |\mathfrak{A}|$. Then $\supset$ is normal implication.

Proof. We first remark that, by Definition 3.12, $B \in D_{\mathfrak{F}}(A, A \supset B)$ and so $B \in Cn_{\mathfrak{F}}(A, A \supset B)$. Thus **MP** holds. For **DT**, let $B \in Cn_{\mathfrak{F}}(\Delta, C)$. We prove by induction on length of proof that $C \supset B \in Cn_{\mathfrak{F}}(\Delta)$. Let $\rho(B) =_{\text{def}} (\mu n)(B \in D_{\mathfrak{F}}^n(\Delta, C))$. We prove by induction on n that

$S(n)$: $(B \in Cn_{\mathfrak{F}}(\Delta, C))(\rho(B) \leq n \;\Rightarrow\; C \supset B \in Cn_{\mathfrak{F}}(\Delta))$

Basis:
$n = 0$. Let $B \in Cn_{\mathfrak{F}}(\Delta, C))$ with $\rho(B) = 0$. Then $B \in \Delta + \{C\}$. Hence (i) $B = C$ or (ii) $B \in \Delta$.

Case (i). $(C \supset (C \supset C)) \supset ((C \supset ((C \supset C) \supset C)) \supset (C \supset C))$ and $C \supset ((C \supset C) \supset C)$ are both theorems by hypothesis. By **MP**, $(C \supset (C \supset C)) \supset (C \supset C)$ is also a theorem. As, by hypothesis, $C \supset (C \supset C) \in Cn_{\mathfrak{F}}(\varnothing)$, again by **MP**, $C \supset C = C \supset B \in Cn_{\mathfrak{F}}(\varnothing) \subseteq C_{\mathfrak{F}}(\Delta)$.

Case (ii). As $B \supset (C \supset B) \in Cn_{\mathfrak{F}}(\varnothing)$ and $B \in Cn_{\mathfrak{F}}(\Delta)$,

$$Cn_{\mathfrak{F}}(B, B \supset (C \supset B)) \subseteq Cn_{\mathfrak{F}}(\Delta, B, B \supset (C \supset B)) = Cn_{\mathfrak{F}}(\Delta).$$

By **MP**, $(C \supset B) \in Cn_{\mathfrak{F}}(\Delta)$. Thus $S(0)$ holds.

Inductive step:
Suppose $S(n)$. Let $B \in Cn_{\mathfrak{F}}(\Delta, C))$ with $\rho(B) = n + 1$. Then $B \in D_{\mathfrak{F}}^{n+1}(\Delta, C) = D_{\mathfrak{F}}(D_{\mathfrak{F}}^n(\Delta, C))$ and $B \notin D_{\mathfrak{F}}^n(\Delta, C)$. Hence there exists an element A such that A, $A \supset B \in D_{\mathfrak{F}}^n(\Delta, C)$. By inductive hypothesis $S(n)$, $C \supset A$, $C \supset (A \supset B) \in Cn_{\mathfrak{F}}(\Delta)$. But $(C \supset A) \supset ((C \supset (A \supset B)) \supset (C \supset B)) \in Cn_{\mathfrak{F}}(\varnothing) \subseteq Cn_{\mathfrak{F}}(\Delta)$.

By two applications of **MP**, $C \supset B \in Cn_{\mathfrak{F}}(\Delta)$. Thus, $S(n + 1)$.

By induction on n, $S(n)$ holds for all n. Hence for all $B \in Cn_{\mathfrak{F}}(\Delta, C)$, $C \supset B \in Cn_{\mathfrak{F}}(\Delta)$. Thus $\mathfrak{F}$ satisfies **DT**. ∎

It follows from this theorem that $\supset$ is normal implication in the classical and intuitionistic propositional logics (Example 3.17(ii)). A kind of converse of theorem 4.4, which holds for a larger class of logics than formal systems, can easily be established.

THEOREM 4.5. *Let $\mathfrak{L} = \langle \mathfrak{A}, Cn \rangle$ be a logic with normal implication $\supset$. Then the theorems of $\mathfrak{L}$ contain all elements $A \supset (B \supset A)$ and $(A \supset B) \supset ((B \supset C) \supset (A \supset C))$ where $A, B, C \in |\mathfrak{A}|$.*

4.2.2 Negation and contradiction

Philosophy of negation Of all the logical operations negation and contradiction have been the most controversial. A philosophical tradition going back to Zeno and Plato, and continued in various forms by idealist philosophers, sees contradiction and negation as prior to formal logic, as constitutive in some way of reality. These concepts are closely connected with problems of self-identity and change.

One aspect of negation is universalized *difference*. The unique individuality of a chosen object (material or abstract) is undermined as soon as its network of relations with other things is seen—they determine each other's nature.

> Thus negation as universal otherness would be difference at the most abstract level of generality, the bare identical character of all differences. On this view reality is not confined to the purely positive... In any given case of 'A is not-B' the exclusion of B as other than A at least partly determines the *nature* of A
>
> Mure 1965, p. 13

Negation and contradiction are thus as immanent in the determination of objects as identity—*omnis determinatio est negatio* (Spinoza). This appears at the level of simple prediction. The process involves a choice of one side of a bifurcation (Chapter 1) (p, n) where p is a witness of P and n is a witness of non-P. The predication procedure involves an orientation of the pair of tokens which is already implied by the reference to n as the witness of non-P. P is an abstraction from a process in which the incorporation of non-P in the constitution of P is suppressed.

For Hegel the presence of 'contradiction' in thought and reality means the presence of mutually opposing tendencies which each moves to complete itself by dominating the total field to the exclusion of opponents. Contradiction is the motive force of development and change. The ubiquity of contradictions in this sense is obvious, though the terminology may not always be useful in particular instances. It is seen in operation in the conceptual transitions in mathematics, for example, in the various movements of completion—from natural numbers, through the integers, rationals, reals to complex numbers, etc.

Though a pupil of Plato, Aristotle took a narrow formal view of negation. In his work on logic negation appears mainly as a *junctor*, an operation on propositions, though there are places where negation determines a name (Bocheński 1961, p. 59). Negation is explicitly a junctor in Megarian–Stoic logic and continued to be so throughout the development of formal logic up to classical logic.

The problem of explaining true negative statements is a recurring one in history, discussed in antiquity by Greek and Indian logicians who spoke of 'negative facts', and in modern times by Frege, Russell, Peirce, etc. The issue is still unresolved.

For Frege a proposition's being negative or positive is an aspect of meaning and is quite different from its truth-value, which is determined by correspondence between facts meant and meaning. A positive statement can be false and a negative statement can be true. Testing a theory by focusing on attempts at refutation shows the importance of negation in the theory of science.

Russell (1903, Chapter 9) developed a theory of how negative sentences can be true and can be known without its being necessary to assume that there are facts which can only be correctly asserted in sentences not containing them.

Vasil'ev, a founder of many-valued logics, claimed that negative predicates are not primitive but are inferred from positive predicates (Kline 1965). Negative propositions concerning perceptions are inferred from propositions about incompatible properties: a denial that an object has property P is founded on the presence of a witness having a property N which excludes P (cf. Chapter 1, §1.3.2). The nature of this incompatibility or exclusion is left unresolved as in Russell's discussion of colour incompatibilities (see, for instance, Pears 1953, Kamlah and Lorenzen 1967, Chapter 6).

Wittgenstein (of the *Tractatus*) considered that negation, as an operation on propositions, was not something in the world, though he came close to seeing it as constitutive of reality in some sense:

> 2.04 The totality of existent atomic facts is the world.
> 2.05 The totality of existent atomic facts also determines which atomic facts do not exist.
> 2.06 The existence and non-existence of atomic facts is the reality.
>   (The existence of atomic facts we also call a positive fact, their non-existence a negative fact.)

p and $\neg p$ lead to the same state of affairs: the state of affairs verifying the asserted proposition falsifies its negation and vice versa.

> 4.1 A proposition represents the existence and non-existence of atomic facts.
>   Wittgenstein 1922

Negation in classical logic is surely the 'standard model' of negation in formal logics. Other examples of negation arise, for a variety of reasons, as deviants. Consider the classical definition of validity in the predicate logic (Definition 2.11) as given by Tarski. This is a symmetrical treatment of the truth-values 0, 1. The essential thing is the inductive nature of the definition: the truth-value of a formula is systematically determined by the truth-values of proper subformulas. Negation is interpreted simply as an operation which interchanges 0 and 1: $[\mathfrak{A}, \alpha](\neg A) = [\mathfrak{A}, \alpha](A)'$ (Definition 2.11(ii)(b)). The special status of the truth-value 1 (truth) first appears in the definition of *satisfaction*, *model*, and *consequence*.

There are two levels at which important philosophical doubt can be expressed in the truth-definition. First, there is the general matter of the effectiveness of the logical operations (including negation), particularly of the quantifiers. The intuitionistic critique of classical logic concerns the notion of existence of mathematical entities: it involves a radical reinterpretation of quantifiers (see Chapter 10).

Second, there are problems in applications concerning the first step in the inductive definition, namely the valuation of atomic formulas. The definition can only be applied to relational systems. These are obtained in applications by a process of idealization: there are reasons, based on consideration of simple predication for considering three truth-values (Chapter 1). The idealized structures should initially be three-valued: how then should negation work? What should it mean at this level and how should it be extended to all formulas? These questions will be treated in Chapters 6 and 8.

Even where the applications are to mathematical structures, there are choices to be made as to which relations are to be taken as primitive, i.e. how to orient the relations. For constructive reasons the relations should be taken positively:

> ... negative concepts are for us even less important than in classical mathematics, whenever possible we replace them by positive concepts.
>
> Heyting 1956, p. 19

Thus, in intuitionist mathematics, inequality between real-number generators is replaced by the apartness relation $\#$. This is a stronger condition than the negation of inequality because $a \# b$ is proved by computation of a witness, whereas $\neg(a = b)$ demands a proof of impossibility in which contradiction plays an essential part. The two notions are not equivalent: equality of real numbers is computationally more complex than apartness.

Negation in formal systems Negation has for long been accorded special treatment in formal systems. Curry (1963, p. 246) has remarked that this separation goes back to Schröder who first developed some axioms which we now recognize as lattice laws (Chapter 5), and afterwards added complementation (negation, in logical terms).

The problem of dealing with negative information in a primarily positive context is a serious and deep question in logic programming (Lloyd 1984, Shepherdson 1988). To fix ideas consider the elementary system $E(\tau)$ (Chapter 2, §2.2.7(iii)). The free-variable Horn clauses are all 'positive' and only Horn clauses are deducible from Horn clauses by *modus ponens* (Definition 2.16). A formula could be considered as *valid* if there is a proof of it. It is natural to say that a formula is *invalid* in the contrary case. Negation defined on the basis of invalidity is called *non-demonstrability* (Curry 1963, p. 255). A rule of inference based on non-demonstrability is called the *closed world assumption* in data-base applications of logic programming. Unfortunately it is not effective.

Another concept, called *refutability* by Curry, is related to our *finite refutability* (Definition 2.21) which itself is intimately connected with *negation as failure* in logic programming. As a rule of inference it is effective, but it is weaker than non-demonstrability.

We now turn attention to logics in which there are operations which are candidates for negation and contradiction. What conditions must be satisfied for them to be so classified? We consider below a number of axioms which are based on variations of conditions which hold in classical and intuitionistic logics.

The 'law of contradiction'—A and not-A cannot both hold—was known to Aristotle and Plato: Aristotle devoted Book Γ of *Metaphysics* to this principle. The principle that every proposition follows from a contradiction appears to have been explicitly enunciated in Scholastic logic (see Bocheński (1961, pp. 148–52) for sources of Scholastic logic). Bocheński quotes the form in which it appeared in Pseudo-Scotus:

> From every proposition implying a contradiction any other formally follows so that there follows for instance: 'Socrates runs and Socrates does not run, therefore you are at Rome'.
>
> Bochénski 1961, 31.25, p. 205

There is also a well-known anecdote concerning Russell on negation:

> Russell is reputed to have been challenged to prove that the (false) hypothesis $2 + 2 = 5$ implies that he was the Pope. Russell replied as follows: 'You admit $2 + 2 = 5$: but I can prove $2 + 2 = 4$: therefore $5 = 4$. Taking two away from both sides we have $3 = 2$: taking one more, $2 = 1$. But you will admit that I and the Pope are two. Therefore I and the Pope are one, qed.'
>
> Birkhoff 1948, p. 194

Hilbert's study of the foundations of mathematics stressed the search for proof of consistency. Prior to 1923 Brouwer dismissed the demand for non-contradiction in the foundations of mathematics as mathematically irrelevant. Brouwer's 'calculus of absurdity' was published in 1923. Thereafter non-contradiction was given the constructive meaning of 'absurdity of absurdity': in Brouwer (1948) he frequently equated the two.

To Brouwer a contradiction is a 'purely linguistic phenomenon'. In a verbal description of a mathematical construction a contradiction cannot arise:

> ... two contradictory theorems cannot be true of a mathematical construction.
>
> Brouwer 1907

Non-contradiction is therefore accepted as a necessary condition for mathematical existence. Contradiction is described as a kind of incompatibility of two mathematical constructions:

> ... the two systems do not fit into each other.
>
> Brouwer 1908

Verbal contradiction reports an obstruction:

> ... the words of your mathematical demonstration are only the accompaniment of a wordless mathematical *building*, and where you pronounce a contradiction, I simply observe that the construction cannot go further, that in the given construction there is no room for a posited structure.
>
> Brouwer 1907

The historical connectives

The word 'posited' here indicates that Brouwer envisaged a verbal hypothesis as the constructive part of a false statement in a hypothesis—a mathematical non-entity which leads to an absurdity like $0 = 1$. This seems to contradict his insistence on the completely mathematical, i.e. languageless, nature of negation.

Heyting maintained that the notion of contradiction is impossible to analyse:

> I think that contradiction must be taken as a primitive notion. It seems very difficult to reduce it to simpler notions, and it is always easy to recognise a contradiction as such. In practically all cases it can be brought to the form $1 = 2$.
> Heyting 1956, p. 98

It is characteristic of the notion of a contradiction in classical and intuitionistic logic that a contradiction implies any proposition. This property can be used in a definition of negation.

In a logic $\mathfrak{L} = \langle \mathfrak{A}, Cn \rangle$ an element f is a *contradiction* (also *absurdity* or *falsehood*) if

F: $\quad Cn(f) = |\mathfrak{A}|$.

(cf. Tarski's Axiom 5, §3.5). Such an element f can be considered as a zero-place connective—anything logically equivalent to it also satisfies **F**. This can be expressed in Qf_2 by the equivalent formulation

$$x \in Cn(f) \qquad (x \in |\mathfrak{A}|).$$

It is sometimes convenient to select an element t which satisfies

T: $\quad t \in Cn(x) \qquad (x \in |\mathfrak{A}|)$.

If such an element exists then

$$Cn(t) = Cn(\emptyset)$$

(cf. Theorem 3.9). t can be read as *theoremhood* or *the true*. Observe that this definition is applicable to logics which are not equipped with negation or contradiction. The motive for this definition is that a theory Δ is usually considered to be inconsistent if it contains a formula A together with its negation $\neg A$. Then, in the usual systems, *every* formula is a consequence of A together with $\neg A$. Such systems satisfy the *law of contradiction*:

N1: $\quad Cn(A, \neg A) = |\mathfrak{A}| \qquad (A \in |\mathfrak{A}|)$.

This argument is due to Post (1921).

N1 holds in intuitionistic logic. The following condition **N2** (Tarski 1956, p. 32) is a version of the *principle of the excluded middle* (PEM):

N2: $\quad Cn(A, \Delta) \cdot Cn(\neg A, \Delta) = Cn(\Delta) \qquad (\Delta \subseteq |\mathfrak{A}|, A \in |\mathfrak{A}|)$.

This holds in classical logic but not in intuitionistic logic (see Chapter 10).

N1 and **N2** are not sufficient to ensure that $\neg$ is a connective. Further conditions are required for this purpose. One such condition, which holds in intuitionistic logic and the three-valued logic of Chapter 8, is

N3.1: $\quad A \in Cn(B, \Delta) \quad \Rightarrow \quad \neg B \in Cn(\neg A, \Delta).$

The three-valued logic of Chapter 8 satisfies **N3.1⁰** (i.e. $\Delta = \emptyset$) but not **N3.1**.

THEOREM 4.6. Let $\mathfrak{L} = \langle \mathfrak{A}, Cn \rangle$ be a logic with a unary operation $\neg$ which satisfies **N3.1**. Then $\neg$ is a strongly logical operation.

Proof. Suppose that $A \leftrightarrow_\Delta B$. Then $A \in Cn(\Delta, B)$ and $B \in Cn(\Delta, A)$. By **N3.1**, $\neg B \in Cn(\Delta, \neg A)$ and $\neg A \in Cn(\Delta, \neg B)$. Thus $\neg A \leftrightarrow_\Delta \neg B$. ∎

Formally related to **N3.1** are the laws

N3.2: $\quad \neg A \in Cn(B, \Delta) \quad \Rightarrow \quad \neg B \in Cn(A, \Delta)$

N3.3: $\quad A \in Cn(\neg B, \Delta) \quad \Rightarrow \quad B \in Cn(\neg A, \Delta),$

of which the first holds for intuitionistic logic and the second holds only in classical logic.

EXAMPLE 4.2. Consider the conditions on logics with a unary function $\neg$:

N4.1: $\quad \neg \neg A \in Cn(A)$

N4.2: $\quad A \in Cn(\neg \neg A).$

Then

$$\begin{array}{rcl}
\textbf{N3.1 \& N4.1} & \Rightarrow & \textbf{N3.2} \\
\textbf{N3.2} & \Rightarrow & \textbf{N4.1} \\
\textbf{N3.1 \& N3.2} & \Leftrightarrow & \textbf{N3.1 \& N4.1} \\
\textbf{N3.1 \& N4.2} & \Leftrightarrow & \textbf{N3.1 \& N3.3} \\
\textbf{N3.1 \& N3.3} & \Rightarrow & \textbf{N3.2}.
\end{array}$$
♦

THEOREM 4.7. If the logic $\langle \Sigma, \neg, Cn \rangle$ satisfies **N3.1 & N3.2** then for all $n \geq 1$,

$$(\neg)^{2n+1} A \leftrightarrow \neg A \quad \text{and} \quad (\neg)^{2n} A \leftrightarrow \neg \neg A$$

and

$$A \in Cn(B_1, \ldots, B_n) \Rightarrow \neg \neg A \in Cn(\neg \neg B_1, \ldots, \neg \neg B_n).$$

Proof. By Example 4.2, **N4.1** can be assumed. Thus $\neg \neg A \in Cn(A)$. By **N3.1**, $\neg A \in Cn(\neg \neg \neg A)$. Further, by **N4.1**, $\neg \neg \neg A \in Cn(\neg A)$. Thus

The historical connectives 117

$\neg\neg\neg A \leftrightarrow \neg A$. As this holds for all formulas A, by induction on n we have $(\neg)^{2n+1}A \leftrightarrow \neg A$. By substituting $\neg A$ for A in this result we then have $(\neg)^{2n}A \leftrightarrow \neg\neg A$. Next, suppose that $A \in Cn(B_1,\ldots,B_n)$. By $2n$ applications of **N3.1**, $(\neg)^{2n}A \in Cn(\neg\neg B_1,\ldots,\neg\neg B_n)$. Hence

$$\neg\neg A \in Cn(\neg\neg B_1,\ldots, B_n). \qquad \blacksquare$$

There are numerous easily established conditions relating negation to absurdity.

THEOREM 4.8. *Let* $\mathfrak{L} = \langle \Sigma, f, \neg, Cn \rangle$ *be a logic wth a non-empty set of theorems which satisfies* **F** *and* **N1**. *Then* $\mathfrak{L}$ *satisfies*

N5: $\quad f \in Cn(A, \Delta) \;\Leftrightarrow\; \neg A \in Cn(\Delta)$

iff it satisfies **N3.1** & **N3.2**.

Proof.
(i) Suppose $\mathfrak{L}$ satisfies **N3.1** & **N3.2**. By **F**, $\neg B \in Cn(f)$ for all $B \in \Sigma$. By **N3.1**, $\neg f \in Cn(\neg\neg B)$. **N4.1** holds by Example 4.2. Hence $\neg f \in Cn(B)$ for all $B \in \Sigma$. Thus $\neg f \in T_{\widetilde{\mathfrak{F}}}$. The set of theorems of $\mathfrak{L}$ is therefore non-empty so, by Theorem 3.9, $\neg f \in Cn(\emptyset)$. Now suppose $f \in Cn(A, \Delta)$. By **N3.1**, $\neg A \in Cn(\neg f, \Delta) = Cn(\Delta)$. Thus

$$f \in Cn(A, \Delta) \Rightarrow \neg A \in Cn(\Delta).$$

Conversely, suppose that $\neg A \in Cn(\Delta)$. Then $Cn(A, \Delta) = Cn(A, \neg A, \Delta)$. By **N1**, $f \in Cn(A, \Delta)$. Thus

$$\neg A \in Cn(\Delta) \;\Rightarrow\; f \in Cn(A, \Delta),$$

i.e. **N5**.
(ii) Suppose **N5** holds.
   ad **N3.1** Suppose that $A \in Cn(B, \Delta)$. Now by **N1**, $f \in Cn(A, \neg A)$. Hence $f \in Cn(B, \neg A, \Delta)$. By **N5**, $\neg B \in Cn(\neg A, \Delta)$. This proves **N3.1**.
   ad **N3.2** Suppose that $\neg A \in Cn(B, \Delta)$. Two applications of **N5** give $\neg B \in Cn(A, \Delta)$. This proves **N3.2**. $\blacksquare$

4.2.3 *Disjunction and conjunction*

Megarian–Stoic logic recognized conjunction and two kinds of disjunction—the complete (exclusive) and the incomplete (non-exclusive) (Bocheński 1961, p. 161). The problems of disjunction were discussed in Scholastic logic. Peter of Spain did not clearly decide between the two interpretations, but Burleigh chose the non-exclusive disjunction and clearly stated the rules

if A and B, then A

if A, then A or B

(Bocheński 1961, p. 197) as a relation between truth-values. The original

118   *Logical operations*

Boolean calculus treated disjunction as exclusive. This was remedied by Jevons and Peirce who rediscovered non-exclusive disjunction.

Whilst classical logic is concerned with reported states of affairs and universal truths, constructive logics stress the discursive nature of logic—it is concerned with rules of correct reasoning, proofs, and truth-claims. Thus intuitionists define the connectives in terms of how a compound proposition is to be established:

> ... $p \wedge q$ can be asserted if and only if both p and q can be asserted.... $p \vee q$ can be asserted if and only if at least one of the propositions p and q *can be asserted*.
>
> Heyting 1956, p. 98

The difference between the constructive and the classical use of conjunction and disjunction was clearly stated by Bishop:

> To prove the statement (P and Q) we must prove the statement P and prove the statement Q, just as in classical mathematics. To prove the statement (P or Q) we must either prove the statement P or prove the statement Q, whereas in classical mathematics it is possible to prove (P or Q) without proving either the statement P or the statement Q.
>
> Bishop 1967, p. 7

Classical disjunction was also criticized from another point of view by the idealists Bradley and Bosanquet (see Chapter 9, §9.1).

Conjunction and disjunction are usually considered as two-place connectives. There is no difficulty, however, in initially considering them to be functions on finite sets. Indeed, there are hints (not to be taken too seriously) that Megarian–Stoic logic applied disjunction to *sets* of propositions (Bocheński 1961, p. 120). Formulating these operations in terms of sets of sentences is a convenient starting point for the study of quantifiers, but they are not expressible in Qf_2. We shall shortly show how two-place connectives can be constructed from them which can then be expressed in Qf_2.

In a logic $\langle \mathfrak{A}, Cn \rangle$, conjunction π and disjunction σ are functions such that for each finite non-empty subset Δ of $|\mathfrak{A}|$, $\pi(\Delta)$ and $\sigma(\Delta)$ are sentences. They satisfy, for all $\Gamma, D \subseteq |\mathfrak{A}|$ and $\Delta \in \hat{\mathbb{P}}_\omega(|\mathfrak{A}|)$,

Cj1: $\qquad\qquad\qquad A \in \Delta \quad \Rightarrow \quad A \in Cn(\Gamma, \pi(\Delta))$

Cj2: $\quad (A \in \Delta)(A \in Cn(\Gamma, D)) \quad \Rightarrow \quad \pi(\Delta) \in Cn(\Gamma, D)$

Dj1: $\qquad\qquad\qquad A \in \Delta \quad \Rightarrow \quad \sigma(\Delta) \in Cn(\Gamma, A)$

Dj2: $\quad (A \in \Delta)(D \in Cn(\Gamma, A)) \quad \Rightarrow \quad D \in Cn(\Gamma, \sigma(\Delta))$.

An operation satisfying **Cj1** and **Cj2** will be called a *normal conjunction* and an operation satisfying **Dj1** and **Dj2** will be called a *normal disjunction*. The designation 'normal' is used because in many important logics, particularly classical logic (Example 2.1(iii)) and intuitionistic logic, conjunction and disjunction give rise to such operations on finite sets of sentences. Weight

The historical connectives

of tradition obliges us to single out these axioms. But the logical operations so defined are fortunately characterized by the existence of upper and lower bounds in the implication orderings (Definition 3.8) (see Chapter 5, §5.6).

The conditions **Cj1**, ..., **Dj2** have been expressed in that particular form because of the connection with quantifiers which will be considered later. However, they can be expressed more simply. Consider the following conditions: for all $\Delta \in \overset{\circ}{\mathbb{P}}_\omega(|\mathfrak{A}|)$, $\Gamma \subseteq |\mathfrak{A}|$

Cj1*: $\Delta \subseteq Cn(\pi(\Delta))$

Dj1*: $\sigma(\Delta) \in \prod \{Cn(A), A \in \Delta\}$

Cj2*: $\pi(\Delta) \in Cn(\Delta)$

Dj2*: $\prod \{Cn(\Gamma, A), A \in \Delta\} \subseteq Cn(\Gamma, \sigma(\Delta))$.

THEOREM 4.9. In any logic $\langle \mathfrak{A}, Cn \rangle$ with operations $\pi, \sigma \colon \overset{\circ}{\mathbb{P}}_\omega(|\mathfrak{A}|) \to |\mathfrak{A}|$, **Cj1**, **Cj2**, **Dj1**, **Dj2** are equivalent to **Cj1***, **Cj2***, **Dj1***, **Dj2*** respectively.

Proof.
  (i) **Cj1** $\Leftrightarrow$ **Cj1***. Suppose **Cj1**. Putting $\Gamma = \varnothing$ gives $\Delta \subseteq Cn(\pi(\Delta))$. Thus **Cj1** $\Rightarrow$ **Cj1***. Conversely, suppose **Cj1***. Let $\Gamma \subseteq |\mathfrak{A}|$. Then $Cn(\pi(X)) \subseteq Cn(\Gamma, \pi(X))$. By **Cj1***, $\Delta \subseteq Cn(\Gamma, \pi(X))$. Thus **Cj1*** $\Rightarrow$ **Cj1**.
  (ii) **Dj1** $\Leftrightarrow$ **Dj1***. Similar to (i).
  (iii) **Cj2** $\Leftrightarrow$ **Cj2***. Suppose **Cj2**. Put $\Gamma = \Delta$ and choose $D \in \Delta$. By **Cj2**, $(A \in \Delta)(A \in Cn(\Delta)) \Rightarrow \pi(\Delta) \in Cn(\Delta)$. But $\Delta \subseteq Cn(\Delta)$. Hence $\pi(\Delta) \in Cn(\Delta)$. Thus **Cj2** $\Rightarrow$ **Cj2***. Conversely, suppose **Cj2***. Let $D \in |\mathfrak{A}|$ and suppose $(A \in \Delta)(A \in Cn(\Gamma, D))$, i.e. $\Delta \subseteq Cn(\Gamma, D)$. Then $Cn(\Delta) \subseteq Cn(\Gamma, D)$. By **Cj2***, $\pi(\Delta) \in Cn(\Gamma, D)$. Thus **Cj2*** $\Rightarrow$ **Cj2**.
  (iv) **Dj2** $\Leftrightarrow$ **Dj2***. **Dj2*** is merely a restatement of **Dj2**. ∎

COROLLARY 4.10. Let $\langle \mathfrak{A}, Cn \rangle$ be a logic with normal conjunction π and normal disjunction σ. Then for all $\Delta \in \overset{\circ}{\mathbb{P}}_\omega(|\mathfrak{A}|)$ and all $\Gamma, D \subseteq |\mathfrak{A}|$

  (i) $Cn(\pi(\Delta)) = Cn(\Delta)$

  (ii) $\sum \{Cn(A), A \in \Delta\} \subseteq Cn(\pi(\Delta))$

  (iii) $\sum \{Cn(A), A \in \Delta\} \subseteq Cn(\Gamma, D) \;\Rightarrow\; \pi(\Delta) \in Cn(\Gamma, D)$

  (iv) $Cn(\Gamma, \sigma(\Delta)) = \prod \{Cn(\Gamma, A), A \in \Delta\}$.

Proof.
  (i) By **Cj1**, $\Delta \subseteq Cn(\pi(\Delta))$ and $\pi(\Delta) \in Cn(\Delta)$. Thus $Cn(\Delta) \subseteq Cn(\pi(\Delta))$ and $Cn(\pi(\Delta)) \subseteq Cn(\Delta)$. Hence $Cn(\pi(\Delta)) = Cn(\Delta)$.
  (ii) This follows directly from (i).
  (iii) This is simply a rewriting of **Cj2**.

(iv) By **Dj1**, $\sigma(\Delta) \in Cn(A)$ for all $A \in \Delta$. So $Cn(\Gamma, A) \supseteq Cn(\Gamma, \sigma(\Delta))$ for all $A \in \Delta$. Hence $\prod \{Cn(\Gamma, A), A \in \Delta\} \supseteq Cn(\Gamma, \sigma(\Delta))$. But by **Dj2***, $\prod \{Cn(\Gamma, A), A \in \Delta\} \subseteq Cn(\Gamma, \sigma(\Delta))$. Hence $\prod \{Cn(\Gamma, A), A \in \Delta\} = Cn(\Gamma, \sigma(\Delta))$. ∎

This corollary enables us to dispense with conjunction and disjunction as functions on finite sets and replace them with two-place operations in the usual sense. First, unit sets can be ignored since for all $A \in |\mathfrak{A}|$, $A \leftrightarrow \sigma(\{A\}) \leftrightarrow \pi(\{A\})$. But more important is the following corollary.

COROLLARY 4.11. Let $\langle \mathfrak{A}, Cn \rangle$ be a logic with normal conjunction π and normal disjunction σ. Then for all $\Delta, \Gamma \in \mathring{\mathbb{P}}_\omega(|\mathfrak{A}|)$

(i) $\pi(\Delta + \Gamma) \leftrightarrow \pi(\{\pi(\Delta), \pi(\Gamma)\})$

(ii) $\sigma(\Delta + \Gamma) \leftrightarrow \sigma(\{\sigma(\Delta), \sigma(\Gamma)\})$.

Proof.

(i) By Corollary 4.10(i).

$$Cn(\pi(\Delta + \Gamma)) = Cn(\Delta + \Gamma) = Cn(\Delta, \pi(\Gamma))$$
$$= Cn(\pi(\Delta), \pi(\Gamma)) = Cn(\pi\{\pi(\Delta), \pi(\Gamma)\}).$$

Hence $\pi(\Delta + \Gamma) \leftrightarrow \pi\{\pi(\Delta), \pi(\Gamma)\}$.

(ii) By Corollary 4.10(ii)

$$Cn(\sigma(\Delta + \Gamma)) = \prod \{Cn(A), A \in \Delta + \Gamma\}$$
$$= \prod \{Cn(A), A \in \Delta\} \cdot \prod \{Cn(A), A \in \Gamma\}$$
$$= Cn(\sigma(\Delta)) \cdot Cn(\sigma(\Gamma)) = Cn(\sigma\{\sigma(\Delta), \sigma(\Gamma)\}),$$

i.e. $\sigma(\Delta + \Gamma) \leftrightarrow \sigma\{\sigma(\Delta), \sigma(\Gamma)\}$. ∎

In the attempt to manufacture connectives in the usual sense we can define operations $\vee, \wedge : |\mathfrak{A}|^2 \to |\mathfrak{A}|$ in terms of π and σ by

$$A \vee B =_{\text{def}} \sigma\{A, B\} \qquad A \wedge B =_{\text{def}} \pi\{A, B\} \qquad (A, B \in |\mathfrak{A}|). \qquad (4.5)$$

By iteration of Corollary 4.10 we have the following corollary.

COROLLARY 4.12.

(i) $\pi\{A_1, \ldots, A_n\} \leftrightarrow (A_1 \wedge \cdots (A_{n-1} \wedge A_n) \cdots)$

(ii) $\sigma\{A_1, \ldots, A_n\} \leftrightarrow (A_1 \vee \cdots (A_{n-1} \vee A_n) \cdots)$.

The historical connectives

The operations $\vee$, $\wedge$ clearly satisfy the following conditions: for all Γ, A, $B \subseteq |\mathfrak{A}|$

Cj1†: $A, B \in Cn(\Gamma, A \wedge B)$

Cj2†: $A, B \in Cn(\Gamma, C) \Rightarrow A \wedge B \in Cn(\Gamma, C)$

Dj1†: $A \vee B \in Cn(\Gamma, A) \cdot Cn(\Gamma, B)$

Dj2†: $C \in Cn(\Gamma, A) \cdot Cn(\Gamma, B) \Rightarrow C \in Cn(\Gamma, A \vee B)$.

This justifies the terms *normal conjunction* and *normal disjunction* to describe two-place operations satisfying **Cj1**†–**Dj2**†. These conditions can obviously be expressed in Qf_2.

We now start with the axioms **Cj1**†–**Dj2**† and show by a roundabout route that normal conjunction and disjunction—as operations on non-empty finite sets—can be constructed from the two-place operations. Operations $\wedge$, $\vee$ satisfying the weaker conditions **Cj1**^{0}, **Cj2**^{0}, **Dj1**^{0}, **Dj2**^{0}, obtained from **Cj1**, ..., **Dj2** respectively by suppression of the set variable Γ, are simply called *conjunction* and *disjunction* respectively.

EXAMPLE 4.3.

(i) if $\wedge$, $\wedge'$ are operations satisfying **Cj1**^{0} & **Cj2**^{0} then $\wedge$ and $\wedge'$ are equivalent in the sense that $A \wedge B \leftrightarrow A \wedge' B$. Similarly, any two operations satisfying **Dj1**^{0} & **Dj2**^{0} are equivalent. Thus conjunction and disjunction are unique up to logical equivalence.

(ii) If a logic has a conjunction (disjunction) then it satisfies the weak conjunction (disjunction) condition **WC** (**WD**), (Chapter 3, §3.10). Conversely, assuming the axiom of choice, if a logic satisfies **WC** (**WD**) then the logic has a conjunction (disjunction). ◆

LEMMA 4.13. *In any logic $\langle \Sigma, \wedge, \vee, Cn \rangle$ with two-place operations $\wedge$, $\vee$*

$$\mathbf{Cj1}^\dagger \ \& \ \mathbf{Cj2}^\dagger \Leftrightarrow Cn(A, B, \Gamma) = Cn(A \wedge B, \Gamma) \qquad (4.6)$$

$$\mathbf{Dj1}^\dagger \ \& \ \mathbf{Dj2}^\dagger \Leftrightarrow Cn(A, \Gamma) \cdot Cn(B, \Gamma) = Cn(A \vee B, \Gamma). \qquad (4.7)$$

Hence

$$\mathbf{Cj1}^0 \ \& \ \mathbf{Cj2}^0 \Leftrightarrow Cn(A, B) = Cn(A \wedge B) \qquad (4.8)$$

$$\mathbf{Dj1}^0 \ \& \ \mathbf{Dj2}^0 \Leftrightarrow Cn(A) \cdot Cn(B) = Cn(A \vee B). \qquad (4.9)$$

COROLLARY 4.14.

(i) Let $\mathfrak{L} = \langle \Sigma, \wedge, Cn \rangle$ be a logic with a binary operation $\wedge$. Then we have the following.

(a) $\mathfrak{L}$ satisfies **Cj1⁰** & **Cj2⁰** iff

$$A \to B \iff A \leftrightarrow A \wedge B \qquad (A, B \in \Sigma).$$

(b) If $\mathbb{L}$ satisfies **Cj1⁰** & **Cj2⁰** then $\wedge$ is a logical operation and if it satisfies **Cj1†** & **Cj2†** then $\wedge$ is a strongly logical operation.

(ii) Let $\mathfrak{L} = \langle \Sigma, \vee, Cn \rangle$ be a logic with a binary operaton $\vee$. Then we have the following.

(a) $\mathfrak{L}$ satisfies **Dj1⁰** & **Dj2⁰** iff

$$A \to B \iff B \Leftrightarrow A \vee B \qquad (A, B \in \Sigma).$$

(b) If $\mathfrak{L}$ satisfies **Dj1⁰** & **Dj2⁰** then $\vee$ is a logical operation and if it satisfies **Dj1†** & **Dj2†** then $\vee$ is a strongly logical operation.

Proof.

(i) (a) This follows from (4.8).

(b) Suppose that $\mathfrak{L}$ satisfies **Cj1⁰** & **Cj2⁰**. Let $A \leftrightarrow A'$ and $B \leftrightarrow B'$. Then $Cn(A) = Cn(A')$ and $Cn(B) = Cn(B')$. By (4.6),

$$Cn(A \wedge B) = Cn(A, B) = Cn(Cn(A), Cn(B))$$
$$= Cn(Cn(A'), Cn(B')) = Cn(A', B') = Cn(A' \wedge B')$$

so that $A \wedge B \leftrightarrow A' \wedge B'$. Thus, $\wedge$ is a logical operation.

(ii) The proofs of (ii)(a) and (ii)(b) are similar to those of (i)(a) and (i((b) respectively. ∎

Consider now the following conditions **L1, L2, L3, L4, L5.1, L5.2** on a logic $\langle \mathfrak{A}, Cn \rangle$ with two-place operations $\wedge$, $\vee$: for all $\Gamma, A, B, C \subseteq |\mathfrak{A}|$

L1($\wedge$): $\quad A \wedge A \leftrightarrow_\Delta A$

L1($\vee$): $\quad A \vee A \leftrightarrow_\Delta A$

L2($\wedge$): $\quad A \wedge B \leftrightarrow_\Delta B \wedge A$

L2($\vee$): $\quad A \vee B \leftrightarrow_\Delta B \vee A$

L3($\wedge$): $\quad A \wedge (B \wedge C) \leftrightarrow_\Delta (A \wedge B) \wedge C$

L3($\vee$): $\quad A \vee (B \vee C) \leftrightarrow_\Delta (A \vee B) \vee C$

L4: $\quad A \wedge (A \vee B) \leftrightarrow_\Delta A$

$\quad\quad\quad A \vee (A \wedge B) \leftrightarrow_\Delta A$

L5.1: $\quad A \wedge (B \vee C) \leftrightarrow_\Delta (A \wedge B) \vee (A \wedge C)$

L5.2: $\quad A \vee (B \wedge C) \leftrightarrow_\Delta (A \vee B) \wedge (A \vee C)$.

These laws, which hold in classical logic (Example 2.1), are closely related to the lattice laws which will be studied more closely in Chapter 5. (Recall here that for a law **M**, **M⁰** is that law derived from **M** but with set variables put to $\emptyset$. Thus **L1($\wedge$)⁰** is $A \wedge A \leftrightarrow A$.)

THEOREM 4.15. Let $\mathfrak{L}$ be a logic.

(i) If $\mathfrak{L}$ has a conjunction $\wedge$ then

$$A \in Cn(B) \quad \Leftrightarrow \quad A \wedge B \leftrightarrow_\Delta B$$

and $\mathbf{L1}(\wedge)^0$, $\mathbf{L2}(\wedge)^0$, and $\mathbf{L3}(\wedge)^0$ hold.

(ii) If $\mathfrak{L}$ has a disjunction $\vee$ then

$$A \in Cn(B) \quad \leftrightarrow \quad A \vee B \leftrightarrow_\Delta A$$

and $\mathbf{L1}(\vee)^0$, $\mathbf{L2}(\vee)^0$ and $\mathbf{L3}(\vee)^0$ hold.

(iii) If $\mathfrak{L}$ has a conjunction and disjunction then $\mathbf{L4}^0$ holds.

(iv) If $\mathfrak{L}$ has a normal conjunction and normal disjunction then $\mathbf{L1}$–$\mathbf{L5}$ hold and

$$A \in Cn(B, \Delta) \quad \Leftrightarrow \quad A \wedge B \leftrightarrow_\Delta B \quad \Leftrightarrow \quad A \vee B \leftrightarrow_\Delta A.$$

Proof. These results follow easily from (4.6)–(4.9).

ad **L5.1** Suppose that $\mathfrak{L}$ has normal conjunction and normal disjunction. Then

$$Cn(A \wedge (B \vee C), \Gamma) = Cn(A, B \vee C, \Gamma) = Cn(A, B, \Gamma) \cdot Cn(A, C, \Gamma)$$
$$= Cn(A \wedge B, \Gamma) \cdot Cn(A \wedge C, \Gamma)$$
$$= Cn((A \wedge B) \vee (A \wedge C), \Gamma).$$

The proof of **L5.2** is similar. ∎

Note that in the above proof the equality $Cn(A, B \vee C, \Gamma) = Cn(A, B, \Gamma) \cdot Cn(A, C, \Gamma)$ is only valid for *normal* disjunction.

Normal conjunction and disjunction can now be reconstructed as operations on finite sets in a logic with a binary operations $\wedge$, $\vee$ as follows. Let $A_1, \ldots, A_n$ be any *sequence* of elements. Define

(i) $\pi(A_1, \ldots, A_n) =_{\text{def}} (A_1 \wedge \cdots (A_{n-1} \wedge A_n) \cdots)$

(ii) $\sigma(A_1, \ldots, A_n) =_{\text{def}} (A_1 \vee \cdots (A_{n-1} \vee A_n) \cdots)$.

COROLLARY 4.16. Let $\mathfrak{L} = \langle \mathfrak{A}, Cn \rangle$ be a logic with two-place operation $\wedge$ satisfying $\mathbf{Cj1}^0$ & $\mathbf{Cj2}^0$. If $\{A_1, \ldots, A_n\} = \{B_1, \ldots, B_m\}$ then $\pi(A_1, \ldots, A_n) \leftrightarrow \pi(B_1, \ldots, B_m)$ so that π is a well-defined operation on finite sets. Further, π satisfies $\mathbf{Cj1}^\dagger$ & $\mathbf{Cj2}^\dagger$. Likewise, if $\mathfrak{L} = \langle \mathfrak{A}, Cn \rangle$ is a logic with two-place operation $\vee$ satisfying $\mathbf{Dj1}^0$ & $\mathbf{Dj2}^0$ then σ is a well-defined operation on finite sets which satisfies $\mathbf{Dj1}^\dagger$ & $\mathbf{Dj2}^\dagger$.

Proof. If $\{A_1, \ldots, A_n\} = \{B_1, \ldots, B_m\}$ then by **L1**, **L2**, and **L3**

$$(A_1 \wedge \cdots (A_{n-1} \wedge A_n) \cdots) \leftrightarrow (B_1 \wedge \cdots (B_{m-1} \wedge B_m) \cdots).$$

Hence $\pi(A_1, \ldots, A_n) \leftrightarrow \pi(B_1, \ldots, B_m)$. Repeated application of the laws shows that **Cj1**† and **Cj2**† are satisfied. Similarly for $\vee$. ∎

THEOREM 4.17. Let $\mathfrak{L} = \langle \mathfrak{A}, Cn \rangle$ be a logic with two-place operations $\wedge$, $\vee$. Then we have the following.

(i) **Cj1**^{0} & **Cj2**^{0} ⇔ **Cj1**† & **Cj2**†.

(ii) **Dj1**† ⇔ **Dj1**^{0}.

(iii) **Dj2**† ⇒ **Dj2**^{0}.

(iv) If $\mathfrak{L}$ is compact, $\wedge$ is normal conjunction and the condition

$$\textbf{D:} \quad (A \wedge B) \vee (A \wedge C) \in Cn(A, B \vee C)$$

is satisfied then **Dj2**^{0} ⇒ **Dj2**†.

Proof.

(i) Putting $\Gamma = \emptyset$ in **Cj1**† and **Cj2**† gives **Cj1**^{0} and **Cj2**^{0}. Thus

$$\textbf{Cj1}^\dagger \,\&\, \textbf{Cj2}^\dagger \;\Rightarrow\; \textbf{Cj1}^0 \,\&\, \textbf{Cj2}^0.$$

Conversely, suppose **Cj1**^{0} & **Cj2**^{0}. Then for every $A, B \in |\mathfrak{A}|$ and $\Gamma \subseteq |\mathfrak{A}|$, $A, B \in Cn(A \wedge B) \subseteq Cn(A \wedge B, \Gamma)$ and $A \wedge B \in Cn(A, B) \subseteq Cn(A, B, \Gamma)$. Thus

$$\textbf{Cj1}^0 \,\&\, \textbf{Cj2}^0 \;\Rightarrow\; \textbf{Cj1}^\dagger \,\&\, \textbf{Cj2}^\dagger.$$

This proves (i).

(ii) and (iii) follow by similar arguments.

(iv) Suppose $\mathfrak{L} = \langle \mathfrak{A}, \wedge, \vee, Cn \rangle$ is compact, $\wedge$ is normal conjunction, and **D** holds. Suppose **Dj2**^{0}. Let $C \in Cn(\Gamma, A) \cdot Cn(\Gamma, B)$. As $\mathfrak{L}$ is compact there exists a finite subset Z of Γ such that $C \in Cn(Z, A) \cdot Cn(Z, B)$. As $\wedge$ is normal conjunction there exists an element t such that $Cn(Z) = Cn(t)$ so $C \in Cn(t \wedge A) \cdot Cn(t \wedge B)$. By **Dj2**^{0}, $C \in Cn((t \wedge A) \wedge (t \wedge B))$. By **D**,

$$C \in Cn(t, A \vee B) = Cn(Z, A \vee B) \subseteq Cn(\Gamma, A \vee B),$$

i.e. **Dj2**† holds. Thus, **Dj2**^{0} ⇒ **Dj2**†. This proves (iv).

This theorem shows that normal conjunction is simply conjunction, but disjunction is not necessarily normal disjunction.

4.3 Relations between connectives

4.3.1 Negation, implication, and contradiction

It is well known that in classical logic all connectives can be defined in terms of f (whose truth-value under every assignment is falsehood) and implication. For instance, Church (1956, p. 72) gives a complete system of the propositional calculus due to Wajsberg with axiom scheme

$$A \supset [B \supset A]$$
$$[A \supset [B \supset C]] \supset [[A \supset B] \supset [A \supset C]]$$
$$[[A \supset f] \supset f] \supset A$$

where the last is the law of double negation when $\neg A$ is taken as an abbreviation of $[A \supset f]$. (The second axiom schema can be replaced by the equivalent schema $[A \supset B] \supset [[A \supset [B \supset C]] \supset [A \supset C]]$.) Negation can also be defined in terms of contradiction and implication in other logics. For the intuitionist,

> ... $\neg p$ can be asserted if, and only if, we possess a construction which from the supposition tht a construction p were carried out, leads to a contradiction.
>
> Heyting 1956, p. 98

We now consider the properties of negation defined in this way.

THEOREM 4.18. Let $\mathfrak{L} = \langle \Sigma, f, \supset, Cn \rangle$ be a logic with absurdity f (satisfying F) and normal implication $\supset$. Define the operation $\neg$ by $\neg A =_{\text{def}} A \supset f$ for all $A \in \Sigma$. Then $f \supset f \in Cn(\emptyset)$, **N1** & **N5** hold, and $Cn(\emptyset)$ includes all elements of the forms $A \supset \neg\neg A$, $(A \supset B) \supset ((A \supset \neg B) \supset \neg A)$ are theorems of $\mathfrak{L}$. If, in addition, $\mathfrak{L}$ satisfies Peirce's law

P: $((A \supset B) \supset A) \supset A) \in Cn(\emptyset) \quad (A, B \in |\mathfrak{A}|)$

then **N4.2** holds and hence

DN: $\neg\neg A \supset A \in Cn(\emptyset)$.

Proof. $f \in Cn(f)$. As implication is normal, $f \supset f \in Cn(\emptyset)$ and $f \in Cn(A, \neg A)$. By **F**, $Cn(A, \neg A) = \Sigma$. Thus **N1** holds **N5** again follows directly from the normality of implication. By Theorem 4.8, **N3.1** & **N3.2** hold. By Example 4.2, $\neg\neg A \in Cn(A)$. Hence $\neg\neg A \supset A \in Cn(\emptyset)$. Further, by Theorem 4.5, $(A \supset B) \supset ((A \supset (B \supset f)) \supset (A \supset f)) \in Cn(\emptyset)$. Hence $(A \supset B) \supset ((A \supset \neg B) \supset \neg A) \in Cn(\emptyset)$.

Suppose $\mathfrak{L}$ satisfies **P**. By **N1**, $A \in Cn(\neg A, \neg\neg A)$. Hence $\neg A \supset A \in Cn(\neg\neg A)$. But by **P**, $(\neg A \supset A) \supset A \in Cn(\emptyset)$ so that $A \in Cn(\neg A \supset A)$. Hence $A \in Cn(\neg\neg A)$. Thus **N4.2** holds and thereby **DN**. ∎

EXAMPLE 4.4. Let $\mathfrak{L}$ be the logic of Theorem 4.19. Then for all $A, B \in \Sigma$,

$$Cn(A) \cdot Cn(B) = Cn(\emptyset) \implies A \in Cn(\neg B).\qquad\blacklozenge$$

4.3.2 Implication, conjunction, and disjunction
Implication and conjunction are related as follows.

THEOREM 4.19. Let $\mathfrak{L} = \langle \mathfrak{A}, Cn \rangle$ be a compact logic with normal conjunction $\wedge$ and a two-place operation $\supset$. Then $\supset$ is normal implication iff for all $A, B, C \in |\mathfrak{A}|$,

$$C \leftrightarrow C \wedge (A \supset B) \Leftrightarrow B \wedge (A \wedge C) \leftrightarrow (A \wedge C). \qquad (4.10)$$

Proof.

(i) Suppose that $\supset$ is normal implication. Then by the normality of conjunction

$$A \supset B \in Cn(C) \Leftrightarrow B \in Cn(A \wedge C) \qquad (4.11)$$

(4.10) now follows by Theorem 4.16(ii).

(ii) Conversely, suppose (4.10). Suppose $A \supset B \in Cn(\Delta)$. As $\mathfrak{L}$ is compact there exists a finite subset Γ of Δ such that $A \supset B \in Cn(\Gamma)$. Then since $\wedge$ is normal conjunction there exists an element C such that $Cn(C) = Cn(\Gamma)$. Thus $A \supset B \in Cn(C)$. By Theorem 4.15, $C \wedge (A \supset B) \leftrightarrow C$. By (4.10), $B \wedge (A \wedge C) \leftrightarrow (A \wedge C)$. By Theorem 4.16, $B \in Cn(A, C) = Cn(A, \Gamma) \subseteq Cn(A, \Delta)$. Thus

$$A \supset B \in Cn(\Delta) \implies B \in Cn(\Delta, A). \qquad (4.12)$$

Next suppose that $B \in Cn(\Delta, A)$. Then, as above, by the compactness of $\mathfrak{L}$ and the normality of $\wedge$, there exists an element C such that $Cn(C) \subseteq Cn(\Delta)$ and $B \in Cn(A, C) = Cn(A \wedge C)$. By Theorem 4.16, $B \wedge (A \wedge C) \leftrightarrow (A \wedge C)$. By (4.91) and Theorem 4.16, $A \supset B \in Cn(D) \subseteq Cn(\Delta)$. Thus

$$B \in Cn(\Delta, A) \implies A \supset B \in Cn(\Delta). \qquad (4.13)$$

(4.12) and (4.13) prove that implication is normal. ∎

EXAMPLE 4.5. Let $\mathfrak{L}$ be a logic with the two-place operations $\wedge, \vee$ together with normal implication. Then $\wedge, \vee$ are normal conjunction and disjunction iff $Th(\mathfrak{L})$ contains all elements of the forms

A3: $A \supset (B \supset (A \wedge B))$
A4: $(A \wedge B) \supset B$
A5: $(A \wedge B) \supset A$
A6: $A \supset (A \vee B)$
A7: $B \supset (A \vee B)$
A8: $(A \supset C) \supset ((B \supset C) \supset ((A \vee B) \supset C))$.

These are all theorems of intuitionistic propositional logic I. Thus $\wedge$, $\vee$, $\supset$ are normal conjunction, disjunction, and implication respectively in I. ◆

4.3.3 Conjunction, disjunction, and negation

The following example connects conjunction and disjunction with truth and contradiction.

EXAMPLE 4.6. Let $\mathfrak{L}$ be a logic with normal conjunction and disjunction.

(i) If $\mathfrak{L}$ has two constants t and f, f satisfies **F** iff

$$\mathbf{LO}(f): \quad f \wedge A \leftrightarrow_\Delta f, \qquad f \vee A \leftrightarrow_\Delta A \qquad (\Gamma, A \subseteq |\mathfrak{A}|)$$

and t satisfies **T** iff

$$\mathbf{LO}(t): \quad t \wedge A \leftrightarrow_\Delta A, \qquad t \vee A \leftrightarrow_\Delta t \qquad (\Gamma, A \subseteq |\mathfrak{A}|).$$

(ii) If $\mathfrak{L}$ has a one-place operation $\neg$ satisfying **N1** then the 'disjunctive syllogism' holds: $B \in Cn(A \vee B, \neg A)$. This law holds in intuitionistic logic (and therefore in classical logic) but not in the three- and four-valued logics of Chapter 8. ◆

THEOREM 4.20. Let $\mathfrak{L} = \langle \mathfrak{A}, Cn \rangle$ be a logic with normal conjunction and disjunction $\wedge$ and $\vee$ and truth and falsehood t, f, and a one-place operation $\neg$ satisfying **N1** and **N5**. Then the following hold.

(i) $\quad \neg A \wedge \neg B \leftrightarrow_\Delta (A \vee B) \qquad (\Delta, A, B \subseteq |\mathfrak{A}|)$

(ii) $\quad \neg(A \wedge B) \in Cn(A \vee B) \qquad (A, B \subseteq |\mathfrak{A}|)$

(iii) $\quad A \wedge \neg A \leftrightarrow_\Delta f \qquad (\Delta, A, B \subseteq |\mathfrak{A}|)$.

If, in addition, **N3.3** holds then so do the following:

(iv) $\quad A \leftrightarrow_\Delta \neg\neg A \qquad (\Delta, A \subseteq |\mathfrak{A}|)$

(v) $\quad \neg(A \wedge B) \leftrightarrow_\Delta \neg A \vee \neg B \qquad (\Delta, A, B \subseteq |\mathfrak{A}|)$

(vi) $\quad A \vee \neg A \leftrightarrow_\Delta t \qquad (\Delta, A \subseteq |\mathfrak{A}|)$.

Proof. Suppose that $\mathfrak{L}$ satisfies **N5**. Then by Theorem 4.1, **N3.1** and **N3.2** also hold.

(i) $A \vee B \in Cn(A, \Delta) \cdot Cn(B, \Delta)$. By **N3.1**, $\neg A, \neg B \in Cn(\neg(A \vee B), \Delta)$. Hence

$$\neg A \wedge \neg B \in Cn(\neg(A \vee B), \Delta). \tag{4.14}$$

By **N1**, $Cn(A, \neg B, \neg A, \Delta) = Cn(B, \neg B, \neg A, \Delta) = |\mathfrak{A}|$. Hence

$$Cn(A, \neg A \wedge \neg B, \Delta) = Cn(B, \neg A \wedge \neg B, \Delta) = |\mathfrak{A}|.$$

By **N5**, $\neg(\neg A \vee \neg B) \in Cn(B, \Delta) \cdot Cn(A, \Delta) = Cn(A \vee B, \Delta)$. By **N3.2**

$$\neg(A \vee B) \in Cn(\neg A \vee \neg B, \Delta). \tag{4.15}$$

(i) follows directly from (4.15) and (4.16).

(ii) $A \in Cn(A, B) = Cn(A \wedge B)$. By **N3.1**, $\neg(A \vee B) \in Cn(\neg A)$. Likewise, $\neg(A \vee B) \in Cn(\neg B)$. Thus, $\neg(A \vee B) \in Cn(\neg A) \cdot Cn(\neg B) = Cn(\neg A \vee \neg B)$.

(iii) By **F**

$$A \wedge \neg A \in Cn(f, \Delta). \tag{4.16}$$

Next, by **N1**,

$$f \in Cn(A, \neg A) \subseteq Cn(A, \neg A, \Delta) = Cn(A \wedge \neg A, \Delta). \tag{4.17}$$

(iii) now follows directly from (4.16) and (4.17).

Suppose now that $\mathfrak{L}$ also satisfies **N3.3**.

(iv) This follows from **N4.1** & **N4.2**.
(v) Thus follows from (i) and (iv).
(vi) This follows by **N1** and Example 4.4. ∎

The conclusions of this theorem with $\Delta = \varnothing$ can be drawn by the same arguments from the hypothesis that $\wedge$, $\vee$ are conjunction and disjunction rather than normal conjunction and disjunction. The theorem applies to classical logic (see Example 2.1(viii)).

The logical identities (i) and (iv) of Theorem 4.20 are known as de Morgan's laws. They were enunciated for the calculus of classes in the form

$$(X \cup Y)' = X' \cap Y'$$

$$(X \cap Y)' = X' \cup Y'$$

by Augustus de Morgan in his *Formal Logic* of 1847. But forms of these laws were recognized in antiquity. Łukasiewicz showed tht they were part of the elementary doctrine of Scholasticism. Bocheński (1961, p. 207) maintained that they first occurred in Ockham and Burleigh.

$(f)^0$ of Theorem 4.21 is **PEM** (also known as *tertium non datur* or *tertium exclusium*). Aristotle devoted a chapter of the *Metaphysics* to it. At least once, however, he restricted its universality by not allowing it to be applied to future contingent events. Neither did the Epicureans accept it. Nevertheless it remained a firm principle of logic until modern times. Łukasiewicz appealed to Aristotle's restriction on **PEM** in founding the study of many-valued logics. The intuitionist critique of the classical notion of mathematical existence also leads to the rejection of the universality of **PEM** (Chapter 10).

4.3.4 Conjunction, disjunction, implication, and contradiction

From Theorem 4.5 and Example 4.5 it is easy to demonstrate the following theorems.

THEOREM 4.21. Let $\mathfrak{L}$ be a logic with normal conjunction $\wedge$ and disjunction $\vee$, normal implication $\supset$, and falsehood f and truth t. Define negation $\neg$

$$\neg x =_{\text{def}} x \supset f.$$

Then $Th(\mathfrak{L})$ contains all sentences of the forms **A0–A9** of Chapter 2, §2.2.7.

4.4 Quantifiers

The quantifiers 'all' and 'some' occur in a rudimentary form in Aristotelian logic. They become more explicit in Scholastic logic, for instance in the discussions of the logical form of quality and quantity. The relation between 'all' and 'some' and negation was clearly stated by Peter of Spain (in his *Summulae Logicales*) in connection with the conversion of propositions:

> Conversion by contraposition is to make the predicate from the subject and conversely, quality and quantity remaining the same, but finite terms being changed to infinite ones. And in this way the universal affirmative is converted into itself and the particular negative into itself, e.g. 'all man is animal', 'all non-animal is non-man'; 'some man is not stone'—'some non-stone is not non-man'.
>
> Bocheński 1961, p. 212

The interpretation of quantifiers as operators forming unbounded conjunctions and disjunctions occurs in Albert of Saxony:

> We proceed to the signs which render (propositions) universal or particular ... Of such signs one is the universal, the other the particular. The universal sign is that by which it is signified that the universal term to which it is adjoined stands copulatively for its suppositum ... The particular sign is that by which it is signified that a universal term stands disjunctively for all its supposita. And I purposely say

copulatively' when speaking of the universal sign, since if one says: 'every man runs' it follows formally: 'therefore this man runs and that man runs, etc.' But of the particular sign I have said that it signifies tht a universal term to which it is adjoined stands disjunctively for all its supposita. This is evident since if one says: 'some man runs' it follows that Socrates or Plato runs, or Cicero runs, and so of each.

Bocheński 1961, p. 234

Quantifiers are represented in the formal calculi—though in the Boolean calculus 'all' and 'some' are expressed as operators on classes—by means of a relation between classes and the universe of discourse, without using the concept of an *individual*.

Consider a quantifier logic $\mathfrak{L} = \langle \Sigma, c_1, \ldots, \exists, \forall, Cn \rangle$ (Definition 3.13) with the set $\mathbb{T}$ of terms and quantifiers $\forall, \exists$, i.e. satisfying Q1–Q4. We shall compromise with tradition by writing $Qx(F)$ instead of $Q(x, F)$ as in Definition 3.11, but it should be borne in mind that the symbol Q denotes a logical operation, and is not in the alphabet of the free-variable logic, and neither is Qx a prefix of some kind, even though in most actual logics the logical operations of quantification are realized by prefixed (see Definition 2.10).

The following axioms are designed as generalizations of the classical results expressed in Theorem 2.14. They are analogous to the axioms **Cj1**, **Cj2**, **Dj1**, **Dj2** for conjunction and disjunction—π corresponds to $\forall$ and σ corresponds to $\exists$.

∀1: For all $A \in \Sigma$, $x \in Fv$ and $t \in \mathbb{T}$

$$A(t/x) \in Cn(\forall x(A))$$

∀2: For all $\Delta, A \subseteq \Sigma$, $x \in Fv - Fv(\Delta)$

$$A \in Cn(\Delta) \quad \Rightarrow \quad \forall x(A) \in Cn(\Delta)$$

∃1: For all $\Delta, A, B \subseteq \Sigma$, $x \in Fv - Fv(\Delta, B)$

$$B \in Cn(\Delta, A) \quad \Rightarrow \quad B \in Cn(\Delta, \exists x(A))$$

∃2: For all $A \subseteq \Sigma$, $x \in Fv$ and $t \in \mathbb{T}$

$$\exists x(A) \in Cn(A(t/x))$$

LEMMA 4.22. *Let $\mathfrak{L} = \langle \mathfrak{A}, Cn \rangle$ be a quantifier logic with the set $\mathbb{T}$ of terms and quantifiers $\forall, \exists$. If $\forall$ satisfies* **∀1** *and* **∀2** *then it is a strong normal quantifier; if $\exists$ satisfies* **∃1** *and* **∃2** *then it is also a strong quantifier* (**Q1***, *Definition 3.13*).

Proof. Suppose $A \leftrightarrow_\Delta B$. Let $x \in Fv - Fv(\Delta)$. $B \in Cn(\forall x(B))$ by **∀1**. Hence $A \in Cn(\Delta, \forall x(B))$. As $x \notin Fv(\Delta, \forall x(B))$ (by **Q3**(a)), then $\forall x(A) \in Cn(\Delta, \forall x(B))$. Similarly, $\forall x(B) \in Cn(\Delta, \forall x(A))$. Thus $\forall x(A) \leftrightarrow_\Delta \forall x(B)$. Thus $\forall$ is a strong quantifier. Likewise ∃ is also a strong quantifier. Let $x \in Fv - Fv(A)$. Then $A \in Cn(\forall x(A))$ by **∀1** and $\forall x(A) \in Cn(A)$ by **∀2**. Thus $A \leftrightarrow_\Delta \forall x(A)$. Thus ∀ is a normal quantifier; similarly so is ∃. ∎

It can easily be seen from the axioms that universal and existential quantifiers are unique. That is, if ∀, ∀′ are universal quantifiers then by **∀1** and **∀2**, $\forall x(A) \leftrightarrow \forall' x(A)$ for all $x \in Fv$, and similarly for existential quantifiers.

Some alternative forms of the axioms **∀1** and **∃2** will prove useful later. They are as follows.

∀1′: For all $\Delta, A, B \subseteq \Sigma$, $x \in Fv$, $t \in \mathbb{T}$

$$B \in Cn(\Delta, A(t/x)) \;\Rightarrow\; B \in Cn(\Delta, \forall x(A))$$

∃2′: For all $\Delta, A \subseteq \Sigma$, $x \in Fv$, $t \in \mathbb{T}$

$$A(t/x) \in Cn(\Delta) \;\Rightarrow\; \exists X(A) \in Cn(\Delta)$$

LEMMA 4.23. *Let $\mathfrak{L} = \langle \mathfrak{A}, Cn \rangle$ be a quantifier logic with the set $\mathbb{T}$ of terms and quantifiers ∀, ∃. Then ∀ satisfies **∀1** iff it satisfies **∀1′** and ∃ satisfies **∃2** iff it satisfies **∃2′**.*

Proof. Let $\Delta, A, B \subseteq |\mathfrak{A}|$, $x \in Fv$, $t \in T(A)$. Suppose that $\mathfrak{L}$ satisfies **∀1** and suppose that $B \in Cn(\Delta, A(t/x))$. By **∀1**, $A(t/x) \in Cn(\forall x(A))$. Hence $B \in Cn(\Delta, \forall x(A))$. This proves **∀1′**. Conversely, suppose that **∀1′** holds. Now $A(t/x) \in Cn(A(t/x))$. By **∀1′** with $B = A(t/x)$ and $\Delta = \emptyset$ we have $A(t/x) \in Cn(\forall x(A))$. This proves **∀1**. The stated relation between **∃2** and **∃2′** is proved similarly. ∎

THEOREM 4.24. *Let $\mathfrak{L} = \langle \mathfrak{A}, Cn \rangle$ be a quantifier logic with the set $\mathbb{T}$ of terms and universal and existential quantifiers ∀, ∃ respectively. Let $x, y \in Fv$. Then*

(i) for $Q \in \{\forall, \exists\}$

$$Qx(Qy(A)) \leftrightarrow_\Delta Qy(Qx(A))$$

(ii) $\forall y(A(y/x)) \in Cn(\forall x(\forall y(A)))$

$$\exists x(\exists y(A)) \in Cn(\exists y(A(y/x)))$$

(iii) $\forall y(\exists x(A)) \in Cn(\exists x(\forall y(A)))$

(iv) $\exists x(A) \in Cn(\forall x(A))$.

Proof.

(i) Consider the case $Q = \forall$. $A \in Cn((\forall x(A))$ and $\forall x(A) \in Cn(\forall y(\forall x(A)))$ by **∀1**. Hence $A \in Cn(\forall y(\forall x(A)))$. But by **Q2**, $x, y \notin Fv(\forall y(\forall x(A)))$. Two applications of **∀2** then give $\forall x(\forall y(A)) \in Cn(\forall y(\forall x(A)))$. The case $Q = \exists$ is proved similarly.

(ii) $A(y/x) \in Cn(\forall x(A))$ and $\forall x(A) \in Cn(\forall y(\forall x(A)))$ by **∀1**. Hence $A(y/x) \in Cn(\forall y(\forall x(A)))$. As $y \notin Fv(\forall y((\forall x(A)))$, by **∀2**, $\forall y(A(y/x)) \in Cn(\forall y(\forall x(A)))$. The stated relationship for $\exists$ is proved similarly.

(iii) $\exists x(A) \in Cn(A)$ by **∃2** and $A \in Cn(\forall y(A))$ by **∀1**. Hence $\exists x(A) \in Cn(\forall y(A))$. As $x \notin Fv(\exists x(A))$, by **∃1**, $\exists x(A) \in Cn(\exists x(\forall y(A)))$. As $y \notin Fv(\exists x(\forall y(A)))$, by **∀2**, $\forall y(\exists x(A)) \in Cn(\exists x(\forall y(A)))$.

(iv) This follows easily from **∃2** and **∀1**. ∎

4.5 Quantifiers and connectives

The interaction of connectives with the universal and existential quantifiers is treated below under three headings: conjunction and disjunction, negation, and implication.

4.5.1 Conjunction and disjunction

The following theorem presents analogues of **Cj1—Dj2**, thus showing a connection between the Aristotelian quantifiers and conjunction and disjunction. Parts (iv)–(vii) relate the quantifiers to the existence of bounds in the implication orderings (Chapter 5, §5.6).

THEOREM 4.25. Let $\mathfrak{L} = \langle \mathfrak{A}, Cn \rangle$ be a quantifier logic with the set $\mathbb{T}$ of terms and universal and existential quantifiers $\forall$ and $\exists$ respectively. Let $\Delta, A, C \subseteq |\mathfrak{A}|$ with $Fv - Fv(\Delta)$ infinite and $x \in Fv - Fv(\Delta)$. Then

(i) $\sum \{Cn(A(t/x)); t \in \mathbb{T}\} \subseteq Cn(\forall x(A))$

(ii) $\sum \{Cn(A(t/x)); t \in \mathbb{T}\} \subseteq Cn(\Delta, C) \Rightarrow \forall x(A) \in Cn(\Delta, C)$

(iii) $Cn(\Delta, \exists x(A)) = \prod \{Cn(\Delta, A(t/x)); t \in \mathbb{T}\}$.

Further let $x\text{-}var(C) =_{\text{def}} \{C(y/x); x = y \text{ or } y \notin Fv(A)\}$. Then

(iv) $\exists x(C) \in Cn(\Delta, B)$ for $B \in x\text{-}var(C)$

(v) $(A \in x\text{-}var(C))(B \in Cn(\Delta, A)) \Rightarrow B \in Cn(\Delta, \exists x(C))$

(vi) $A \in Cn(\Delta, \forall x(C))$ for $A \in x\text{-}var(C)$

(vii) $(A \in x\text{-}var(C))(A \in Cn(\Delta, B)) \Rightarrow \forall x(C) \in Cn(\Delta, B)$.

Proof.

(i) This is simply a rewriting of **∀1**.
(ii) Suppose that $\sum \{Cn(A(t/x)); t \in \mathbb{T}\} \subseteq Cn(\Delta, C)$. Then $A \in Cn(\Delta, C)$. As $Fv - Fv(\Delta)$ is infinite there exists a free variable $x \in Fv - Fv(\Delta, C)$. By **∀2**, $\forall x(A) \in Cn(\Delta, C)$.
(iii) By **∃2**, $Cn(\exists x(A)) \subseteq Cn(A(t/x))$ for all $t \in \mathbb{T}$. Hence

$$Cn(\Delta, \exists x(A)) \subseteq \prod \{Cn(\Delta, A(t/x)); r \in \mathbb{T}\}. \qquad (4.18)$$

Now let $B \in \prod \{Cn(\Delta, A(t/x)); t \in \mathbb{T}\}$. Then $B \in Cn(\Delta, A)$. As $x \in Fv - Fv(\Delta)$, by **∃1**, $B \in Cn(\Delta, \exists x(A))$. Hence

$$\prod \{Cn(\Delta, A(t/x)); t \in \mathbb{T}\} \subseteq Cn(\Delta, \exists x(A)). \qquad (4.19)$$

The required equality follows from (4.19) and (4.20).

(iv) Let $B \in x\text{-}var(C)$. Then $B = C(y/x)$, say, for some $y \in \{x\} + (Fv - Fv(C))$. By **∃2** it follows that $\exists x(C) \in Cn(B) \subseteq Cn(\Delta, B)$.
(v) Suppose $B \in Cn(\Delta, A)$ for all $A \in x\text{-}var(C)$. As $Fv - Fv(\Delta)$ is infinite there can be found a free variable y not in $Fv(\Delta, B)$. Then $C(y/x) \in x\text{-}var(C)$ so that $B \in Cn(\Delta, C(y/x))$. By **∃1**, $B \in Cn(\Delta, \exists y(C(y/x)))$ so, by **Q3**, $B \in Cn(\Delta, \exists x(C))$.
(vi) and (vii) These are similar to (iv) and (v) respectively, but using **∀1** and **∀2** instead of **∃1** and **∃2**. ∎

EXAMPLE 4.7.

(i) $B \in Cn(\Delta, \forall x(A), A(t/x)) \Rightarrow B \in Cn(\Delta, \forall x(A))$.

(ii) If $\vee$ is disjunction (i.e. satisfies **Dj1⁰** & **Dj2⁰**) then

$$\exists x(A) \vee A(t/x) \in Cn(\Delta) \Rightarrow \exists x(A) \in Cn(\Delta). \qquad \blacklozenge$$

In view of Theorem 4.25 it is not surprising that where the logic has conjunction and disjunction, universal quantifiers commute with conjunction and existential quantifiers commute with disjunction.

THEOREM 4.26. Let $\mathfrak{L} = \langle \mathfrak{A}, Cn \rangle$ be a quantifier logic with universal and existential quantifiers $\forall$ and $\exists$ respectively, normal conjunction π, and normal disjunction σ. Let $x \in Fv$ and let Δ be a finite set of formulas. Then

(i) $\forall x(\pi(\Delta)) \leftrightarrow \pi\{\forall x(A); A \in \Delta\}$

(ii) $\exists x(\sigma(\Delta)) \leftrightarrow \sigma\{\exists x(A); A \in \Delta\}$.

Proof.

(i) Let $x \in Fv - Fv(\Delta)$ and $A \in \Delta$. Then by **∀1**, $A \in Cn(\forall x(A)) \subseteq Cn(\{\forall x(A); A \in \Delta\})$. By **Cj2**, $\pi(\Delta) \in Cn(\{\forall x(A); A \in \Delta\})$. By **∀2** and **Q2**

$$\forall x(\pi(\Delta)) \in Cn(\{\forall x(A); A \in \Delta\}). \qquad (4.20)$$

Next, by **Cj1** and **∀1**, $A \in Cn(\pi(A)) \subseteq Cn(\forall x(\pi(\Delta)))$. Hence, by **∀2**, $\forall x(A) \in Cn(\forall x(\pi(\Delta)))$. Thus $\forall x(A) \in Cn(\forall x(\pi(\Delta)))$. Then by (4.20),

$$\forall x(\pi(\Delta)) \leftrightarrow \pi\{\forall x(A); A \in \Delta\}.$$

(ii) This is proved similarly but using the laws for disjunction and existential quantification. ∎

Similar commutation relations hold if the logic has normal binary conjunction and disjunction. These are proved using **Cj1†–Dj2†**.

THEOREM 4.27. Let $\mathfrak{L} = \langle \mathfrak{A}, Cn \rangle$ be a quantifier logic with universal and existential quantifiers $\forall$ and $\exists$ respectively, normal conjunction $\wedge$, and normal disjunction $\vee$. Let $A, B \in |\mathfrak{A}|$ and $x \in Fv$. Then

(i) $\forall x(A) \wedge \forall x(B) \leftrightarrow \forall x(A \wedge B)$

(ii) $\exists x(A) \vee \exists x(B) \leftrightarrow \exists x(A \vee B)$.

If, in addition, $x \notin Fv(A)$ then

(iii) $A \wedge \forall x(B) \leftrightarrow \forall x(A \wedge B)$

(iv) $A \vee \exists x(B) \leftrightarrow \exists x(A \vee B)$

(v) $A \wedge \exists x(B) \leftrightarrow \exists x(A \wedge B)$

(vi) $A \vee \forall x(B) \leftrightarrow \forall x(A \vee B)$.

4.5.2 Negation

The interchange of universal with existential quantifiers under the action of negation is familiar in classical logic (Example 2.9(ii)). The following results hold for intuitionistic logic and therefore also for classical logic.

THEOREM 4.28. Let $\mathfrak{L} = \langle \mathfrak{A}, Cn \rangle$ be a quantifier logic with universal and existential quantifiers $\forall, \exists$ and negation $\neg$ satisfying **F** and **N5**. Let $A \in |\mathfrak{A}|$ and $x \in Fv$. Then

(i)(a) $\neg \forall x(\neg A) \in Cn(\exists x(A))$

(b) $\neg\neg \exists x(A) \in Cn(\neg \forall x(\neg A))$

(ii)(a) $\neg \exists x(\neg A) \in Cn(\forall x(A))$

(b) $\neg\neg \forall x(A)) \in Cn(\neg \exists x(\neg A))$.

Proof.

(i) (a) Let $x \in Fv - Fv(A)$. Then $x \notin Fv(\neg A)$. By **∀1**, $\neg A \in Cn(\forall x(\neg A))$. By **N5**, $f \in Cn(A, \forall x(\neg A))$. As $x \notin Fv(\forall x(\neg A))$, by **∃1** and **Q4**, $f \in Cn(\exists x(A), \forall x(\neg A))$. By **N5**, $\neg \exists x(A) \in Cn(\forall x(\neg A))$. By Theorem 4.8, **N3.1** and **N3.2** hold. Hence $\neg \forall x(\neg A) \in Cn(\exists x(A))$.

(b) Let $x \in Fv - Fv(A)$. Then by **∃2**, $\exists x(A) \in Cn(A)$. By **N3.1**, $\neg A \in Cn(\neg \exists x(A))$. But $x \notin Fv(\neg \exists x(A))$. By **∀2** and **Q2** $\forall x(\neg A) \in Cn(\neg \exists x(A))$ and so by **N3.1** $\neg \neg \exists x(A) \in Cn(\neg \forall x(\neg A))$.

(ii) This is similar to (i). ∎

4.5.3 *Implication*

The results listed here hold for intuitionistic logic

THEOREM 4.29. Let $\mathfrak{L} = \langle \mathfrak{A}, Cn \rangle$ be a quantifier logic with universal and existential quantifiers ∀ and ∃ respectively and normal implication ⊃. Let $A, B \in |\mathfrak{A}|$ and $x \in Fv$.

(i) if $x \notin Fv(A)$ then $\forall x(A \supset B) \leftrightarrow A \supset \forall x(B)$.
(ii) If $x \notin Fv(B)$ then $\forall x(A \supset B) \leftrightarrow \forall x(A) \supset B$.
(iii) If $x \notin Fv(A)$ then $A \supset \exists x(B) \in Cn(\exists x(A \supset B))$.
(iv) If $x \notin Fv(B)$ then $\forall x(A) \supset B \in Cn(\exists x(A \supset B))$.
(v) $\exists x(A) \supset \exists x(B) \in Cn(\exists x(A \supset B))$.

Proof.

(i) As ⊃ is normal, $\forall x(B) \in Cn(A, A \supset \forall x(B))$. By **∀1**, $B \in Cn(\forall x(B))$. Hence $B \in Cn(A, A \supset \forall x(B))$. Thus $A \supset B \in Cn(A \supset \forall x(B))$. Now $x \notin Fv(A, \forall x(B))$ so by **∀2**

$$\forall x(A \supset B) \in Cn(A \supset \forall x(B)). \tag{4.21}$$

By **∀1**, $A \supset B \in Cn(\forall x(A \supset B))$. Hence $B \in Cn(A, \forall x(A \supset B))$. By **∀2**, $\forall x(B) \in Cn(A, \forall x(A \supset B))$. Thus

$$A \supset \forall x(B) \in Cn(\forall x(A \supset B)). \tag{4.22}$$

(i) now follows directly from (4.21) and (4.22).

The proofs of (ii)–(v) similarly involve appeal to the quantifier axioms and axiom (4.1) for normal implication. ∎

5

Order and lattices

5.1 Introduction

As in algebraic systems, the subsystems (closed sets) of logic are ordered by the relation of *inclusion* and in Chapter 3, §3.10, this relation was represented by 'lattice diagrams'. This itself is sufficient reason for developing a theory about this ordering. But, more importantly, the logical consequence of a logic itself defines an order relation, *implication* (Definition 3.8), which has familiar properties (Example 3.4(ii)). The notion of *logical equivalence* is defined in terms of this relation.

The presence of certain connectives in a logic, particularly conjunction and disjunction, endows implication with *least upper bounds* and *greatest lower bounds* (Theorem 5.14)—operations which are best described by the theory of *lattices*. Thus, associated with various logics—classical, intuitionist, three-valued, etc.—are corresponding types of lattices—Boolean algebras, I-lattices, quasi-Boolean algebras respectively. Conversely, logics can naturally be constructed (Chapter 6) from various classes of lattices, with the lattice operations becoming the connectives. For example, one form of quantum logic can be constructed from ortho-modular lattices (Chapter 6, §6.8).

Therefore the aim of this chapter is to develop such parts of the elementary theory of relations, order, and lattices as are necessary for the analysis and synthesis of logics. The story of lattices begins with the idea of a *relation* (§§5.2–5.4), particularly binary relations, and proceeds via the notion of *quasi-ordering* and the associated *partial ordering* (§§5.5–5.6) to the definition of lattices (§5.7). Under the *lub* and *glb* operations lattices become algebras. These algebras can then be classified according to the laws governing the operations (§§5.8–5.10).

Further information on lattices can be found in the classic text by Birkhoff (1948). Chapter 4 of Curry (1963) is an excellent survey of such parts of lattice theory as are useful to logicians and contains historical notes.

5.2 Relations

The institutionalization of Aristotle's doctrines ensured their remarkable persistence. This was the case, in particular, with Aristotle's ('disastrous'—Geach) subject–predicate logic which endured into the modern era.

Aristotle's syllogistic concerned sentences of the forms

(a) all P are Q
(i) some P are Q
(e) no P is Q
(o) not all P are Q.

where P, Q are predicates: there are no proper names in these sentences. (The symbols a, i, e, and o are of medieval origin.) Aristotle himself usually wrote such sentences as a string beginning with P and ending with Q. If, therefore, 'all P are Q' is written as PaQ, the symbols a, i, e, o can be interpreted as binary relations between predicates. In later developments of the syllogistic, the predicates were represented as areas enclosed by circles and the relations between them illustrated by intersections and inclusions of such regions. This device, though usually attributed to Euler, was also used by Leibniz with more diverse shapes to reveal more complicated relations between predicates. It is clear even in Aristotle that relational arguments (i.e. at the 'metalevel') were used to reveal the workings of the syllogistic.

The syllogisms were typically of the form

$$PrQ \ \& \ QsR \ \Rightarrow \ PrsR$$

where r and s are relations between predicates, involving the product relation (Definition 5.1(ii)). The interpretation of Aristotle's syllogisms as products of relations was first given by de Morgan in 1847. Peirce regarded de Morgan as the founder of the logic of relations:

> The real founder of the modern logic of relations is de Morgan of whom Peirce, himself a great logician, said that he 'was one of the best logicians that ever lived and unquestionably the father of the logic of relative'.
>
> Bocheński 1961, p. 347

de Morgan defined the concepts of relative product of binary relations, converse of a binary relation and transitivity. The subject of binary relations was developed by Peirce in his articles on the algebra of logic (Peirce 1880, 1885).

The inability of Aristotle's logic to give a correct syllogistic proof of a correct argument such as 'Because a horse is an animal, the head of a horse is the head of an animal' was recognized by the logician Joachim Jungius (1587–1657), who had attempted to classify them. His works were known to Leibniz (1646–1716). Jungius argued that logic (as then understood) should be augmented to cover relational arguments.

Leibniz's early education was in a German-Protestant Aritotelian tradition. He first learned logic at school from a textbook by Johannes Rhenius. He evinced an early interest in logic—his youthful work *De Arte Combinatoria* contained the basic ideas of his logic to which he adhered for the rest of his life, that every proposition has a subject and a predicate and that all truths are reducible to identical propositions. These ideas became deeply embedded in his later philosophy.

Leibniz was acutely aware of the importance of relations, but he argued that by supplementing traditional logic with 'rational grammar' relational arguments could be transformed to forms amenable to the syllogistic treatment. Leibniz did not fully develop this notion. However, his work on particular kinds of binary relations is remarkably modern (Parkinson 1966). For instance, Leibniz defined a notion of *coincidence* of terms (a 'term' is the subject or predicate of a categorical proposition).

> Those terms are 'the same' or 'coincident' of which either can be substituted for the other whenever we please without loss of truth . . . .

He proved symmetry (Proposition 1) and transitivity (Proposition 3) of this relation (Parkinson 1966, Chapter 16). In the same text there is defined a notion of *containment* of terms:

> That A 'is in' L or that L 'contains' A is assumed to be coincident with several terms taken together, among which is A.

The transitivity and reflexiveness of this relation was proved in Propositions 15 and 7 respectively, though these concepts were not explicitly defined.

The modern extensional view of a binary relation as a class of pairs was adopted by Peirce (Bocheński 1961, p. 377). It is interesting to observe that Russell at first adopted an intensional view of relations:

> In addition to the defects of the old symbolic logic, their (i.e. Peirce, Schröder) method suffers from the fact that they regard a relation as essentially a class of couples, thus requiring elaborate formulae of summation for dealing with single relations. This view is derived, I think, probably unconsciously from a philosophical error: it has always been customary to suppose relational propositions less ultimate than class propositions (or subject–predicate propositions, with which class propositions are habitually confounded), and this had led to a desire to treat relations as a kind of classes. However this may be it was certainly from the opposite philosophical belief . . . that I was led to a different formal treatment of relations.
>
> Russell 1903, p. 24

In a later passage Russell explains how this intensional viewpoint avoids taking the notion of *couple* (ordered pair) as a primitive concept:

> There is a temptation to regard a relation as definable in extension as a class of couples . . . it is necessary to give sense to the couple, to distinguish the referent from the relatum: thus a couple become essentially distinct from a class of two terms, and must itself be introduced as a primitive idea. It would seem, viewing the matter philosophically, that the sense can only be derived from some relational proposition, and that the assertion that a is reference and b relatum already involves a purely relational proposition in which a and b are terms, though the relation asserted is only the general one of referent to relatum . . . It seems

Relations

therefore more correct to take an intensional view of relations, and identify them with class concepts than with relations.

Russell 1903, p. 99

The *Principia*, in effect, restates the necessity of using a relational proposition to establish order in a couple, but asserts an extensional view of the identity of relations:

> A *relation* as we shall use the word, will be understood in extension: It may be regarded as the class of couples for which some given function $\psi(x, y)$ is true.
>
> Russell 1910, Vol. 1, p. 200

At first sight it might seem that the use of the proposition to order the class of couples can be avoided by using set-theoretical definitions of the notion of ordered pair. Thus Wiener (1912) represented the ordered pair $\langle x, y \rangle$ by the construction

$$\langle x, y \rangle =_{\text{def}} \iota'(\iota'\iota'x \cup \iota'\Lambda) \cup \iota'\iota'\iota'y. \tag{5.1}$$

Wiener remarked in the same paper:

> It will be seen that what we have done is practically to revert to Schröder's treatment of a relation as a class of ordered couples. The complicated apparatus of the ι's and Λ's of which we have made use is simply and solely devised for the purpose of constructing a class which shall depend only on an ordered pair of values of x, y and which shall correspond to only one such pair. The method selected for doing this is a matter of choice; for example, I might have substituted V or any other constant class not a unit class, and existing in every type of classes in every place I have written ι..

Hausdorff gave a definition of ordered pair $\langle x, y \rangle$ as $\{\{x, 1\}, \{y, 2\}\}$ where 1 and 2 are two distinct objects differing from x and y. Kuratowski (1921) employed a similar device

$$\langle x, y \rangle =_{\text{def}} \{\{x\}, \{x, y\}\}$$

which satisfies the fundamental property (5.1). This is the current fashion (van Heijenoort 1967b, pp. 224–7).

These artifices, as Wiener remarked, are somewhat arbitrary. They are hardly an *analysis* of the notion of ordered pair; rather they are convenient *representations* in a structure which allows escalation of types, and there are many ways of doing this (e.g. Quine 1945). In order to represent a particular relation by one of these tricks, it still has to be decided which of the arguments x, y is to be taken as the referent and which the relatum—the ordering of the pair is actually used in the construction of its representation.

5.3 Equivalence relations

As in the logical papers of Leibniz, certain kinds of binary relations have proved to be of exceptional importance in mathematics and logic. The following definition is necessary for dealing with such relations.

DEFINITION 5.1. Let D be a non-empty set

(i) The *identity relation* on D, id_D, is defined by
$$id_D =_{def} \{(x, x); x \in D\}.$$

(ii) Let R, S be binary relations on D. The *product* RS and *intersection* $R \cdot S$ of R and S are defined by
$$RS =_{def} \{(x, z); (Ey)(xRy \ \& \ ySz)\}$$
and
$$R \cdot S =_{def} \{(x, y); xRy \ \& \ xSy\}$$
respectively. The *converse* $\breve{R}$ of R is defined by
$$\breve{R} =_{def} \{(x, y); yRx\}.$$

(iii) The *powers* R^n of R are defined inductively by
$$R^0 =_{def} id_D \qquad R^{n+1} =_{def} RR^n$$
The *transitive closure* R^* of R is
$$R^* =_{def} \sum \{R^n; n \in \mathbb{N}\}.$$

(iv) The *domain* $D'R$, the *converse domain* $\mathcal{A}'R$, and the *field* $C'R$ of a binary relation R are defined by
$$D'R =_{def} \{x; (Ey)(xRy)\}$$
$$\mathcal{A}'R =_{def} \{y; (Ex)(xRy)\}$$
$$C'R =_{def} D'R + \mathcal{A}'R.$$

Since relations are considered as sets, the inclusion relation between set induces an inclusion relation between binary relations: write $R \subseteq S$ if R is included in S.

It is quite easy to establish the following relations between binary relations.

EXAMPLE 5.1. For binary relations R, S, T,
$$R \cdot S = S \cdot R, \qquad R(S \cdot T) = RS \cdot RT, \qquad (S \cdot T)R = SR \cdot TR$$
$$\breve{RS} = \breve{S}\breve{R}, \qquad \breve{\breve{R}} = R$$
$$R \subseteq S \ \Rightarrow \ \breve{R} \subseteq \breve{S} \ \& \ TR \subseteq TS \ \& \ RT \subseteq ST$$

These results are in Peirce (1880, 1885). ◆

Equivalence relations

The notions concerning the concept of *order* are given in the following definition.

DEFINITION 5.2. Let D be a non-empty set and let R be a binary relation on D (i.e. $R \subseteq D \times D$). Then R is said to be:

(i) *reflexive* if $id_D \subseteq R$, i.e. xRx ($x \in D$);
(ii) *symmetric* if $\tilde{R} = R$, i.e. $xRy \Rightarrow yRx$ ($x, y \in D$);
(iii) *transitive* if $RR \subseteq R$, i.e.

$$xRy \ \& \ yRz \ \Rightarrow \ xRz \qquad (x, y, z \in D);$$

(iv) *anti-symmetric* if $R \cdot \tilde{R} \subseteq id_D$, i.e. $xRy \ \& \ yRx \Rightarrow x = y$;
(v) *asymmetric* if $R \cdot \tilde{R} = \emptyset$, i.e.

$$\text{not-}(R(x, y) \ \& \ R(y, x)) \qquad (x, y \in D);$$

(vi) an *equivalence relation* on D if it is reflexive, symmetric, and transitive;
(vii) *connected* if xRy or yRx ($x, y \in D$).

EXAMPLE 5.2.

(i) The relation $D \times D$ is an equivalence relation on D. It makes every element of D equivalent to every other element. Every equivalence relation on D is included in this relation. id_D is an equivalence relation on D and it is included in every equivalence relation on D.
(ii) The intersection of a class of equivalence relations on D is again an equivalence relation on D.
(iii) The product of two equivalence relations on D is also an equivalence relation on D.
(iv) For any relation R on D, R^* is transitive and reflexive. If R is a symmetric relation then R^* is an equivalence relation. ◆

The term 'transitive' in the above definition appears to have been first used by de Morgan (1856). The property of reflexiveness was so-called by Peano (1894) who showed that it cannot be inferred from symmetry and transitivity.

EXAMPLE 5.3. Let $\mathfrak{L} = \langle \Sigma, Cn \rangle$ be a logic. Implication relations imp_Δ, for $\Delta \subseteq \Sigma$, were defined in Definition 3.8; imp_Δ is reflexive and transitive (see Example 3.4(ii)). Logical equivalence Cn_Δ^e is exactly $imp_\Delta \cdot \widetilde{imp}_\Delta$ (Definition 3.10); it is an equivalence relation. Further, the relations imp_Δ (considered as sets of ordered pairs) reflect the ordering of the closed sets $Cn(\Delta)$ by inclusion:

$$A \ imp_\Delta \ B \ \Leftrightarrow \ Cn_\Delta(A) \supseteq Cn_\Delta(B)$$

$$imp_\Gamma \supseteq imp_\Delta \ \Leftrightarrow \ Cn(\Gamma) \supseteq Cn(\Delta) \qquad (\Delta, \Gamma \subseteq \Sigma). \qquad ◆$$

EXAMPLE 5.4.

(i) For any binary relation R, $R\breve{R}$ is symmetric.
(ii) Let S be a connected relation. If R is a relation such that $S \subseteq R$, then $S \cdot \breve{S} \subseteq R \cdot \breve{R}$.
(iii) If R is symmetric and transitive then R is reflexive on its field. ♦

Particular kinds of equivalence relations have long been known in mathematics. Similarity and congruence of triangles were known to ancient Greek geometers. A theory of identity was invented by Aristotle (Bocheński 1961, p. 92). Leibniz appreciated similarity and congruence as organizing principles in geometry. He was first influenced to reform geometry on Greek foundations, as against the Cartesian project, by his reading of Hobbes. These equivalence relations were fundamental principles of Leibniz's synthesis:

> We will reduce equality to congruence and ratio to similarity. Two things are equal if one can be transformed into the other.
> Couturat 1961, p. 152; Loemker 1969, p. 254

5.4 Abstraction

The importance of symmetric transitive relations derives from the following well-known and easily proved fact.

THEOREM 5.1. Let R be a symmetry transitive relation. For each $x \in C`R$ define $[x] =_{\text{def}} \{y; xRy\}$. Then we have the following.

(i) $x \in [x]$ $(x \in C`R)$.
(ii) For each $x, y \in C`R$ either $[x]$ and $[y]$ are disjoint ($[x] \cdot [y] = \varnothing$) or they coincide ($[x] = [y]$).
(iii) For each $x, y \in C`R$

$$xRy \Leftrightarrow [x] = [y].$$

The sets $[x]$ in the above theorem are traditionally known as *equivalence classes*. This result underlies the process which Peano called *definition by abstraction* (Peano 1894, p. 5, Russell 1903, p. 219). Given the symmetric transitive relation R, to each element x of the field of R one assigns a new entity $[x]$ which is identical with $[y]$ iff xRy. If $C`R$ is considered as a set of *points* then $[x]$ can be considered as a point in a new space formed by *identifying* all the poings in $[x]$. Thus the relation $R$ is represented as the relation of membership between an element of $C`R$ and its equivalence class and the relation of identity between equivalence classes (Theorem 5.2).

The mapping $x \to [x]$ is variously called the 'projection map' or 'canonical projection', and the set of equivalence classes is called the *quotient set* of $C`R$ and is denoted by $C`R/R$ (i.e. $C`R$ reduced by R). It is sometimes convenient to write $x = y\,mod(R)$ instead of xRy when R is an equivalence relation. This convention is particularly useful where the equivalence relation is compounded from other such relations. For instance, if S is also an equivalence relation then

$$x = y\,mod(R \cdot S) \quad \Leftrightarrow \quad x = y\,mod(R)\ \&\ x = y\,mod(S).$$

THEOREM 5.2. *Let R be a symmetric transitive relation. Let I be the identity relation on $C`R/R$ and $M$ the relation $\varepsilon$ of membership between an element $x$ of $C`R$ and its equivalence class $[x]$. Then $R = \varepsilon I \breve{\varepsilon}$.*

Proof. By Theorem 5.1(iii),

$$xRy \quad \Rightarrow \quad x \in [x]\ \&\ [x] = [y]\ \&\ y \in [y] \quad \Rightarrow \quad x(\varepsilon I \breve{\varepsilon})y.$$

Conversely, suppose that $x(\varepsilon I \breve{\varepsilon})y$. Then there exist equivalence classes $[s]$, $[t]$ such that $x \in [s]$, $[s] = [t]$, $y \in [t]$. Again by Theorem 5.1(iii), $x \in [s]$, sRt, and $y \in [t]$. Then by the definition of equivalence classes, xRs, sRt, and tRy. By the transitivity of R, xRy. Thus

$$x(\varepsilon I \breve{\varepsilon})y \quad \Rightarrow \quad xRy. \quad \blacksquare$$

An early use of abstraction was in Eudoxus' theory of ratio (Euclid, Books v, vi). Eudoxus did not define 'ratio'. He only said that it is a relation without magnitude. He then went on to define 'equality of ratio'. In modern terms, he constructed ordered pairs $a:b$ of magnitudes and then defined the relation of equality of ordered pairs. This relation is symmetric and transitive. He also defined an order relation $>$ between ratios, or, more exactly, a relation $>$ between ordered pairs which induces an order relation on the equivalence classes, i.e. order and equality satisfy the condition

$$x:y = x':y'\ \&\ x':y' > w:z\ \&\ w:z = w':z' \quad \Rightarrow \quad x:y > w':z'.$$

Eudoxus did not proceed to form equivalence classes.

Leibniz defined *space* by a formally similar method, first defining the symmetric and transitive relation of *same place* and then defining *place* by abstraction. Finally, space is the class of places.

Place is that, which we say is the same to A and to B, when the relation of coexistence of B with C, E, F, G etc. supposing there has been no cause of change in C, E, F, G etc. It may be said also, without entering into any further particularity, that place is that which is the same in different moments to different existent things, when their relations of coexistence with certain other existents, which are supposed to continue fixed from one of those moments to the other, agree entirely together.... lastly space is that which results from places taken together.... I have done much like Euclid, who, not being able to make his readers understand what ratio is absolutely in the sense of geometricians, defines what is the same ratios.

> Thus in like manner, in order to explain what place is, I have been content to define what is the *same place*.
>
> Alexander 1956, p. 69

Definitions by abstraction abound in modern mathematics—Frege's definition of cardinal and ordinal numbers, construction of integers from natural numbers, rational numbers from integers, real numbers from rationals, etc. These examples do not concern us here. However, the more algebraic examples—factor group, quotient rings, etc.—are relevant to our work. Abstraction from equivalence relations which are compatible in some sense with the relations of a relational system or the operations of an algebraic system preserves certain properties of the system.

DEFINITION 5.3.

(i) Let $\mathfrak{A} = \langle |\mathfrak{A}|, c_1, \ldots, c_n \rangle$ be an algebra. A binary relation $\sim$ on $|\mathfrak{A}|$ is a *congruence* if for each i and all $x_1, x_2, \ldots, y_1, y_2, \ldots \in |\mathfrak{A}|$

$$x_1 \sim y_1 \ \& \ x_2 \sim y_2 \ \& \ \cdots \ \Rightarrow \ c_i(x_1, x_2, \ldots) = c_i(y_1, y_2, \ldots).$$

If $\sim$ is a congruence on $|\mathfrak{A}|$ then the *reduced algebra* (*quotient algebra*) $\mathfrak{A}/\sim$ is the algebra $\mathfrak{A}/\sim \ =_{\text{def}} \langle |\mathfrak{A}|, f_1, \ldots, f_n \rangle$ where

$$f_i([x_1], [x_2], \ldots) =_{\text{def}} [c_i(x_1, x_2, \ldots)].$$

(It is easily shown that the f_i are well defined.)

(ii) Let $\mathfrak{D} = \langle |\mathfrak{D}|, R_1, R_2, \ldots \rangle$ be a relational system. A binary relation $\sim$ on $|\mathfrak{D}|$ is an *r-congruence* on $\mathfrak{D}$ if, for each i and all $x_1, x_2, \ldots, y_1, y_2, \ldots \in |\mathfrak{D}|$,

$$R_i(x_1, x_2, \ldots) \ \& \ x_1 \sim y_1 \ \& \ x_2 \sim y_2 \ \& \ \cdots \ \Rightarrow \ R_i(y_1, y_2, \ldots).$$

If $\sim$ is an equivalence relation on $|\mathfrak{D}|$ then the *reduced system* (*quotient system*) $\mathfrak{D}/\sim$ is the relational system $\langle |\mathfrak{D}|/\sim, S_1, S_2, \ldots \rangle$ where the relations $S_1, S_2, \ldots$ are defined by

$$S_i([x_1], [x_2], \ldots) \Leftrightarrow_{\text{def}} (Ey_1 \in [x_1])(Ey_2 \in [x_2]) \cdots R_i(y_1, y_2, \ldots)$$

for $i = 1, 2, \ldots$.

EXAMPLE 5.5.

(i) It is well known that for an algebra $\mathfrak{A}$ on which there is a congruence $\sim$ the canonical projection is a homomorphism of $\mathfrak{A}$ onto the reduced algebra $\mathfrak{A}/\sim$.
(ii) If in the relational system $\mathfrak{D}$ the relations $R_1, R_2, \ldots$ are binary relations then a necessary and sufficient condition that an equivalence relation E on $|\mathfrak{D}|$ be an r-congruence is $ER_iE \subseteq R_i$ for $i = 1, 2, \ldots$. ◆

A connection between the concept of congruence on an algebra and r-congruence on a relational system is established by replacing each operation of the *algebra* $\mathfrak{A}$ by its graph to obtain the *relational system* $Gr(\mathfrak{A})$. Thus, if

$\mathfrak{A} = \langle |\mathfrak{A}|, c_1, c_2, \ldots \rangle$ is an algebra, the graph of c_i is the relation R_i defined by

$$R_i =_{\text{def}} \{(x_1, x_2, \ldots, y); c_i(x_1, x_2, \ldots) = y\}.$$

Then $Gr(\mathfrak{A}) =_{\text{def}} \langle |\mathfrak{A}|, R_1, R_2, \ldots \rangle$. The next theorem follows from the definition of *congruence* and from Definition 5.3.

THEOREM 5.3. Let $\mathfrak{A} = \langle |\mathfrak{A}|, c_1, c_2, \ldots \rangle$ be an algebra. If $\sim$ is an r-congruence on $Gr(\mathfrak{A})$ then it is a congruence on $\mathfrak{A}$ such that, for $1 \leq i \leq k$,

$$x_1 \sim y_1 \ \& \ x_2 \sim y_2 \ \& \ \cdots \ \Rightarrow \ c_i(x_1, x_2, \ldots) = c_i(y_1, y_2, \ldots) \quad (5.2)$$

Thus, r-congruences correspond to special kinds of congruences, namely those which satisfy (5.2)—notice the equality sign on the right-hand side.

EXAMPLE 5.6. The converse of Theorem 5.3 is false. For example, let $\mathfrak{A} = \langle \{a, b, c\}, f \rangle$ where f is a unary function such that $f(a) = f(b) = f(c) = b$. Then $Gr(\mathfrak{A}) = \langle \{a, b, c\}, R \rangle$ where $R = \{(a, b), (b, b), (c, b)\}$. Consider the equivalence relation $\sim$ on $\{a, b, c\}$ whose equivalence classes are $\{a, b\}$ and $\{c\}$. Then $\sim$ is a congruence on $\mathfrak{A}$. But $R(c, b)$, $c \sim c$, $a \sim b$ holds but not $R(c, a)$, so that $\sim$ is not an r-congruence. ♦

5.5 Order

The importance of the concept of order in modern mathematics was clearly recognized by Russell. Indeed, one of the six parts of his *Principles of Mathematics* was devoted to the subject:

> The importance of order, from a purely mathematical standpoint, has been immeasurably increased by many modern developments ... Moreover, the whole philosophy of space and time depends on the view we take of order. Thus a discussion of order has become essential to any understanding of the foundations of mathematics.
>
> Russell 1903, p. 200

Russell concluded that all order depends upon transitive asymmetric relations:

> The minimum ordinal proposition, which can always be made whenever there is order at all, is of the form: 'y is between x and z'; and this proposition means: 'There is some asymmetrical transitive relation which holds between x and y and between y and z.'
>
> Russell 1903, p. 217

The most favoured way of studying order nowadays is via the related notion of the *partially ordered set* (Definition 5.4), though the wider notion of *quasi-ordering* is often more appropriate.

DEFINITION 5.4.

(i) A binary relation R on a non-empty set D is a *quasi-order* if R is transitive and reflexive; $\langle D, R \rangle$ is then a *quasi-ordered set* (*qoset*).
(ii) A binary relation R on a non-empty set D is a *partial ordering* if R is transitive, reflexive, and antisymmetric. $\langle D, R \rangle$ is then a *partially ordered set* (*poset*).

EXAMPLE 5.7. A binary relation on a set D is said to have the *Church–Rosser property* if, for all $a, b, c \in D$,

$$aRc \ \& \ bRc \ \Rightarrow \ (Ed \in D)(dRa \ \& \ dRb).$$

This is equivalent to the condition $R\tilde{R} \subseteq \tilde{R}R$. If R is a quasi-ordering which has the Church–Rosser property then $\tilde{R}R$ is symmetric and transitive, so by Example 5.4(iii) it is an equivalence relation on its field. This result is the essence of 'Newman's principle'. (See Börger 1989, Chapter A, Part I, §1.) ◆

The difference between Russell's concept of order (transitive, assymetric relation) and the concept of partial ordering is trivial as is easily proved in the following theorem.

THEOREM 5.4. Let R be a binary relation on a non-empty set D. Define binary relations S, T on D by

$$S = \{(x, y); R(x, y) \vee x = y\}$$
$$T = \{(x, y); R(x, y) \ \& \ x \neq y\}.$$

Then we have the following.

(i) If R is asymmetric, then S is antisymmetric.
(ii) If R is antisymmetric then T is asymmetric.
(iii) If R is transitive then so are S and T.

The term 'partially ordered set' (*teilweise geordnete Menge*) seems to have been first used by Hausdorff (1914, Chapter VI). Leibniz, Peirce, and Schröder studied particular examples (see Birkhoff 1948, p. 1n) of transitivity, reflexiveness, and antisymmetry, but they did not possess the general notion of ordering in the abstract sense of the above definition.

It was remarked previously that Leibniz proved the reflexiveness and transitivity of the containment relation between the terms of categorical propositions. He also proved the following properties of this relation, where in his notation $A = B$ means 'A coincides with B':

Prop. 5. If A is in B and $A = C$ then C is in B.
Prop. 6. If C is in B and $A = B$ then C is in A.
Prop. 7. A is in A.
Prop. 8. If $A = B$, A is in B.
Prop. 15. If A is in B and B is in C then A is in C.
Prop. 17. If A is in B and B is in A then $A = B$.

Parkinson 1966, Chapter 16

Proposition 17 looks like the antisymmetric law. Proposition 7 is the reflexive law and Proposition 15 states transitivity. It almost looks as if the relation of containment is a partial ordering—but not quite. Coincidence, however, can hold between terms which are not absolutely identical. Propositions 5 and 6 show that coincidence is an r-congruence on the system of terms ordered by containment. The quotient system is easily shown to be a partial ordering in which the containment relation becomes absolute equality so that Proposition 17 is the antisymmetric law.

Early in the history of the calculus of classes, Peirce studied the relation of inclusion of classes and saw that equality of classes could be defined in terms of inclusion:

Inclusion in or *being as small as* is a transitive relation. The consequence holds that

$$\text{if } x \prec y$$
$$\text{and } y \prec z$$
$$\text{then } x \prec z.$$

Equality is the conjunction of being as small as its converse. To say that $x = y$ is to say that $x \prec y$ and $y \prec x$.

Peirce 1870

Later, Schröder (1890) based his study of the inclusion relation on the reflexiveness and transitivity of this relation. Like Peirce, Schröder *defined* equality in terms of inclusion. From these two principles and the definition of equality it follows that equality is an r-congruence on the sets ordered by inclusion (Schröder 1890, Theorems 1–4, p. 184).

An easily proved, more general, statement relating equality and quasi-ordering is as follows.

THEOREM 5.5. Let $\mathfrak{D} = \langle D, R \rangle$ be a qoset and E an r-congruence on it such that $R \cdot \breve{R} \subseteq E$, i.e.

$$xRy \ \& \ yRx \ \Rightarrow \ xEy \qquad (x, y \in D). \tag{5.3}$$

Then $\mathfrak{D}/E$ is a poset.

Every qoset can be reduced to a poset because an r-congruence E satisfying (5.3) can always be constructed from R. This theorem was first proved by Schröder (1890, p. 184) (see also Birkhoff 1948, p. 4).

THEOREM 5.6. Let $\mathfrak{D} = \langle D, R \rangle$ be a qoset. Then $R \cdot \check{R}$ is an r-congruence on $\mathfrak{D}$ and $\mathfrak{D}/R \cdot \check{R}$ is a poset.

Proof.

(i) It follows from Definition 5.2 that $R \cdot \check{R}$ is an equivalence relation.
   (a) $R \cdot \check{R}$ is reflexive. For $id_D \subseteq R$. Hence $id_D = \widetilde{id_D} \subseteq \check{R}$. So $id_D \subseteq R \cdot \check{R}$.
   (b) $R \cdot \check{R}$ is symmetric. For $\widetilde{R \cdot \check{R}} = \check{R} \cdot \widetilde{\check{R}} = \check{R} \cdot R = R \cdot \check{R}$.
   (c) $R \cdot \check{R}$ is transitive. Since R is transitive, $RR \subseteq R$. Hence $\check{R}\check{R} \subseteq \check{R}$. So $(R \cdot \check{R})(R \cdot \check{R}) \subseteq RR \cdot \check{R}\check{R} = R \cdot \check{R}$.
(ii) $R \cdot \check{R}$ is an r-congruence since $(R \cdot \check{R})R(R \cdot \check{R}) \subseteq RRR \subseteq R$.
(iii) $R \cdot \check{R}$ satisfies (5.3) trivially. Hence by Theorem 5.5 $\mathfrak{D}/R \cdot \check{R}$ is a poset.

For any qoset $\mathfrak{D} = \langle D, R \rangle$, $R \cdot \check{R}$ will be called the *natural congruence*. If $\mathfrak{D} = \langle D, R \rangle$ is a poset then the natural congruence is id_D.

A graphical representation of finite qosets can be obtained from the notion of 'covering'. This notion was used for posets by Birkhoff who attributed it to Dedekind (1900).

DEFINITION 5.5. Let $\langle D, R \rangle$ be a qoset.

(i) Let $a, b, c \in D$. If aRb, bRc, $b \neq a$, and $b \neq c$, then b is between a and c.
(ii) When bRa, $b \neq a$, and no member of D is between a and b, it is said that *a covers b*.

Let $\langle D, R \rangle$ be a finite qoset: associate with each element $a \in D$ a distinct point $P(a)$ in the plane. Then if a covers b there is a directed line from $P(a)$ to $P(b)$. The points can be so arranged that if a covers b and b does not cover a then $P(a)$ is above $P(b)$ and there is a directed line from $P(a)$ to $P(b)$ (Fig. 5.1a). If a covers b and b covers a then the points $P(a)$ and $P(b)$ are on the same horizontal level, and the two lines from $P(a)$ to $P(b)$ and from $P(b)$ to $P(a)$ can be drawn coinciding (Fig. 5.1(b)). We remark that such a

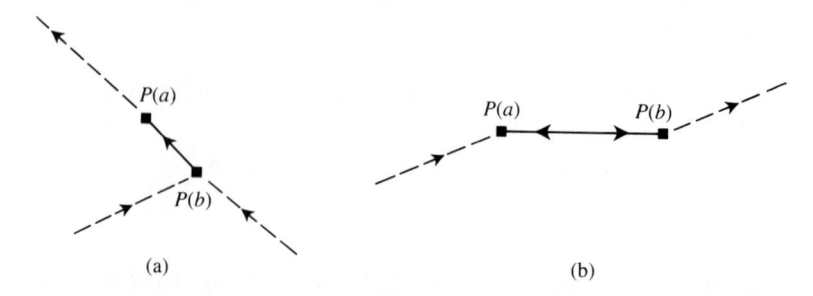

Fig. 5.1. The covering relation: (a) a covers b, b does not cover a; (b) a covers b, b covers a.

diagram of a poset cannot have two distinct points $P(a)$ and $P(b)$ on the same horizontal level joined by a line segment. For by the preceding definition of the diagram it would indicate that aRb and bRa, which by the antisymmetry law for partial ordering implies $a = b$. The highest and lowest points in these diagrams of qosets are in an obvious sense the maximal and minimal elements. The formal definitions are as follows.

DEFINITION 5.6. An element m of a qoset $\langle D, \leqslant \rangle$ is said to be:

(i)   a *maximal* element   if $m \leqslant x \Rightarrow x \leqslant m$ $(x \in D)$;

(ii) a *maximum* element   if $x \leqslant m$ $(x \in D)$;

(iii) a *minimal* element   if $x \leqslant m \Rightarrow m \leqslant x$ $(x \in D)$;

(iv) a *minimum* element   if $m \leqslant x$ $(x \in D)$.

EXAMPLE 5.8. Let $\mathfrak{L} = \langle \Sigma, Cn \rangle$ be a logic.

(i) The implication relation $imp_\varnothing$ is a quasi-ordering by Example 5.1. If $\mathfrak{L}$ has truth t (or falsehood f) then t (f respectively) is a maximum (minimum respectively) element in this ordering.

(ii) The passage from the set-to-set function Cn to the implication relation can be reversed to some extent if $\mathfrak{L}$ has conjunction so that the conjunction operation π (Chapter 4, §4.2.3) can be defined (Corollary 4.16). Now suppose that $\leqslant$ is a quasi-ordering on Σ and let $\equiv$ be the natural congruence. Suppose also that on Σ there is defined a two-place function $\wedge$ satisfying $\mathbf{L1}(\wedge)^0$, $\mathbf{L2}(\wedge)^0$, $\mathbf{L3}(\wedge)^0$ of Chapter 4, §4.2.3, and $A \wedge B \leqslant A$ $(A, B \in \Sigma)$. Define $Cn\colon \mathring{\mathbb{P}}_\omega(\Sigma) \to \mathbb{P}_\omega(\Sigma)$ by

$$Cn(\Gamma) =_{\text{def}} \{A; \pi(\Gamma) \leqslant A\} \qquad (\Gamma \in \mathring{\mathbb{P}}_\omega(\Sigma)).$$

Then $\langle \Sigma, Cn^\blacktriangledown \rangle$ is a logic in which $\leqslant$ coincides with the implication relation and $\wedge$ is conjunction.

(iii) Define the relation $\leqslant$ on $\mathbb{P}(\Sigma)$ by

$$\Delta \leqslant \Gamma \quad \Leftrightarrow_{\text{def}} \quad Cn(\Delta) \subseteq Cn(\Gamma).$$

Then $\langle \mathbb{P}(\Sigma), \leqslant \rangle$ is a qoset whose natural congruence is logical equivalence of sets (Definition 3.10). This qoset, reduced by logical equivalence, is a poset which is isomorphic to the partial ordering of closed sets. Figures 5.2, 5.3, and 5.4 each show the diagram of a finite logic as in Chapter 3, §3.10, the corresponding qoset, and the poset formed from this qoset by reduction with the natural congruence. In Fig. 5.2 there are three closed sets in the logic—$\{a, b, c\}$, $\{a, b\}$, and $\{a\}$. Further, $Cn(\varnothing) = Cn(a)$, $Cn(a, b) = Cn(b)$ and $Cn(a, b, c) = Cn(b, c) = Cn(a, c) = Cn(c)$. Therefore there should be horizontal lines connecting all the points corresponding to $\{a, b, c\}$, $\{b, c\}$, $\{a, c\}$, and $\{c\}$. We have omitted lines joining the points for $\{a, c\}$, $\{b, c\}$ and $\{a, b, c\}$, $\{c\}$, and adopted the subterfuge of a curved line connecting the points for $\{a, c\}$ and $\{c\}$.

(iv) The strong and the weak reduction relations between logics, defined in Chapter 3, §3.12, are quasi-orderings. ◆

150   Order and lattices

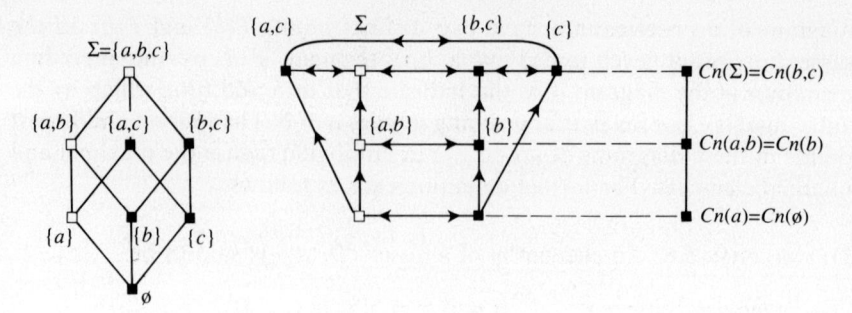

Fig. 5.2.

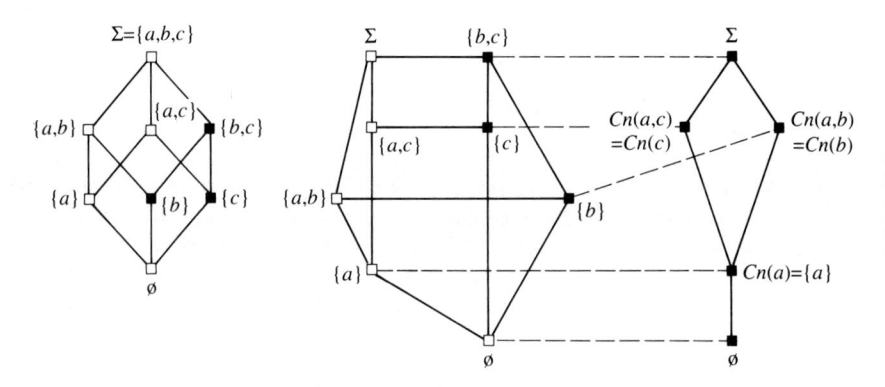

Fig. 5.3.

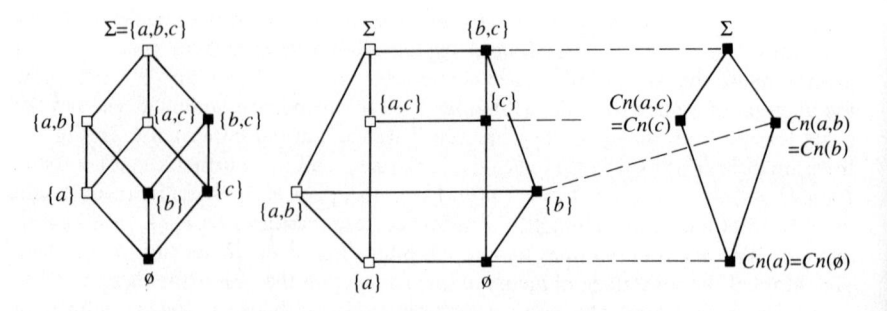

Fig. 5.4.

Upper and lower bounds

The relation between maximal and maximum elements is defined in the following easily proved statement.

THEOREM 5.7. Let $\mathfrak{D} = \langle D, \leqslant \rangle$ be a qoset. Then we have the following.

(i) Every maximum (minimum) element of $\mathfrak{D}$ is maximal (minimal).
(ii) If $m, n \in D$ are maximum (or minimum) elements then $n \leqslant m$ and $m \leqslant n$.
(iii) Let $m \in D$. If $\sim (s \leqslant m)$ for all $s \in D$, then m is minimal; if $\sim (m \leqslant s)$ for all $s \in D$, then m is maximal.
(iv) If D is finite then $\mathfrak{D}$ has a maximal element and a minimal element.
(v) If $\mathfrak{D}$ is a poset then there is at most one maximum element and at most one minimum element.

Note that not all infinite qosets have minimal or maximal elements the set of positive and negative integers under the usual ordering has neither maximal nor minimal elements. Figure 5.5 is the diagram of a qoset with

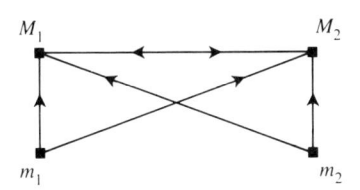

Fig. 5.5..

four elements M_1, M_2, m_1, m_2 in which M_1, M_2 are maximal elements and m_1, m_2 are minimal elements.

5.6 Upper and lower bounds

Many qosets that occur in logics have additional structure conferred on them by connectives. Conjunction and disjunction and Aristotelian quantifiers are concerned with the presence in the qoset of upper and lower bounds.

DEFINITION 5.7. Let $\langle D, \leqslant \rangle$ be a qoset. Let $X \subseteq D$.

(i) An element $c \in D$ is a *lower bound* of X if $c \leqslant x$ for all $x \in X$. $L(\leqslant : X)$ denotes the set of lower bounds of X. An element $d \in D$ is an upper bound of X if $x \leqslant d$ for all $x \in X$. $U(\leqslant : X)$ denotes the set of upper bounds of X.
(ii) An element $d \in D$ is a *least upper bound* of X if $d \in U(\leqslant : X)$ and d is a minimum element of $U(\leqslant : X)$, i.e. $d \leqslant y$ for all $y \in U(\leqslant : X)$. $\text{lub}(\leqslant : X)$ denotes the set of least upper bounds of X. Similarly an element $c \in D$

is a *greatest lower bound* of X if $c \in L(\leqslant : X)$ and c is a maximum element of $L(\leqslant : X)$, i.e. $y \leqslant c$ for all $y \in L(\leqslant : X)$. $glb(\leqslant : X)$ denote the set of least upper bounds of X. Where a fixed ordering $\leqslant$ is under consideration we write $L(X)$, $glb(X)$, etc. instead of $L(\leqslant : X)$, $glb(\leqslant : X)$ respectively.

EXAMPLE 5.9.

(i) Replacing the relation $\leqslant$ by its converse $\geqslant$ interchanges upper and lower bounds, i.e. $L(\geqslant : X) = U(\leqslant : X)$, $U(\geqslant : X) = L(\leqslant : X)$.
(ii) $U(\leqslant : \emptyset) = L(\leqslant : \emptyset) = D$ so that $lub(\leqslant : \emptyset)$ is the set of minimum elements and $glb(\leqslant : \emptyset)$ is the set of maximum elements. ◆

It follows immediately from these definitions that

$$c \in glb(\leqslant : X) \Leftrightarrow (x \in X)((c \leqslant x)$$
$$\& \ (y \in D)[(x \in X)(y \leqslant x) \Rightarrow y \leqslant c]) \quad (5.4)$$

$$d \in lub(\leqslant : X) \Leftrightarrow (x \in X)((x \leqslant d)$$
$$\& \ (y \in D)[(x \in X)(x \leqslant y) \Rightarrow d \leqslant y]). \quad (5.5)$$

EXAMPLE 5.10. Let $\mathfrak{L} = \langle \Sigma, Cn \rangle$ be a logic. Consider the qosets $\langle \Sigma, imp_\Delta \rangle$ where $\Delta \subseteq \Sigma$ are imp_Δ is the implication relation $\rightarrow_\Delta$ (Definition 3.8). Then for all $\Gamma \subseteq \Sigma$

$$C \in L(imp_\Delta : \Gamma) \Leftrightarrow (A \in \Gamma)(A \in Cn(\Delta, C))$$
$$D \in U(imp_\Delta : \Gamma) \Leftrightarrow (A \in \Gamma)(D \in Cn(\Delta, A)).$$

Further, by (5.2) and (5.3) respectively,

$$C \in glb(imp_\Delta : \Gamma) \Leftrightarrow (A \in \Gamma)(A \in Cn(\Delta, C))$$
$$\& \ [(A \in \Gamma)(A \in Cn(\Delta, B) \Rightarrow C \in Cn(\Delta, B)]$$

$$D \in lub(imp_\Delta : \Gamma) \Leftrightarrow (A \in \Gamma)(D \in Cn(\Delta, A))$$
$$\& \ [(A \in \Gamma)(B \in Cn(\Delta, A) \Rightarrow B \in Cn(\Delta, D)]. \quad \blacklozenge$$

The relations (5.4) and (5.5) become particularly simple when Γ is a set of two elements as is the case when conjunction and disjunction as two-place connectives are considered. Let $b, c \in D$. Then

$$d \in glb(\leqslant : \{b, c\}) \Leftrightarrow (a \in D)[a \leqslant b \ \& \ a \leqslant c \Leftrightarrow a \leqslant d] \quad (5.6)$$

$$d \in lub(\leqslant : \{b, c\}) \Leftrightarrow (a \in D)[b \leqslant a \ \& \ c \leqslant a \Leftrightarrow d \leqslant a]. \quad (5.7)$$

This enables the order relation of a qoset to be defined in terms of least upper bounds and greatest lower bounds.

THEOREM 5.8. Let $\langle D, \leqslant \rangle$ be a qoset. Let $b, c \in D$. Then

(i) $b \leqslant c \Leftrightarrow b \in glb(\leqslant : \{b, c\})$

(ii) $b \leqslant c \Leftrightarrow c \in lub(\leqslant : \{b, c\})$.

Proof.

(i) Suppose $b \in glb(\leqslant : \{b, c\})$. By (5.6)

$$a \leqslant b \,\&\, a \leqslant c \Leftrightarrow a \leqslant b \qquad (a \in S). \qquad (5.8)$$

In particular, putting $a = b$ in (5.8) we obtain $b \leqslant c$. Conversely, suppose $b \leqslant c$. Clearly, for all $a \in D$, $a \leqslant b \,\&\, a \leqslant c \Rightarrow a \leqslant b$. But suppose that $a \leqslant b$. As $b \leqslant c$, we have $a \leqslant c$. Thus $a \leqslant b \Rightarrow a \leqslant b \,\&\, a \leqslant c$. Hence (5.8) holds for all $a \in D$. By (5.4), $b \in glb(\leqslant : \{b, c\})$.

(ii) Similar to (i). ∎

THEOREM 5.9. Let $\langle D, \leqslant \rangle$ be a qoset and let $\equiv$ be the natural congruence. Then for each $X \subseteq D$, either $lub(\leqslant; X)$ is empty or is a $\equiv$-class. Likewise either $glb(\leqslant; X)$ is empty or is a $\equiv$-class.

Proof. It is sufficient to prove (5.9) and (5.10):

$$x, y \in glb(\leqslant : \{b, c\}) \Rightarrow x \equiv y \qquad (5.9)$$

$$x \in glb(\leqslant : \{b, c\}) \,\&\, y \equiv x \Rightarrow y \in glb(\leqslant : X). \qquad (5.10)$$

ad (5.9) Suppose $x, y \in glb(\leqslant : \{b, c\})$. Then $z \leqslant x$ for all $z \in L(\leqslant : X)$. But $y \in L(\leqslant : X)$. Hence $y \leqslant x$. Likewise $x \leqslant y$ and so $x \equiv y$.

ad (5.10) Suppose $x \in glb(\leqslant : X)$ and $y \equiv x$. Then $x \leqslant z$ for all $z \in X$. But as $y \equiv x$, $y \leqslant x$. As $\leqslant$ is transitive, $y \leqslant z$ for all $z \in L(\leqslant : X)$.

Next, let $z \in L(\leqslant : X)$. Then $z \leqslant x$ as $x \in glb(\leqslant : X)$. But as $y \equiv x$, $x \leqslant y$. Hence, as $\leqslant$ is transitive, $z \leqslant y$. Thus $z \leqslant y$ for all $z \in L(\leqslant : X)$. So $y \in glb(\leqslant : X)$. This proves (5.10). ∎

COROLLARY 5.10. Let $\langle D, \leqslant \rangle$ be a poset. Then for each $X \subseteq D$, $lub(\leqslant : X)$ and $glb(\leqslant : X)$ each contain at most one member.

Theorem 5.9 and Corollary 5.10 raise the question of the relation between upper and lower bounds in a qoset and upper and lower bounds in the system reduced by an r-congruence. It is quite easy to prove that upper and lower bounds are preserved by this process of reduction. The proof depends on the following lemma.

LEMMA 5.11. Let $\mathfrak{D} = \langle D, R \rangle$ be a qoset and $\sim$ an r-congruence on $\mathfrak{D}$. Let $\bar{R}$ be the corresponding relation on the quotient system $\mathfrak{D}/\sim$. Then for each $A, B \in D$

$$ARB \Leftrightarrow [A]\bar{R}[B].$$

Then directly from the definition of upper and lower bounds we have the following theorem.

THEOREM 5.12. Let $\mathfrak{D} = \langle D, R \rangle$ be a qoset and $\sim$ an r-congruence on $\mathscr{D}$. Let $\mathscr{D}/\sim = \langle D/\sim, \bar{R} \rangle$. Then for every $X \subseteq D$,

$$c \in L(R: X) \Leftrightarrow c \in L(\bar{R}: X/\sim)$$
$$d \in U(R: X) \Leftrightarrow d \in U(\bar{R}: X/\sim)$$
$$c \in glb(R: X) \Leftrightarrow c \in glb(\bar{R}: X/\sim)$$
$$d \in lub(R: X) \Leftrightarrow d \in lub(\bar{R}: X/\sim).$$

Example 5.8 raises the possibility of expressing quasi-orderings algebraically. Under certain conditions this can be done.

THEOREM 5.13. Let $\mathfrak{A} = \langle \Sigma, \cdot \rangle$ be an algebra and $\equiv$ a congruence on $\mathfrak{A}$ such that $x \equiv x \cdot x$, $x \cdot y \equiv y \cdot x$, $x \cdot (y \cdot z) \equiv (x \cdot y) \cdot z$ for all $x, y, z \in \Sigma$. The relation $\geqslant$ is defined on Σ by

$$x \geqslant y \Leftrightarrow_{def} y \equiv x \cdot y.$$

Then $\langle \Sigma, \geqslant \rangle$ is a qoset such that $x \cdot y \cdot \cdots \cdot z \in glb(\geqslant : \{x, y, \ldots, z\})$ and $\equiv$ is the natural congruence.

Proof. That $\langle \Sigma, \geqslant \rangle$ is a qoset follows easily from the identities. Let $w \in L(\{x, y, \ldots, z\})$. Then $x, y, \ldots, z \geqslant w$. Thus $w \equiv x \cdot w \equiv y \cdot w \cdots \equiv z \cdot w$. Since $\equiv$ is a congruence, by the identities, $w \equiv (x \cdot y \cdot \cdots \cdot z) \cdot w$. By the definition of the order relation, $x \cdot y \cdot \cdots \cdot z \geqslant w$. Thus $x \cdot y \cdot \cdots \cdot z \in glb(\geqslant : \{x, y, \ldots, z\})$. Finally, suppose that $x \geqslant y$ and $y \geqslant x$. Then by the identities, $y \equiv x \cdot y \equiv y \cdot x \equiv x$. Thus $\equiv$ is the natural congruence ∎

COROLLARY 5.14. Let $\mathfrak{L} = \langle \Sigma, \wedge, Cn \rangle$ be a logic with conjunction $\wedge$. The relation $\leqslant$ is defined on Σ by

$$A \geqslant B \Leftrightarrow_{def} B \leftrightarrow A \wedge B.$$

Then $\langle \Sigma, \geqslant \rangle$ is a qoset in which $A \wedge B \wedge \cdots \wedge C \in glb(\{A, B, \ldots, C\})$, $\geqslant$ coincides with $imp_\varnothing$ and the natural congruence coincides with logical equivalence. Further, for any $\Gamma \in \mathbb{P}_\omega(\Sigma)$

$$A \in Cn(\Gamma) \quad \Leftrightarrow \quad A \in U(\geqslant : \Gamma).$$

Clearly, normal conjunction and disjunction in a logic are closely connected with upper and lower bounds via the implication relations.

THEOREM 5.15. Let $\mathfrak{L} = \langle \Sigma, Cn \rangle$ be a logic and π, σ two functions on Σ such that, for each finite non-empty set $\Delta \leqslant \Sigma$, $\pi(\Delta), \sigma(\Delta) \in \Sigma$. Then π is normal conjunction iff **Cj3** holds, where

Cj3: $\pi(\Gamma) \in lub(imp_\Delta : \Gamma) \quad (\Gamma \in \overset{\circ}{\mathbb{P}}_\omega(D), \Delta \subseteq D)$.

σ is normal disjunction iff **Dj3** holds, where

Dj3: $\sigma(\Gamma) \in glb(imp_\Delta : \Gamma) \quad (\Gamma \in \overset{\circ}{\mathbb{P}}_\omega(D), \Delta \subseteq D)$.

Proof. By Example 5.8, **Cj3** is a rewording of **Cj1** and **Cj2** and **Dj3** is a rewording of **Dj1** and **Dj2**. ∎

The analogue of Theorem 5.15 for universal and existential quantifiers follows directly from Theorem 4.25(iv)–(vii), (5.4), and (5.5):

THEOREM 5.16. Let $\langle \mathfrak{A}, \exists, \forall, Cn \rangle$ be a quantifier logic with existential and universal quantifiers $\exists, \forall$ respectively. Let $\Delta \subseteq |\mathfrak{A}|$ be such that $Fv - Fv(\Delta)$ is infinite and let $x \in Fv - Fv(\Delta)$. Then for any $F \in |\mathfrak{A}|$,

$$\exists x(F) \in lub(imp_\Delta : x\text{-}var(F))$$

$$\forall x(F) \in glb(imp_\Delta : x\text{-}var(F)).$$

5.7 Lattices

Lattices are posets in which least upper bounds and greatest lower bounds of finite sets of elements always exist. Consideration of the almost trivial way in which logics can be manufactured from lattices is deferred to the next chapter. Here, we develop that part of lattice theory which enables us to see how lattices arise in logics.

DEFINITION 5.8. A poset $\langle D, \leqslant \rangle$ in which $lub(\leqslant : \{a, b\}) \neq \emptyset$ for all $a, b \in D$ is an *upper semi-lattice* and a poset $\langle D, \leqslant \rangle$ in which $glb(\leqslant : \{a, b\}) \neq \emptyset$ for all $a, b \in D$ is a *lower semi-lattice*. A poset $\langle D, \leqslant \rangle$ in which both $lub(\leqslant : \{a, b\}) \neq \emptyset$ and $glb(\leqslant : \{a, b\}) \neq \emptyset$ is a *lattice*.

It follows from Theorem 5.9 that if $\langle D, \leqslant \rangle$ is a lattice then $lub(\leqslant : \{a, b\})$ and $glb(\leqslant : \{a, b\})$ are both unit sets for all $a, b \in D$. Hence there exist two functions $\cup, \cap : D \times D \to D$ such that for all $a, b \in D$, $a \cup b$ and $a \cap b$ are the unique members of $lub(\leqslant : \{a, b\})$ and $glb(\leqslant : \{a, b\})$ respectively. The functions $\cup$ and $\cap$ are often called *join* and *meet* respectively. They enable the *lub* and *glb* of any finite set of elements to be computed:

$$x_1 \cup (x_2 \cup \cdots \cup (x_{n-1} \cup x_n) \cdots) \in lub(\leqslant : \{x_1, \ldots, x_n\})$$
$$x_1 \cap (x_2 \cap \cdots \cap (x_{n-1} \cap x_n) \cdots) \in glb(\leqslant : \{x_1, \ldots, x_n\}).$$

It should be remarked here that, according to our definition, *lub* and *glb* are sets of elements. This was forced on us because of the application to qosets. Hence *lub* and *glb* are unit sets when defined on posets, whereas in the traditional treatments of lattice theory *lub* and *glb* are initially defined for posets and hence are defined as elements, namely the unique members of what we have termed *lub* and *glb*. When dealing with posets we shall henceforth revert to traditional usage and write $x = lub(X)$ instead of $x \in lub(X)$.

For brevity of expression we shall express the meet operation simply by juxtaposition and use the convention that meet is a stronger binding than join. Thus $xy \cup zw$ stands for $(x \cap y) \cup (z \cup w)$. The reader can easily restore the symbol $\cap$ with its attendant brackets. This notation, however, tends to obscure the symmetry between join and meet; where this needs to be emphasized $\cap$ will be used.

DEFINITION 5.9. Let $\mathfrak{D}$ be a collection of subsets of a given set M. If $\mathfrak{D}$ is closed under intersection and union then $\mathfrak{D}$ is said to be a *ring of sets*; if $\mathfrak{D}$ is a ring of sets and is also closed under complementation with respect to M then it is said to be a *field of sets*.

EXAMPLE 5.11. Every ring of sets is a lattice in which meet and join are defined by $X \cap Y =_{def} X \cdot Y$ and $X \cup Y =_{def} X + Y$ respectively. ◆

EXAMPLE 5.12.

(i) Let $\mathfrak{L} = \langle \Sigma, Cn \rangle$ be a logic and D its set of closed sets. Thus for $\Delta \in D$, $Cn(\Delta) = \Delta$. Let $\subseteq$ denote the set-inclusion relation on D. Then $cl(\mathfrak{L}) =_{def} \langle D, \subseteq \rangle$ is a lattice (cf. Examples 3.7 and 5.8(iii)) which we call the *lattice of closed sets* of $\mathfrak{L}$. It can easily be checked that, for $\Delta, \Gamma \in D$,

$$\Delta \cdot \Gamma = glb(\subseteq : \{\Delta, \Gamma\}) \qquad Cn(\Delta, \Gamma) = lub(\subseteq : \{\Delta, \Gamma\}).$$

Thus $\Delta \cdot \Gamma = \Delta \cap \Gamma$ and $Cn(\Delta, \Gamma) = \Delta \cap \Gamma$. The maximum element of $cl(\mathfrak{L})$ is Σ and the minimum element is $Cn(\emptyset)$. The behaviour of the connectives in $\mathfrak{L}$ has considerable influence on the structure of $cl(\mathfrak{L})$ (Corollary 5.30).

(ii) Let $\langle \Sigma, \wedge, \vee, Cn \rangle$ be a logic with normal conjunction and disjunction. Then for every $\Gamma \subseteq D$, $\langle \Sigma, imp_\Gamma \rangle$ is a qoset (Example 5.8, Theorem 5.15) in which

Lattices

$A \vee B \in glb(imp_\Gamma : \{A, B\})$ and $A \wedge B \in lub(imp_\Gamma : \{A, B\})$. By Theorem 5.6 $\langle \Sigma, imp_\Gamma \rangle / imp_\Gamma \cdot \widetilde{imp}_\Gamma$ is a lattice (recall that by Example 5.3, $imp_\Delta \cdot imp_\Gamma = Cn_\Gamma^e$). Let $[A]_\Gamma$ denote the logical equivalence class $mod(Cn_\Gamma^e)$ of A. Thus

$$B \in [A]_\Gamma \Leftrightarrow Cn(\Gamma, A) = Cn(\Gamma, B).$$

Hence in this lattice

$$[A]_\Gamma \cap [B]_\Gamma = [A \wedge B]_\Gamma \qquad [A]_\Gamma \cup [B]_\Gamma = [A \vee B]_\Gamma.$$

Thus normal disjunction and normal conjunction determine least upper bounds and greatest lower bounds in the lattices $\langle \Sigma, imp_\Gamma \rangle / imp_\Gamma \cdot \widetilde{imp}_\Gamma$ for $\Gamma \subseteq D$. ◆

EXAMPLE 5.13. Figure 5.6 shows the diagrams of some finite lattices. ◆

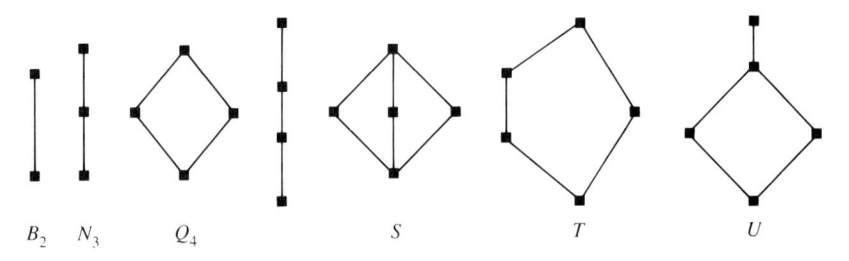

$B_2 \qquad N_3 \qquad Q_4 \qquad\qquad S \qquad\qquad T \qquad\qquad U$

Fig. 5.6.

We now examine some identities satisfied by the lattice operations $\cup$ and $\cap$. From (5.6), (5.7), and Theorem 5.8 we have the following lemma.

LEMMA 5.17. Let $\langle L, \leqslant \rangle$ be a lower semi-lattice. Then for all $x, y, p \in L$

(i)(a) $y \leqslant x \Leftrightarrow xy = y$

  (b) $xy \leqslant x, y$

  (c) $p \leqslant x, y \Rightarrow p \leqslant xy$

(ii)(a) $x \leqslant y \Leftrightarrow x \cup y = y$

  (b) $x, y \leqslant x \cup y$

  (c) $x, y \leqslant p \Rightarrow x \cup y \leqslant p$.

Let $\geqslant$ denote the converse of $\leqslant$. Then comparison of the form of the statements (i)(a) with (ii)(a), (i)(b) with (ii)(b), ... reveals that (i)(a) can be obtained from (ii)(a), (i)(b) from (ii)(b), ... by replacing each occurrence of the symbols $\cup$, $\cap$, and $\leqslant$ by $\cap$, $\cup$, and $\geqslant$ respectively. This is an example of a very general phenomenon in lattice theory known as *duality*. It is of great significance but is based on quite trivial observations: in fact, by immediate derivation from the definitions of qoset, lub, glb, etc.

THEOREM 5.18 (Duality). Let $\mathfrak{S} = \langle L, \leqslant \rangle$ and $\breve{\mathfrak{S}} = \langle L, \geqslant \rangle$ where $\leqslant$ is the converse of $\geqslant$.

(i) $\mathfrak{S}$ is a qoset iff $\breve{\mathfrak{S}}$ is a qoset.
(ii) if $\mathfrak{S}$ is a qoset then for every $X \subseteq L$

$$lub(\leqslant : X) = glb(\geqslant : X) \qquad \text{and} \qquad glb(\leqslant : X) = lub(\geqslant ; X).$$

(iii) $\mathfrak{S}$ is a lattice iff $\breve{\mathfrak{S}}$ is a lattice.
(iv) If x is a maximal (maximum) element of $\mathfrak{S}$ then x is a minimal (minimum) element of $\breve{\mathfrak{S}}$; if x is a minimal (minimum) element of $\mathfrak{S}$ then x is a maximal (maximum) element of $\breve{\mathfrak{S}}$.

There follows the useful Duality principle:

If $S(\cup, \cap, \leqslant)$ is a statement which holds for all lattices then so is $S(\cap, \cup, \geqslant)$.

$S(\cup, \cap, \leqslant)$ is the *dual statement* of $S(\cap, \cup, \geqslant)$. Observe that duality of statements is a symmetric relation.

The weakness of this enunciation of the duality principle is that the term 'statement' is undefined. We need to define the term in the context of a language. There are many such languages that can be used to define conditions on lattices—quantifier-free first-order languages, first-order predicate languages, higher-order languages, etc. As we shall mostly be concerned with identities we shall assume that the non-formally expressed conditions are expressed in a suitable formal language. Thus, the statements (i)(a), (i)(b), and (i)(c) of Lemma 5.17 hold in all lattices. The dual statements (ii)(a), (ii)(b), and (ii)(c) respectively therefore also hold in all lattices.

The essential connection between albebraic operations and order has already been stated in Theorem 5.14. Its extension to lattices yields the following theorem.

THEOREM 5.19. Let $\langle L, \leqslant \rangle$ be a lattice with *glb* and *lub* operations $\cap, \cup$ respectively. Then the algebra $\langle L, \cap, \cup \rangle$ satisfies

K0: $x \leqslant y \Leftrightarrow x = x \cap y \Leftrightarrow y = x \cup y$

K1: $x \cap y = y \cap x, \qquad x \cup y = y \cup x$

K2: $x \cap (y \cap z) = (x \cap y) \cap z, \qquad x \cup (y \cup z) = (x \cup y) \cup z$

K3: $x \cap (x \cup y) = x, \qquad x \cup (x \cap y) = x$

K4: $x = x \cap x, \qquad x = x \cup x.$

If, further, the lattice has a maximum element 1 and a minimum element 0 then

K5: $0 \cap x = 0, \quad 0 \cup x = x, \quad 1 \cap x = x, \quad 1 \cup x = 1.$

Proof. **K0–K5** follow from Lemma 5.17. The second identity in each pair **K1–K4** follows by duality from the first.

ad **K0** This follows from (i)(a) and (ii)(a).

ad **K1** $yz \leqslant x, y$ by (i)(b) and $yx \leqslant xy$ by (i)(c). Hence $xy \leqslant yx$. Thus $yx = xy$.

ad **K2** By (i)(b), $x(yz) \leqslant x$, yz and $yz \leqslant z$. Hence $x(yz) \leqslant x, y, z$. But then, by (c), $x(yz) \leqslant (xy)$ and $z \leqslant (xy)z$. By a similar argument $(xy)z \leqslant x(yz)$. Hence $x(yz) = (xy)z$.

ad **K3** By (ii)(b) $x \leqslant x \cup y$. (i)(c) gives $x \leqslant x(x \cup y)$. Also by (i)(b), $x(x \cup y) \leqslant x$ so that $x = x(x \cup y)$.

ad **K4** By (i)(a).

ad **K5** If 0 is the minimum element then $0 \leqslant x$ for all $x \in L$. Thus, $0 = glb(\leqslant : \{x, 0\})$ and $x = lub(\leqslant : \{x, 0\})$. Hence $0 = x0$ and $x = x \cup 0$. By duality, $1 = x \cup 1$ and $x = x1$. ∎

The identities **K1–K4** characterize lattices, and the relation **K0** enables the order relation of the lattice to be defined in terms of the **glb** and **lub** operations.

THEOREM 5.20. *Let* $\langle L, \cap, \cup \rangle$ *be an algebra (Appendix, §A) of type* $\langle 2, 2 \rangle$ *satisfying* **K1–K4**. *Define the relation* $\leqslant$ *on* L *by*

$$y \leqslant x \Leftrightarrow_{def} x \cap y = y.$$

Then $\langle L, \leqslant \rangle$ *is a lattice with* $x \cap y = glb(\leqslant : \{x, y\})$ *and* $x \cup y = lub(\leqslant : \{x, y\})$.

Proof. First, it follows from **K3** that $xy = y \Rightarrow x \cup y = x$. Thus the ordering defined by meet coincides with that defined by join. The theorem then follows from Theorem 5.13. ∎

Theorems 5.19 and 5.20 justify our referring to an algebra $\langle L, \cap, \cup \rangle$ of type $\langle 2, 2 \rangle$ as a lattice if it satisfies **K1–K4**. An algebra $\langle L, \cap, \cup, 0, 1 \rangle$ of type $\langle 2, 2, 0, 0 \rangle$ satisfying **K1–K5** will be called a *lattice with maximum and minimum elements*.

EXAMPLE 5.14. Let $\mathfrak{L} = \langle \Sigma, \wedge, \vee, Cn \rangle$ be a logic with normal conjunction $\wedge$ and normal disjunction $\vee$. By Theorem 4.15 the Lindenbaum algebra (Example 3.10(ii)), $\mathfrak{L}\P = \langle \Sigma, \wedge, \vee \rangle / Cn^e$ satisfies the lattice laws

$$[A] \cap [A] = [A], \qquad [A] \cup [A] = [A]$$
$$[A] \cap [B] = [B] \cap [A], \qquad [A] \cup [B] = [B] \cup [A]$$
$$[A] \cap ([B] \cap [C]) = ([A] \cap [B]) \cap [C]$$
$$[A] \cup ([B] \cup [C]) = ([A] \cup [B]) \cup [C]$$
$$[A] \cap ([A] \cup [B]) = [A], \qquad [A] \cup ([A] \cap [B]) = [A]$$

where $[A]$ denotes the logical equivalence class of A. The Lindenbaum algebra is therefore a lattice. The lattice operations and relation are related to Cn by $A \in Cn(B) \Leftrightarrow A \wedge B \leftrightarrow B$. Further, $A \in Cn(B) \Leftrightarrow [A] \geq [B]$. It is a peculiarity of normal conjunction and disjunction that, corresponding to **L5.1** and **L5.2** of Chapter 4, §4.2.3, the Lindenbaum algebra also satisfies the *distributive* laws:

$$[A] \cap ([B] \cup [C]) = ([A] \cap [B]) \cup ([A] \cap [C])$$
$$[A] \cup ([B] \cap [C]) = ([A] \cup [B]) \cap ([A] \cup [C]). \qquad \blacklozenge$$

The identities **K1–K5** can be taken as axioms for lattices. Other formulations of the axioms are possible. Thus a structure $\langle L, \cap, \cup, \leq \rangle$ is a lattice iff $\leq$ is a reflexive, transitive, and antisymmetric relation and the operations $\cup$ and $\cap$ satisfy the conditions

$$x \leq y \,\&\, x \leq z \;\Leftrightarrow\; x \leq y \cap z$$
$$y \leq x \,\&\, z \leq x \;\Leftrightarrow\; y \cup z \leq x.$$

These axioms were formulated by Peirce (1880, 1885) and Schröder (1890). But neither author was concerned with the abstract concept of ordered sets. They were both intent on axiomatizing the classical two-valued logic. Peirce's axioms were intended to characterize the classical propositional calculus: the relation symbol $\prec$ in his system was an implication *connective* (Peirce 1880, p. 21). Schröder, however, was studying the form of 'categorical judgements' (Schröder 1890, p. 132) which are of the form 'Subject $\in$ Predicate' (Schröder used $\in$ where we would use $\leq$):

> Here, so we assert, we can always regard Subject and Predicate and represent the logical content of a judgement by interpreting it as an assertion: the subject class is wholly contained in the predicate class.
>
> Schröder 1890, p. 147

Schröder therefore developed his algebra of logic by formulating axioms for the 'identity calculus with regions of a manifold', i.e. for subsets of a given set (Schröder 1890, p. 157). An *abstract* definition of 'lattice' goes back to the work of Dedekind (1897) who first conceived of modules and ideals of a ring as elements of a lattice in order to obtain an exact view of their decomposition properties. Dedekind recorded how his theory was influenced by Schröder's *Verlesungen über die Algebra der Logik* (Dedekind 1897, p. 112), but whereas Schröder was concerned with classical logic, Dedekind used **K1–K4** as axioms—any system of objects with operations $\cap, \cup$ satisfying these axioms was to be called a *dual group*. The modern German term *Verband* is due to Fritz Klein. The English term 'lattice' has been attributed to Birkhoff (Hermes and Köthe 1939).

Dedekind's work on lattices remained unnoticed for about 30 years. Without knowledge of Dedekind's work, Klein (1932) again put forward the idea of lattices as a basic notion of similar significance to that of groups.

Lattice theory has indeed found many important applications in mathematics, as is clear from the earlier survey article by Hermes and Köthe.

5.8 Complete lattices

There exist lattices in which greatest lower bounds and least upper bounds exist not only of finite sets of elements but of arbitrary sets. These are the *complete lattices*.

DEFINITION 5.10. An upper semi-lattice is said to be *complete* if every set of its elements has a *lub*. Dually, a lower semi-lattice is said to be *complete* if every set of elements has a *lub* or a *glb*.

A complete upper semi-lattice $\mathfrak{L}$ has a maximum element $1 = lub(|\mathfrak{L}|)$. Dually, a complete lower semi-lattice $\mathfrak{L}$ has a minimum element $0 = glb(|\mathfrak{L}|)$.

THEOREM 5.21. Every complete lower (upper) semi-lattice with maximum (minimum) element is a complete lattice, i.e. every set of elements has a *lub* (*glb*).

Proof. Let $\mathfrak{L} = \langle |\mathfrak{L}|, \geqslant \rangle$ be a complete lower semi-lattice. Let X be a non-empty subset of $|\mathfrak{L}|$ and let $U = U(X)$, i.e.

$$x \in U \iff (y \in X)(x \geqslant y).$$

$U \neq \emptyset$. Hence $a = glb(U)$ exists. Then $a = lub(X)$. For suppose that $b \in X$. Then $x \in U \Rightarrow x \geqslant b$. Thus $b \in L(U)$. Hence $a \geqslant b$. Thus $a \in U(X)$. Now let $z \in U$ and suppose that $z \in U(X)$. Then $z \geqslant y$ for all $y \in X$. Then $z \geqslant a$ as $a \in L(U)$. Hence $a = lub(X)$. ∎

EXAMPLE 5.15.
  (i) For a given set X, $\mathbb{P}(X)$, ordered by inclusion, is a complete lattice.
  (ii) The closed subsets (dually, open) of any topological space form a complete lattice (Appendix, §T).
  (iii) The subalgebras of a given algebra form a complete lattice. Likewise, the congruence relations of an algebra form a complete lattice. (The congruence relations, considered as sets of ordered pairs, are ordered by inclusion.) Thus the set of normal subgroups of a given group form a lattice. Likewise the set of ideals of a given ring form a lattice. These lattices embody the decomposition properties of normal subgroups and ideals respectively (Cohn 1965, p. 86).
  (iv) The closed subspaces of a Hilbert space form a lattice.
  (v) Every finite lattice is complete.
  (vi) The lattice of closed sets of a logic $\mathfrak{L}$ is complete (Theorem 3.7). ◆

Order and lattices

The followig three results concern fixed points of an isotone funcion f from a complete lattice $\mathfrak{L}$ to itself. An isotone function f on $\mathfrak{L}$ is one which satisfies the condition

$$x \leqslant y \Rightarrow f(x) \leqslant f(y) \qquad (x, y \in |\mathfrak{L}|).$$

Let $\Omega(f)$ denote the set of fixed points of f, i.e. $\Omega(f) = \{a; a = f(i)\}$.

THEOREM 5.22 (Fixed-point theorem). $\Omega(f) \neq \emptyset$.

Proof. Let $K = \{x; x \leqslant f(x)\}$ and let $a = lub(K)$. Since $0 \leqslant f(0)$, a exists. Since f is isotone, $a \geqslant x$ for all $x \in K$. Thus

$$f(a) \geqslant f(x) \geqslant x \qquad (x \in K).$$

Hence $f(a) \geqslant lub(K) = a$. As f is isotone, $f(f(a)) \geqslant f(a)$. Hence $f(a) \in K$ and so, as $a = lub\ K$, $a \geqslant f(a)$. Thus $a = f(a)$. ∎

THEOREM 5.23. If $X \subseteq \Omega(f)$ then $glb(X) \in \Omega(f)$.

Proof. $glb(X)$ exists as the lattice is complete. As f is monotone $f(glb\ X) \leqslant f(x) = x$ $(x \in X)$, i.e. $f(glb(X)) \in L(X)$. So $f(glb(X)) \leqslant glb(X)$. As f is isotone, $glb(X) \leqslant f(glb(X)$. Hence $f(glb(X)) = glb(X)$. Thus $glb(X) \in \Omega(f)$. ∎

COROLLARY 5.24. f has a least fixed point.

We now give a thumbnail sketch of Scott's (1970) theory of data types (see also Scott 1976b, Stoy 1977): complete lattices play an important part. Data types and mappings between them are axiomatized as follows.

AXIOM 1. A data type is a poset.

The order relation $x \leqslant y$ is understood to mean that y is *consistent with* x and *more accurate than* x, and it is said that x *approximates* y though this usage should not necessarily be taken in the usual metric sence of *approximation*, i.e. of x being within ε of y. And if x, y approximate the same entity then y gives more *information* about it than x does.

Let f be a mapping of the data type D (with approximation ordering $\leqslant$) into the data type $D^{\neq}$ (with approximation ordering $\leqslant^{\neq}$). If f were defined by a computation procedure then the more accurate (in the above sense) the argument of f the more accurate would be the value, i.e. if $x \leqslant y$ then $f(x) \leqslant^{\neq} f(y)$. Thus f is monotone with respect to partial ordering.

Various kinds of lattices

AXIOM 2. Mappings between data types are monotonic.

If $x_0 \leqslant x_1 \leqslant \cdots \leqslant x_n \leqslant x_{n+1} \leqslant \cdots$ is an infinite sequence of 'approximations', then it is assumed that it actually approximates something, i.e. the sequence tends to a *limit* y which is a member of the data type. Thus $y = lub\{x_n; n \in \mathbb{N}\}$. For mathematical simplicity it is assumed that *every* subset of the data type has a *lub*. Thus, we have the following axiom (see Theorem 5.23).

AXIOM 3. A data type is a complete lattice under its partial ordering.

The maximum element 1 of the complete lattice is regarded as the *most overdetermined element* so that $x \cup y = 1$ means that x and y are *inconsistent*. The minimum element 0 is seen as the *most underdetermined element*: $xy = 0$ means that x and y are *unconnected* in the sense that there is no overlap of information between them.

Topological concepts are introduced via the idea of a *directed set*. A subset X of the data type D is *directed* if every finite subset of X has an upper bound in X (note that every directed set is non-empty). The *limit* of the directed set is defined as $lub(X)$. Then a mapping f which preserves limits is *continuous* (Appendix, §T), i.e. $f(lub(X)) = lub\{f(x); x \in X\}$.

AXIOM 4. Mappings between data types are continuous.

A topology is defined on a data type D by specifying the open sets. $U \subseteq D$ is open under the conditions:

(i) $x \in U$ & $x \leqslant y$ $\Rightarrow$ $y \in U$;

(ii) if $X \subseteq D$ is directed and $lub(X) \in U$ then $X \cdot U \neq \emptyset$.

Then $F: D \to D^{\neq}$ is continuous in the topological sense iff it is continuous in the above sense of preserving limits.

5.9 Various kinds of lattices

We have seen that a lattice can be considered as an algebra of type $\langle 2, 2 \rangle$ satisfying the identities **K1**–**K4**, and that a lattice with maximum and minimum elements can be considered as an algebra of type $\langle 2, 2, 0, 0 \rangle$ satisfying **K1**–**K5** (Theorem 5.20). Various kinds of lattice can be defined by adding further conditions to this list of axioms, for example, the distributive laws in Example 5.14. These are the most important type from the logical point of view (§5.9.1): the meet and join operations are comparable to normal conjunction and disjunction in logics.

Other kinds of lattices can be constructed by equipping them with additional operators which are constrained to behave in certain ways relating to join and meet. The particular concern of the following sections is the unary operation of *complementation*, though one type of it will be derived from a binary operation of *relative complementation* (§5.10.2). This form of complementation leads to Skølem's concepts of *I-lattices* and *classical I-lattices* (§5.10.2) which are related to intuitionistic logic and classical logic respectively. Another form of complementation leads to *Boolean algebras* (§5.10.6), which are again related to classical logic, and to *quasi-Boolean algebras*, which are similarly related to the three- and four-valued logics discussed in Chapter 8.

Occasionally the additional conditions are most naturally expressed in more complicated terms than identities. In such cases it is of some interest, in view of Birkhoff's theorem (Appendix, §A), to be able to replace the logically more complex conditions by conditions of least logical complexity, i.e. identities. This cannot always be done, but it is possible where the condition is expressible in quantifier-free conditional form:

$$x = t \Rightarrow E$$

where t is a term in the appropriate language, x is a free variable, and E is a quantifier-free formula. Lemma 5.25 follows from the definition of satisfaction (Chapter 2).

LEMMA 5.25. Let x_i, t_i for $1 \leqslant i \leqslant n$ be distinct free variables and terms respectively in a language of type τ and let E be a formula of this language. Then the conjunction of the identities $x_i = t_i$ implies that the formula $x_1 = t_1 \ \& \ \cdots \ x_n = t_n \Rightarrow E$ is equivalent to $E(t_1, \ldots, t_n / x_1, \ldots, x_n)$.

THEOREM 5.26. Let $t_1, \ldots, t_{n+m}$ be terms and $x_1, \ldots, x_{n+m}$ be distinct free variables. Then **K1**–**K5** imply that, for every formula E, the formula

F: $\quad x_1 \leqslant t_1 \ \& \ \cdots \ \& \ x_n \leqslant t_n \ \& \ x_{n+1} \leqslant t_{n+1} \ \& \ \cdots \ x_{n+m} \leqslant t_{n+m} \Rightarrow E$

is equivalent to the formula

G: $\quad E(x_1 t_1, \ldots, x_n t_n, x_{n+1} \cup t_{n+1}, \ldots, x_{n+m} \cup t_{n+m} / x_1, \ldots, x_{n+m})$.

Proof. **K1**–**K5** are the lattice axioms. Hence they imply $x_i = x_i t_i$ for $1 \leqslant i \leqslant n$ and $x_i = x_i \cup t_i$ for $n + 1 \leqslant 1 \leqslant n + m$. The equivalence of **F** and **G** follows from lemma 5.25. ∎

EXAMPLE 5.16. A necessary and sufficient condition for a lattice to satisfy the condition

$$x(y \cup z) \leqslant xy \cup z \qquad (5.11)$$

is that, for all elements x, y, z of the lattice,

$$xy \leqslant z \ \& \ x \leqslant y \cup z \ \Rightarrow \ x \leqslant z. \qquad (5.12)$$

This follows from Theorem 5.27 because the result of substituting $x(y \cup z)$ for x and $xy \cup z$ for z in $x \leqslant z$ is (5.11). (5.12) is, in fact, equivalent to the distributive law (Lorenzen 1951), which is the subject of the next section. ♦

5.9.1 Distributive lattices

Dedekind (1897) regarded the following lattice identities as particularly important.

M: $\quad x(y \cup z) \cup yz = (x \cup yz)(y \cup z)$

K6.1: $\quad xy \cup xz = x(y \cup z)$

K6.2: $\quad (x \cup y)(x \cup z) = x \cup yz$

He called **M** the 'modular law' (*Modulgesetz*) because all submodules of an R-module are modular, i.e. satisfy **M**. He called **K6.1**, now known as the 'distributive law', the 'ideal law' (*Idealgesetz*) as the ideals of a finite field satisfy **K6.1**. The modular law **M**, as expressed by Dedekind, is expressed differently nowadays (e.g. Birkhoff 1948, Cohn 1965, p. 65).

Dedekind counted the modular law as more significant than the distributive laws **K6.1** and **K6.2**. However, the distributive laws are more important in logic. For instance, by Example 5.14 they hold in logics with normal conjunction and disjunction. We therefore proceed with the development of the theory of distributive lattices, i.e. lattices satisfying **K6.1**.

First observe that every lattice satisfies the 'semi-distributive laws'.

THEOREM 5.27. Every lattice satisfies the conditions

K7.1: $\quad xy \cup xz \leqslant x(y \cup z)$

K7.2: $\quad (x \cup y)(x \cup z) \geqslant x \cup yz$

Proof. By Lemma 5.17, $xy, xz \leqslant x(y \cup z)$. Hence $xy \cup xz \leqslant x(y \cup z)$. **K7.2** is proved by duality. ∎

Not every lattice is distributive: the lattices S and T of Fig. 5.6 are not distributive. Moreover **K6.1** is equivalent to **K6.2** so that they can be referred to simply as **K6**.

166   *Order and lattices*

THEOREM 5.28. A lattice satisfies **K6.1** iff it satisfies **K6.2**.

Proof. Suppose that a lattice $\mathfrak{L}$ satisfies **K6.1**. Then, for any elements x, y, z of $\mathfrak{L}$,

$$\begin{aligned}(x \cup y)(x \cup z) &= (x \cup y)(x \cup y)xz & &\text{by } \mathbf{K3}\\&= x \cup (x \cup y)z & &\text{by } \mathbf{K1}\ \&\ \mathbf{K3}\\&= x \cup (xz \cup yz) & &\text{by } \mathbf{K6.1}\\&= (x \cup xz) \cup yz & &\text{by } \mathbf{K2}\\&= x \cup yz & &\text{by } \mathbf{K3}.\end{aligned}$$

Thus $\mathfrak{L}$ satisfies **K6.2**. Hence **K1** & **K2** & **K3** & **K6.1** implies **K6.2**. By duality, **K1** & **K2** & **K3** & **K6.2** implies **K6.1**. ∎

DEFINITION 5.11. A distributive lattice (DL) is a lattice which satisfies **K6.1** (or, by Theorem 5.28, **K6.2**).

The distributive laws **K6.1**, **K6.2** generalize to the 'infinite distributive laws': for arbitrary sets Y of elements

K$^\infty$6.1: $x \cap lub(Y) = lub(\{x \cap y; y \in Y\})$

K$^\infty$6.2: $x \cup glb(Y) = glb(\{x \cup y; y \in Y\})$

DL's form a smaller class than modular lattices: the lattice T of Fig. 5.6 is modular but non-distributive. We remark here that a lattice is distributive iff it is modular and does not contain a sublattice isomorphic to T (e.g. Cohn 1965, p. 69).

EXAMPLE 5.17. A strengthening of Example 5.11: every ring of sets is a DL. ◆

The internal structure of a logic $\mathfrak{L}$ affects the lattice $cl(\mathfrak{L})$ of closed sets via the following lemma.

LEMMA 5.29. Let $\mathfrak{L} = \langle \Sigma, Cn \rangle$ be a compact logic. If either of (i), (ii), or (iii) hold then for all $\Delta, \Gamma, \Omega \subseteq \Sigma$

$$Cn(\Delta) \cdot Cn(\Gamma, \Omega) \subseteq Cn(Cn(\Delta) \cdot Cn(\Omega), Cn(\Delta) \cdot Cn(\Gamma)).$$

(i) $\mathfrak{L}$ has a normal implication.
(ii) $\mathfrak{L}$ has normal conjunction and normal disjunction.
(iii) $\mathfrak{L}$ is strongly compact.

Various kinds of lattices

Proof. By Theorem 5.29 it suffices to prove in each case that for all closed sets a, b, c of $\mathfrak{L}$, $a(b \cup c) \leq ab \cup ac$. Join and meet in $cl(\mathfrak{L})$ have been defined in Example 5.12. Thus we prove that, for all $\Delta, \Gamma, \Omega \subseteq L$,

$$A \in Cn(\Delta) \cdot Cn(\Gamma, \Omega) \;\Rightarrow\; A \in Cn(Cn(\Delta) \cdot Cn(\Omega), Cn(\Delta) \cdot Cn(\Gamma)).$$

Suppose that $A \in Cn(\Delta) \cdot Cn(\Gamma, \Omega)$. We prove in each case that $A \in Cn(Cn(\Delta) \cdot Cn(\Gamma), Cn(\Delta) \cdot Cn(\Gamma))$.

(i) Let $\supset$ be the normal implication in $\mathfrak{L}$. As $\mathfrak{L}$ is compact there exist finite sets $\Delta', \Gamma', \Omega'$ such that $\Delta' \subseteq \Delta$, $\Gamma' \subseteq \Gamma$, $\Omega' \subseteq \Omega$, and $A \in Cn(\Delta')$, $A \in Cn(\Gamma', \Omega')$. It may be assumed that Γ' is non-empty: suppose that $\Gamma' = \{b_1, \ldots, b_n\}$. Let $\Gamma' \supset A$ stand for $(b_1 \supset \cdots (n_{n-1} \supset A)) \cdots)$. As implication is normal, $\Gamma' \supset A \in Cn(\Omega')$. As $A \in Cn(\Delta')$, then $A \in Cn(\Delta', \Omega')$ so that $\Gamma' \supset A \in Cn(\Delta')$. Thus

$$\Gamma' \supset A \in Cn(\Delta') \cdot Cn(\Omega') \subseteq Cn(\Delta) \cdot Cn(\Omega). \tag{5.13}$$

Further, as implication is normal, $A \in Cn(\Gamma', \Gamma' \supset A)$. Hence

$$(\Gamma' \cap A) \supset A \in Cn(\Gamma') \subseteq Cn(\Gamma).$$

Next, as $A \in Cn(\Delta)$, $A \in Cn(\Gamma' \supset A, \Delta)$. Hence $(\Gamma' \supset A) \supset A \in Cn(\Delta)$ so that

$$(\Gamma' \supset A) \supset A \in Cn(\Delta) \cdot Cn(\Gamma), \tag{5.14}$$

But $A \in Cn(\Gamma' \supset A, (\Gamma' \supset A) \supset A)$. So by (5.13) and (5.14), $A \in Cn(Cn(\Delta) \cdot Cn(\Omega), Cn(\Delta) \cdot Cn(\Gamma))$.

(ii) Let $\wedge, \vee$ be the normal conjunction and disjunction respectively. As $\mathfrak{L}$ is compact there exist finite sets $\Delta', \Gamma', \Omega'$ such that $\Delta' \subseteq \Delta$, $\Gamma' \subseteq \Gamma$, $\Omega' \subseteq \Omega$, and $A \in Cn(\Delta')$, $A \in Cn(\Gamma', \Omega')$. Let F, G, H be conjunctions of all members of $\Delta', \Gamma', \Omega'$ respectively. Then $Cn(F) = Cn(\Delta')$, $Cn(G) = Cn(\Gamma')$, $Cn(H) = Cn(\Omega')$, $Cn(F \wedge G) = Cn(\Delta', \Gamma')$, and $Cn(F \vee G) = Cn(\Delta') \cdot Cn(\Gamma')$ etc. as conjunction and disjunction are normal. Thus $A \in Cn(F \vee (G \wedge H))$. But by Theorem 4.16(iv),

$$F \vee (G \wedge H) \;\leftrightarrow\; (F \vee G) \wedge (F \vee H).$$

Hence

$$\begin{aligned}
A \in Cn((F \vee G) \wedge (F \vee H)) &= Cn(F \vee G, F \vee H) \\
&= Cn(Cn(\Delta') \cdot Cn(\Gamma'), Cn(\Delta') \cdot Cn(\Omega')) \\
&= Cn(Cn(\Delta) \cdot Cn(\Gamma), Cn(\Delta) \cdot Cn(\Omega)).
\end{aligned}$$

(iii) Suppose that $\mathfrak{L}$ is strongly compact. Then $A \in Cn(\Delta)$ and $A \in Cn(y)$ for some $y \in \Gamma + \Omega$. So $y \in \Gamma$ or $y \in \Omega$. If $y \in \Gamma$ then $y \in Cn(\Gamma)$ so that $A \in Cn(\Delta) \cdot Cn(\Gamma)$. Thus $A \in Cn(Cn(\Delta) \cdot Cn(\Gamma), Cn(\Delta) \cdot Cn(\Omega))$. Similarly for $y \in \Omega$. ∎

168   *Order and lattices*

COROLLARY 5.30. If $\mathfrak{L}$ is compact and one of (i), (ii), or (iii) in Lemma 5.29 holds then $cl(\mathfrak{L})$ is distributive.

The distributive laws **K6.1, K6.2** are independent of the lattice axioms **L1, L2, L3**. This was first proved by Schröder (1890) who compared his work with that of Beltrami, Cayley, and Klein in establishing the unprovability of the parallel axiom from the remaining axioms of Euclidean geometry, namely by constructing a 'non-standard' model of the Euclidean axioms (less the parallel axiom) in which the parallel axiom failed. Schröder's axioms for logic were, in effect, our **K1–K4**. His 'standard model' was 'identity calculus with regions of a manifold'; his non-standard model in which **K0–K5** held but the distributive laws failed was called the 'logical calculus with groups'. It is described in his *Algebra der Logik*, Vol. I, *Anhang* 3 (Schröder 1890, p. 609). The objects of investigation are to be 'manifolds' of sentences and the 'regions' are to be 'formal groups', i.e. sets of sentences extended by all their logical consequences (Schröder 1890, p. 622). Thus, Schröder had in mind what we can now recognize as the closed sets of a logic though he did not have the abstract notion of a logic. In particular, he investigated the subgroups of a given group and constructed a group whose subgroups failed to satisfy the distributive law. Schröder (1890, p. 291) concluded that there were really two calculi of logic—one in which only one side of the distributive law held (Theorem 5.27), the other in which both sides held (i.e. **K6.1** and **K6.2**). The first one should be called the 'identity calculus' and the second—the 'properly logical' one—the 'calculus with groups'. Both, however, belong to the 'algebra of logic'.

Dedekind was evidently impressed by Schröder's proof of the independence of the distributive law (Schröder 1890, pp. 112–3). Schröder's idea of applying the logical-algebraic laws derived from his axioms for logic probably led Dedekind to their application to ideals and submodules and thence to his general notion of lattice.

Schröder's construction of a group whose lattice of subgroups is non-distributive is cumbersome. A simpler construction is provided by the following.

EXAMPLE 5.18. The lattice S of Fig. 5.6 is the lattice of subgroups of the group whose table is

	e	a	b	c
e	e	a	b	c
a	a	e	c	b
b	b	c	e	a
c	c	b	a	e

Various kinds of lattices

This lattice is non-distributive. In fact, if $\mathfrak{G}$ is a non-trivial group, then the lattice of subgroups of $\mathfrak{G} \times \mathfrak{G}$ is non-distributive. ◆

Dedekind (1900) proved that the normal subgroups of any group form a modular lattice (Birkhoff 1948, p. 65). The condition for the lattice of subgroups of a group $\mathfrak{G}$ to be distributive can be expressed entirely in terms of the internal structure of $\mathfrak{G}$. A group $\mathfrak{G}$ is said to be *generalized cyclic* if for all $a, b \in \mathfrak{G}$ there exists $c \in \mathfrak{G}$ and integers m, n such that $a = c^m, b = c^n$. Ore (1938) proved that the lattice is distributive iff $\mathfrak{G}$ is generalized cyclic (see also Birkhoff 1948, p. 96).

Peirce, working in the context of classical propositional calculus, had claimed that the distributive laws for the propositional calculus are derivable from his axioms (which were equivalent to lattice axioms), 'but the proof is too tedious to give' (Peirce 1880, p. 22) Schröder communicated his non-derivability result to Peirce (Schröder 1890, p. 291) who acknowledged this result but did not deny the validity of the distributive laws for the propositional calculus. Peirce (1885, p. 185) gave a truth-table type of argument to establish the 'fifth icon', $((x \prec y) \prec x) \prec x$, adding in a footnote:

> It is interesting to observe that this reasoning is dilemmatic. In fact the dilemma involves the fifth icon. The dilemma was only introduced into logic from rhetoric by the humanists of the renaissance: and at that time logic was studied with so little accuracy that the peculiar nature of this reasoning escaped notice. I was thus led to suppose that the whole non-relative logic was derivable from the principles of the ancient syllogistic, and this error was involved in Chapter II of my paper in the third volume of this journal. My friend, Professor Schröder, detected the mistake and showed that the distributive formulae
>
> $$(x + y)z \prec (x + y)z$$
> $$xy + z \prec (x + z)(y + z)$$
>
> could not be deduced from syllogistic principles. I had myself independently discovered and virtually stated the same thing (*Studies in Logic*, p. 189). There is some disagreement as to the definition of the dilemma (see Keynes's excellent *Formal Logic* p. 241), but the most useful definition would be a syllogism depending on the above distributive formulae. The distributive formulae
>
> $$xy + yz \prec (x + y)z$$
> $$xy + z \prec (x + z)(y + z)$$
>
> are strictly syllogistic.
>
> Peirce 1885, p. 190n

Peirce's proof of the distributive law appeared for the first time in a paper by Huntington (1904). It is a derivation from 'the second set of postulates' which are axioms for what we now call Boolean algebra. In a footnote

Huntington (1904, pp. 300–1n) quoted from a letter (dated 14 February 1904) in which Peirce explained the circumstances in which his proof remained hidden:

> I venture to opine that it fully vindicates my characterisation of it as tedious..

It has been asserted that Peirce believed that all lattices are distributive (Hermes and Köthe 1939, p. 13, n. 2) and that Peirce first retracted this claim under Schröder's criticism and later (as quoted by Huntington (1904)) 'boldly defended his original view' (Birkhoff 1948, p. 133). These claims are open to doubt. Peirce did not have Dedekind's abstract conception of lattice. He was concerned entirely with the axioms for the algebra of classical logic. Further remarks on the work of Peirce appear in Curry (1963, p. 160).

5.9.2 *Representation of distributive lattices*

A major structural role in lattices, particularly DL's, is played by *ideals*. It is not necessary to go far in this direction, but they will be used for the construction of DL's in Chapter 6 and 7.

DEFINITION 5.12. Let $\mathfrak{L} = \langle L, \cup, \cap \rangle$ be a lattice.

(i) An *ideal* of $\mathfrak{L}$ is a subset J of L such that

$$x, y \in J \Rightarrow x \cup y \in J$$
$$y \leqslant x \; \& \; x \in J \Rightarrow y \in J.$$

(ii) If $J \subseteq L$ then (J) denotes the ideal *generated by* J, i.e. the smallest ideal of $\mathfrak{L}$ which includes J. (Thus (J) consists of all elements $j_1 l_1 \cup \cdots \cup j_n l_n$, where $j_i \in J$ and $l_i \in L$.)

EXAMPLE 5.19.

(i) For ideals J, K of a lattice $\mathfrak{L}$, $J \cdot K = \{xy; x \in J, y \in L\}$. Further, the ideals of a lattice are closed under arbitrary intersection and therefore themselves form a complete lattice.

(ii) The ideals of a distributive lattice $\mathfrak{L}$ form a DL. For ideals J, K, $J \cup K =_{\text{def}}$ lub$(J + K)$, and $J \cap K =_{\text{def}}$ glb$(J + K)$. Then $J \cup K = \{x \cup y; x \in J, y \in K\}$ and $J \cap K = J \cdot K$. By Theorem 5.27 it suffices to prove that, for any ideals H, J, K of $\mathfrak{L}$, $H \cap (J \cap K) \subseteq (H \cap J) \cup (H \cap K)$. Now every element of $H \cap (J \cap K)$ is of the form $h(j \cup k)$ where $h \in H$, $j \in J$, and $k \in K$. But $h(j \cup k) = hj \cup hk \in (H \cap J) \cup (H \cap K)$.

(iii) Let $\mathfrak{L}_i = \langle L_i, \cup, \cap, 0_i \rangle$ for $i = 1, 2$ be lattices with minimum elements. Let θ be a *join-homomorphism* of $\mathfrak{L}_1$ to $\mathfrak{L}_2$, i.e. a mapping such that $\theta(x \cup y) = \theta(x) \cup \theta(y)$ for all $x, y \in L_1$. Then $\{x \in L_1; \theta(x) = 0_2\}$ is an ideal of $\mathfrak{L}_1$. ◆

Various kinds of lattices 171

EXAMPLE 5.20. The closed sets of a topological space X, including the empty set, form a DL $\mathbb{L}(X)$ under union and intersection. The set of closed boundary sets of X are an ideal of $\mathbb{L}(X)$ (Appendix, §T). ◆

The representation theory of DL's can be approached via a representation of posets.

THEOREM 5.31. *Let $P = \langle S, \leqslant \rangle$ be a poset. For each $x \in S$ put $M(x) = \{y; y \leqslant x\}$. Let $S' = \{M(x); x \in S\}$ and $P' = \langle S', \subseteq \rangle$ where $\subseteq$ denote inclusion of sets. Then P is isomorphic to P'. If, in addition, P is a lattice then the glb operation in P corresponds to intersection of sets in S'.*

Proof. M is an order-preserving map of P to P', i.e.

$$x \leqslant y \quad \Rightarrow \quad M(x) \subseteq M(y).$$

For suppose $x \leqslant y$. Let $z \in M(x)$. Then $z \leqslant x$. Hence $z \leqslant y$, i.e. $z \in M(y)$. Hence $M(x) \subseteq M(y)$.

Next, M has an inverse since $M(x) = M(y) \Rightarrow x = y$. For suppose $M(x) = M(y)$. As $y \in M(y)$, $y \in M(x)$ so that $y \leqslant x$. Similarly, $x \leqslant y$. Hence $x = y$.

Finally, suppose that P is a lattice. Let $\cap$ denote the *glb* operation. By Lemma 5.17, $z \leqslant xy \Leftrightarrow z \leqslant x \ \& \ z \leqslant y$. By the definition of M, this gives $z \in M(xy) \Leftrightarrow z \in M(x) \ \& \ z \in M(y)$, i.e. $M(xy) = M(x) \cdot M(y)$. ∎

The above theorem can be applied to the construction of logics from finite lattices, a process which is of some importance for the construction of logics from truth-values.

THEOREM 5.32. *Let $P = \langle \Sigma, \leqslant \rangle$ be a finite lattice. Define $v: \mathbb{P}(\Sigma) \to \Sigma$ by*

$$v(X) =_{\text{def}} (\iota x)(x \in lub(X)) \qquad (X \subseteq \Sigma).$$

Then for all $X, Y, y \subseteq \Sigma$

$$y \in X \quad \Rightarrow \quad y \leqslant v(X)$$
$$X \subseteq Y \quad \Rightarrow \quad v(X) \leqslant v(Y)$$
$$y = v(M(y)).$$

A simple corollary of this theorem is that every finite lattice is isomorphic to the lattice of closed sets of some logic (Example 5.12(i)).

COROLLARY 5.33. *Let $P = \langle \Sigma, \leqslant \rangle$ be a finite lattice. Define $Cn: \mathbb{P}(\Sigma) \to \mathbb{P}(\Sigma)$ by*

$$Cn(X) = M(v(X)) \qquad (X \subseteq \Sigma)$$

Then $\mathfrak{L} = \langle \Sigma, Cn \rangle$ is a logic and $cl(\mathfrak{L})$ is isomorphic to P.

It follows from the above result and Theorem 5.29 that there exist logics on which neither normal implication nor normal conjunction and disjunction can be defined.

A representation theorem for DL's can be obtained by modifying the function M of Theorem 5.31 (Campbell 1943). The modification requires the notion of *join-irreducibility*.

DEFINITION 5.13. An element a of a lattice is said to be *join-irreducible* if for all $x, y \in \mathfrak{L}$

$$x \cup y = a \quad \Rightarrow \quad x = a \text{ or } y = a.$$

LEMMA 5.34. In a finite lattice $\mathfrak{L} = \langle L, \leqslant \rangle$ every element is the joint of join-irreducible elements.

Proof. As $\mathfrak{L}$ is finite its elements $m_1, \ldots, m_k$ can be so ordered that if $m_i \leqslant m_j$ and $m_i \neq m_j$ in $\mathfrak{L}$ then $i < j$ in the standard ordering of $\mathbb{N}$. Suppose that m_r is not join-irreducible. Then there exist elements $x, y \in L$ such that $x \cup y = m_r$ and $x < m_r$ and $x < m_r$. Thus, for all r, where $1 \leqslant r \leqslant k$, either m_r is join-irreducible or there exists an $i < r$ such that $m_r = m_i \cup m_j$. It follows by induction on r that m_r is the join of join-irreducible elements. ∎

For each element x of a lattice $\mathfrak{L}$ let $N(x)$ be the collection of all join-irreducible elements y such that $y \leqslant x$. Lemma 5.34 shows that $N(x) \neq \emptyset$ when $\mathfrak{L}$ is finite.

THEOREM 5.35. Let $\mathfrak{L} = \langle L, \cup, \cap \rangle$ be a finite DL. Let $L'' =_{\text{def}} \{N(x); x \in L\}$ and $\mathfrak{L}'' =_{\text{def}} \langle L'', +, \cdot \rangle$ where $+, \cdot$ are union and intersection of sets respectively. Then $\mathfrak{L}$ is isomorphic to $\mathfrak{L}''$.

Proof. It suffices to prove that

$$N(x) = N(y) \quad \Rightarrow \quad x = y \tag{5.15}$$

$$N(xy) = N(x) \cdot N(y) \tag{5.16}$$

$$N(x \cup y) = N(x) + N(y). \tag{5.17}$$

(5.15) and (5.16) are proved as in Theorem 5.31.

ad (5.17) Let $z \in N(x \cup y)$, where $N(x) = \{x_1, \ldots, x_k\}$ and $N(y) = \{y_1, \ldots, y_n\}$, say. Then z is join-irreducible and $z \leqslant x_1 \cup \cdots \cup x_k \cup y_1 \cup \cdots \cup y_m$. Hence $z = z(x_1 \cup \cdots \cup x_k \cup y_1 \cup \cdots \cup y_m)$. As $\mathfrak{L}$ is distributive, $z = zx_1 \cup \cdots \cup zx_k \cup zy_1 \cup \cdots \cup zy_m$. But z is join-irreducible so $z = zx_i$ for some i such that $1 \leqslant i \leqslant k$, or $z = zy_j$ for some j such that $1 \leqslant j \leqslant m$. Hence $z \leqslant x_1$ or $\ldots$ or $z \leqslant y_m$. By Lemma 5.34, $x = x_1 \cup \cdots \cup x_k$ and $y = y_1 \cup \cdots \cup y_m$. Thus $z \leqslant x$ or $z \leqslant y$. This proves that

$$N(x \cup y) \subseteq N(x) + N(y). \tag{5.18}$$

Various kinds of lattices

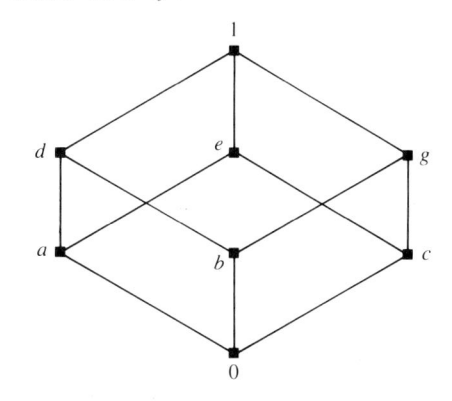

x	$N(x)$
0	$\{0\}$
a	$\{a,0\}$
b	$\{b,0\}$
c	$\{c,0\}$
d	$\{a,b,0\}$
e	$\{a,c,0\}$
g	$\{b,c,0\}$
1	$\{a,b,c,0\}$

Fig. 5.7.

Next, let $z \in N(x)$. Then z is join-irreducible and $z \leq x$. So $z \leq x \cup y$. Thus $z \in N(x \cup y)$ and so $N(x) \subseteq N(x \cup y)$. Likewise $N(y) \subseteq N(x \cup y)$. So
$$N(x) \cup N(y) \subseteq N(x \cup y). \tag{5.19}$$
(5.22) follows from (5.18) and (5.19). ∎

EXAMPLE 5.21. Consider the eight-element lattice $\mathfrak{L}$ of Fig. 5.7. The join-irreducible elements are $0, a, b, c$. The sets $N(x)$ can easily be determined. They are given in the table in Fig. 5.7. The sets $N(x)$, ordered by inclusion form the lattice $\mathfrak{L}''$, are shown in Fig. 5.8. $\mathfrak{L}$ is obviously isomorphic to $\mathfrak{L}''$. ◆

A more general representation theorem applicable to arbitrary DL's can be obtained via Birkhoff's theorem (Appendix §A). It is therefore necessary to identify the subdirectly irreducible DL's.

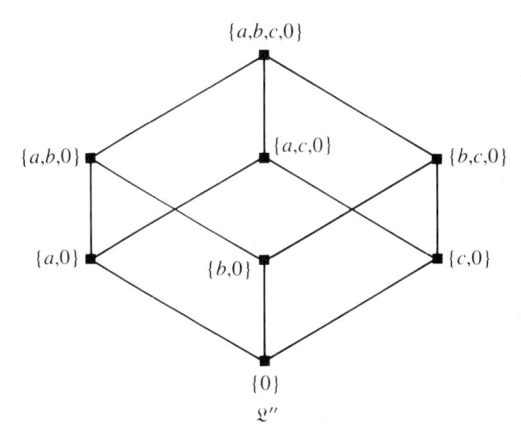

Fig. 5.8.

LEMMA 5.36. Let $\mathfrak{L} = \langle L, \cup, \cap, 0, 1 \rangle$ be a DL with maximum and minimum elements. If L contains an element $x \neq 0$ and $x \neq 1$ then there exist congruences θ, ψ on $\mathfrak{L}$ such that neither θ nor ψ is the identity id_L but their intersection is id_L.

Proof. Let $x \in L$ where $x \neq 0$ and $x \neq 1$. Define relations θ, ψ on L by

$$u = v(\theta) \Leftrightarrow_{\text{def}} ux = vx$$
$$u = v(\psi) \Leftrightarrow_{\text{def}} u \cup x = v \cup x.$$

(i) By **K2** and **K6** it easily follows that θ and ψ are congruences on $\mathfrak{L}$.
(ii) $\theta, \psi \neq id_L$. For by the definitions of θ and ψ $x = 1(\theta)$ and $x = 0(\psi)$. As, by hypothesis, $x \neq 1$, $\theta \neq id_L$. Likewise as $x \neq 0$, $\psi \neq id_i$.
(iii) $\theta \cdot \psi = id_L$. For suppose $u = v(\theta\psi)$. Then $u = v(\theta)$ and $u = v(\psi)$. By the definitions of θ, ψ and **K1**–**K6** it follows that $u = v$. ∎

COROLLARY 5.37. The only subdirectly irreducible DL's are the one-element lattice $\mathbb{1}$ and the two-element lattice $\mathbb{2}$ (Fig. 5.6, B_2).

Proof. $\mathbb{1}$ and $\mathbb{2}$ are subdirectly irreducible (Appendix, §A). Lemma 5.36 shows that there are no other DL's. ∎

COROLLARY 5.38. Every DL except $\mathbb{1}$ is isomorphic to a subdirect product of replicas of $\mathbb{2}$.

This result, first proved by Birkhoff (1933), follows trivially from Corollary 5.37 and Birkhoff's theorem (Appendix, §A). The representation theorem for DL's now follows from Corollary 5.38.

THEOREM 5.39. Every DL is isomorphic to a ring of sets.

Proof. Let $\mathfrak{L}$ be a DL. If $\mathfrak{L}$ is isomorphic to $\mathbb{1}$ then it is isomorphic to the ring of subsets of the empty set. Suppose $\mathfrak{L}$ has more than one element. By Corollary 5.38 there is an isomorphism § from $\mathfrak{L}$ onto a subdirect product of a family $\{\mathbb{2}_i; i \in I\}$ where $\mathbb{2}_i$ is isomorphic to $\mathbb{2}$. The minimum and maximum elements of $\mathbb{2}_i$ are 0_i and 1_i, say. Then for each element m of $\mathfrak{L}$, §$[m]$ is a function such that for each $i \in I$

$$\S[m](i) \in \{0_i, 1_i\}$$
$$\S[m_1 \cup m_2](i) = \S[m_1](i) \cup \S[m_2](i)$$
$$\S[m_1 m_2](i) = \S[m_1](i)\S[m_2](i).$$

Complemented lattices 175

For each element m of $\mathfrak{L}$ define $X(m) = \{i \in I; \S[m](i) = 1_i\}$. Then

$$i \in X(m_1 \cup m_2) \Leftrightarrow \S[m_1 \cup m_2](i) = 1_i = \S[m_1](i) \cup \S[m_2](i)$$
$$\Leftrightarrow \S[m_1](i) = 1_i \text{ or } \S[m_2](i) = 1_i$$
$$\Leftrightarrow i \in X(m_1) + X(m_2).$$

Thus $X(m_1 \cup m_2) = X(m_1) + X(m_2)$. Similarly, $X(m_1 m_2) = X(m_1) \cdot X(m_2)$.
Let $\mathfrak{L}'$ be $\{X(m); m \in |\mathfrak{L}|\}$ under the operations of union and intersection. Then X maps $\mathfrak{L}$ homomorphically onto $\mathfrak{L}'$ so that $\mathfrak{L}'$ is a ring of sets. It remains to prove that X is an isomorphism. For this it is sufficient to prove that

$$X(m_1) = X(m_2) \Rightarrow m_1 = m_2 \qquad (5.20)$$

Suppose $X(m_1) = X(m_2)$. Then for all $i \in I$, $\S[m_1](i) = 1_i$ iff $\S[m_2](i) = 1_i$. Since 2 has only two elements, $\S[m_1](i) = \S[m_2](i)$ for all $i \in I$. Hence $\S[m_1] = \S[m_2]$. As $\S$ is an isomorphism, $m_1 = m_2$. ∎

5.10 Complemented lattices

Further kinds of lattice can be defined by equipping them with additional operations. Traditionally these operations are derived from some concept of complementation which has an obvious importance in the calculus of classes and is related to negation in classical logic. Other kinds of complementation arise in intuitionistic logic and many-valued logics. We begin with a weak form of complementation to which further conditions will be added later.

5.10.1 Proto-complementation

A primitive feature of most kinds of complementation is order-reversal which is closely connected with the de Morgan laws (Chapter 4, §4.3.3) and, in particular, with Theorem 4.21. The idea of order-reversal is embodied in the definition of proto-complementation.

DEFINITION 5.14. A unary operation $'$ on a lattice $\langle L, \leqslant \rangle$ is proto-complementation if the following holds:

PC1: $x \leqslant y \Rightarrow x' \geqslant y'$ $(x, y \in L)$
PC2: $x \leqslant x''$ $(x \in L)$.

Then x' is the *proto-complement* of x. A lattice with proto-complementation is said to be *proto-complemented*.

PC1 and PC2 are clearly lattice versions of the logic laws **N3.1°** and **N4.1** (Chapter 4, §4.2.2). The various kinds of complementation we consider below satisfy **PC1** and **PC2** and some of the identities.

In: $\quad x'' = x$

PC1': $\quad x' \geq y' \Rightarrow x \leq y$

DM1: $\quad (x \cup y)' = x'y'$.

DM2: $\quad (xy)' = x' \cup y'$.

DM1 and **DM2** are the de Morgan laws which are a distinctive feature of Boolean algebras and quasi-Boolean algebras.

The following theorem relates join and meet to proto-complementation. It follows directly from the lattice laws and Definition 5.14.

THEOREM 5.40.

(i) In a proto-complemented lattice $\mathfrak{L}$ the following hold:

$$(x \cup y)' \leq x'y' \qquad (5.21)$$

$$x' \cup y' \leq (xy)' \qquad (5.22)$$

$$x''' = x'. \qquad (5.23)$$

If $\mathfrak{L}$ has a minimum element 0, then $0'$ is the maximum element.

(ii) If, in addition, $\mathfrak{L}$ satisfies **In** then it also satisfies **PC1'**, **DM1**, and **DM2**. And if $\mathfrak{L}$ has a minimum element 0 (maximum element 1) then it has a maximum element 1 (minimum element 0) and $0x = 0$, $0 \cup x = x$, $1 \cup x = 1$, $1x = x$, $1' = 0$ and $0' = 1$.

It easily follows from **K0** that the order-reversal properties follow from **In** and the de Morgan laws: *if a lattice has a unary operation ' which satisfies* **In**, **DM1**, *and* **DM2**, *then* **PC1** *and* **PC1'** *also hold*.

The first variant of proto-complementation—*pseudo-complementation*—is derived from a two-place operation which is related to the implication connective in classical and intuitionistic logic.

5.10.2 *Relative complementation*

The notion of the *implicative lattice* appears to have been formulated by Skølem (1919a). The operation $\rightarrow$, as will be obvious from the following theory, is analogous to the implication connective. Birkhoff (1948, p. 147) uses the term 'relative pseudo-complement'.

Complemented lattices

DEFINITION 5.15.

(i) Let a, b be elements of a lattice $\mathfrak{L}$ and let $R(a, b) =_{\text{def}} \{x \in |\mathfrak{L}|; ax \leqslant b\}$. If $R(a, b)$ contains a greatest element, then that element is denoted by $a \to b$ and is called the *pseudo-complement of a relative to b*. An *I-lattice* (or *implicative lattice*) is an algebra $\langle L, \cup, \cap, \to \rangle$ such that $\langle L, \cup, \cap \rangle$ is a lattice and, for all $a, b \in L$, $a \to b$ is the pseudo-complement of a relative to b.

(ii) An I-lattice which satisfies the identities $x \cup (x \to y) = 1$ and $(x \to y) \to x \leqslant x$ is called a *classical implicative lattice*. (This term was introduced by Curry (1963, p. 149). The second of the identities is obviously analogous to the 'Peirce law' **P** (Theorem 4.19).)

EXAMPLE 5.22.

(i) In the lattice of Fig. 5.7, the pseudo-complement of a relative to b does not exist since $\{x; ax \leqslant b\} = \{0, b, c\}$ which clearly does not contain the maximum element.
(ii) If the lattice $\mathfrak{L}$ has a minimum element 0, say, and if $0 \to 0$ exists then $0 \to 0$ is the maximum element. Every element of the lattice N_3 of Fig. 5.6 is pseudo-complemented but N_3 is not an I-lattice since $n \cup (n \to 0) = n$.
(iii) In an I-lattice $R(a, a) = L$ for all $a \in L$. Hence $a \to a$ is the maximum element. Thus every I-lattice has a maximum element.
(iv) The Lindenbaum algebra of the classical propositional calculus H (Chapter 2, §2.2.5) is a classical I-lattice with minimum element.
(v) Let V be a set. $\mathbb{P}(V)$ is closed under union, intersection, and complementation with respect to V. Define the two-place operation $\to$ on $\mathbb{P}(V)$ by $X \to Y =_{\text{def}} X^c + Y$ for all $X, Y \in \mathbb{P}(V)$, where X^c is the complement of X with respect to V. Then $\langle \mathbb{P}(V), +, \cdot, \to \rangle$ is an I-lattice.
(vi) I-lattices naturally occur in topologies (see Tarski 1956, Chapter 17). Let S be a topological space and O the collection of open sets of S (Appendix, §T). Then O is closed under union and intersection. Hence operations $\wedge$ and $\vee$ can be defined on O by

$$X \vee Y =_{\text{def}} X + Y, \qquad X \wedge Y =_{\text{def}} X \cdot Y \qquad (X, Y \in O).$$

The operation $\to$ on O can be defined using the closure operation C of the topology (Example 3.2(iv)):

$$X \to Y =_{\text{def}} (C(X \cdot Y^c))^c.$$

Then $\langle O, \vee, \wedge, \to \rangle$ is an I-lattice. To prove this it is sufficient to show from the properties of the closure operation that

$$Z \subseteq X \to Y \Leftrightarrow Z \cdot X \subseteq Y \qquad (X, Y, Z \in O), \qquad \blacklozenge$$

It easily follows from Definition 5.15 that an element c of a lattice $\mathfrak{L}$ is the pseudo-complement of a relative to b iff c satisfies

$$x \leqslant c \Leftrightarrow x \leqslant b \qquad (x \in \mathfrak{L}). \tag{5.24}$$

Thus $\langle A, \cup, \cap, \rightarrow \rangle$ is an I-lattice iff it satisfies **KO–K4** and the condition

$$x \leq a \rightarrow b \quad \Leftrightarrow \quad ax \leq b. \tag{5.25}$$

It follows from Theorem 5.27 that (5.30) is equivalent in lattice theory to the two identities

K8.1: $x(x \rightarrow y) \leq y$

K8.2: $z \leq x \rightarrow (y \cup xz)$.

Thus the class of I-lattices is a variety and so is closed under homomorphic images, subsystems, and direct products (Appendix, §A).

The following facts about pseudo-complementation are proved by Rasiowa and Sikorski (1963, Chapter 1, §12): they are derived more or less directly from the lattice laws, Definition 5.15, **K8.1**, and **K8.2**. By Example 5.21(iii), every I-lattice has a maximum element which, as usual, will be denoted by 1.

THEOREM 5.41. Let $\mathfrak{L}$ be an I-lattice. Then for all $a, b, c \in |\mathfrak{L}|$:

(i) $\quad a \rightarrow a = 1 \qquad a \rightarrow 1 = 1 \qquad 1 \rightarrow a = a$;

(ii) $\quad a \rightarrow b = 1 \Leftrightarrow a \leq b$;

(iii)
$$b \leq a \rightarrow b$$
$$a \leq b \rightarrow ab$$
$$c \rightarrow a \leq (c \rightarrow (a \rightarrow b)) \rightarrow (c \rightarrow b)$$
$$(a \rightarrow b)(b \rightarrow c) \leq a \rightarrow c$$
$$a \rightarrow b \leq ((b \rightarrow c) \rightarrow (a \rightarrow c))$$
$$a \rightarrow (b \rightarrow c) \leq ((a \rightarrow b) \rightarrow (a \rightarrow c))$$
$$(a \rightarrow c) \cup (b \rightarrow c) \leq (ab \rightarrow c);$$

(iv)
$$ab \cup ac = a(b \cap c)$$
$$a(a \rightarrow b) = ab$$
$$b(a \rightarrow b) = b$$
$$(a \rightarrow b)(a \rightarrow c) = a \rightarrow bc$$
$$(a \rightarrow c)(b \rightarrow c) = ((a \cup b) \rightarrow c)$$
$$c(ca \rightarrow cb) = c(a \rightarrow b)$$
$$a \rightarrow (b \rightarrow c) = ab \rightarrow c = (b \rightarrow (a \rightarrow c));$$

(v) $\quad x \leq y \Rightarrow y \rightarrow b \leq x \rightarrow b$

$\qquad x \leq y \Rightarrow a \rightarrow x \leq a \rightarrow y.$

It follows from (iv) of this theorem that every I-lattice is distributive. It is therefore appropriate to ask whether the converse is true, i.e. whether every distributive lattice can be expanded to an I-lattice.

THEOREM 5.42.

(i) Every finite distributive lattice can be expanded to an I-lattice.
(ii) There exists an infinite distributive lattice which cannot be expanded to an I-lattice.

Proof.

(i) Let $\mathfrak{L} = \langle L, \cap, \cup \rangle$ be a finite distributive lattice and let $a, b \in L$. As L is finite it has a minimum element 0. Clearly $0 \in R(a, b)$. Then for $x, y \in R(a, b)$ there holds $axy \leq ax \leq b$ so that $xy \in R(a, b)$. Further, as $ax, ay \leq b$ there holds $ax \cup ay \leq b$. As $\mathfrak{L}$ is a distributive lattice, $axy \leq b$, i.e. $x \cup y \in R(a, b)$. Thus, $R(a, b)$ is a finite lattice and so has a maximum element. By Definition 5.15, this maximum element is the pseudo-complement of a relative to b. The operation $\to$ on L can therefore be defined by $a \to b =_{\text{def}} (\iota x)(x \in lub(R(a, b)))$ for all $a, b \in L$. $\langle L, \cap, \cup, \to \rangle$ is therefore an I-lattice.

(ii) Let $V = \mathbb{P}_\omega(\mathbb{N}) + \{\mathbb{N}\}$. V is a DL under the operations of union and intersection of sets. Let $a, b \in \overset{\circ}{\mathbb{P}}_\omega(\mathbb{N})$ where a is not a subset of b. Then $\mathbb{N} \notin R(a, b)$. Let $m = \max(a \cup b) + 1$. Then for all $n \geq m$, $a\{n\} = \emptyset \subseteq b$. Thus $n \geq m \Rightarrow \{n\} \in R(a, b)$. If $R(a, b)$ had a maximum element, X say, then X would contain all numbers exceeding m and so would be infinite. But the only infinite set in V is $\mathbb{N}$. This contradicts $\mathbb{N} \notin R(a, b)$. Thus $R(a, b)$ cannot have a maximum element. Hence relative complementation cannot be defined in V. ∎

Pseudo-complementation plays a part in the lattice $cl(\mathfrak{L})$ of closed sets (Example 3.9) and the Lindenbaum algebra $\mathfrak{L}\P$ of a logic $\mathfrak{L}$. It enables Corollary 5.30 to be strengthened (cf. Tarski 1956, Chapter 12).

For a logic $\mathfrak{L} = \langle \Sigma, Cn \rangle$ and closed sets Δ, Γ

$$\Delta \to \Gamma =_{\text{def}} Cn(\sum \{\Theta \in |cl(\mathfrak{L})|; \Theta \cdot \Delta \subseteq \Gamma\}).$$

LEMMA 5.43. Suppose that $\mathfrak{L}$ is compact and for all $\Delta, \Gamma, \Theta \subseteq \Sigma$

$$Cn(\Delta) \cdot Cn(\Gamma, \Theta) \subseteq Cn(Cn(\Delta) \cdot Cn(\Theta), Cn(\Delta) \cdot Cn(\Gamma)). \quad (5.26)$$

Then for closed sets $\Delta, \Gamma, \Delta \to \Gamma$ is the pseudo-complement in $cl(\mathfrak{L})$ of Δ relative to Γ.

Proof. Let $\Delta, \Gamma_1, \ldots, \Gamma_n$ be closed sets. Then by (5.26)

$$\Delta \cdot Cn(\Gamma_1, \ldots,) \subseteq Cn(\Delta \cdot \Gamma_1, \ldots, \Delta \cdot \Gamma_n). \tag{5.27}$$

Now let $A \in (\Delta \to \Gamma) \cdot \Delta$. As $\mathfrak{L}$ is compact, by the definition of $\Delta \to \Gamma$ there exist closed sets $\Gamma_1, \ldots, \Gamma_n$, say, such that $\Gamma_i \cdot \Delta \subseteq \Gamma$ for $1 \leqslant i \leqslant n$ and $A \in Cn(\Gamma_1, \ldots, \Gamma_n)$. By (5.27), $A \in Cn(\Delta \cdot \Gamma_1, \ldots, \Delta \cdot \Gamma_n) \subseteq (\Gamma)$. Thus $(\Delta \to \Gamma) \cdot \Delta \subseteq \Gamma$. Clearly, $\Delta \to \Gamma$ is the maximum element of $R(\Delta, \Gamma)$. By Definition 5.15, $\Delta \to \Gamma$ is the pseudo-complement of Δ relative to Γ. ∎

COROLLARY 5.44. *If $\mathfrak{L}$ is compact and satisfies* (i), (ii), *or* (iii) *of Lemma 5.29 then $cl(\mathfrak{L})$ is an I-lattice with minimum element.*

THEOREM 5.45. *Let $\mathfrak{L} = \langle \Sigma, \supset, f, Cn \rangle$ be a logic with normal implication $\supset$ and contradiction f. For any set $\Delta \subseteq \Sigma$,*

$$\sim \Delta =_{\text{def}} \prod \{Cn(\neg A); A \in \Delta\}.$$

Then $\sim \Delta$ is a closed set and $\Delta \to Cn(\varnothing) \subseteq \sim \Delta$. If, in addition, $\mathfrak{L}$ satisfies $\mathbf{N2^0}$, *then* $\Delta \to Cn(\varnothing) = \sim \Delta$.

Proof. Let $A \in \Delta \to Cn(\varnothing)$. Then $Cn(A) \cdot Cn(\Delta) = Cn(\varnothing)$. Hence, for all $B \in \Delta$, $Cn(A) \cdot Cn(B) = Cn(\varnothing)$. By Example 4.7, $A \in Cn(\neg B)$ so that $A \in \sim \Delta$. Thus $\Delta \to Cn(\varnothing) \subseteq \sim \Delta$.

Now suppose that $\mathbf{N2^0}$ holds. Let $A \in \Delta \cdot \sim \Delta$. Then $A \in Cn(\neg B)$ for all $B \in \Delta$. In particular, $A \in Cn(\neg A)$. Thus $A \in Cn(A) \cdot Cn(\neg A)$. By $\mathbf{N2^0}$, $A \in Cn(\varnothing)$. So $\Delta \cdot \Delta \subseteq Cn(\varnothing)$. Hence $\sim \Delta \subseteq \Delta \to Cn(\varnothing)$. ∎

THEOREM 5.46.

(i) *Let $\mathfrak{L} = \langle L, \wedge, \vee, \supset, Cn \rangle$ be a logic with normal implication $\supset$, normal conjunction $\wedge$, and normal disjunction $\vee$. Then the Lindenbaum algebra $\mathfrak{L}\P$ is an I-lattice. If, in addition, $\mathfrak{L}$ has falsehood f, satisfies P, and has negation defined by $\neg A =_{\text{def}} A \supset f$ satisfies $\mathbf{N2}$, then $\mathfrak{L}\P$ is a classical implicative lattice.*
(ii) *The Lindenbaum algebra $I\P$ of the intuitionistic propositional logic $\mathbb{I}$ is an I-lattice with a minimum element.*

Proof.

(i) Since implication and disjunction are normal $A \supset B \in Cn(D) \Leftrightarrow B \in Cn(A, D)$, i.e.

$$Cn((A \supset B) \wedge D) = Cn(D) \quad \Leftrightarrow \quad Cn(A \wedge B \wedge D) = Cn(A \wedge D).$$

Thus, by Example 5.14,

$$[(A \supset B) \wedge D] = [D] \quad \Leftrightarrow \quad [A \wedge B \wedge D] = [A \wedge D]$$

so that

$$[A \supset B] \cap [D] = [D] \Leftrightarrow [A] \cup [B] \cap [D] = [A] \cap [D].$$

Hence $[A \supset B] \geq [D] \Leftrightarrow [B] \geq [A] \cap [D]$. Since $\supset$ is a connective the operation $\to$ on equivalence classes can be defined by $[A \supset B] =_{\text{def}} [A] \to [B]$. Hence $[A] \to [B] \geq [D] \Leftrightarrow [B] \geq [A][D]$. By (5.25), $\mathfrak{L}\P$ is an I-lattice.

Suppose now that **P** and **N2** are satisfied. $[f]$ is the minimum element of $\mathfrak{L}$ by **F**. Let $1 =_{\text{def}} [f \supset f]$: this is the maximum element and $f \supset f \in Cn(\emptyset)$. By **P**, $A \in Cn((A \supset B) \supset A)$. Hence $(A \supset B) \supset A \in L(imp_\emptyset : A)$ and so $([A] \to [B]) \to [A] \leq [A]$. Further, by **N2**, $Cn(A) \cdot Cn(A \supset f) = Cn(\emptyset)$. As $\supset$ is normal, $f \in Cn(A, A \supset f)$. By **F**, $B \in Cn(A, A \supset f)$. Hence $Cn(A \supset B) \subseteq Cn(A \supset f)$. Thus $Cn(A) \cdot Cn(A \supset B) = Cn(\emptyset)$. As $\vee$ is normal, $A \vee (A \supset B) \in Cn(\emptyset)$. Thus

$$[A] \vee ([A] \to [B]) = 1.$$

Hence $\mathfrak{L}\P$ is a classical implicative lattice.

(ii) This is derived from (i) and Example 4.4. ∎

Theorem 5.41 can be used to deepen the connection between I and I-lattices which was stated in Theorem 5.46(ii).

THEOREM 5.47. *Let G be a formula of I, the intuitionistic propositional calculus. Then G is a theorem of I iff for every I-lattice $\mathfrak{L}$ with minimum element and every valuation $h: \mathbb{F} \to |\mathfrak{L}|$ of the formulas $\mathbb{F}$ of I, $h(G)$ is the maximum element of $\mathfrak{L}$.*

Proof. Let $\mathfrak{L} = \langle |\mathfrak{L}|, \wedge, \vee, \to \rangle$ be an I-lattice with minimum element 0, say. Then it also has a maximum element 1, say. Let G be a theorem of I. Then it has a derivation $D_0, D_1, \ldots, D_n = G$, say. We prove by induction on $r \leq n$ that

$S(r): \quad h(D) = 1 \quad$ for $s \leq r$

where h is a valuation of $\mathbb{F} = F(\{f, \vee, \wedge, \supset\}, P)$ (cf. Chapter 2, §2.2.7) in $\mathfrak{L}$.

Basis:
$S(0) = 0$. D_0 is an axiom. Consider the cases **A0–A8** (Chapter 2, §2.2.7).

A0 $D_0 = t$. Then $h(D_0) = 1$.
       $D_0 = f \supset A$. Then $h(D_0) = 0 \to h(A)$. By Theorem 5.41(iii), $h(D_0) = 1$.

A1 $D_0 = A \supset (B \supset A)$. Then $h(D_0) = h(A) \to (h(B) \to h(A))$. By Theorem 5.41(i), (iv), $h(D_0) = 1$.

A2 $D_0 = (A \supset B) \supset ((A \supset (B \supset C)) \supset (A \supset C))$. By Theorem 5.41(iii)

$$(h(A) \to h(B)) \leqslant (h(A) \to (h(B) \to h(C))) \to (h(A) \to h(C)).$$

By Theorem 5.41(iii), $h(D_0) = 1$.

The remaining cases similarly use parts of Theorem 5.41.

Inductive step:
Suppose $S(r)$ and consider D_{r+1}. Then either (a) D_{r+1} is a theorem or (b) $D_m = D_k \supset D_{r+1}$ for some $m, k \leqslant r$.

(a) As for the basis of induction, $h(D_{r+1}) = 1$.
(b) By inductive hypothesis, $h(D_k) = h(D_m) = 1$. By Theorem 5.41, $1 = h(D_k) \leqslant h(D_{r+1})$. As 1 is the maximum element of $\mathfrak{L}$, $h(D_{r+1}) = 1$.

Thus $S(r)$ implies $S(r+1)$. It now follows by induction that $S(r)$ holds for all $r \leqslant n$. In particular $h(G) = h(D_n) = 1$.

Conversely, suppose that $h(G) = 1$ for every I-lattice $\mathfrak{L}$ with minimum element 0 and every valuation $h\colon \mathbb{F} \to |\mathfrak{L}|$ of the formulas $\mathbb{F}$. Let $h(A) =_{\text{def}} [A]$, where $[A]$ is the logical equivalence class in I of the formula. Then h is a valuation of $\mathbb{F}$ in $I\P$ which by Theorem 5.46(ii) is an I-lattice with minimum element. By hypothesis, $h(G) = [G] = 1$. Hence G is a theorem. ∎

The relation between classical I-lattices with minimum element and the classical propositional calculus is analogous to the relation between I-lattices with minimum element and intuitionistic propositional calculus.

THEOREM 5.48. *Let G be a formula of H. Then G is a theorem of H iff for every I-lattice $\mathfrak{L}$ with minimum element and every valuation $h\colon \mathbb{F} \to |\mathfrak{L}|$ of formulas $\mathbb{F}$ of H, $h(G)$ is the maximum element of $\mathfrak{L}$.*

Proof. As in Theorem 5.47, but here there is an additional axiom schema

A10: $\neg\neg A \supset A$

to include in the case analysis.

Let $\mathfrak{L}$ be an I-lattice with minimum element 0 and maximum element $1_{\mathfrak{L}}$ and let $h\colon \mathbb{F} \to |\mathfrak{L}|$ be a valuation of $\mathbb{F}$ in $\mathfrak{L}$. Now $\neg\neg A \supset A$ is equivalent to $((A \supset f) \supset f) \supset A$. Hence $h(\neg\neg A \supset A) = ((h(A) \to 0) \to h(A)$. But as $\mathfrak{L}$ is a classical I-lattice, $(h(A) \to 0) \to 0 \leqslant h(A)$. By Theorem 5.41(ii), $h(\neg\neg A \supset A) = 1_{\mathfrak{L}}$. ∎

Note that if, for elements b, c of a lattice $\mathfrak{L}$, there exist elements c, d such that for all $x \in |\mathfrak{L}|$

$$x \leqslant c \Leftrightarrow ax \leqslant b \quad \text{and} \quad x \leqslant d \Leftrightarrow ax \leqslant b$$

Complemented lattices

then $c = d$. Hence relative complementation is uniquely defined by the first order condition (5.25). Thus the class K of lattices which can be expanded to I-lattices is that class of lattices satisfying the condition

$$(a)(b)(Ec)(x \leq c \Leftrightarrow ax \leq b).$$

This condition cannot be replaced by a set of identities, i.e. K is not an equational class. For $\mathbb{P}(\mathbb{N})$, under the operations of union and intersections of sets, is in K but this lattice has as a subsystem the lattice defined in Theorem 5.42(ii) which cannot be expanded to an I-lattice. Thus K is not hereditary, so by Birkhoff's theorem (Appendix, §A) K is not an equational class.

5.10.3 Pseudo-complementation

A type of complementation—a unary operation—can be derived from the binary operation of relative complementation of §5.10.2. It is analogous to the definition of negation from contradiction and implication in intuitionistic logic.

DEFINITION 5.16. Let a be an element of a lattice $\mathfrak{L}$ with minimum element 0. If $a \to 0$ exists it is denoted by $a^\dagger$ and is called the *pseudo-complement* of a.

Thus in an I-lattice with minimum element, every element is pseudo-complemented. It should be remarked that it is possible for every element of a lattice to be pseudo-complemented without there being a relative complementation.

The following theorem follows from Definition 5.16 and Theorem 5.41. Parts (i) and (ii) show that pseudo-complementation is a proto-complementation and so appeal can be made to Theorem 5.43.

THEOREM 5.49. In an I-lattice with minimum element 0, $1 = 0^\dagger$ and the following identities and relations hold:

(i) **PC1**: $x \geq y \Rightarrow y^\dagger \geq x^\dagger$
(ii) **PC2**: $x \leq x^{\dagger\dagger}$
(iii) $x^\dagger = x^{\dagger\dagger\dagger}$
(iv) **DM1**: $x^\dagger y^\dagger = (x \cup y)^\dagger$
(v) $x^\dagger \cup y^\dagger \leq (xy)^\dagger$
(vi) $0^\dagger = 1, \quad 1^\dagger = 0$
(vii) $xx^\dagger = 0, \quad 1 = (x^\dagger \cup x)^{\dagger\dagger}$
(viii) $x \to y \leq y^\dagger \to x^\dagger.$

COROLLARY 5.50. In an I-lattice with minimum element 0, if the identity **In** holds then so also do the identities **DM1, DMS**, and

$$x \cup x^\dagger = 1. \tag{5.28}$$

THEOREM 5.51. An I-lattice with minimum element 0 is a classical implicative lattice iff the identities **In** (§5.10.1) and

$$x \cup x^\dagger = 1 \qquad x^{\dagger\dagger} = x \tag{5.29}$$

hold.

Proof. This proof relies on the identities of Theorem 5.41. Let $\mathfrak{L}$ be a classical implicative lattice. By Definition 5.15 the identities $x \cup (x \to y) = 1$ and $(x \to y) \to x \leqslant x$ hold in $\mathfrak{L}$. Putting $y = 0$ in these gives $x \cup (x \to 0) = 1$, i.e. $x \cup x^\dagger = 1$, and $x^\dagger \to x \leqslant x$ respectively. Hence $x^\dagger \to 0 \leqslant x^\dagger \to x$. Thus $x^{\dagger\dagger} \leqslant x$. But by Theorem 5.48, $x \leqslant x^{\dagger\dagger}$. Hence $x = x^{\dagger\dagger}$.

Conversely, suppose $\mathfrak{L}$ is an I-lattice with minimum element 0 which satisfies (5.29). By Theorem 5.41(iv) $\mathfrak{L}$ is distributive. Hence, since $x \cup x^\dagger = 1$, again by Theorem 5.41(iv),

$$x^\dagger \to x = (x \cup x^\dagger)(x^\dagger \to x) = x(x^\dagger \to x) \cup x^\dagger(x^\dagger \to x)$$
$$= x \cup xx^\dagger = x.$$

But $x^\dagger \leqslant x \to y$ by Theorem 5.41(v). Also by Theorem 5.41(v), $(x \to y) \to x \leqslant x^\dagger \to x = x$. By Definition 5.15(ii) $\mathfrak{L}$ is a classical I-lattice. ∎

COROLLARY 5.52. Every classical I-lattice satisfies **DM1** and **DM2**.

5.10.4 Complementation

Another concept of complementation is given in Definition 5.17. It is clearly related to negation in classical logic and to complementation as it occurs in the classical calculus of classes.

DEFINITION 5.17. Let $\mathfrak{L}$ be a lattice with maximum and minimum elements of 0 and 1 respectively.

(i) Let $a \in |\mathfrak{L}|$. An element x of $\mathfrak{L}$ is a complement of a if $ax = 0$ and $a \cup x = 1$. An element is said to be *complemented* if it has a complement. $cm(x)$ denotes the set of complements of x.
(ii) An algebra $\langle A, \cup, \cap, 0, 1, ' \rangle$ of type $\langle 2, 2, 0, 0, 1 \rangle$ is a *complemented lattice* if $\langle A, \cup, \cap, 0, 1 \rangle$ is a lattice and, for all $x \in A$, $x' \in cm(x)$.

Schröder (1890, Chapter 7) has a similar definition of negation though he was defining the algebra of classical logic and not the general notion of 'lattice'.

Complemented lattices

EXAMPLE 5.23. The most familiar example of a complemented lattice is the classical algebra of classes. Let V be a non-empty set. The complement X' of a subset X of V is defined as usual: $X' =_{\text{def}} \{x \in V; x \notin V\}$. The algebra $\langle \mathbb{P}(V), \cup, \cap, V, ' \rangle$ is a complemented lattice in which the empty set $\emptyset$ is the minimum element and V is the maximum element. This algebra has a subsystem whose elements are those subsets of V which are finite or cofinite (i.e. finite complement). Any field of sets (Definition 5.9) is a complemented lattice. ◆

EXAMPLE 5.24. Consider the classical propositional calculus H as defined in Chapter 2, §2.2. Let $\mathbb{F} = \mathbb{F}(\{\vee, \wedge, \neg\}, P)$. Consider the algebra of formulas $\mathfrak{F} = \langle \mathbb{F}, \wedge, \vee, \neg \rangle$. A definition of logical consequence which applies to this algebra of formulas is given in Definition 2.4. From Example 2.1 it follows that $\mathfrak{F}/\leftrightarrow$ is a complemented distributive lattice. ◆

EXAMPLE 5.25. Complementation does not necessarily reverse order as in proto-complementation (**PC1**). In the complemented lattice of Fig. 5.9, where $a' = A$, $b' = B$,

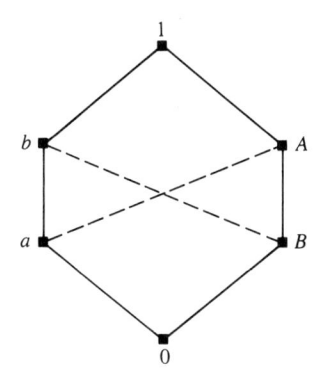

Fig. 5.9.

$0' = 1$, $1' = 0$, $a \geqslant b$ and $a' = A \geqslant B = b'$ hold. Thus, order is not reversed. Further, complements are not necessarily unique. For instance in the above lattice both A and B are complements of a. ◆

However, complements are unique in DL's.

THEOREM 5.23. Let $\mathfrak{L}$ be a DL. Then we have the following.

(i) No element of $\mathfrak{L}$ can have more than one complement. If $\mathfrak{L}$ has a maximum element 1 and a minimum element, 0, then $0 \in cm(1)$ and $1 \in cm(0)$.
(ii) For $x, y \in |\mathfrak{L}|$, if $x' \in cm(x)$ and $y' \in cm(y)$ then $x' \cup y' \in cm(xy)$ and $x'y' \in cm(x \cup y)$. Thus the complemented elements of $\mathfrak{L}$ form a sublattice of $\mathfrak{L}$.
(iii) If $x \in cm(y)$ then $y \in cm(x)$.

Proof.

(i) Suppose that $x, y \in cm(i)$. Then
$$ax = 0 = ay \qquad a \cup x = 1 = a \cup y. \qquad (5.30)$$

Hence $y(a \cup x) = y$. As $\mathfrak{L}$ is distributive, $ya \cup yx = y$. By (5.30), $xy = y$ and $yx = x$. Hence $x = y$. Thus a has just one complement. (This argument was given by Schröder (1890, Chapter 7) who attributed it to R. Grassmann.) Further, since $0 \cup 1 = 1$ and $01 = 0$, $0 \in cm(1)$ and $1 \in cm(0)$.

(ii) Since $\mathfrak{L}$ is distributive and $x \cup x' = y \cup y' = 1$,
$$(x \cup y) \cup x'y' = (x \cup y \cup x')(x \cup y \cup y') = 11 = 1.$$

Also, $(x \cup y)x'y' = xx'y \cup yx'y' = 00 = 0$ as $xx' = yy' = 0$. Thus, $x'y' \in cm(x \cup y)$ and $x' \cup y' \in cm(xy)$. By (i), $x'y'$ is the (unique) complement of $x \cup y$ and $x' \cup y'$ is the unique complement of xy. Hence the complemented elements of $\mathfrak{L}$ form a sublattice.

(iii) This follows immediately from Definition 5.17. ∎

The next theorem indicates some conditions under which complementation and pseudo-complementation coincide.

THEOREM 5.54.

(i) Every I-lattice with a minimum element which satisfies the identity $x^{\dagger\dagger} = x$ is a complemented lattice.
(ii) Every classical I-lattice with a minimum element is complemented.

Proof.

(i) By Theorems 5.49(vii) and 5.51, $x^\dagger$ is the complement of x.
(ii) Let $\mathfrak{L}$ be a classical I-lattice. By Theorem 5.51 the identities $y \cup y^\dagger = 1$ and $y^{\dagger\dagger} = y$ hold in $\mathfrak{L}$. By Theorem 5.49(vii), $xx^\dagger = 0$ and $1 = x^\dagger \cup x$ hold in $\mathfrak{L}$. By Definition 5.17, $x^\dagger \in cm(x)$. ∎

COROLLARY 5.55. In a complemented DL, complementation is order-reversing and the following identities hold: **In, DM1, DM2**, and $0' = 1, 1' = 0$.

5.10.5 Involution and ortho-complemented lattices

The last form of complementation to be dealt with is derived from proto-complementation by strengthening **PC2** to an equality. The notion of *involution* defined below is of importance in quasi-Boolean algebras and the logics discussed in Chapter 8.

Complemented lattices

DEFINITION 5.18.

(i) Let $P = \langle A, \leqslant \rangle$ be an ordered set. An order-reversing map $x \to x^*$ of period 2 is an *involution*, i.e. the operation * satisfies the laws **PC1** and **In** (§5.10.1).
(ii) An algebra $\langle A, \cup, \cap, * \rangle$ of type $\langle 2, 2, 1 \rangle$ is an *involuted* lattice if $\langle A, \cup, \cap \rangle$ is a lattice and * is an involution on it.
(iii) An *ortho-complemented lattice* (*OCL*) or *ortho-lattice* is an algebra $\langle A, \cup, \cap, 0, 1, * \rangle$ such that $\langle A, \cup, \cap, 0, 1 \rangle$ is a lattice with maximum and minimum elements 1,0 respectively, and * is an involution and complementation.
(iv) An *ortho-modular lattice* (OML) is an OCL satisfying the *ortho-modular* identity

$$a \leqslant b \quad \Rightarrow \quad a \cup ba^* = b.$$

Two elements a, b of an ortho-modular lattice are said to *commute* (written aCb) when $a = ab \cup ab^*$.

EXAMPLE 5.26.

(i) The three element set $\{1, n, 0\}$ with the ordering defined by $1 \geqslant n \geqslant 0$ and the operation *, where $1^* = 0$, $0^* = 1$, $n^* = n$ is an involuted lattice (i.e. N_3 of Fig. 5.6).
(ii) By Theorem 5.40 proto-complementation † satisfying **In** is an involution. By Example 5.25 complementation is not necessarily an involution.
(iii) Let $\mathfrak{L}$ be a DL with maximum and minimum elements 1, 0 respectively, and let J be an ideal of $\mathfrak{L}$ (Definition 5.12(i) and Example 5.19). Define

$$L =_{\text{def}} \{(x, y); x, y \in |\mathfrak{L}|, xy \in J, x \cup y = 1\}$$

and define operations $\cup$, * on L by

$$(x, y)(u, v) =_{\text{def}} (xu, y \cup v)$$
$$(x, y) \cup (u, v) =_{\text{def}} (x \cup u, yv)$$
$$(x, y)^* =_{\text{def}} (y, x).$$

Then the algebra $(\mathfrak{L}, J) =_{\text{def}} \langle L, \cup, \cap, *, \mathbf{0}, \mathbf{1} \rangle$, where $\mathbf{0} =_{\text{def}} (0, 1)$, $\mathbf{1} =_{\text{def}} (1, 0)$, is an involuted DL with maximum and minimum elements in which there hold **DM1**, **DM2**, **In**, and $ss^* \leqslant t \cup t^*$ for all $s, t \in L$. ◆

The notion of involution was defined by Birkhoff (1948, p. 4). In the same work (p. 124) the notion of ortho-complementation was defined for modular lattices only. It is closely connected with de Morgan's laws **DM1** and **DM2** via Theorem 5.40.

LEMMA 5.56.

(i) If * is a mapping of period 2 on a lattice $\mathfrak{L}$ such that **DM1** (with ' interpreted as *) holds in $\mathfrak{L}$, then * is an involution and **DM2** also holds in $\mathfrak{L}$.
(ii) If * is an involution on a lattice, then **DM1** and **DM2** hold in it.

COROLLARY 5.57.

(i) An involuted lattice is a lattice with a unary operation * such that **DM1**, **DM2**, and **In** hold.
(ii) An OCL is a complemented lattice in which **DM1** and **DM2** hold.

EXAMPLE 5.27.

(i) In a complete involuted lattice

$$(lub(X))^* = glb(X^*) \qquad (glb(X))^* = lub(X^*)$$

hold.
(ii) The lattice of Fig. 5.9, in which $a' = A$, $A' = a$, $b' = B$, $B' = b$, $0' = 1$, and $1' = 0$, is an OCL but is not ortho-modular since $a \cup ba' = a \cup 0 = a \neq b$ and $a \leqslant b$.
(iii) The smallest non-modular irreducible OML is shown in Fig. 5.10.
(iv) The lattice of all closed subspaces of a Hilbert space H is complete and ortho-modular. It is modular iff H is finite-dimensional. It can be shown that two projections P, Q commute, i.e. $PQ = QP$, iff the associated subspaces N_P, N_Q commute in the sense of Definition 5.18(iv). ♦

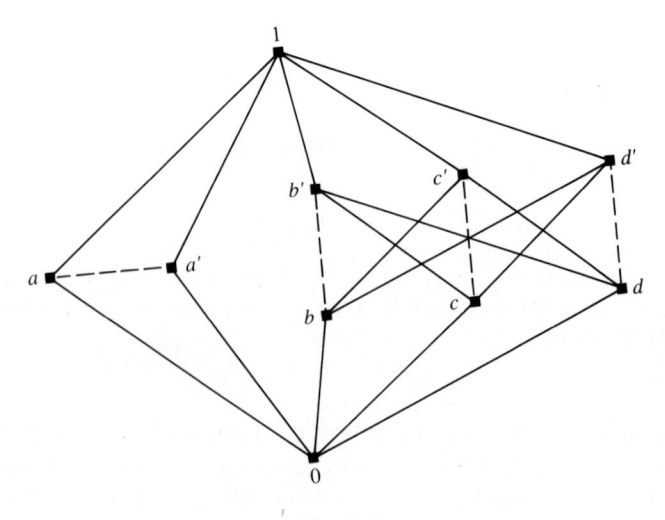

Fig. 5.10..

OMLs are thought by some to be 'fundamental' to quantum logic in the same way that Boolean algebra is to classical logic. One of the principal mathematical instruments in the study of quantum mechanics is the interpretation of observables as operators in Hilbert spaces and the study of lattices of projection operators. Birkhoff and von Neumann (1936) employed this idea in an attempt to construct lattice-theoretical models of the 'logic of quantum mechanics'. They tried to discover:

> ... what logical structure one may hope to find in physical theories which, like quantum mechanics, do not conform to classical logic. Our main conclusion, based on admittedly heuristic arguments, is that one can reasonable expect to find a calculus of propositions which is formally indistinguishable from the calculus of linear subspaces [of a Hilbert space] with respect to *set products, linear sums* and *orthogonal complements*—and resembles the usual calculus of propositions with respect to *and, or* and *not*.

A defect of this approach was the use of the modular law which holds only in the case of finite-dimensional spaces. Later, Husimi (1937) proposed instead to study the ortho-modular law which holds for projection operators of arbitrary Hilbert spaces. However, not every OML is isomorphic to the lattice of linear subspaces of some Hilbert space (Sasaki 1954).

5.10.6 *Boolean algebras and related structures*

The very brief introduction to Boolean algebra given here will suffice for the applications given later. The reader who wishes to pursue the subject further is referred to the abundance of literature on the subject ('The amount of literature produced in this field today probably exceeds that produced in all of mathematical logic' (Hanf 1974)), but particularly Rasiowa and Sikorski (1963) and Birkhoff (1948).

DEFINITION 5.19. A complemented DL is a *Boolean lattice* (or *Boolean algebra* when considered as an algebra).

Boole's mathematical logic developed from attempts to broaden the Aristotelian syllogistic. Bocheński (1961, p. 296) cites Augustus de Morgan here. Boole's calculus was of historical importance because of its interpretation as both propositional logic and class calculus. The system was developed by Peirce, Schröder, and Jevons, reaching its peak in Schröder's *Vorlesung über die Algebra der Logik* (1890–1910). Huntington's (1904) investigation of postulates for Boolean algebra is particularly important.

Numerous examples of Boolean algebras can be given.

EXAMPLE 5.28.

(i) Let V be a non-empty set. The algebra $\langle \mathbb{P}(V), \cup, \cap, \emptyset, V, ' \rangle$ is a Boolean algebra.
(ii) Let V be an infinite set. Let $\mathbb{P}_{\text{cof}}(V)$ denote the set of finite subsets of V together with their complements with respect to V. $\mathbb{P}_{\text{cof}}(V)$ is closed under union, intersection, and complementation. Thus, $\langle \mathbb{P}_{\text{cof}}(V), \cup, \cap, \emptyset, V, ' \rangle$ is a subalgebra of $\langle \mathbb{P}(V), \cup, \cap, \emptyset, V, ' \rangle$ and is likewise a Boolean algebra.
(iii) Every field of sets is a Boolean algebra.
(iv) By Theorem 5.53, the complemented elements of a DL with maximum and minimum elements form a Boolean algebra.
(v) Let $\langle L, \cap, \cup, \rightarrow, 0 \rangle$ be a classical I-lattice with minimum element. Define the operation $'$ by $x' =_{\text{def}} x \rightarrow 0$. Then it follows from Theorem 5.54(ii) that $\langle L, \cap, \cup, ', 0 \rangle$ is a Boolean algebra. ◆

EXAMPLE 5.29.

(i) Every Boolean algebra is an OCL and every ortho-modular DL is a Boolean algebra.
(ii) Let $\mathfrak{L}$ be an OML. It can be proved that the commutation relation C (Definition 5.18(iv)) is a reflexive and symmetric relation, that $x \leqslant y \Rightarrow xCy$, and that if M is a subset of $|\mathfrak{L}|$ such that xCy for all $x, y \in M$ then the sublattice generated by M is a Boolean algebra (Kalmbach 1983). ◆

Boolean algebras have already been encountered in the guise of I-lattices in the following sense.

THEOREM 5.58.

(i) Let $\mathfrak{L} = \langle L, \cup, \cap, \rightarrow, 1, 0 \rangle$ be a classical I-lattice. Define the operation $'$ on L by $x' =_{\text{def}} x \rightarrow 0$. Then $\langle L, \cup, \cap, 1, 0, ' \rangle$ is a Boolean algebra.
(ii) Let $\langle L, \cup, \cap, 1, 0, ' \rangle$ be a Boolean algebra. Define the two-place operation $\rightarrow$ on L by $x \rightarrow y =_{\text{def}} x' \cup y$. Then $\langle L, \cup, \cap, \rightarrow, 1, 0 \rangle$ is a classical I-lattice.

As with the case of DL's (Corollary 5.38) a structure theorem can be obtained by identifying the subdirectly irreducible algebras and then using Birkhoff's theorem.

THEOREM 5.59. *The only join-irreducible Boolean algebra is the two-element Boolean algebra B_2 (Fig. 5.6).*

Proof. Let $\mathfrak{B} = \langle B, \cup, \cap, 0, 1, ' \rangle$ be a Boolean algebra. Suppose B contains an element $x \neq 0, 1$. Then there exist congruences θ, ψ on $\mathfrak{B}$ such that neither θ nor ψ is the identity on B but $\theta \cdot \psi$ is the identity.

Complemented lattices

As in Lemma 5.36, define the relations θ, ψ on B by

$$u = v(\theta) \Leftrightarrow ux = vx$$
$$u = v(\psi) \Leftrightarrow u \cup x = v \cup x.$$

Then by the distributive and de Morgan laws

$$u = v(\theta) \ \& \ u_1 = v_1(\theta) \Rightarrow u \cup u_1 = v \cup v_1(\theta) \ \& \ uu_1 = vv_1(\theta)$$
$$u = v(\theta) \Rightarrow u' = v'(\theta).$$

Thus θ is a congruence on $\mathfrak{B}$. Similarly, ψ is also a congruence on $\mathfrak{B}$.

As in the proof of Lemma 5.36, neither θ nor ψ is the identity on B but $\theta \cdot \psi$ is the identity. Hence (Appendix, §A) the two-element Boolean algebra is the only subdirectly irreducible Boolean algebra. Theorem 5.60 follows from Birkhoff's theorem. ∎

THEOREM 5.60. *Every Boolean algebra is isomorphic to a subdirect product of replicas of B_2.*

The representation theorem for Boolean algebras follows from this and the definition of 'field of sets'. The proof is similar to that of Theorem 5.39.

THEOREM 5.61. *Every Boolean algebra is isomorphic to a field of sets.*

The structure theorem for finite DL's (Theorem 5.38) was based on the notion of join-irreducible elements. A similar result holds for Boolean algebras, but this depends on the notion of an 'atom'.

DEFINITION 5.20. *An element a of a poset $\langle A, \leqslant \rangle$ with minimum element is an* atom *if every element x of A such that $x \leqslant a$ is either 0 or a itself.*

LEMMA 5.62. *An element x of a Boolean algebra such that $x \neq 0$ is join-irreducible iff it is an atom.*

Recall that, by the structure theorem for finite DL's (Theorem 5.35), associated with each element of a DL there is a set $N(x)$ of all join-irreducible elements y such that $y \leqslant x$. For an element x of a Boolean algebra we now define $At(x)$ to be the set of atoms a such that $a \leqslant x$. Then $N(x) = \{0\} \cup At(x)$. Further, $At(0) = \emptyset$ and $At(1)$ is the set of all atoms in the Boolean algebra.

THEOREM 5.63. *For any finite Boolean algebra $\mathfrak{B} = \langle B, \cap, \cup, ', 0, 1 \rangle$ the sets $At(x)$, where $x \in B$, are closed under union, intersection, and complementation with respect to $At(1)$; they form a Boolean algebra isomorphic to B.*

Proof. From Theorem 5.35, (5.15)–(5.17) hold. Hence

$$N(x) \cdot N(x') = N(xx') = N(0) = \{0\}$$
$$N(x) + N(x') = N(x \cup x') = N(1).$$

From this

$$At(x) + At(y) = At(x \cup y)$$
$$At(x) \cdot At(y) = At(xy)$$
$$At(x) = At(y) \Rightarrow x = y$$
$$At(x) \cdot At(x') = \emptyset$$
$$At(x) + At(x') = A(1).$$

The theorem follows from these relations. ∎

EXAMPLE 5.30. The lattice of Fig. 5.11 is distributive. The operation ′ defined by $a' = A$, $A' = a$, $b' = B$, $B' = b$, $c' = C$, $C' = c$, $0' = 1$, $1' = 0$ makes this lattice into a Boolean algebra. The atoms are a, b, c. Thus, $At(a) = \{a\}$, $At(b) = \{b\}$, $At(c) = \{c\}$. Hence $At(A) = \{b, c\}$, $At(B) = \{a, c\}$, $At(1) = \{a, b, c\}$, and $At(0) = \emptyset$. This Boolean algebra is isomophic to the Boolean algebra of subsets of $\{a, b, c\}$. ◆

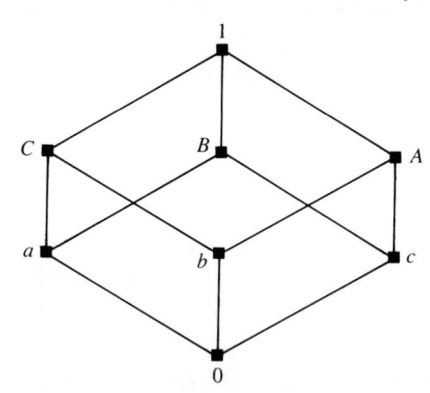

Fig. 5.11.

6

Constructing logics

6.1 Introduction

An important class of logics, typified by classical logic (Chapter 2), is based primarily on a concept of *truth*. The logical consequence functions of such logics are derived from the interpretation of a language by structures of truth-values. These logics could therefore be called *semantic logics*.

The purpose of this chapter is to treat semantic logics by a (classical) generalization of the principles on which classical logic is founded. Such logics, if they purport to represent anything at all, are therefore *classical representations*—of various grades of utility depending on applications—of states of affairs. They cannot totally replace classical or constructive logic. In the first section of this chapter the background is set with a brief discussion of many-valued logics. The remainder of the chapter is divided into three parts, in the first of which (§§6.3, 6.4) it will be shown how ordered truth-values can be organized as logics—the 'closure method'—particularly when the truth-values form a lattice. In the second part (§§6.5–6.9) a general method of constructing propositional and quantifier logics from logics of truth-values is defined, which is a generalization of the construction of classical logic from the two truth-values (Chapter 2, §2.2.2, §2.3.2). This method is applied to logics whose truth-values form classes of lattices. It gives rise, for instance, to 'quantum logic' (§6.9) and three-valued logics (Chapter 8). Quantum logics are treated very briefly in §6.9 as an application of these construction principles. An adequate treatment of the issues raised in the subject is beyond the scope of this book: the reader is referrred to the prolific output of the quantum industry, a minute fraction of which is quoted in the text. Finally (§§6.10, 6.11) there is a brief treatment of temporal logic and Kripke's model structures.

6.2 Many-valued logics

A standard technical device for investigating the independence of the axioms and rules of a system of propositional calculus is to use as truth-values the natural numbers $1, 2, \ldots, M$, the first s of these being selected as *designated* truth-values. One of these truth-values is assigned to each of the constants of the formal language of the calculus, and a *truth-table* in these truth-values is associated with each of the junctors. The value of a formula for given values of the propositional variables can then be computed. A formula is a *tautology* if it takes a designated value for each assignment of truth-values to its variables. If every rule of inference preserves tautologies and all axioms except one are tautologies, then the axiom is independent, i.e. it cannot derived from the remaining axioms.

Amongst the connectives which are frequently used in many-valued logics are *conjunction* $\wedge$, *disjunction* $\vee$, and the *conditional* $\supset$. Their truth-tables are defined by (Rosser and Turquette 1952)

$$p \wedge q = \max(p, q) \qquad p \vee q = \min(p, q) \qquad p \supset q = \max(1, q - p + 1)$$

for $p, q \in \{1, 2, \ldots, M\}$. Under the operations of conjunction and disjunction the truth-values acquire a lattice ordering. For this reason we shortly investigate the construction of logics and ordered sets of truth-values.

The employment of many-valued truth-tables for the purpose of proving independence of the axioms of the propositional calculus was introduced by P. Bernays in his *Habilitationsschrift* of 1918, but this was not published until 1926. This method was known, prior to its publication, by Łukasiewicz, who, independently of Bernays and following a suggestion of Tarski (Tarski 1956, p. 8), first applied his many-valued logics to proofs of independence and subsequently defined the general method.

The general notion of a three-valued logic was probably first discussed in depth by N. A. Vasil'ev in four papers published between 1910 and 1913 (see Kline 1965), the idea of 'non-Aristotelian' logic being obtained by abandoning the law of contradiction. But credit for the development of three-valued formal logics is usually accorded to Łukasiewicz, Post and Tarski. The n-valued systems were discovered in 1922 and reported in 1929. Independently of Łukasiewicz, Post, in his thesis of 1920 (published in 1921), studied many-valued logics from a purely abstract standpoint. In contrast with the logics of Łukasiewicz, Post allowed more than one designated value.

Since the pioneering work of Łukasiewicz, many-valued logic has developed into a major, albeit esoteric, industry based on philosophical confusion and, until recent events in computer science, lack of concrete application. Quite recently there has been a renewed interest in the subject arising from computer chip technology (IEEE 1976). For instance, the increased reliability of miniaturized integrated circuits has made many-valued logic circuits distinctly feasible if not desirable. Many-valued digital signals increase the flow per ip–op pin so that the pin limitation problem of integrated circuit chips can be overcome.

Further information on the development of many-valued logics is to be found in Church (1956, §29) and Tarski (1956, Chapter 4). Rescher (1969) gives a 49 page bibliography. See also Wolf (1977) and Urquhart (1986).

It can legitimately be asked whether many-valued logics have any use except as artefacts for manufacturing independence results. Many logicians think that many-valued logics have no ready interpretation and so can hardly count as logics. Beth (1959, p. 231) remarked:

> ... at present the various languages of modal logic and many-valued logics are merely studied, they are hardly ever used.

See also Scott (1972, p. 266). Łukasiewicz justified his own systems of many-valued logics by appeal to future contingent events. This reasoning is not regarded very seriously. Chwistek (1948, p. 132) wrote that:

> ... the philosophical arguments which Łukasiewicz employs and in particular the Aristotelian classification of events into contingent and necessary are quite naive and antagonize the reader.

Nevertheless there is a three-valued logic which does have a clear and down-to-earth application. This logic will be discussed in Chapter 8. It is constructed by organizing the three truth-values as a lattice and then constructing a logic by a general principle from the lattice.

Generally, the study of many-valued 'logics' is tautology-oriented. The rules of inference are designed so that only tautologies can be derived from tautologies. Łukasiewicz seemed not to have been interested in the notion of logical consequence. It has been suggested (Smiley 1962, p. 435) that many-valued logics should be so organized that the truth-values themselves form a logic in the sense of Definition 3.7. This approach was called the 'closure method' by Smiley, who pointed out that if K is a set of truth-values with a subset D of designated values then a consequence function Cn can be defined by

$$Cn(S) =_{\text{def}} \begin{cases} D & \text{if } S \subseteq D \\ K & \text{otherwise} \end{cases} \qquad (6.1)$$

Thus

$$x \in Cn(S) \quad \Leftrightarrow \quad (S \subseteq D \Rightarrow x \in D) \qquad (6.2)$$

Then $\langle K, Cn \rangle$ is a logic whose theorems $Cn(\emptyset)$ are exactly the set D of designated values (see Example 3.10(iii)). This approach gives a finer analysis than can be obtained by concentrating on tautologies. Smiley has remarked that the same truth-values and truth-tables as were used by Łukasiewicz can be employed to define at least three logics whose theorems are the same but which differ in such matters as the validity of *modus ponens* and the equivalence of a conjunction to the joint force of its components.

6.3 Logics from truth-values

Consider the general problem of defining a logical consequence function on a structured set of truth-values.

Let $\mathfrak{A}$ be a structure of type τ (relational system) of truth-values. Consequence functions can be defined on $\mathfrak{A}$ in many ways. A particularly flexible instrument in framing such definitions is the first-order language with equality $L'(\tau)_=$, of $\mathfrak{A}$ containing, in addition, the predicate symbols corresponding to the relations of $\mathfrak{A}$, a one-place predicate symbol S. Let $\Psi(S, x)$ be a formula of this language with at most one free variable x (see Chapter 2, §2.3.2). Then from $\Psi(S, x)$ the set-to-set function $Cn_\Psi \colon \mathbb{P}(|\mathfrak{A}|) \to \mathbb{P}(|\mathfrak{A}|)$ can be defined by

$$a \in Cn_\Psi(Y) \quad \Leftrightarrow \quad (\mathfrak{A}, Y, a) \vDash \Psi(S, x). \tag{6.3}$$

(On the right-hand side, the predicate symbol S is interpreted as the set Y and the individual variable x is interpreted as the element a.) The requirement that Cn_Ψ be a consequence function imposes some conditions on $\Psi(S, x)$. It would be of some interest to find necessary and sufficient syntactical conditions on $\Psi(S, x)$ that this be the case. However, we can define a particularly simple syntactical form for which it is easy to prove that it defines a consequence function. Theorem 6.1 easily follows from Example 3.10(iii).

THEOREM 6.1. Let $\mathfrak{A} = \langle A, o_1, o_2, \ldots \rangle$ be an algebra of type τ and let Ψ be a formula of $L(\tau)_=$. Let $\bar{y}$ stand for the string of variables $y_1, \ldots, y_n$. Define the equivalence formula $E(z, x)$ to be $(\bar{y})(\psi(\bar{y}, x) \equiv \psi(\bar{y}, z))$ and, for $i = 1, 2, \ldots$, the congruence formula con_i to be $(E(x_1, z_1) \,\&\, E(x_2, z_2) \,\&\, \cdots \Rightarrow E(o_i(x_1, x_2, \ldots), o_i(z_1, z_2, \ldots)))$. Let ψ be a formula of $L'(\tau)_=$ (Chapter 2, §2.2.5) with free variables $y_1, \ldots, y_n, z$ and not containing the predicate symbol S. Then the formula

$$\Psi(S, x) = (\bar{y})(Ez)[(S(z) \,\&\, \sim\!\psi(\bar{y}, z)) \vee \psi(\bar{y}, x)]$$

defines a consequence function Cn_Ψ on $\mathfrak{A}$ and E defines the associated logical equivalence. Further, if $\mathfrak{A}$ satisfies $con_1 \,\&\, con_2 \,\&\, \cdots$ then the operations of $\mathfrak{A}$ are connectives in the logic $\langle \mathfrak{A}, Cn_\Psi \rangle$.

Corollary 5.15 shows that if the chosen logic of truth-values is equipped with a normal conjunction then the truth-values have a quasi-ordering which is the converse of the implication relation. Thus we start by assuming that the truth-values form a qoset.

Consider now the problem of constructing a consequence function on a qoset $\mathfrak{D} = \langle T, \leqslant \rangle$ of truth-values. Our definition of a consequence function on $\mathfrak{D}$ is motivated by an established notion of *compactness* for elements of a complete lattice. This notion was first introduced by Nachbin (1949) and Büchi (1952). It has some connection with chain conditions.

DEFINITION 6.1.
(i) A poset P such that each of its non-empty subsets contains a maximal element is said to satisfy the *ascending chain condition* (ACC). If each

non-empty subset of P contains a minimal element then P satisfies the *descending chain condition* (DCC).
(ii) An element c of P is said to be *compact* if, whenever $c \geqslant glb(\leqslant : S)$, there exists a finite subset Y of S such that $c \geqslant glb(\leqslant : Y)$ (Definition 5.7).

P satisfies the ACC iff it contains no infinite sequence of elements $a_1, a_2, a_3, \ldots$ such that $a_1 < a_2 < a_3 < \cdots$, and it satisfies the DCC iff it contains no infinite sequence $a_1, a_2, a_3, \ldots$ such that $a_1 > a_2 > a_3 > \cdots$ (Birkhoff 1948, p. 37). It can easily be shown (Crawley and Dilworth 1973, p. 14) that every element of P is compact iff P satisfies the DCC.

For our purpose of investigating logics the reference to completeness, except where quantifiers are concerned (i.e. greatest lower bounds of arbitrary sets), can be avoided by observing that in a given qoset

$$c \geqslant glb(S) \quad \Leftrightarrow \quad (y \in L(S))(c \geqslant y)$$

(cf. Definition 5.7). Thus

$$c \geqslant glb(S) \quad \Leftrightarrow \quad L(c) \supseteq L(S).$$

This enables compactness to be defined for elements of an arbitrary qoset.

DEFINITION 6.2. Let $\langle Q, \leqslant \rangle$ be a qoset. An element $c \in Q$ is *compact* if, for all $S \subseteq Q$, whenever $L(c) \supseteq L(S)$ there exists a finite subset Y of S such that $L(c) \supseteq L(Y)$. $\langle Q, \leqslant \rangle$ is said to be *compact* if every element of it is compact.

Once the completeness condition has been dropped the connection between DCC and compactness is no longer so transparent, as it is not too difficult to construct qosets which are compact but which do not satisfy DCC and there are qosets which satisfy DCC but are not compact. However, the following theorem holds.

THEOREM 6.2. Let $\langle P, \leqslant \rangle$ be a poset in which $glb(\{x, y\}) =_{def} xy$ exists for all $x, y \in P$. If $\langle P, \leqslant \rangle$ satisfies DCC then it is compact.

Proof. Let $S \subseteq P$ and let x be an element of P such that

$$L(x) \supseteq L(S). \tag{6.4}$$

We prove that there exists a finite subset Y of S such that

$$L(x) \supseteq L(Y). \tag{6.5}$$

For a finite set $Y = \{y_1, y_2, \ldots, y_n\}$ let $\pi(Y) = y_1 y_2 \cdots y_n$. Then

$$L(\pi(Y)) = \pi(\{L(y_i), 1 \leqslant i \leqslant n\}) = L(Y). \tag{6.6}$$

Let $S' = \{\pi(Y), Y \in \mathring{\mathbb{P}}_\omega(S)\}$. Then since $\langle P, \leqslant \rangle$ satisfies DCC, S' has a minimal element $\pi(Y)$, say. Thus, for all $Z \in \mathring{\mathbb{P}}_\omega(S)$, $\pi(Z) \leqslant \pi(Y) \Rightarrow \pi(Z) = \pi(Y)$. By (6.6), $L(Z) \subseteq L(Y) \Rightarrow L(Z) = L(Y)$. But $L(Z + Y) = L(Z) \cdot L(Y) \subseteq L(Y)$. Hence $L(Z) \cdot L(Y) = L(Y)$, i.e. $L(Z) \supseteq L(Y)$. So $L(S) = L(Y)$. (6.5) now follows from (6.4). ∎

We return to the problem of making a logic out of a qoset of truth-values.

DEFINITION 6.3. Consider the qoset $\mathfrak{D} = \langle T, \leqslant \rangle$. $Cn_\mathfrak{D}$ is the set-to-set function on T defined by

$$Cn_\mathfrak{D}(X) =_{\text{def}} U(\leqslant : L(\leqslant : X)) \qquad (X \subseteq T). \qquad (6.7)$$

(cf. Definition 5.7). $\mathfrak{D}^\ddagger$ is the prelogic $\langle T, Cn_\mathfrak{D} \rangle$.

EXAMPLE 6.6.

(i) It follows immediately from the definitions of U and L that

$$x \in Cn_\mathfrak{D}(X) \Leftrightarrow x \geqslant y \qquad (y \in L(X)).$$

Hence

$$x \in Cn_\mathfrak{D}(X) \Leftrightarrow L(x) \supseteq L(X) \qquad (6.8)$$

and, in particular,

$$x \in Cn_\mathfrak{D}(y) \Leftrightarrow y \leqslant x. \qquad (6.9)$$

(ii) For $\mathfrak{D}$ a complete lattice, $x \in Cn_\mathfrak{D}(X) \Leftrightarrow x \geqslant glb(X)$.
(iii) If $\mathfrak{D}$ has a maximum element 1 and a minimum element 0 then from (6.8) and Chapter 5, §5.6, we have

$$Cn_\mathfrak{D}(\varnothing) = Cn_\mathfrak{D}(1) = \{1\} \qquad Cn_\mathfrak{D}(0) = T. \qquad (6.10)$$

◆

Example 6.1(i) establishes the relation between the consequence function and the definition of *compact element* in Definition 6.2. The right-hand side of (6.8) can be expressed by the first-order formula

$$(z)((y)(y \in X \supset y \leqslant z) \supset z \leqslant x)$$

or, in prenex form, $(z)(Ey)[y \in X \wedge \neg(z \leqslant y) \vee z \leqslant x]$. This formula is of the kind considered in Theorem 6.1 (with $z \in X$ in place of $S(z)$). Thus logical equivalence is expressed by the formula $(z)(z \leqslant x \equiv z \leqslant y)$, i.e. $x \leqslant y \wedge y \leqslant x$, and so is the natural congruence (Chapter 5, §5.5). ◆

Logics from lattices

COROLLARY 6.3. $\mathfrak{D}^{\ddagger}$ is a logic in which the relation of logical equivalence is the natural congruence on $\mathfrak{D}$. Moreover, $\mathfrak{D}^{\ddagger}$ is a simple logic (Definition 3.7) if $\mathfrak{D}$ is a poset. Further, it is a compact logic iff $\mathfrak{D}$ is a compact qoset (Definition 6.2).

EXAMPLE 6.2. By (6.8) and the definition of compactness, if $\mathfrak{D}$ is a compact lattice then $Cn_{\mathfrak{D}}$ is the smallest ideal including X. ◆

Corollary 6.3 together with (6.9) enable the lattice conditions of Chapter 5 which hold in $\mathfrak{D}$ to be translated into equivalent logical conditions of Chapter 3 which hold in the associated logic $\mathfrak{D}^{\ddagger}$ given by Definition 6.3. (The compactness condition in this statement follows from (6.8).)

EXAMPLE 6.3. Consider the algebra $\mathbb{T}v = \langle \{0, 1\}, ', \cap, \cup, \supset, 0, 1 \rangle$ of truth-values for the classical logic (Chapter 2, §2.2.2). This is a Boolean algebra and, by (6.8),

$$Cn_{\mathbb{T}v}(\varnothing) = Cn_{\mathbb{T}v}(1) = \{1\} \qquad Cn_{\mathbb{T}v}(0) = Cn_{\mathbb{T}v}(0, 1) = \{0, 1\}. \qquad ◆$$

Where the ordered set T of truth-values has a non-empty set M of maximum elements, these can be regarded as a set of designated values and a consequence function Cn can be defined on T by Smiley's definition (6.2):

$$x \in Cn(X) \Leftrightarrow (X \subseteq M \Rightarrow x \in M). \qquad (6.11)$$

It can easily be shown that Cn is coarser then $Cn_{\mathfrak{D}}$ in the sense that $Cn_{\mathfrak{D}}(X) \subseteq Cn(X)$ for all $X \subseteq T$. This is not surprising since the order relation enters into the construction of Cn only in so far as it determines the designated values M, whereas the full force of the order relation is used in $Cn_{\mathfrak{D}}$. Moreover, Cn entails a non-simple logic, thus putting some unequal truth-values logically equivalent.

6.4 Logics from lattices

Where the ordered set $\mathfrak{D}$ is a poset, it follows from (6.9) that logical equivalence as determined by $Cn_{\mathfrak{D}}$ is the identity. Thus, if $\mathfrak{D}$ is a lattice then its associated logic (Definition 6.3) $\mathfrak{D}^{\ddagger}$ is a simple logic and every lattice operation of $\mathfrak{D}$ is therefore a logical operation. We consider now the problem of strongly logical operations.

Throughout this subsection we consider a fixed lattice $\mathfrak{D} = \langle T, \cap, \cup \rangle$ - sometimes with distinguished elements such as maximum and minimum elements or with further operation such as complementation—and its related logic $\mathfrak{D}^{\ddagger} = \langle T, \cap, \cup, Cn_{\mathfrak{D}} \rangle$.

6.4.1 Contradiction and theoremhood

It follows from (6.8) that an element of the logic is contradiction in the logic iff it is the maximum element of $\mathfrak{D}$, and it is theoremhood in the logic iff it is the minimum element of $\mathfrak{D}$.

6.4.2 Conjunction and disjunction

By Theorem 4.16 and Example 5.12(ii), the Lindenbaum algebra of a logic with conjunction and disjunction is a lattice. Moreover, if these connectives are normal then they are strongly logical operations and the lattice is distributive. Consider now the reverse process of constructing a logic from a lattice.

THEOREM 6.4. Let $\mathfrak{D}$ be a lattice. Then we have the following.

(i) Meet is a strong connective of $\mathfrak{D}^{\ddagger}$. Also, join is a strong connective iff $\mathfrak{D}$ is distributive.
(ii) **Cj1**†, **Cj2**†, **Dj1**† hold in $\mathfrak{D}^{\ddagger}$.
(iii) If $\mathfrak{D}$ is compact then **Dj2**† holds in $\mathfrak{D}^{\ddagger}$ iff $\mathfrak{D}$ is distributive.

Proof.

(i) To show that meet is a strongly logical operation it suffices to prove that for all $x, y, t \in T$ and $\Delta \subseteq T$

$$x \leftrightarrow_\Delta y \quad \Rightarrow \quad xt \leftrightarrow_\Delta yt \tag{6.12}$$

where $Cn_\mathfrak{D}$ is defined by (6.7) and $\leftrightarrow_\Delta$ denotes $Cn^e_{\mathfrak{D},\Delta}$, the logical equivalence relation determined by it (Definitions 3.2(ii) and 3.10). First,

$$\begin{aligned} x \in Cn_{\mathfrak{D},\Delta}(y) &\Leftrightarrow x \in Cn_\mathfrak{D}\Delta, y) \\ &\Leftrightarrow L(\Delta, y) \subseteq L(x) \quad \text{by (6.8)} \\ &\Leftrightarrow L(\Delta) \cdot L(y) \subseteq L(x) \end{aligned} \tag{6.13}$$

and

$$t \in L(y) \Leftrightarrow t \leqslant y. \tag{6.14}$$

Next, suppose that $x \leftrightarrow_\Delta y$. Let (6.14),

$$L(\Delta) \cdot L(y) \subseteq L(x). \tag{6.15}$$

We show that

$$L(\Delta) \cdot L(yt) \subseteq L(xt). \tag{6.16}$$

Suppose that $z \in L(\Delta) \cdot L(yt)$. Then $z \in L(\Delta)$ and, by (6.14), $z \leqslant yt$. Hence $z \leqslant y$ and $z \leqslant t$. So by (6.15) and (6.16), $z \leqslant x$. As $z \leqslant t$, $z \leqslant xt$, i.e. $z \in L(xt)$.

This proves (6.16). By (6.13), $xt \in Cn_{\mathfrak{D},\Delta}(yt)$. Similarly $yt \in Cn_{\mathfrak{D},\Delta}(xt)$. Thus $xt \leftrightarrow_\Delta yt$. This proves (6.12).

Consider now the join operation. Suppose that $\mathfrak{D}$ is distributive. We show that for all $x, y, t \in T$ and $\Delta \subseteq T$,

$$x \leftrightarrow_\Delta y \quad \Rightarrow \quad x \cup t \leftrightarrow_\Delta y \cup t. \tag{6.17}$$

Suppose that $x \leftrightarrow_\Delta y$. We first prove that

$$L(\Delta) \cdot L(y \cup t) \subseteq L(x \cup t). \tag{6.18}$$

Let $z \in L(\Delta) \cdot L(y \cup t)$. By (6.14), $z \in L(\Delta)$ and $z \leqslant y \cup t$. Now $zy \leqslant y$ and $zy \leqslant z$. Hence $zy \in L(\Delta)$ and $zy \in L(y)$. By (6.12), (6.14), and $x \leftrightarrow_\Delta y$, we have $zy \leqslant x$. Hence $zy \cup t \in L(x \cup t)$. Then by the distributive law, $zy \cup t \geqslant z(y \cup t)$. As $z \leqslant y \cup t$, $z = z(y \cup t)$. Hence $z \in L(x \cup t)$. This proves (6.18). By (6.13), $y \cup t \in Cn_{\mathfrak{D},\Delta}(x \cup t)$. Similarly, $x \cup t \in Cn_{\mathfrak{D},\Delta}(y \cup t)$. Hence $x \cup t \leftrightarrow_\Delta y \cup t$. This proves (6.17).

Conversely, suppose (6.17). We prove that the distributive law holds in $\mathfrak{D}$. Let $z \in T$. By (6.13)

$$x \leftrightarrow_\Delta y \quad \Leftrightarrow \quad L(z) \cdot L(x) = L(t) \cdot L(y).$$

But $L(z) \cdot L(x) = L(zx)$. Hence

$$x \leftrightarrow_\Delta y \quad \Leftrightarrow \quad xz = zy. \tag{6.19}$$

By (6.18) and (6.16)

$$zx = zy \quad \Rightarrow \quad z(x \cup t) = z(y \cup t). \tag{6.20}$$

Putting $y = zx$ in (6.20) gives the identity $z(x \cup t) = z(zx \cup t)$. Substituting t for x and zx for t in this gives $z(t \cup xz) = z(zt \cup zx)$ which in turn yields

$$z(x \cup t) = z(zt \cup zx) \leqslant zt \cup zx.$$

By Theorem 5.29, the distributive law **K6** holds.

(ii) Consider **Cj1**†. For any elements $x, y \in T$, $x \geqslant xy$ by the lattice laws. By (6.10), $x \in Cn_{\mathfrak{D}}(xy)$. By Corollary 6.3, $x \in Cn_{\mathfrak{D}}(xy, \Delta)$ for every $\Delta \subseteq T$. Similarly, $y \in Cn_{\mathfrak{D}}(xy, \Delta)$. Hence **Cnj1**† holds. The proofs that **Cnj2**†, **Dj1**† hold follow similarly from the lattice laws and the properties of the consequence function $Cn_{\mathfrak{D}}$.

(iii) Suppose that $\mathfrak{D}$ is compact. By Corollary 6.3, $\mathfrak{D}^\ddagger$ is also compact. Suppose now that $\mathfrak{D}$ is distributive. Let $d \in Cn_{\mathfrak{D}}(\Delta, x) \cdot Cn_{\mathfrak{D}}(\Delta, y)$. Then by compactness of $\mathfrak{D}^\ddagger$, there exist finite subsets Δ', Γ' of Δ such that $d \in Cn_{\mathfrak{D}}(\Delta', x) \cdot Cn_{\mathfrak{D}}(\Gamma', y)$. Let $k = \text{glb}(\Delta' + \Gamma')$. But $Cn_{\mathfrak{D}}(k, x) \supseteq Cn_{\mathfrak{D}}(\Delta', x)$ and $Cn_{\mathfrak{D}}(k, y) \supseteq Cn_{\mathfrak{D}}(\Gamma', y)$. Hence $d \in Cn_{\mathfrak{D}}(k, x) \cdot Cn_{\mathfrak{D}}(k, y)$. By (6.9), $d \geqslant kx$ and $d \geqslant ky$. Thus $d \geqslant kx \cup ky$. As the lattice is distributive, $d \geqslant k(x \cup y)$. Thus $L(d) \supseteq L(k, x \cup y) = L(\Delta' + \Gamma', x \cup y) \supseteq L(\Delta, x \cup y)$. By (6.8), $d \in Cn_{\mathfrak{D}}(\Delta, x \cup y)$. Thus **Dj2**† holds. Conversely, suppose that **Dj2**† holds in $\mathfrak{D}^\ddagger$. Let $x, y, z \in T$. Then by the lattice laws, $xy \cup yz \geqslant xz$ and $xy \cup yz \geqslant yz$.

By (6.9), $xy \cup yz \in Cn_{\mathfrak{D}}(x, y) + Cn_{\mathfrak{D}}(y, z)$. But by **Dj2**†, $xy \cup yz \in Cn_{\mathfrak{D}}(x \cup z, y) = Cn_{\mathfrak{D}}(y(x \cup z))$. By (6.9), $xy \cup yz \geqslant y(x \cup z)$. Hence the lattice is distributive. ∎

6.4.3 Implication

It was proved in Theorem 5.47 that in a logic $\mathfrak{D} = \langle S, \wedge, \vee, \supset, Cn \rangle$ with normal conjunction, disjunction, and implication, the Lindenbaum algebra $\mathfrak{D}\P$ (Example 3.9) is an I-lattice. As a kind of converse we have the following theorem.

THEOREM 6.5. Let $\mathfrak{D} = \langle T, \cap, \cup, \rightarrow \rangle$ be a DL with a binary operation $\rightarrow$. Then $\mathfrak{D}$ is an I-lattice iff $\rightarrow$ is normal implication in the associated logic $\mathfrak{D}^\ddagger$.

Proof. By (6.9) the condition $x \leqslant a \rightarrow b \Leftrightarrow ax \leqslant b$ is equivalent to $a \rightarrow b \in Cn_{\mathfrak{D}}(x) \Leftrightarrow b \in Cn_{\mathfrak{D}}(ax)$. ∎

6.4.4 Negation

Various kinds of complementation in lattices give rise to negation (Chapter 4, §4.2.2) in the associated logics.

THEOREM 6.6. Let $\mathfrak{D} = \langle T, \cap, \cup, 0, 1, ' \rangle$ be a lattice with minimum element 0, maximum element 1, and a unary operation '.

 (i) $\mathfrak{D}$ is pseudo-complemented iff $\mathfrak{D}^\ddagger$ satisfies **N5**.
 (ii) ' is complementation in $\mathfrak{D}$ iff $\mathfrak{D}^\ddagger$ satisfies **N1** and **N2**^{0}.
 (iii) ' is ortho-complementation in $\mathfrak{D}$ iff $\mathfrak{D}^\ddagger$ satisfies **N1**, **N2**^{0}, **N3.1**^{0}, **N4.1**, and **N4.2**.
 (iv) Then $\mathfrak{D}$ is a Boolean algebra iff meet and join are normal conjunction and disjunction respectively in $\mathfrak{D}^\ddagger$ and $\mathfrak{D}^\ddagger$ satisfies **N1**, **N3.3**, and **N5**.

Proof. (6.9) provides the connection between the logical laws and the algebraic relations in the lattice. For example, in (i), the condition (Definition 5.16) $x \leqslant y' \Leftrightarrow yx = 0$ is equivalent, by Definition 6.3, to $y' \in Cn_{\mathfrak{D}}(x) \Leftrightarrow 0 \in Cn_{\mathfrak{D}}(x, y)$ which is **N5**. (ii), (iii), and (iv) are proved similarly. ∎

6.5 Logical consequence in classical logic

The concept of logical consequence in the classical two-valued logic comes from the notion of satisfaction which itself depends on the truth-table interpretation of the propositional junctors (see Chapter 2, §2.2.2).

The thinkers of the Megarian–Stoic school established correct truth-matrices for the most significant junctors. The idea of truth-values as the values of functions was first described by Frege. His doctrine was derived from his principle according to which every proposition is a name for truth or falsehood. This idea is expounded by Church (1956, p. 23ff), though it has not been generally followed. Definitions of propositional junctors by means of tables occur in Boole, though with no explicit reference to truth-values, and later in Frege's *Begriffschrift*. Peirce expressed the idea explicitly in 1880. The same idea was systematically developed by Lukasiewicz, Post, and Wittgenstein in about 1920.

The basic assumptions behind classical logic are, first, that logic concerns propositions with well-defined truth-values (true, false) and that the junctors by which new propositions are formed from given ones are such that the truth-value of a compound formula depends only on the truth-values of its components. The historical basis of the pre-eminence of the connectives of negation, conjunction, and disjunction lies in the Greek tradition—Kreisel and Krivine (1967, p. 4) call them the 'Aristotelian connectives'.

Tarski (1956, p. 413) remarked that the first attempt to formulate a precise definition of the notion of logical consequence was due to Carnap (1935). He added that this attempt was connected rather closely with the particular properties of the formalized language concerned. Carnap's definition can be formulated as follows:

> The sentence X *follows logically* from the sentences of the class K if, and only if, the class consisting of all the sentences of K and of the negation of X is contradictory.
>
> Tarski 1956, p. 413

Tarski observed that the decisive element in this definition is the term 'contradictory'. It should be added that the concept is not applicable to languages without negation (e.g. positive logic) or where negation behaves in an unusual fashion (e.g. the three-valued logic discussed in Chapter 8). Tarski's definition is:

> The sentence X *follows logically* from the sentences of the class K if, and only if, every model of the class K is also a model of the sentence X.
>
> Tarski 1956, p. 417

Tarski acknowledged the analogy between this definition and one suggested 100 years previously by Bolzano (Tarski 1956, p. 417n).

The definition of logical consequence given in Definition 2.11 proceeds from the notion of *model*. A model of a set Δ of formulas is a valuation $v \in \mathbb{V}_{\text{pred}}$ which makes every formula of Δ true, i.e. $v(A) = 1$ for every $A \in \Delta$ (this can also be expressed as $v(\Delta) \subseteq \{1\}$). Thus, using the logical consequence function $Cn_{\mathbb{T}v}$ on the algebra $\mathbb{T}v$ of truth-values (Example 6.1), we have

$$A \in Cn(\Delta) \quad \Leftrightarrow \quad (v \in \mathbb{V}_{\text{pred}})(v(A) \in Cn_{\mathbb{T}v}(\Delta)). \tag{6.21}$$

This shifts the emphasis away from the *algebra* of truth-values towards truth-values as a *logic*.

6.6 Logical consequence in semantic logics

We now consider propositional calculi interpreted in arbitrary logics of truth-values. Let $\mathbb{T} = \langle T, o_1, o_2, \ldots, Cn \rangle$ be a logic of type τ with connectives $o_1, o_2, \ldots$. (Think of $\mathbb{T}$ as being the logic of the truth-values.) A propositional logic based on this structure of truth-values is constructed in an analogous way to the construction of the classical propositional calculus in Chapter 2, §2.2.1.

A formal language L is defined as follows. The primitive symbols of the language L are the pairwise disjoint alphabets P, J, G where P and G are as in Chapter 2, §2.2.1, and J is the list of symbols (one for each of the operations o_i)

$$c_1, c_2, c_3, \ldots$$

called *junctors*.

The class $\mathbb{F}(J, P)$ of *formulas* is the subset of L^* defined inductively by the following rules (cf. Definition 2.1):

(F1) $P \subseteq \mathbb{F}(J, P)$
(F2) for $i = 1, 2, 3, \ldots$, if $F_1, \ldots, F_n \in \mathbb{F}(J, P)$ and a_i is an n-place operation then $c_i[F_1, \ldots, F_n] \in \mathbb{F}(J, P)$.

Note that $\mathbb{F}(J, P)$ is the set of terms (Definition 2.10) in a first-order language $L(\tau)$ with P as the set of free variables. As in Chapter 2, §2.2.1, the junctors can be considered as operations on $\mathbb{F}(J, P)$. Then $\mathbb{F} = \langle \mathbb{F}(J, P), c_1, c_2, \ldots \rangle$ is the *algebra of formulas*.

Every mapping of P into T can be uniquely extended to homomorphism of $\mathbb{F}$, the algebra of formulas, into $\mathbb{T}$, the algebra of truth-values. Such homomorphisms are the *valuations* (Definition 2.3). Let $\mathbb{V}$ be the set of valuations. (6.21) can then be used to construct a set-to-set function $Cn[\mathbb{T}]$ on $\mathbb{F}$: for any set Δ of formulas

$$A \in Cn[\mathbb{T}](\Delta) \Leftrightarrow_{\text{def}} (v \in \mathbb{V})(v(A) \in Cn(v(\Delta))).$$

Thus

$$Cn[\mathbb{T}](\Delta) = \prod \{v^{-1} Cn(v(\Delta)); v \in \mathbb{V}\} \tag{6.22}$$

By Theorem 3.3, every SDPC which holds in the logic of truth-values holds in the prelogic $\mathfrak{L}[\mathbb{T}] = \langle \mathbb{F}(J, P), c_1, c_2, \ldots, Cn[\mathbb{T}] \rangle$. Hence we have the following theorem.

Logical consequence in classical logic 205

THEOREM 6.7. $\mathfrak{L}[\mathbb{T}]$ is a logic and every SDPC which holds in $\mathbb{T}$ also holds in $\mathfrak{L}[\mathbb{T}]$.

Thus every logical identity which holds in $\mathbb{T}$ also holds in $\mathfrak{L}[\mathbb{T}]$. For this reason we call $\mathfrak{L}[\mathbb{T}]$ the *propositional logic of T*. Its formulas under the relation of logical equivalence are sometimes called the *polynomials over* $\mathbb{T}$, and when $\mathbb{T}$ is a lattice they are *lattice polynomials*.

EXAMPLE 6.4. If $\mathbb{T}$ is the two-element Boolean algebra B_2 then $\mathfrak{L}[\mathbb{T}]$ is the classical propositional logic. By Theorem 6.7, $\mathfrak{L}[\mathbb{T}]\P$ is itself a Boolean algebra. ◆

It may be asked whether the process of constructing a new logic $\mathfrak{L}[\mathbb{T}]$, from the given logic $\mathbb{T}$ can repreated to obtain yet another new logic. This, however, is not the case.

THEOREM 6.8. $\mathfrak{L}[\mathfrak{L}[\mathbb{T}]] = \mathfrak{L}[\mathbb{T}]$.

Proof. $Cn[\mathfrak{L}[\mathbb{T}]](\varDelta) = \prod \{u^{-1} Cn[\mathbb{T}](u(\varDelta)); \ v \in \mathbb{V}'\}$ where $\mathbb{V}'$ is the set valuation in $\mathbb{F}(J, P)$, i.e. homomorphisms of $\mathbb{F}(J, P)$ into $\mathbb{F}(J, P)$. Further, $Cn[\mathbb{T}](u(\varDelta) = \prod \{v^{-1} Cn(v(u(\varDelta))); \ v \in \mathbb{V}\}$. Hence

$$Cn[\mathfrak{L}[\mathbb{T}]](\varDelta) = \prod \{(u \circ v)^{-1} Cn(u \circ v(\varDelta)); \ u \in \mathbb{V}', \ v \in \mathbb{V}\}.$$

But $\{u \circ v; u \in \mathbb{V}, v \in \mathbb{V}\} = \mathbb{V}$. Thus $Cn[\mathfrak{L}[\mathbb{T}]](\varDelta) = Cn[\mathbb{T}](\varDelta)$. ∎

The above construction can be widened still further by interpreting formulas in a class of similar logics. Thus let K be such a class. Define the logical consequence function $Cn[K]$ on the algebra of formulas by

$$Cn[K](\varDelta) =_{\text{def}} \prod \{Cn[\mathbb{T}](\varDelta); \ \mathbb{T} \in K\}. \tag{6.23}$$

By Theorem 3.3, every SDPC which holds in every member of K also holds in the prelogic $\mathfrak{L}[K] =_{\text{def}} \langle \mathbb{F}(J, P), c_1, c_2, \ldots, Cn[K] \rangle$. In particular, $\mathfrak{L}[K]$ is a logic such that every logical identity which holds in every member of K also holds in $\mathfrak{L}[K]$.

Now suppose that K is the class of models of a set Θ (so that Θ is a set of closed formulas in the first-order language $L(\tau)_=$). Let ψ be a formula of $L(\tau)_=$ such that the formulas con_i of Theorem 6.1 are consequences of Θ. Then by Theorem 6.1 the operations of each algebra $\mathfrak{A}$ in K are logical operations with respect to the logical consequence function defined by $\Psi(S, x)$ on $\mathfrak{A}$. Thus K can be considered as a class of similar logics.

THEOREM 6.9. $\mathfrak{L}[K]$ is compact.

Proof. Let $\varDelta, A \subseteq \mathbb{F}(J, P)$. Then by (6.17) and (6.19), recalling that $\mathbb{F}(J, P)$ is the set of terms of $L(\tau)$, $A \in Cn[K](\varDelta)$ iff for each $A \in K$ and each valuation

α of $\mathbb{F}(J, P)$ in $\mathfrak{A}$ (i.e. homomorphism of the algebra of formulas into $\mathfrak{A}$)

$$(\mathfrak{A}, \alpha(\Delta), \alpha) \vDash (\bar{z})((Ey)(S(y) \ \& \sim\psi(\bar{z}, y)) \vee \psi(\bar{z}, A)). \tag{6.24}$$

Since Δ is a set of terms, (6.24) is equivalent to

$$(\mathfrak{A}, \alpha) \vDash (\bar{z})(\bigvee_{t \in \Delta} \sim \psi(\bar{z}, t) \vee \psi(\bar{z}, A)). \tag{6.25}$$

($\bigvee$, $\bigwedge$ denote arbitrary disjunctions and conjunctions respectively, cf. Chapter 2, §2.2.2.)

To prove compactness it is sufficient to prove that if $A \notin Cn[K](\Gamma)$ for all $\Gamma \in \mathring{P}_\omega(\Delta)$ then $A \notin Cn[K](\Delta)$. So suppose that $A \notin Cn[K](\Gamma)$ for all $\Gamma \in \mathring{P}_\omega(\Delta)$. Then by (6.18), for each $\Gamma \in \mathring{P}_\omega(\Delta)$ there exists $\mathfrak{A} \in K$ and a valuation α of $\mathbb{F}(J, P)$ in $\mathfrak{A}$ such that

$$(\mathfrak{A}, \alpha) \vDash (E\bar{z})(\bigwedge_{t \in \Gamma} \psi(\bar{z}, t) \ \& \sim\psi(\bar{z}, A)).$$

Thus the set of $L(\tau)_=$ formulas $\Theta + \{\psi(\bar{z}, t); t \in \Gamma\} + \{\sim\psi(\bar{z}, A)\}$ is consistent. By the compactness theorem of classical predicate logic (Theorem 2.29) $\Theta + \{\psi(\bar{z}, t); t \in \Delta\} + \{\sim\psi(\bar{z}, A)\}$ is consistent. Thus there exists a $\mathfrak{B} \in K$ and a valuation β of $\mathbb{F}_{prop}(\alpha)$ in $\mathfrak{B}$ such that

$$(\mathfrak{B}, \beta) \vDash \bigwedge_{t \in \Delta} \psi(\bar{z}, t) \ \& \sim\psi(\bar{z}, A).$$

By (6.24) and (6.25) this contradicts $A \in Cn[K](\Gamma)$. Hence $A \notin Cn[K](\Gamma)$. ∎

COROLLARY 6.10. *Let K be the class of lattices which satisfy a given set of first-order sentences. Then with $\psi(x, y) =_{def} x \leqslant y$, $\mathfrak{L}[K]$ is compact.*

EXAMPLE 6.5.

(i) The propositional logics of lattices, I-lattices, modular lattices, distributive lattices, complemented lattices, ortho-complemented lattices, and Boolean algebras are compact.
(ii) The propositional logic of a finite lattice is compact.   ◆

Where K consists of a single finite logic $\mathbb{T}$ the following proof of this last result can be readily extended to more complex logics such as quantifier logics, tense logics, etc.

THEOREM 6.11. *$\mathfrak{L}[\mathbb{T}]$ is compact.*

Proof. With each propositional variable $p \in P$ we associate new propositional variables p^a for $a \in T$. Let P^{cl} be the set of all such variables and let $\mathbb{F}_{cl}$ be the set of formulas of the classical propositional calculus constructed from them (Chapter 2, §2.2.1). Then the $\mathbb{T}$-valued formulas of $\mathbb{F}(J, P)$ can be translated into classical formulas of $\mathbb{F}_{prop}$ (Definition 2.1) as follows: for every

$F \in \mathbb{F}(J, P)$ and every $a \in T$ we construct inductively, following the formation rules of $\mathbb{F}_{\text{prop}}$, a formula $F^a \in \mathbb{F}_{cl}$:

$T(\mathfrak{L})_1$: $(p)^a = p^a$ for $p \in P$.

$T(\mathfrak{L})_2$: $(c_i[A, B, \ldots])^a = \bigvee \{A^p B^q \cdots;\ o_i(p, q, \ldots) = a\}$.

(The disjunction of an empty set of formulas is defined as 0, the constant false proposition.)

Next, valuations v of $\mathbb{F}(J, P)$ in $\mathbb{T}$ are 'translated' into (classical) valuations $\bar{v}$ of $\mathbb{F}_{cl}$:

$$\bar{v}(p^a) = \begin{cases} 1 & \text{if } v(p) = a \\ 0 & \text{otherwise.} \end{cases}$$

Then the proposition below easily follows from these definitions.

PROPOSITION

(i) For every valuation v of $\mathbb{F}(J, P)$ in $\mathbb{T}$, $\bar{v} \vDash M$ (i.e. $\bar{v}$ is a classical propositional model of M, Definition 2.4(i)) where $M = \{V_1, V_2, \ldots\}$ and V_i is the $\mathbb{F}_{cl}$ formula

$$\bigvee \{p_i^a\ \&\ \bigwedge \{\sim p_i^b;\ a \neq b,\ b \in T\};\ a \in T\}.$$

(V_i says classically that p_i takes on exactly one of the truth-values in T.)

(ii) For every classical propositional model u of M (i.e. $u \in \mathbb{V}_2$) there is a valuation v of $\mathbb{F}(J, P)$ in $\mathbb{T}$ such that $\bar{v} = u$.

(iii) For all $F \in \mathbb{F}(J, P)$, $a \in T$, and valuations v of $\mathbb{F}(J, P)$ in $\mathbb{T}$,

$$v(F) = a \quad \Leftrightarrow \quad \bar{v} \vDash F^a.$$

(iv) Let $A, B_1, \ldots, B_n \in \mathbb{F}(J, P)$ and let G_n be the formula

$$\bigvee \{A^a B_1^b \cdots B_n^d;\ a \notin Cn(b, \ldots, d)\}.$$

Then for any valuation v of $\mathbb{F}(J, P)$ in $\mathbb{T}$,

$$v(A) \in Cn(v(B_1), \ldots, v(B_n)) \quad \Leftrightarrow \quad \bar{v} \vDash \sim G_n.$$

The compactness of $\mathfrak{L}[\mathbb{T}]$ now follows from the compactness of the classical logic by the above translation procedure. Let M and G_n be as defined in the above proposition and let Δ be an infinite set of $\mathbb{F}(J, P)$ formulas: $\Delta = \{B_1, B_2, \ldots\}$, say. Suppose that $A \notin Cn[\mathbb{T}](B_1, \ldots, B_n)$ for all n. We deduce that $A \notin Cn[\mathbb{T}](\Delta)$.

There exists a valuation v such that $v(A) \notin Cn(v(B_1), \ldots, v(B_n))$, i.e. $\bar{v} \vDash M + \{G_n\}$. But as C is a consequence function, if $n > m$ then $M, G_n \to G_m$. Hence $M + \{G_n;\ n = 1, 2, 3, \ldots\}$ is finitely satisfiable. By the compactness theorem (Theorem 2.6) there exists a model u of $M + \{G_n;\ n = 1, 2, 3, \ldots\}$.

Hence there exists a valuation v of $\mathbb{F}(J, P)$ in $\mathbb{T}$ such that $u = \bar{v}$. Thus, for all n, $v(A) \notin C(v(B_1), \ldots, v(B_n))$.

Now suppose that $v(A) \in Cn(v(\Delta))$. Since T is finite, for some n, $v(\Delta) = \{v(B_1), \ldots, v(B_n)\}$ so that $v(A) \in Cn(v(B_1), \ldots, v(B_n))$. This is a contradiction. Hence $v(A) \notin Cn(v(\Delta))$. Thus $A \notin Cn[\mathbb{T}](\Delta)$. ∎

THEOREM 6.12. Let R be a binary relation on $\mathbb{T}$. Let $\Delta, \Gamma \subseteq \mathbb{F}(J, P)$. If for all valuations v in $\mathbb{T}$ there exist $A \in \Delta$, $B \in \Gamma$ such that $v(A) R v(B)$ then there exists a finite set $\{(A_1, B_1), \ldots, (A_n, B_n)\}$ of pairs (A_i, B_i) with $A_i \in \Delta$ and $B_i \in \Gamma$ for $1 \leq i \leq n$ such that for all valuations v in $\mathbb{T}$, $\bigvee_{1 \leq i \leq n} v(A_i) R v(B_i)$.

Proof. Using the above proposition the condition $v(A) R v(B)$ can be expressed classically. The theorem then follows from Theorem 2.5(ii) (with $\Theta = M$). ∎

COROLLARY 6.13. Let $\mathbb{T} = \mathfrak{D}^{\ddagger}$ where $\mathfrak{D}$ is a lattice (Definition 6.3) and let $\Delta, \Gamma \subseteq \mathbb{F}(J, P)$ such that for all valuations v in D there exist $\Delta' \in \mathbb{P}_\omega(\Delta)$, $\Gamma' \in \mathbb{P}_\omega(\Gamma)$ with $v(\bigwedge \Delta') \leq v(\bigvee \Gamma')$. Then there exist $\Delta'' \in \mathbb{P}_\omega(\Delta)$, $\Gamma'' \in \mathbb{P}_\omega(\Gamma)$ such that, for all valuations v in $\mathfrak{D}$, $v(\bigwedge \Delta'') \leq v(\bigvee \Gamma'')$.

Proof. From Theorem 6.12 with R as the order relation $\leq$ on D it follows that there exist $\Delta'_1, \ldots, \Delta'_n \in \mathbb{P}_\omega(\Delta)$ and $\Gamma'_1, \ldots, \Gamma'_n \in \mathbb{P}_\omega(\Gamma)$ such that for all valuations v in D, $v(\bigwedge \Delta'_i) \leq v(\bigvee \Gamma'_i)$ for some i. Let $\Delta'' =_{\text{def}} \sum \{\Delta_i; 1 \leq i \leq n\}$ and $\Gamma'' =_{\text{def}} \sum \{\Gamma_i; 1 \leq i \leq n\}$. Since D is a lattice,

$$v(\bigwedge \Delta'') \leq v(\bigwedge \Delta'_i) \leq v(\bigvee \Gamma'_i) \leq v(\bigvee \Gamma'').$$ ∎

6.7 Functionally free algebras

In this section we enquire into the possibility of an effective test of logical equivalence of two formulas in the logic $\mathfrak{L}[K] = \langle \mathbb{F}_{\text{prop}}, c_1, c_2, \ldots, Cn[K] \rangle$, where K is a class of simple logics. The key idea is the notion of *functionally free algebra* formulated by Białynicki-Birula and Rasiowa (1957).

DEFINITION 6.4. Let K be a class of similar algebras and $\mathbb{F}(J, P)$ be the set of formulas of that type. Let $\mathfrak{A} \in K$, $A, B \in \mathbb{F}(J, P)$ and V be the set of homomorphisms of $\mathbb{F}(J, P)$ into $\mathfrak{A}$. $\mathfrak{A}^e$ is the equivalence relation on $\mathbb{F}(J, P)$ defined by

$$A = B \; mod(\mathfrak{A}^e) \quad \Leftrightarrow_{\text{def}} \quad h(A) = h(B) \quad (h \in V).$$

$K^e =_{\text{def}} \prod \{\mathfrak{A}^e; \mathfrak{A} \in K\}$. Thus K^e is the equivalence relation on $\mathbb{F}(J, P)$ such that

$$A = B \; mod(K^e) \quad \Leftrightarrow \quad A = B \; mod(\mathfrak{A}^e) \quad (\mathfrak{A} \in K).$$

𝔄 is *functionally free over* K if, for every $A, B \in \mathbb{F}(J, P)$,

$$A = B \, mod(K^e) \quad \Leftrightarrow \quad A = B \, mod(\mathfrak{A}^e).$$

Thus if 𝔄 is functionally free over K, then to test whether $A = B \, mod(K^e)$ we need only test whether $A = B \, mod(\mathfrak{A}^e)$, and if 𝔄 is finite this test is effective.

THEOREM 6.14. *If every member of K is a subdirect power of 𝔄 and $\mathfrak{A} \in K$ then 𝔄 is functionally free over K.*

Proof. The proof follows easily from the following proposition.

PROPOSITION. *If 𝔅 is a subsystem of 𝔄, a direct power of 𝔄, or a homomorphic image of 𝔄, then*

$$A = B \, mod(\mathfrak{A}^e) \quad \Rightarrow \quad A = B \, mod(\mathfrak{B}^e).$$

Clearly, if $A = B \, mod(K^e)$ then, as $\mathfrak{A} \in K$, $A = B \, mod(\mathfrak{A}^e)$.

Conversely, suppose $A \neq B \, mod(K^e)$. Then $A \neq B \, mod(\mathfrak{B}^e)$ for some $\mathfrak{B} \in K$. But 𝔅 is a subdirect power of 𝔄. Hence $A \neq B \, mod(\mathfrak{A}^e)$. ∎

EXAMPLE 6.6. Every Boolean algebra is a subdirect power of B_2. Hence to test for any $F, G \in \mathbb{F}(J, P)$ whether $F = G \, mod(\mathfrak{B}^e)$ for all Boolean algebras 𝔅, it suffices to test whether $F = G \, mod(B_2^e)$. As B_2 has only two elements, this test is effective. Further examples are given in the next chapter. ◆

COROLLARY 6.15. *Let K be a class of lattices and suppose Q is functionally free over K. Then $\mathfrak{L}[K] = \mathfrak{L}[Q]$.*

6.8 The logic of quantum mechanics

Birkhoff and von Neumann (1936) proposed that experimental propositions of quantum mechanics were governed by laws that differed somewhat from the familiar laws of classical logic by abandonment of the distributive law **K6** (Chapter 5, §5.9.1). This idea has been developed by a number of authors, but the subject continues to be controversial.

> ... quantum logic is not a closed subject. Although penetrating investigations have been made there exists no universally accepted theory. The basic framework has not been established to the satisfaction of even a majority of researchers.
> Greechie and Gudder 1975

Some authors (e.g. Garden 1984) claim that the apparent failure of the distributive law arises because the quantum-mechanical description of reality is not capable of being in a unique correspondence with reality itself. Garden concludes that Birkhoff and von Neumann

> do not establish that the operations in the calculus of quantum probabilities are non-distributive. Rather what first appears to be a failure of distribution is understood as a case in which the operations cannot be defined ... [We therefore] have no reason to suppose that the propositional logic of quantum theories, which we use to reason about microscopic things is radically non-classical..

Indeed, a significant subsidiary of the quantum logic company manufactures partial Boolean algebras. There is a case here for considering 'logics' with partially defined connectives.

The diversity of opinion and the extent of agreement was defined in 1966:

> The yes-or-no questions about a physical system (example of such a proposition: is the energy greater than 6 ergs) are partially ordered by 'implication' and have the natural orthocomplementation wherein the orthocomplement of a proposition a is its negation not-a. In classical theory, these propositions constitute a Boolean lattice (Boolean algebra). But in quantum theory this is no longer true. The distributivity appears to be inconsistent with the essential new features of quantum mechanics, notably the Heisenberg uncertainty principle. There seems to be general agreement among all investigators that these propositions constitute an 'orthomodular partially ordered set'; but there the agreement ends. There seems to be no clear-cut reason why they should form a lattice, and indeed one may argue the question both ways.
>
> Holland 1975, pp. 490–1

The status of OMLs and the associated logics (quantum logics) is quite clear. The apparatus of classical analysis applied to the Schrödinger partial differential equation singles out the closed subspaces of Hilbert spaces H. By normal classical reasoning these are proved to be complete OMLs. They also have other properties:

> However, as with completeness these features are not equationally definable and accordingly they are not regarded as fundamental to the purely logical structure of concrete quantum logic. Thus, at least until a smaller variety containing Hilbert lattices is discovered, the study of abstract quantum logic is subsumed under the study of general OMLs.
>
> Hardegree 1976, p. 57

Standard mathematical construction and reasoning then delivers a logic from the class of ortho-modular lattices (Theorem 6.7). This logic bears the title of 'the logic of quantum mechanics', which misleadingly suggests that the subject of quantum mechanics is somehow 'founded' on its offspring:

> But from our point of view a logic of quantum mechanics is simply an attempt to give a systematic account of the semantic relations among the elementary

The logic of quantum mechanics   211

statements of that theory. And these semantic relations are *to be deduced* from the quantum theory—*that* is the sense in which this logic is a quantum logic. It is not meant to be the basis for a formalization of the theory for a new, non-standard *Principia*.

van Fraassen 1975, p. 600

There have been various attempts at constructing logics from OMLs. Kalmbach (1983, Chapter 15) constructs a theorem-oriented logic with connectives of conjunction, disjunction, negation, and 'implication'. The first three of these are interpreted as the usual lattice operations meet, join, and complementation, whilst implication is interpreted as a two-place operation defined in terms of them. The logic has 15 axioms and one rule of inference—*modus ponens*. A completeness theorem holds, i.e. the semantic consequences coincide with syntactical deduction.

There seems to be considerable controversy (Hardegree 1975) over the 'implication' in ortho-modular logic, much of it occasioned, as in Birkhoff and von Neumann (1936), by identifying the logic with the lattices themselves, whereas Kalmbach's method is, in effect, to treat the class of OMLs, as in our general method, as a class of truth-value structures. Some exponents of quantum logic regard an 'implication' as of prime importance in a class of lattices underlying a logic (see Greechie and Gudder 1973) and maintain that without accepting a conditional in quantum logic it is questionable whether the lattice of general quantum mechanics is a logic. Birkhoff and von Neumann (1936) required of a definition of a conditional in terms of meet, join, and complementation (i.e. a *polynomial p* in two variables) that it satisfy

$$p(a, b) = 1 \quad \Leftrightarrow \quad a \leqslant b$$

in every OML. In fact, there are five such polynomials (Kalmbach 1983, Chapter 15):

$$a'b' \vee a'b' \vee a(a' \vee b)$$
$$a'b \vee ab \vee (a' \vee b \vee b')$$
$$a' \vee ab$$
$$b \vee a'b',$$
$$a'b \vee ab \vee a'b.$$

The whole problem of the conditional rests on the confusion of 'implication' as an operation on propositions (which is more correctly termed 'conditional') and 'implication' as a metarelation between propositions (i.e. logical consequence). In our treatment of logic there is no necessity for

212   Constructing logics

lattices, or any set of truth-value structures, to be equipped with a conditional. Later, in Chapter 10, we shall exhibit a general method of extending a logic to include an implication operation.

6.9 Predicate logics

Classical predicate logic is concerned with relational systems. The relations of such structures can be identified with their characteristic functions, i.e. mappings into $\{0, 1\}$. One way of generalizing this situation is to replace such characteristic functions by functions with values in an arbitrary set of truth-values (Appendix, §R).

6.9.1 *Lattice-valued sets*

Let X be a non-empty set. A subset V of X can be identified with its characteristic function, i.e. $V: X \to (0, 1\}$. Thus V is a member of the direct power (Appendix, §A) B_2^X. The operations on this direct power are inherited from the Boolean operations of B_2. Thus the calculus of classes—strictly speaking, the subsets of X—is isomorphic to B_2^X. Now replace B_2 by an algebra $\mathfrak{L}$. $\mathfrak{L}^X$ is the calculus of $\mathfrak{L}$-valued sets. By Theorem A1 (Appendix, §A), every identity which holds in $\mathfrak{L}$ also holds in $\mathfrak{L}^X$. Where $\mathfrak{L}$ is a lattice with maximum element 1 and minimum element 0 and a unary operation $'$ such that $0' = 1$, $1' = 0$ and such that $\{0, 1\}$ is a subalgebra of $\mathfrak{L}$, then the classical calculus of classes B_2^X is a subsystem of $\mathfrak{L}^X$. This idea will be developed further in Chapter 7 and 8.

6.9.2 *Lattice-valued predicate logics*

We now consider many-valued structures, the values being taken from a lattice. The idea of the definition is to replace an n-ary relation R on a set X by its characteristic function ϕ:

$$R(x_1, \ldots, x_n) \Leftrightarrow \phi(x_1, \ldots, x_n) = 1$$
$$\text{not} \sim R(x_1, \ldots, x_n) \Leftrightarrow \phi(x_1, \ldots, x_n) = 0.$$

ϕ is a function with values in $\{0, 1\}$. Now replace $\{0, 1\}$ by the set L to give an *L-valued relation* (Appendix, §R).

DEFINITION 6.5. Let R, Fn (Chapter 2, §2.3.1) and $\tau: R + Fn \to \mathbb{N}$ be as in Chapter 2, §2.2.1. An *L-valued structure* of type τ is a structure

$$\langle A, g_1, g_2, \ldots, R_1, R_2, \ldots \rangle$$

Predicate logics 213

such that, for each i,

(i) if $\tau(f_i) = n$ then $g_i: A^n \to A$
(ii) if $\tau(P_i) = n$ then $R_i: A^n \to L$.

If $\mathfrak{L}$ is a logic then an $\mathfrak{L}$-*valued structure of type* τ is an $|\mathfrak{L}|$-valued structure.

We confine attention here to lattice logics because otherwise it is not at all clear how to interpret universal and existential quantifiers. Moreover, we must assume that the lattice is complete in order to effect this. Hence $\mathfrak{L}$ is assumed to have a maximum element 1 and a minimum element 0 (Chapter 5, §5.8) which can be identified with the classical 0 and 1 of the calculus of classes.

A language for discussing $\mathfrak{L}$-valued structures differs from the classical predicate logic of the appropriate type only by the provision of junctors which are appropriate for $\mathfrak{L}$. Thus the primitive symbols of the language are the pairwise disjoint alphabets Fv, Bv, R, Fn, Q, G of Chapter 2, §2.3.1, but the set J of junctors is now the list of symbols $\vee, \wedge, t, f, c_3, c_4, \ldots$ corresponding to the operations of $\mathfrak{L}$ and the maximum and minimum elements respectively. Admitting operations $o_3, o_4, \ldots$ and corresponding junctors in the language enables us to deal generally with the various forms of complementation and pseudo-complementation.

The classes $\mathbb{T}$ of terms and the class At of atomic formulas of the language are as defined in Definitions 2.10 and 2.11 except that the symbols t, f must now be counted at atomic formulas. The definition of the class $\mathbb{F}_{pred}$ of formulas differs from Definition 2.12 only in the clause (F2) which concerns the junctors. This clause is replaced by the following.

(F2′) If $A_1, \ldots, A_n \in \mathbb{F}_{pred}$ and $\tau(o_i) = n$ then $c_i[A_1, \ldots, A_n] \in \mathbb{F}_{pred}$. If $A, B \in \mathbb{F}_{pred}$ the $[A \vee B]$ and $[A \wedge B]$ are also in $\mathbb{F}_{pred}$. $t, f \in \mathbb{F}_{pred}$.

A *tf-free formula* is one in which there is no occurrence of t or f.

The semantical notions for this language follow closely Definition 2.13 of the classical case. The notions of $\mathfrak{L}$-valued *realization* and *assignment* are as in Definitions 2.11(i) and 2.11(ii) respectively. Likewise the notions related to *canonical realization* (Definition 2.15) can be appropriated for the present case: $Can(X)$ is the set of $\mathfrak{L}$-valued realizations on $\mathbb{T}(X)$ and $Can(\emptyset)$ is the set of $\mathfrak{L}$-valued Herbrand realizations. The computation of truth-values, however, has to be generalized from the use of the two truth-values of B_2 to the use of values in $\mathfrak{L}$. Thus clause (iii) of Definition 2.13 must now read as follows.

(iii)′ Given a realization $\mathfrak{A}$ and an assignment α to $\mathfrak{A}$ there is a function $[\mathfrak{A}, \alpha]$—a *valuation*—which assigns truth-values in $\mathfrak{L}$ defined by

$$[\mathfrak{A}, \alpha](P_i(t_1, \ldots, t_n)) = R_i(\alpha(t_1), \ldots, \alpha(t_n))$$

when $\tau(P_i) = n$, and $[\mathfrak{A}, \alpha](f)$, $[\mathfrak{A}, \alpha](t)$ are respectively the minimum and maximum elements of $\mathbb{T}$. (Recall that $R_i\colon |\mathfrak{A}|^n \to \mathbb{T}$). $\mathbb{V}_{pred}$ is the set of all such valuations.

The $[\mathfrak{A}, \alpha]$ extends to all $\mathbb{F}_{pred}$ by the following inductive definition.

(iii) (a) If $A \in At$ then $[\mathfrak{A}, \alpha](A)$ has already been defined.
 (b) If $A, B \in \mathbb{F}_{pred}$ then

$$[\mathfrak{A}, \alpha]([A \vee B]) = [\mathfrak{A}, \alpha](A) \cup [\mathfrak{A}, \alpha](B)$$

$$[\mathfrak{A}, \alpha]([A \wedge B]) = [\mathfrak{A}, \alpha](A) \cap [\mathfrak{A}, \alpha](B).$$

If o_i is an n-place operation (i.e. $\tau(o_i) = n$) in $\mathfrak{L}$ and $A_1, \ldots, A_n \in \mathbb{F}_{pred}$ then

$$[\mathfrak{A}, \alpha](c_i[A_1, \ldots, A_n]) = o_i([\mathfrak{A}, \alpha](A_1), \ldots, [\mathfrak{A}, \alpha](A_n)).$$

(c) If $A \in \mathbb{F}_{pred}$, $x \in Fv$, $a \in Bv - Bv(A)$, then

$$[\mathfrak{A}, \alpha](\forall a A(a/x)) = glb\{[\mathfrak{A}, \alpha[s/x]](A);\ s \in |\mathfrak{A}|\}$$

$$[\mathfrak{A}, \alpha](\exists a A(a/x)) = lub\{[\mathfrak{A}, \alpha[s/x]](A);\ s \in |\mathfrak{A}|\}$$

(note that the least upper bounds and greatest lower bounds exist as $\mathfrak{L}$ is complete).

It must now be verified that $[\mathfrak{A}, \alpha]$ is a *function*, i.e. it assigns a *unique* truth-value to each formula. This is trivial for a formula of the form $c_i[A_1, \ldots, A_n]$ because it is constructed in only one way from its subformulas $A_1, \ldots, A_n$. However, a quantified formula $\forall a A(a/x)$, say, can be constructed from A or from another formula A' which differs from A only by the occurrence of a different free variable y in exactly those places where x occurs in A. Thus $A' = A(y/x)$ where $y \notin Fv(A)$ and $\forall a A(a/x) = \forall a A'(a/y)$. But $[\mathfrak{A}, \alpha](A)$ depends only on the free variables of F, i.e. if β is an assignment such that $\alpha(z) = \beta(z)$ for all $z \in Fv(A)$ then $[\mathfrak{A}, \alpha](A) = [\mathfrak{A}, \beta](A)$. Thus, for every $s \in |\mathfrak{A}|$,

$$[\mathfrak{A}, \alpha[s/x]](A) = [\mathfrak{A}, \alpha[s/y]](A').$$

Thus, by (iii)(c),

$$[\mathfrak{A}, \alpha[s/x]](\forall a A(a/x)) = [\mathfrak{A}, \alpha[s/y]](\forall a A'(a/y)).$$

Hence $[\mathfrak{A}, \alpha]$ assigns a unique truth-value to $(\forall a A(a/x))$. Similarly for $\exists a A(a/x)$.

A logical consequence function $Cn_{\mathfrak{L}}$ can now be defined on $\mathbb{F}_{pred}$ as in Example 6.1 which, by Theorem 3.3, makes $\langle \mathbb{F}_{pred}, \vee, \wedge, c_3, c_4, \ldots, Cn_{\mathfrak{L}} \rangle$ into a quantifier logic $Pred[\mathfrak{L}]$ in which every logical identity which holds in $\mathfrak{L}$ also holds in $Pred[\mathfrak{L}]$ (Theorem 6.7). We shall write $\Delta \to F$ as an abbreviation of $F \in Cn_\mathfrak{L}(\Delta)$. Then

$$\Delta \to F \quad \Leftrightarrow \quad ([\mathfrak{A}, \alpha] \in \mathbb{V}_{pred})([\mathfrak{A}, \alpha](F) \in Cn_\mathfrak{L}([\mathfrak{A}, \alpha](\Delta))). \quad (6.26)$$

We now investigate the quantifier symbols. First, $Pred[\mathfrak{L}]$ is a free-variable logic. The only clause of Definition 3.13 which needs justifying is (iii). Let σ be the substitution operator (t/x) where $t \in \mathbb{T}$ and $x \in Fv$. Then

$$\Delta \to F \quad \Rightarrow \quad \Delta\sigma \to F\sigma. \quad (6.27)$$

Suppose $\Delta \to F$. Let $[\mathfrak{A}, \alpha] \in \mathbb{V}_{pred}$. Then $[\mathfrak{A}, \alpha[\alpha(t)/x]] \in \mathbb{V}$. By (6.26), $[\mathfrak{A}, \alpha[\alpha(t)/x]](F) \in Cn_\mathfrak{L}([\mathfrak{A}, \alpha[\alpha(t)/x]](\Delta))$. But for any formula G, $[\mathfrak{A}, \alpha[\alpha(t)/x]](G) = [\mathfrak{A}, \alpha](G(t/x))$ (cf. (2.16)). Hence $[\mathfrak{A}, \alpha](F\sigma) \in Cn_\mathfrak{L}([\mathfrak{A}, \alpha](\Delta\sigma))$. This holds for every evaluation. By (6.26), $\Delta\sigma \to F\sigma$.

It follows easily from the definition of *valuation* and *logical consequence* that $\forall, \exists$ satisfy the quantifier axioms **∀1,∀2, ∃1, ∃2** of Chapter 4, §4.4 (cf. Theorem 2.14).

THEOREM 6.16. Let F be a formula $a \in Bv - Bv(F)$, and $x \in Fv$.

(i) For $a, b \in Bv - Bv(F)$, $QaF(a/x) = QbF(b/x) \, mod(Cn_\mathfrak{L}^e)$ for $Q = \exists, \forall$.

(ii) (a) (cf. **∀2**) For $\Delta \subseteq \mathbb{F}_{pred}$ suppose that $x \notin Fv(\Delta)$. Then

$$\Delta \to F \quad \Rightarrow \quad \Delta \to \forall aF(a/x).$$

(b) For $G \in \mathbb{F}_{pred}$, suppose that $x \notin Fv(G)$. Then

$$F \to G \quad \Rightarrow \quad \exists aF(a/x) \quad \Rightarrow \quad G.$$

(c) (cf **∃1**) If $\mathfrak{L}$ satisfies the 'infinite distributive laws' (Chapter 5, §5.9.1) then for $\Delta, G \subseteq \mathbb{F}_{pred}$, and $x \notin Fv(\Delta, G)$,

$$\Delta, F \to G \quad \Rightarrow \quad \Delta \quad \exists aF(a/x) \to G.$$

(iii) (cf. **∀1, ∃2**) For all $t \in \mathbb{T}$

$$\forall aF(a/x) \to F(t/x) \qquad F(t/x) \to \exists aF(a/x).$$

COROLLARY 6.17. For $F, G \in \mathbb{F}_{pred}$, and any $x \in Fv$ and bound variables $a, b \notin Bv(F, G)$

$$F = G \, mod(Cn_\mathfrak{L}^e) \quad \Rightarrow \quad QaF(a/x) = QbG(b/x) \, mod(Cn_\mathfrak{L}^e)$$

where Q is $\forall$ or $\exists$ and $Cn_\mathfrak{L}^e$ is the logical equivalence of $\mathfrak{L}$ (Definitions 3.2(i) and 3.10).

From the symbols $\exists, \forall$ we can define quantifiers in the sense of Definition 3.13. First consider $\forall$. It can be turned into a function $\forall: \mathbb{F}_{pred} \times Fv \to \mathbb{F}_{pred}$

by the definition

$$\forall(x, F) =_{def} \forall a F(a/x)$$

where a is the first bound variable which does not occur in F. Then, by Corollary 6.15, condition **Q1** is satisfied. **Q3**, **Q3**, and **Q4** follow from the formation rules for quantified expressions. Further, it follows from Theorem 6.16 (ii)(c) that $[\mathfrak{A}, \alpha](\forall a F(a/x)) = [\mathfrak{A}, \alpha](F)$ for all evaluations $(\mathfrak{A}, \alpha)$. Thus if $x \notin Fv(F)$ then $\forall(x, F) = F \, mod(Cn_{\mathfrak{L}}^e)$. Hence **Q2** holds. So $\forall$, and likewise $\exists$, are quantifiers.

From Theorem 6.16 we also have $\forall$ is a universal quantifier, i.e. it satisfies **∀1** and **∀2** (Chapter 4, §4.11). Also, $\exists$ is a weak existential quantifier, i.e. it satisfies **∃1°** and **∃2**; it is an existential quantifier (i.e. it satisfies **∃1** and **∃2**) if $\mathfrak{L}$ satisfies the infinite distributivity laws. Thus the above construction gives a predicate logic $Pred[\mathfrak{L}]$ with universal and existential quantifiers.

As in the classical logic there is a close connection between canonical realizations and propositional valuations of atomic formulas. In fact, Theorem 2.18 can be taken over entirely to the $\mathfrak{L}$-valued logic by reading $Can(Y)$ and V_{prop} as $\mathfrak{L}$-valued canonical realizations and propositional valuations in $\mathfrak{L}$ respectively. This enables a version of the uniformity theorem (Theorem 2.24) to be proved for the case of finite $\mathfrak{L}$.

THEOREM 6.18. *Let $\mathfrak{L}$ be a finite lattice, and let A, B be quantifier-free formulas such that $\forall a A(a/x) \to \exists b B(b/y)$. Then there exist ground instances $A_1, \ldots, A_n$ of A and $B_1, \ldots, B_m$ of B such that $\bigwedge_{1 \leq i \leq n} A_i \to_{prop} \bigvee_{1 \leq j \leq m} B_j$.*

Proof. Let v be a propositional valuation of $\{A(s/x); s \in \mathbb{T}(\emptyset)\}$. By Theorem 2.18 ($\mathfrak{L}$-valued version) there is an $\mathfrak{L}$-valued Herbrand realization $\mathfrak{A}$ such that $v(A(s/x)) = [\mathfrak{A}, \imath](A(s/x))$. Now as $\mathfrak{L}$ is finite there exist $U, V \in \mathbb{P}_\omega(\mathbb{T}(\emptyset))$ such that

$$[\mathfrak{A}, \imath](\forall a A(a/x)) = glb\{[\mathfrak{A}, \imath]A(s/x); s \in U\}$$
$$= [\mathfrak{A}, \imath](\bigwedge \{A(s/x); s \in U\})$$
$$[\mathfrak{A}, \imath](\exists b B(b/x)) = lub\{[\mathfrak{A}, \imath]B(t/x); t \in V\}$$
$$= [\mathfrak{A}, \imath](\bigvee \{B(t/x); t \in U\}).$$

But as $\forall a A(a/x) \to \exists b B(b/y)$,

$$[\mathfrak{A}, \imath](\forall a A(a/x)) \leq [\mathfrak{A}, \imath](\exists b B(b/y)).$$

Hence

$$v(\{A(s/x); s \in U\}) \leq v(\{B(t/x); t \in U\}).$$

By Corollary 6.13, there exist $s_1, \ldots, s_n, t_1, \ldots, t_m \in \mathbb{T}(\emptyset)$ such that $\bigwedge_{1 \leq i \leq n} A(s_i/x) \to_{prop} \bigvee_{1 \leq i \leq m} B(t_i/y)$. ∎

Predicate logics

If, instead of a single complete lattice $\mathfrak{L}$, there is a class K of complete lattices, a logical consequence function can be defined on $\mathbb{F}_{\text{pred}}$ by (6.23) which gives the quantifier logic $Pred[K]$. For instance, if K is the class of lattices of closed subspaces of Hilbert spaces, this construction gives a predicate logic which could be called the *Hilbert space logic*.

6.9.3 Compactness of quantifier logics

As with the propositional logic, the $\mathfrak{L}$-valued predicate logic inherits compactness from the corresponding predicate logic. The burden of the proof rests on establishing this correspondence.

THEOREM 6.19. *If $\mathfrak{L}$ is finite then $Pred[\mathfrak{L}]$ is compact.*

Proof. The $\mathfrak{L}$-valued logic can be translated into classical logic by following the procedure of Theorem 6.11.

With each $\mathfrak{L}$-valued structure (Definition 6.5) $\mathfrak{A} = \langle \mathfrak{B}, R_1, R_2, \ldots \rangle$, where $\mathfrak{A}$ is an algebra of type τ, there is associated the relational system $\overline{\mathfrak{A}} = \langle \mathfrak{B}, R_1^e, R_1^b, \ldots, R_1^c, R_2^a, R_2^b, \ldots, R_2^c, \ldots \rangle$ formed from $\mathfrak{B}$ by replacing each $\mathfrak{L}$-valued function R_i by the sequence $R_i^a, R_i^b, \ldots, R_i^c$ of $\beta(i)$-place relations, where $a, b, \ldots, c \in |\mathfrak{L}|$. Clearly, $\mathfrak{B}$ satisfies the condition (informal language) $(\bar{x}) \bigvee_\xi \{R_i^\xi(\bar{x}) \ \& \ \bigwedge_{\mu \neq \xi} \sim R_i^\mu(\bar{x})\}$ for $i = 1, 2, \ldots$. Let M be the set of such sentences. (This set plays a similar role to the set M in Theorem 6.11.) Then for any structure $\mathfrak{C}$ (of the appropriate type) satisfying M there is an $\mathfrak{L}$-valued structure $\mathfrak{A}$ such that $\mathfrak{C} = \overline{\mathfrak{A}}$.

Next, for each $F \in \mathbb{F}_{\text{pred}}$ and $\xi \in |\mathfrak{L}|$ there is a formula F^ξ of the classical predicate logic such that for all valuations $[\mathfrak{A}, \alpha]$ in $\mathfrak{L}$

$$[\mathfrak{A}, \alpha](F) = \xi \quad \Rightarrow \quad (\overline{\mathfrak{A}}, \alpha) \vDash F^\xi.$$

This is proved by induction on the construction of F.

Basis:
F is the atomic formula $R_i(\bar{t})$. Then F^ξ is $R_i^\xi(\bar{t})$.

Inductive step:
(i) F is $c_i(A, B, \ldots)$. Then F^ξ is $\bigvee \{A^\mu B^\nu \cdots; o_i(\mu, \nu, \ldots) = \xi\}$. Since $|\mathfrak{L}|$ is finite this formula is a finite disjunction of previously defined formulas $A^\mu, B^\nu, \ldots$.
(ii) F is $\exists a G(a/x)$. Then for each valuation $[\mathfrak{A}, \alpha]$,

$$[\mathfrak{A}, \alpha](\exists a G(a/x)) = \xi \quad \Leftrightarrow \quad \xi = \text{lub}\{[\mathfrak{A}, \alpha[z/x]](G); z \in |\mathfrak{A}|\}.$$

Now $\xi \in U(\{[\mathfrak{A}, \alpha[z/x]](G); z \in |\mathfrak{A}|\})$ iff $(\overline{\mathfrak{A}}, \alpha) \vDash H^\xi$ where $H(\xi)$ is $\forall a \bigvee_{\mu \leq \xi} G^\mu(a/x)$. Thus ξ is the least upper bound if $\sim H(\mu)$ holds whenever $\mu < \xi$, i.e. $(\overline{\mathfrak{A}}, \alpha) \vDash \bigwedge_{\mu < \xi} \sim H(\mu)$. Thus F^ξ is

$$H(\xi) \ \& \ \bigwedge_{\mu < \xi} \sim H(\mu).$$

(iii) F is $\forall a G(a/x)$. Let $K(\xi)$ be the formula $\forall a \bigvee_{\mu \geq \xi} G^b(a/x)$. Then F^ξ is $K(\xi)$ & $\bigwedge_{\mu > \xi} \sim K(\mu)$.

Thus we have the following proposition.

PROPOSITION

(i) For every valuation $[\mathfrak{A}, \alpha]$ of $\mathbb{F}_{\text{pred}}$ in $\mathfrak{L}$, $(\mathfrak{A}, \alpha) \vDash M$.
(ii) For every classical model $(\mathfrak{B}, \alpha)$ of M there is a valuation $[\mathfrak{A}, \alpha]$ of $\mathbb{F}_{\text{pred}}$ in $\mathfrak{L}$ such that $\mathfrak{B} = \bar{\mathfrak{A}}$.
(iii) For all $F \in \mathbb{F}_{\text{pred}}$, $\tau \in \mathfrak{L}$, and valuation $[\mathfrak{A}, \alpha]$ of $\mathbb{F}_{\text{pred}}$ in $\mathfrak{L}$, $v(F) = \tau \Leftrightarrow \bar{v} \vDash F^\tau$.
(iv) Let $A, B_1, \ldots, B_n \in \mathbb{F}_{\text{pred}}$ and let G_n be the formula

$$\bigwedge \{A^a B_1^b \cdots B_n^d; a \notin Cn_{\mathfrak{L}}(b; c, \ldots)\}.$$

Then for any valuation $[\mathfrak{A}, \alpha]$ of $\mathbb{F}_{\text{pred}}$ in $\mathbb{T}$,

$$[\mathfrak{A}, \alpha](A) \in Cn_{\mathfrak{L}}([\mathfrak{A}, \alpha](B_1), \ldots, [\mathfrak{A}, \alpha](B_n)) \quad \Leftrightarrow \quad (\bar{\mathfrak{A}}, \alpha) \vDash \sim G_n.$$

The remainder of the proof now follows as in Theorem 6.11. ∎

6.10 Temporal logics

The standard formulations of classical and intuitionistic propositional or predicate logic make no special reference to propositions which relate to time. The aim of temporal logic—also known as 'tense logic' or 'change logic'—is to systematize discourse of propositions involving the concepts embodied in the various tenses of the verbs such as are expressed by the various temporal modes of an object possessing a property: 'A has the property P', 'A had P', 'A will have P', 'A will have had P', etc. In such propositions there is essential reference to the relationship before–after or past–present–future. The usual treatments of temporal logic are founded on classical logic despite the difficulty of maintaining that all tensed propositions are truth-definite.

A rudimentary theory of temporal quantifiers (such as 'sometimes', 'always', etc.) and a theory of temporalized modalities were developed by the Megarians and Stoics, particularly Diodorus Cronus (Kneale 1962, Chapter 3, §2, Rescher and Urquhart 1971, Chapter 1). Aristotelian and Stoic logic of temporal relations was further developed by some medieval Arabic logicians. There has been a major revival of interest in this subject since the late 1940's, in which various theorem-oriented systems have been propounded (see references in Rescher and Urquhart (1971, p. 12)). The reader is referred to Rescher and Urquhart (1971) and McArthur (1976) for a fuller discussion of the technical and philosophical issues raised by temporal logic.

Quite recently researchers in artificial intelligence have shown interest in the subject (e.g. Gabbay et al. 1980, McDermott 1982). Tense logics also arise in computer science by regularizing discourse about the ongoing behaviour of computer programs (Pneuli 1977).

We first restrict attention to temporal propositional calculus, particularly to the *minimum system* K_t of tense logic due to E. J. Lemmon. This is a classical calculus to which is adjoined the *temporal operators* (i.e. one-place connectives) F, P which are interpreted as follows.

FA: A is true at some future time.

PA: A was true at some past time.

The formation rules for formulas are as in Chapter 2, §2.2.1, except that the symbols F and P are adjoined to the list J of junctors and the following clause is added to (F2):

If $A \in \mathbb{F}_{prop}$ then $FA \in \mathbb{F}_{prop}$ and $PA \in \mathbb{F}_{prop}$.

Let G be an abbreviation of $\neg F \neg$ and H an abbreviation of $\neg P \neg$. Then GA can be read informally as 'A will be true at all future times' and HA as 'A has always been true in the past'. The axiom schemas of K_t are as follows.

T1: $G[A \supset B] \supset [GA \supset GB]$

T2: $H[A \supset B] \supset [HA \supset HB]$

T3: $A \supset HFA$

T4: $A \supset GPA$

The rules of inference are as follows.

R1: If A is a classical tautology then $\vdash A$

R2: If A is an axiom then $\vdash A$

R3: If $\vdash A$ then $\vdash HA$ and $\vdash GA$

MP: If $\vdash A$ and $\vdash [A \supset B]$ then $\vdash B$

The interpretation of formulas allows that a given formula can be true (or false) at one time and false (or true) at a later or earlier instant. It is based on the following.

(i) A model of time which is at last a set T of instants on which there is a binary relation $\ll$ of temporal precedence—before–after, or earlier–later.
(ii) A valuation (Definition 2.3) $\mathfrak{A}_t$ for each instant $t \in T$ which determines a (classical, i.e. in B_2) truth-value of each propositional variable in P and thereby of each formula of $\mathbb{F}_{prop}$.

Thus the structure in which formulas are interpreted is a non-empty set T of instants, a binary relation $\ll$ on T, and a set $\{\mathfrak{A}_t; t \in T\}$ of valuations indexed by T. The structure $\mathfrak{G} = \langle T, \ll, \{\mathfrak{A}_t; t \in T\}\rangle$ is called a *tense structure* and $\mathfrak{A}_t$ is called the *t-section* of $\mathfrak{G}$ and defines the 'state of the universe' at time t. Our definition is slightly different from those given by Rescher and Urquhart and by McArthur. It is framed in this way to bring out the similarity with Kripke's model structures.

Thus a given a tense structure $\mathfrak{G}$ determines a mapping $\phi: T \times P \to \{0, 1\}$ defined by $p_t = \mathfrak{A}_t(p)$. Then ϕ can be extended to a mapping $T \times \mathbb{F}_{\text{prop}} \to \{0, 1\}$ by the interpretation of the classical junctors as given in Chapter 2, §2.2.2, and the intended interpretation of F and P. In this, $t_1 \ll t_2$ is the representation of 'instant t_2 follows (is after) t_1' or 'instant t_1 precedes (is before) t_2'. The *future of* t is $f(t) = \{s \in T; t \ll s\}$ and the *past of* t is $p(t) = \{s \in T; s \ll t\}$. Then ϕ is determined by the inductive definition

$$p_t = \mathfrak{A}_t(p) \qquad \text{for } p \in P$$
$$[A \supset B]_t = (A_t)' \cup B_t$$
$$(\neg A)_t = (A_t)'$$
$$(FA)_t = \text{lub}\{A_s; s \in f(t)\}$$
$$(PA)_t = \text{lub}\{A_s; s \in p(t)\}.$$

We write $(\mathfrak{G}, t) \vDash F$ if $F_t = 1$ and say that F is *true in* $\mathfrak{G}$ *at time* t. F is said to be *true in* $\mathfrak{G}$ if it is true in $\mathfrak{G}$ for all $t \in T$. F is *valid* if it is true in all tense structures.

K_t can be proved complete with respect to this semantics: a formula of K_t is valid iff it is a theorem of K_t. A proof based on semantic tableaux is given by Rescher and Urquhart and a Henkin-type proof is given by McArthur.

Several other temporal logics have been studied by placing further conditions on temporal succession. For instance, a linear tense logic is one whose model of time is a line. Such logics arise in the logic of computer programs. For instance, Emerson and Sistla (1984) have defined a temporal propositional logic which, as well as linear time operators G ('always'), F ('sometimes'), X ('next time'), and U ('until'), also contains quantifiers over computation paths: A ('for all paths') and E ('for some paths').

There are several ways of effecting linearity of time. If T is a line then $f(s) \subseteq f(t)$ when $s \in f(t)$. Hence if FFA holds at t then so does FA. Thus $FFA \supset FA$ is valid. This is clearly related to the transitivity of $\ll$.

Linearity also ensures that if FA and FB hold at t then at least one of $F(AB)$, $F(A \wedge FB)$, $F(FA \wedge B)$ is also true at B and similarly for P. A linear tense logic was formulated by Cocchiarella (1966) with the following axioms.

T5: $FFA \supset FA$

T6: $PPA \supset PA$

T7: $(FA \wedge FB) \supset (F(A \wedge B) \vee (FA \wedge B) \vee F(A \wedge FB))$

T8: $(PA \wedge PA) \supset (P(A \wedge B) \vee (PA \wedge B) \vee P(A \wedge PB))$

Temporal logics

The semantics of this system differs from that of K_t by imposing the conditions of transitivity **P1**, right-linearity **P2**, and left-linearity **P3** on $\ll$.

P1: $(x, y, z)[x \ll y \land y \ll z. \supset x \ll z]$

P2: $(x, y, z)[x \ll y \land x \ll z. \supset (x = y \lor y \ll z \lor z \ll y)]$

P3: $(x, y, z)[y \ll x \land z \ll x. \supset (y = z \lor z \ll y \lor y \ll z)]$

A *linear tense structure* $\mathfrak{G} = \langle T, \ll, \{\mathfrak{A}_t; t \in T\}\rangle$ is a tense structure where $\ll$ satisfies **P1**, **P2**, and **P3**. This system is also complete.

There have been attempts to construct many-valued temporal logics, e.g. Rescher and Urquhart (1971, Chapter 18). As many-valued logics without theorems can easily be constructed, it would appear more natural to abandon the theorem-oriented approach and to study the notion of logical consequence. One might hope to axiomatize this concept by a Gentzen type of sequent calculus.

Let us take a lattice $\mathfrak{L}$ as the system of truth-values. Then a tense structure $\mathfrak{G} = \langle T, \ll, \{\mathfrak{A}_t; t \in T\}\rangle$ based on $\mathfrak{L}$ differs from the above two-valued case only in that the t-sections $\mathfrak{A}_t$ are now valuations in $\mathfrak{L}$: we call this structure an $\mathfrak{L}$-tense-structure. The interpretation of the junctors F (P) depends on truth-values at all future (past) times. Therefore $\mathfrak{L}$ must be complete. The connectives built into the logic of $\mathfrak{L}$ will be conjunction and disjunction, corresponding to join and meet in $\mathfrak{L}$, together with junctors corresponding to further operations with which $\mathfrak{L}$ is equipped. There is considerable choice here (see Chapter 4). Let us choose a lattice with a proto-complementation $'$ (Definition 5.14). There is another choice to be made, namely which notion of logical consequence shall be used. Here we shall use Definition 6.3 (and therefore (6.9)).

Then the formal language of the tense logic is built from an infinite list P of propositional variables and two-place junctors $\land$, $\lor$ for conjunction and disjunction, and one-place junctors $\neg$, F, P for negation and the tense operators F, P. Let $\mathbb{F}_{\text{prop}}$ be the set of formulas. Then from the $\mathfrak{L}$-tense-structure $\mathfrak{G} = \langle T, \ll, \{\mathfrak{A}_t; t \in T\}\rangle$ there can be constructed a valuation of all formulas, i.e. an $\mathfrak{L}$-valued function on $T \times \mathbb{F}_{\text{prop}}$, defined by induction as follows.

If $A \in P$ then $A_t = \mathfrak{A}_t(A)$.

If $A, B \in \mathbb{F}_{\text{prop}}$ then $[A \land B]_t = A_t \cup B_t$ and $[A \lor B]_t = A_t \cap B_t$.

If $A \in \mathbb{F}_{\text{prop}}$ then

$$(\neg A)_t = (A_t)'$$
$$(FA)_t = \text{lub}\{A_s; s \in f(t)\}$$
$$(PA)_t = \text{lub}\{A_s; s \in p(t)\}.$$

For a set $\Delta \subseteq \mathbb{F}_{\text{prop}}$ put $\Delta_t =_{\text{def}} \{A_t; A \in \Delta\}$.

A logical consequence function $Cn_{\mathfrak{G},t}$ on F can now be defined (Definition 6.3) by

$$A \in Cn_{\mathfrak{G},t}(\Delta) \Leftrightarrow A_t \geqslant glb(\Delta_t). \tag{6.28}$$

This is a logical consequence depending on a particular $\mathfrak{L}$-tense-structure and a particular $t \in T$. Then the logical consequence $Cn_{\mathfrak{L}}$ for the tense logic is defined by

$$A \in Cn_{\mathfrak{L}}(\Delta) \Leftrightarrow A \in Cn_{\mathfrak{G},t}(\Delta))$$

for all $t \in T$ and all $\mathfrak{L}$-tense-structures $\mathfrak{G}$. We write $\Delta \to A$ instead of $A \in Cn_{\mathfrak{L}}(\Delta)$ and $A \leftrightarrow B$ for logical equivalence with respect to this consequence function.

THEOREM 6.20. For all $A, B \in \mathbb{F}_{\text{prop}}$ the following hold.

(i) If $A \to B$ then $OA \to OB$ where $O \in \{F, P, G, H\}$.

(ii) For $O \in \{HF, GP\}$, $\neg\neg A \to OA$,

(iii) For $O \in \{F, P\}$, $OA \vee OB \leftrightarrow O\langle A \vee B)$.

(iv) For $O \in \{G, H\}$, $OA \vee OB \to O(A \vee B)$.

(v) For $O \in \{F, P, G, H\}$, $O(AB) \to OA \wedge OB$.

(vi) For $O \in \{PG, FH\}$, $OA \to \neg\neg A$.

Proof.

(i) Suppose $A \to B$. Let $\mathfrak{G}$ be an $\mathfrak{L}$-tense-structure. Then by (6.28), $B_t \geqslant A_t$ for all $t \in T$. Hence $lub\{B_s; s \in f(t)\} \geqslant lub\{A_s; s \in f(t)\}$ so that $(FB)_t \geqslant (FA)_t$. By (6.28), $FB \in Cn_{\mathfrak{G},t}(FA))$. Thus $FA \to FB$. Similarly, $PA \to PB$. Now by **PC1** and (6.28), $\neg B \to \neg A$. Hence $F\neg B \to F\neg A$. Thus $\neg F\neg A \to \neg F\neg B$, i.e. $GA \to GB$. Similarly, $HA \to HB$.

(ii) PROPOSITION. In any $\mathfrak{L}$-tense-structure,

$$s \in p(t) \Rightarrow A_t \leqslant (FA)_s$$

$$s \in f(t) \Rightarrow A_t \leqslant (PA)_s.$$

For the first implication, $A_t \in \{A_w : w \in f(s)\}$ if $s \in P(t)$. Thus $A_t \leqslant lub\{A_w : w \in f(s)\} = (TA)_s$. The second of these relations is proved similarly. To prove (ii), suppose that $s \in p(t)$. By the proposition $(\neg A)_t \geqslant (\neg FA)_s$. Hence $(\neg A)_t \geqslant lub\{(\neg FA)_s; s \in p(t)\} = (P\neg FA)_t$. Thus

$$P\neg FA \to \neg A. \tag{6.29}$$

Hence $\neg\neg A \to HFA$. Similarly, $\neg\neg A \to GPA$.

(iii) $F(A \vee B)_t = lub\{(A \vee B)_s; s \in f(t)\}$
$= lub\{lub\{A_s, B_s\}\}; s \in f(t)\}$
$= lub\{lub\{A_s; s \in f(t)\}, lub\{B_s; s \in f(t)\}\}$
$= (F(A) \vee F(B))_t.$

Thus, $F(A \vee B) \leftrightarrow F(A) \vee F(B)$. Similarly, $P(A \vee B) \leftrightarrow P(A) \vee P(B)$.
(iv) As $\mathfrak{L}$ is proto-complemented, by (5.26) and (6.9), $\neg(A \vee B) \to \neg A \wedge \neg B \to \neg A, \neg B$. By (i), $F(\neg(A \vee B)) \to F\neg A \wedge F\neg B$. Hence

$$\neg(F\neg A \wedge F\neg B) \to \neg F\neg(A \vee B) = G(A \vee B).$$

But, by (5.27) and (6.9), $\neg F\neg A \vee \neg F\neg B \to \neg(F\neg A \wedge F\neg B)$. Hence $GA \vee GB \to G(A \vee B)$. Similarly, $HA \vee HB \to H(A \vee B)$.
(v) $AB \to A, B$. By (i), $O(AB) \to OA, OB$ for $O \in \{F, P, G, H\}$. Hence $O(AB) \to OA \wedge OB$.
(vi) From (6.29), $P\neg F\neg A \to \neg\neg A$, i.e. $PGA \to \neg\neg A$. Likewise, $FGA \to \neg\neg\neg A$. ∎

THEOREM 6.21. *For a linear $\mathfrak{L}$-tense structure:*

(i) $OOA \to OA$ for $O \in \{F, P\}$;
(ii) *if $\mathfrak{L}$ is finite and is itself linearly ordered (e.g. N_3 or B_2 of Fig. 5.6, but not Q_4) then*

T'7: $FA \wedge FB \to F(AB) \vee F(A \wedge FB) \vee F(FA \wedge B)$

T'8: $PA \wedge PB \to P(AB) \vee P(A \wedge PB) \vee P(PA \wedge B).$

Proof.

(i) First, suppose that $s \in f(t)$. Then by linearity, $f(s) \subseteq f(t)$. Hence $(FA)_s \leq (FA)_t$. Thus $s \in f(t) \to (FA)_s \leq (FA)_t$. But $(FFA)_t = lub\{(FA)_s; s \in f(t)\}$. Hence $(FFA)_t \leq lub\{(FA)_t; s \in f(t)\} = (FA)_t$. Thus $FFA \to FA$. Similarly, $PPA \to PA$.
(ii) To prove **T'7** use will be made of the following proposition.

PROPOSITION. *If $\mathfrak{L}$ is a finite linearly ordered lattice, then for some $s \in f(t)$, $A_s = (FA)_t.$*

For as $\mathfrak{L}$ is finite, then $\{A_s: s \in f(t)\}$ is a finite subset of $\mathfrak{L}$. As $\mathfrak{L}$ is linearly ordered, for some $s \in f(t)$, $A_s = lub\{A_w; w \in f(t)\} = (FA)_t$. Now put $a = (FA)_t$ and $b = (FB)_t$. Then by the proposition there exist $u, v \in f(t)$ such that $a = A_u$ and $b = B_v$. By linearity, exactly one of the cases (a) $u = v$, (b) $u \in f(v)$, or (c) $v \in f(u)$ holds.

(a) $u = v$. Then
$$(F(AB))_t = \text{lub}\{(AB)_s; s \in f(t)\} \geq (AB)_w = a \cap b.$$

(b) $u \in f(v)$. Then
$$(F(B \wedge FA))_t \geq (B \wedge FA)_v = B_v(FA)_v \geq B_v \cap A_w$$

by the proposition. Thus $(F(B \wedge FA))_t \geq ab$.

(c) $v \in f(u)$. Then as in (b), $(F(A \wedge FB))_t \geq a \cap b$.

From each of these cases we have $(F(AB) \vee F(A \wedge FB) \vee F(FA \wedge B))_t \geq (F(AB))_t$. Hence

$$FA \wedge FB \to F(AB) \vee F(A \wedge FB) \vee F(FA \wedge B).$$

T′8 is proved similarly. ∎

EXAMPLE 6.7. The condition that the lattice ordering be linear in Theorem 6.21(ii) cannot be omitted as a Q_4-valued tense structure can be constructed which is a counterexample to **T′7**. ◆

THEOREM 6.22. *If $\mathfrak{L}$ is finite then the $\mathfrak{L}$-tense logic is compact.*

Proof. As in Theorem 6.11, the proof method consists of translating the tense logic into a suitable classical logic. Corresponding to the $\mathfrak{L}$-tense structure $\mathfrak{G} = \langle T, \ll, \{\mathfrak{A}_t; t \in T\}\rangle$ there is a relational system $\bar{\mathfrak{G}} = \langle T, \ll, \ldots R_i^\xi \cdots\rangle$ wherein to each propositional variable p_i of $\mathbb{F}_{\text{prop}}$ and each $\xi \in |\mathfrak{L}|$ there is a one-place predicate R_i^ξ on T. We shall express conditions holding in such relational systems in the appropriate informal language.

$\bar{\mathfrak{G}}$ satisfies the set M of (classical) formulas $(t) \bigvee_\xi \{R_i^\xi(t) \& \bigwedge_{\xi \neq \mu} \sim R_i^\mu(t)\}$ for $i = 1, 2, 3, \ldots$. Conversely, for every relational system satisfying M there exists an $\mathfrak{L}$-tense structure $\mathfrak{H}$ such that $\bar{\mathfrak{H}} = \mathfrak{H}$.

Each tense logic formula $D \in \mathbb{F}_{\text{prop}}$ translates to classical formulas $D^\xi(t)$ with one free variable t for each $\xi \in |\mathfrak{L}|$ such that

$$\mathfrak{A}_t(D) = \xi \iff \bar{\mathfrak{H}} \vDash D^\xi(t).$$

The procedure for affecting the translation follows inductively the formation rules and the definition of valuation.

(i) If D is p_i then D^ξ is $R_i^\xi(t)$.
(ii) If D is $c_i(A, B, \ldots)$ then D^ξ is $\bigvee \{A^\mu(t)B^\nu(t)\cdots; o_i(\mu, \nu, \ldots) = \xi\}$.
(iii) If D is FA then $(FA)_t = \text{lub}(\Phi_t(A))$ where $\Phi_t(A) = \{A_s; s \in f(t)\}$. Then ξ is an upper bound of $\Phi_t(A)$ iff for all $s \gg t$, $A_s \leq \xi$, i.e. iff $\bar{\mathfrak{G}} \vDash H^\xi$ where H^ξ is $(s)(t \ll s \Rightarrow \bigvee_{\mu \leq \xi} A^\mu(t))$. Hence ξ is the least upper bound of $\Phi_t(A)$ iff ξ is an upper bound and $\mu \geq \xi$ whenever μ is an upper bound. Thus $(FA)_t = \xi$ iff $\bar{\mathfrak{G}} \vDash D^\xi$ where D^ξ is $H^\xi(t) \& \bigwedge_{\mu < \xi} H^\mu(t)$.

(iv) If D is PA, define $J^\xi(t)$ to be the formula $(s)(s \ll t \Rightarrow \bigwedge_{\mu \leqslant \xi} A^\mu(t))$. Then D^ξ is $J^\xi(t)$ & $\bigwedge_{\mu < \xi} \sim J^\xi(t)$.

The remainder of the proof is as in Theorem 6.11. ∎

EXAMPLE 6.8. Linear $\mathfrak{L}$-tense logic is compact when $\mathfrak{L}$ is finite. For a proof of this, adjoin to M the conditions required for a tense structure to be linear. Any first-order specification of conditions on tense structures will correspondingly give a compact propositional $\mathfrak{L}$-tense logic. ◆

There are considerable delicate and deep philosophical problems in temporal quantifier logics concerning the range of quantification—what is to be counted as an individual for the purposes of quantification? The main problems concern the temporal integrity of individuals (e.g. Pivčević 1990).

> Things—the 'real' things of the physical world—of course exist within the temporal framework, be they tables, clouds, men, palm trees or pyramids. All such things have a history: they come into being, they pass away and they change during the course of their lifespan, so much rudimentary metaphysics is needed to guide the logician.
>
> Rescher and Urquhart 1971

We refer the reader to the discussion of these problems by Rescher and Urquhart (1971, Chapter 20). But it is clear that each definition of connectives and quantification will generate its own particular temporal logic. There is no unique temporal quantifier logic—it depends on the application and philosophical presuppositions.

6.11 Model structures

Structures similar to tense structures have been exploited by Kripke to interpretations of a range of logics from modal to formal intuitionistic logics.

By a *model structure* for a propositional logic formulated as in Chapter 2, §2.2, but with additional one-place connectives, we mean a structure $\mathfrak{G} = \langle T, R, \{\mathfrak{A}_t; t \in T\}, \tau \rangle$ where τ is a distinguished member of T and, for each $t \in T$, $\mathfrak{A}_t$ is a valuation in $\{0, 1\}$ and is called a 'possible world'. The binary relation R (the *world relation*) and the way in which the valuations are extended to all formulas depend on the logic concerned. Thus for modal systems there is at least one modal operator $\square$. With a fixed model structure let us write F_t as an abbreviation for $\mathfrak{A}_t(F)$. Then the valuations given in the model structure enable truth-values to be assigned to each modal formula for each t-section.

For the connectives

$$[A \vee B]_t = A_t \cup B_t$$
$$[A \wedge B]_t = A_t \cap B_t$$
$$[A \supset B]_t = A'_t \cup B_t$$
$$(\neg A)_t = A'_t.$$

For the modal operator,

$$(\Box A)_t = glb\{A_s; sRt\}.$$

Then a model structure $\mathfrak{G}$ is a *model* of a formula F if $F_\tau = 1$ in $\mathfrak{G}$ and we write $\mathfrak{G} \vDash F$. Logical consequence can then be defined in the usual way for classical logic: $\Delta \to F$ if all tense structures of the appropriate sort which are models of every member of Δ are also models of F. A formula F is *valid* if every model structure of the appropriate type is a model of F.

Various modal logics are obtained by imposing conditions on R. If R, we obtain models of a modal system M (also called T by Cressell and Hughes (1968, pp. 30ff); this book also contains definitions of various modal systems—S4, S5, etc.). If $\ll$ is transitive we have the system S4, if $\ll$ is transitive and symmetric we have S5, and if $\ll$ is symmetric we have the Brouwersche system (Cresswell and Hughes 1968, p. 58). Completeness proofs can be given for many such modal systems with respect to the associated model structures. See Cresswell and Hughes (1968) for copious references and historical remarks on the modal industry.

For the intuitionistic propositional calculus the relation R is a quasi-ordering and the assignment of truth-values to atomic formulas satisfies the condition

$$\mathfrak{A}_t(p_i) \leq \mathfrak{A}_s(p_i) \qquad (sRt).$$

The assignment of truth-values to formulas compounded with the logical junctors $\wedge$, $\vee$, $\wedge$, $\neg$ also depends on R:

$$[A \wedge B]_t = A_t \cap B_t,$$
$$[A \vee B]_t = A_t \cup B_t$$
$$(\neg A)_t = glb\{A'_s; sRt\}$$
$$[A \supset B]_t = glb\{A'_s \cup B_s; sRt\}.$$

Kripke (1965) proved that a formula is valid iff it is provable in the formal intuitionistic calculus. A similar interpretation based on the notion of *forcing* introduced by Cohen (1963) was made independently by Grzegorczyk (1964).

EXAMPLE 6.9. Consider the model structure $\langle \{\tau, \mu\}; R, \{\mathfrak{A}_\tau, \mathfrak{A}_\mu\}, \tau \rangle$ where $\mathfrak{A}_\tau(p) = 0$ for all $p \in P$ and $\mathfrak{A}_\mu(p) = 1$ if $p = p_1$ and is 0 otherwise. This is a model structure for the intuitionistic propositional calculus. It is not a model of $[\neg\neg p_1 \supset p_1]$ since $\mathfrak{A}_t([\neg\neg p_1 \supset p_1]) = 0$. Hence $[\neg\neg p_1 \supset p_1]$ is not provable in the intuitionistic propositional calculus. ♦

As with temporal logics, the model structures and the associated logics can be translated into classical terms and this enables compactness proofs to be manufactured.

Quantificational model structures can be constructed for modal logics and for intuitionistic predicate calculus. Once more the concept of validity can be translated into classical terms, thereby enabling compactness proofs to be given. As with temporal logic there are severe metaphysical problems concerned with identification of individuals in different sections (possible worlds).

Quantificational structures for intuitionistic predicate calculus can also be defined. Kripke (1965) has proved the completeness with respect to such models. The metaphysical problems here, however, are quite different from the problems surrounding temporal and modal logic. Suffice it to say that these structures are classical *representations* of intuitionistic logic.

7

Quasi-Boolean algebras and empirical continuity

7.1 Introduction

A class of lattices called *quasi-Boolean algebras* (QBA's) is examined in this chapter. These lattices include Boolean algebras, but the relation between the two classes is stronger than mere inclusion, as is shown in §7.5. The main reason for studying QBA's is that they are structures underlying Körner's theory of inexactness and empirical continuity (§§7.7, 7.8) and his three-valued logic of inexactness discussed in Chapter 8.

In pursuing this study some topological terminology has been introduced, such as *boundary*, *continuity*, and *connectedness* (Appendix, §T). The connection of elementary notions of point-set topology with QBA's is derived from the well-known fact that the closed sets of a topological space together with the empty set form a distributive lattice (Birkhoff 1948, p. 51). These notions are particularly suitable for Körner's theory of empirical continua.

The main results of this chapter were first developed in Cleave (1976).

7.2 Quasi-Boolean algebras

DEFINITION 7.1. A *quasi-Boolean algebra* (QBA) is an involuted distributive lattice. A QBA$^+$ is a QBA with a maximum or a minimum element. A *normal quasi-Boolean algebra* (NQBA) is a QBA which satisfies the identity

NQ: $x \cap x' \leqslant y \cup y'$.

A QBA is therefore an algebra $\langle A, \cap, \cup, ' \rangle$ satisfying the *QBA identities* **K1–K6**, **In**, **DM1**, and **DM2**. Thus QBA's are an equational class, so by Theorem A1 (Appendix, §A) every subsystem, every homomorphic image of a QBA is a QBA and every direct product of QBA's is a QBA.

If 1 is the maximum element (0 the minimum element) of a QBA $\mathfrak{L}$, then $\mathfrak{L}$ also has a minimum element 0 (maximum element 1) such that $1' = 0$, $0' = 1$.

The QBA's with two, three, or four elements are shown in Fig. 7.1. The action of the involution is shown by the dotted line joning an element x to x'. B_2, N_3, N_4, and B_4 are NQBA's. Q_4 is not an NQBA. Note that the

Quasi-Boolean algebras

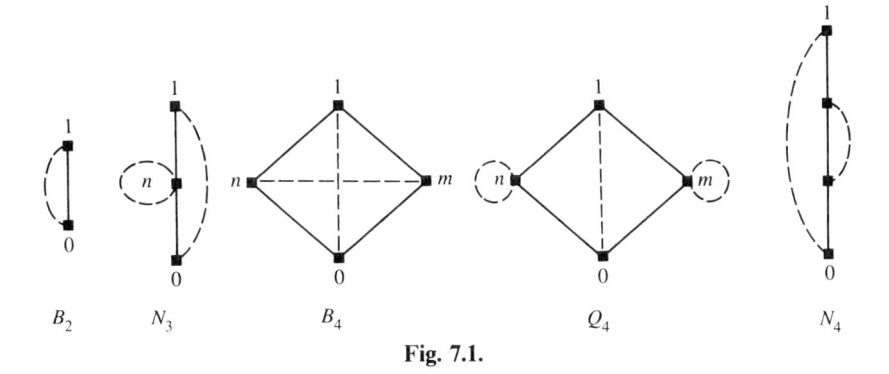

Fig. 7.1. B_2, N_3, B_4, Q_4, N_4

involution ′ of a QBA $\mathfrak{L}$ becomes a complementation, and thus $\mathfrak{L}$ becomes a Boolean algebra (BA) if the identities

BA: $\quad xx' = 0 \quad\quad x \cup x' = 1$

hold in $\mathfrak{L}$. Thus every BA is an NQBA. B_2 and B_4 are BA's; N_3 and N_4 are NQBA's but not BA's.

QBA$^+$'s were also called 'de Morgan algebras' by Monteiro (1960), and in that paper he used the term 'de Morgan lattices' for QBA's. Various authors (e.g. Rasiowa 1974) have studied QBA's: it seems that QBA's were first introduced—and so termed—by Białnicki-Birula and Rasiowa (1957). de Morgan lattices were introduced and studied under the term 'distributive i-lattices', and distributive i-lattices satisfying **K9** were introduced under the term 'normal distributive i-lattice' by Kalman (1958). Monteiro named them 'Kleene lattices', reserving the name 'Kleene algebras' for NQBA's.

EXAMPLE 7.1. Let $K(N)$ be the set of factors of a positive integer N ordered by divisibility. Then for all $m, n \in K(N)$, $glb\{m, n\} = gcd\{m, n\}$ and $lub\{m, n\} = lcm\{m, n\}$. Putting $n' = N/n$ defines ′ as an order-reversing map satisfying $n'' = n$, i.e. an involution. Hence $K(N)$ is a QBA with maximum element N and minimum element 1. ◆

EXAMPLE 7.2. (Finite subdivisions of the continuum). Our treatment of empirical continuity in §7.8 is guided by the following construction. Let D be a finite subset—including 0 and 1 of the closed interval $[0, 1]$ of the real line. Let J denote the set of unions of closed intervals $[d_1, d_2]$ where $d_1, d_2 \in D$. By convention $\emptyset \in J$. Thus J is closed under union and intersection. The non-empty members of J are closed sets whose boundary points are in D. For each $U \in J$ there is at least one set $V \in J$ such that $U + v = [0, 1]$ and $U \cdot V \subseteq D$. We think of U and V as complementary parts of the unit interval and so are led to consider the ordered pair (U, V). Let

$$F = \{(U, V); U, V \in J, U + V = [0, 1], U \cdot V \subseteq D\}.$$

F includes the special cases $O = (\emptyset, [0, 1])$, $I = ([0, 1], \emptyset)$. Operations $\cup, \cap, '$ are defined on F in an obvious way:

$$(U, V) \cup (W, Y) = (U + W, V \cdot Y)$$
$$(U, V) \cap (W, Y) = (U \cdot W, V + Y) \quad (7.1)$$
$$(U, V)' = (V, U).$$

Clearly, F is closed under these operations and so we have an algebra

$$K_D = \langle F, 0, I, \cup, \cap, ' \rangle.$$

It is easy to verify that K_D is an NQBA. This example can be generalized. Let K be a simplicial complex (e.g. Hilton and Wylie 1967, p. 18) and let K^+ be K augmented by the empty simplex $\emptyset$. Let J be the collection of finite unions of the closed simplexes of K^+. Then J is a DL under the operations of union and intersection since two closed simplexes of K^+ intersect in a closed face of each. The construction of an NQBA from J then proceeds as above. ◆

7.3 Constructing quasi-Boolean algebras

There are many ways of constructing QBA's besides direct products, subsystems, and homomorphic images which are characteristic of equational classes. The construction which we now describe gives 'free products' of QBA's. It is an important idea applicable particularly to arbitrary equational classes and is treated in much greater generality by Grätzer (1968, §28). A particular case of free products of QBA's will be applied to the logic of relevance in Chapter 9.

Let I be a non-empty index set and, for each $i \in I$, let $Q_i = \langle |Q_i|, \cup, \cap, ' \rangle$ be a QBA. Suppose that the domains $|Q_i|$ are pairwise distinct. Put $D = \sum \{|Q_i|; i \in I\}$. Treat D as a set of propositional letters and define the set $\mathbb{F}$ of formulas on D using the two-place junctors $\wedge, \vee$ and the one-place junctor $\neg$ (cf. Chapter 2, §2.2.1, and Chapter 6, §6.6) as follows.

F1: $D \subseteq \mathbb{F}$

F2: If $A, B \in \mathbb{F}$ then $\neg A, [A \vee B], [A \wedge B] \in \mathbb{F}$

The algebra of formulas has type $\langle 2, 2, 1 \rangle$; there should be no confusion in using $\mathbb{F}$ to denote this algebra as well as its domain.

We next define a congruence E on $\mathbb{F}$: $\mathbb{F}/E$ is then the free product and is denoted by $\prod * \{Q_i; i \in I\}$, and if I is finite, $I = \{1, 2, \ldots, n\}$ say, $\prod * \{Q_i; i \in I\}$ will also be written as $Q_1 * Q_2 * \cdots * Q_n$. ... E is defined using the notion of an *instance of a QBA identity* which we define by the following example. Consider **K3** (Theorem 5.20): $x \cap (x \cup y) = x$, $x \cup (x \cap y) = x$. Let $A, B \in \mathbb{F}$. Then the two ordered pairs $([A \wedge [A \vee B]], [A \vee [A \wedge B]])$, $([A \vee [A \wedge B]], [A \wedge [A \vee B]])$ are instances of **K3**. The congruence E is

Constructing quasi-Boolean algebras

determined by the following inductively defined set of ordered pairs. The initial elements are given by (i).

(i) (a) For $A, B \in |Q_i|$ for some i,

$$(\neg A, A'), \quad ([A \vee B], A \cup B), \quad ([A \wedge B], A \cap B) \in E.$$

(b) For all $A \in \mathbb{F}$, $(A, A) \in E$.
(c) Every instance of the QBA identities is in E.

The closure specification is defined by (ii):

(ii) For all $A, B, C \in \mathbb{F}$:

(a) $(A, B) \in E \Rightarrow (B, A) \in E$

(b) $(A, B), (B, C) \in E \Rightarrow (A, C) \in E$

(c) $(A, B) \in E \Rightarrow (\neg A, \neg B),$

$([A \vee C], [B \vee C]),$

$([A \wedge C], [A \wedge C]) \in E.$

These clauses make E a congruence on $\mathbb{F}$ and ensure that $\mathbb{F}/E$ satisfies the QBA identities, i.e. is a QBA.

Consider the case where $Q_i = \bar{1} =_{\text{def}} \langle \{i\}, \cup, \cap, ' \rangle$ is the unit QBA, i.e. its domain consists of one element i and $i = i \cup i = i \cap i = i'$. The diagrams of $\bar{1} * \bar{2}$ and $\bar{1} * \bar{2} * \bar{3}$ are shown in Fig. 7.2 with the elements labelled with representatives of the relevant E-classes. Thus, in $\bar{1} * \bar{2}$ (which is clearly isomorphic to Q_4) the elements are represented by the formulas $[1 \vee 2]$ (the maximum element), $1, 2$ and $[1 \wedge 2]$ (the minimum element). Here $\neg 1 = 1 \ mod(E), \ \neg 2 = 2 \ mod(E), \ \text{and} \ [1 \vee 2] = \neg\neg[1 \vee 2] = \neg[\neg 1 \wedge \neg 2] \ mod(E)$, etc.

A construction of a lattice $(\mathfrak{L}, J)$ from a given DL $\mathfrak{L}$ and an ideal J of $\mathfrak{L}$ (Definition 5.12(i)) was given in Example 5.26(iii): this lattice can now be recognized as an NQBA. The artifice employed in this example can be used to associate an NQBA with a given topological space X. It motivates the use of topological terms—particularly the notion of connectedness—in the theory of QBA's in the following section.

It is well known that the closed sets of X, including the empty set $\emptyset$, form a DL $\mathbb{L}(X)$ under union and intersection and that the set B_0 of closed boundary sets is an ideal of it (Appendix, §T and Kuratowski 1961, p. 125). $\mathfrak{N}(X) =_{\text{def}} (\mathbb{L}(X), B_0)$ is the *principal NQBA of* X. Let Cl denote the class of closed sets of X. Let $\pi: \mathbb{L}(X) \times \mathbb{L}(X) \to Cl$ be the projection onto the first member, i.e. $\pi(x, y) = x$.

THEOREM 7.1. B_0 is the smallest ideal B of $\mathbb{L}(X)$ such that $\pi(\mathbb{L}(X), B) = Cl$.

Quasi-Boolean algebras

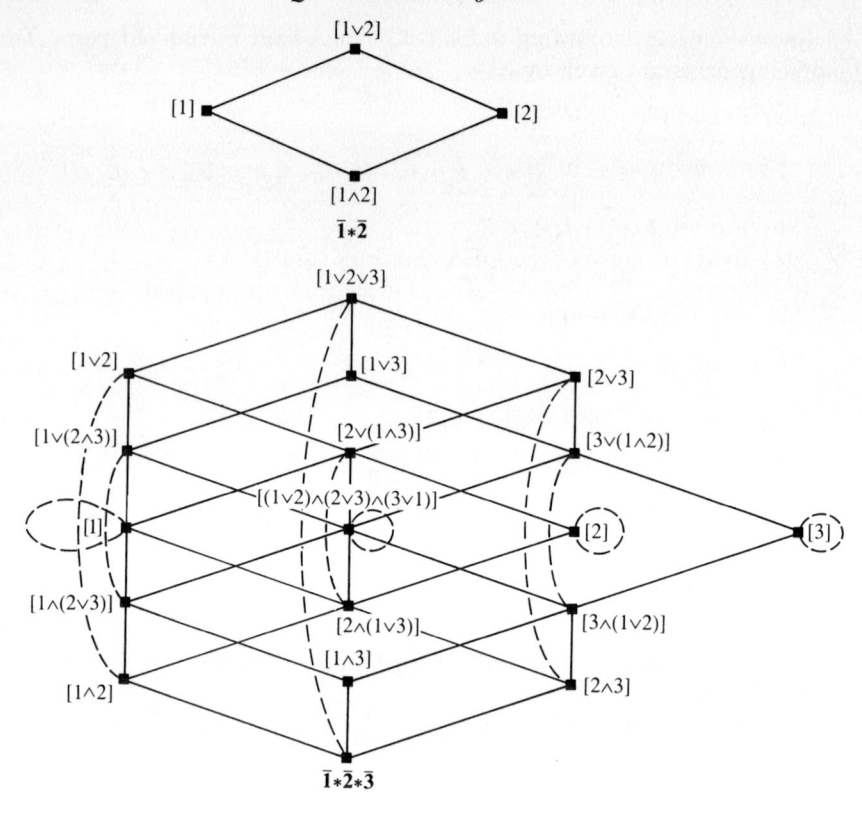

Fig. 7.2.

Proof. Let B be an ideal such that $\pi(\mathbb{L}(X), B) = Cl$. We first show that

$$B \supseteq B_0. \tag{7.2}$$

For let $p \in B_0$. As $\pi(\mathbb{L}(X), B) = Cl$, there exists a closed set x such that $(p, x) \in \pi(\mathbb{L}(X), B)$, i.e. $p + x = X$ and $p \cdot x \in B$. Then $x \supseteq X - p$ so that $C(x) \supseteq C(X - p)$. (C denotes topological closure.) But p is a boundary set so that $X - p$ is dense in X. Thus $C(X - p) = X$. Further, as x is closed, $C(x) = x$. Thus $x = X$. Hence $p = p \cdot X \in B$. This proves (7.2).

Next,

$$\pi(\mathbb{L}(X), B_0) = Cl. \tag{7.3}$$

For let $p \in Cl$. Then $p + C(X - p) = X$ and $p \cdot C(X - p)$ is a closed boundary set. Hence $(p, C(X - p))$ is an element of $(\mathbb{L}(X), B_0)$ and $\pi(p, C(X - p)) = p$. This proves (7.3).

7.4 The boundary elements of a quasi-Boolean algebra

Let X be a topological space. $\mathfrak{N}(X) = (\mathbb{L}(X), B_0)$ is its principlal NQBA. Let $a = (x, y) \in \mathfrak{N}(X)$ so that $x + y = X$ and $x \cdot y \in B_0$. Then $a \leq a' \Leftrightarrow (x \cdot y, x + y) = (x, y) \Leftrightarrow x \in B_0$ and $y = X$. The right-hand side of this equivalence justifies the appellation *boundary element* for such elements a. Thus we have a purely algebraic characterization—an element a of $\mathfrak{N}(X)$ is a boundary element if $a \leq a'$, i.e. $a = aa'$. Clearly, the boundary elements of $\mathfrak{N}(X)$ are precisely the elements bb' for $b \in \mathfrak{N}(X)$. We therefore make the following definition.

DEFINITION 7.2. Let $\mathfrak{N}$ be a QBA.
(i) $\partial \mathfrak{N} =_{\text{def}} \{a \in \mathfrak{N}; a \leq a'\}$, $\delta \mathfrak{N} =_{\text{def}} \{a \in \mathfrak{N}; a' \leq a\}$.
(ii) For all $x \in \mathfrak{N}$, $\delta x = x \cup x'$, $\partial x = xx'$. ∂x is the *boundary* of x and δx is the *coboundary* of x. x is a *boundary element* if $x = \partial x$ and a *coboundary element* if $x = \delta x$.

Coboundary and *boundary* are clearly dual concepts.
In the topological applications of QBA's ∂x is the 'limit' or 'extremity' of x, to use some metaphors of Aristotle (1936).
Some trivial facts about boundaries and coboundaries are recorded in the following lemma.

LEMMA 7.2. For all $x \in \mathfrak{N}$:
(i) $xx' \in \partial \mathfrak{N}$ $x \cup x' \in \delta \mathfrak{N}$;
(ii) $\partial x = \partial \partial x = \partial \delta x = \partial(x') = (\delta x)'$
$\delta x = \delta \delta x = \delta \partial x = \delta(x') = (\partial x)'$;
(iii) $\partial(xy), \partial(x \cup y) \leq \partial x \cup \partial y$
$\delta(xy), \delta(x \cup y) \geq \delta x \delta y$
$\partial x \leq \delta y$;
(iv) $\partial(y \partial x) = y \partial x$
$\delta(y \cup \delta x) = y \cup \delta x$.

EXAMPLE 7.3. Let $\mathfrak{L}$ be a DL and B an ideal of $\mathfrak{L}$. Then using the construction of Example 5.26(iii), $\partial(\mathfrak{L}, B) = B \times \{1\}$, $\delta(\mathfrak{L}, B) = \{1\} \times B$ ◆

The condition that a QBA be normal can be given in terms of the boundary elements since **NQ** (Definition 7.1) can be expressed as

$$\partial x \leqslant \delta y. \tag{7.4}$$

THEOREM 7.3. *Let $\mathfrak{N}$ be a QBA. Then $\mathfrak{N}$ is normal iff $\partial\mathfrak{N}$ is an ideal.*

Proof. Suppose that $\mathfrak{N}$ is normal. Let $p, q \in \partial K$. Then $p = \partial p$, $q = \partial q$, and $\delta p = p'$, $\delta q = q'$. By (7.4) $\partial p \leqslant \delta q = q'$ and $\partial q \leqslant \delta p = p'$. Now by the de Morgan laws and distributivity, $\partial(p \cup q) = q'\partial p \cup p'\partial q$. Hence $\partial(p \cup q) = \partial p \cup \partial p$. Thus $\mathfrak{N}$ is closed under $\cup$. Now let $p \in \partial\mathfrak{N}$. Then by Lemma 7.2(iv), for any q, $\partial(qp) = qp$. Hence $qp \in \partial\mathfrak{N}$. Thus $\partial\mathfrak{N}$ is an ideal.

Conversely, suppose that $\partial\mathfrak{N}$ is an ideal. We show that (7.4) holds. Let $x, y \in \mathfrak{N}$. Then $\partial x, \partial y \in \partial\mathfrak{N}$. As $\partial\mathfrak{N}$ is an ideal, $\partial x \cup \partial y \in \partial\mathfrak{N}$. Thus, $\partial x \cup \partial y \leqslant (\partial x \cup \partial y)'$. By Lemma 7.2(ii),

$$\partial x \leqslant \partial x \cup \partial y \leqslant (\partial x \cup \partial y)' = \delta x \delta y \leqslant \delta y. \qquad \blacksquare$$

COROLLARY 7.4. *Let $\mathfrak{N}$ be a QBA. Then $\mathfrak{N}$ is normal if $\delta\mathfrak{N}$ is a dual ideal.*

COROLLARY 7.5. *Let $\mathfrak{N}$ be a QBA such that $n = lub(\partial\mathfrak{N})$ exists. Then $\mathfrak{N}$ is normal iff $n \leqslant n'$ (i.e. $n \in \partial\mathfrak{N}$).*

Proof. Suppose that $\mathfrak{N}$ is normal. Then by (7.4), $\partial p \leqslant \partial q$ for all $p, q \in \mathfrak{N}$. Thus, each member of δK is an upper bound of $\partial\mathfrak{N}$. Hence n is a lower bound of δK. But $n' = glb(\delta\mathfrak{N})$. Hence $n \leqslant n'$. Hence $n \in \partial\mathfrak{N}$.

Conversely, suppose that $n \leqslant n'$. Then $\partial p \leqslant n \leqslant n' \leqslant \delta q$ for all $p, q \in \mathfrak{N}$. Thus (7.4) holds and so $\mathfrak{N}$ is normal. $\blacksquare$

$lub(\partial\mathfrak{N})$ always exists in a finite QBA and so Corollary 7.5 provides a convenient test of normality.

EXAMPLE 7.4. In Q_4, $\partial(Q_4) = \{n, m, 0\}$ and $lub\,\partial(Q_4) = 1$. As $1 \not\leqslant 0$, Q_4 is non-normal. ◆

LEMMA 7.6. *Let $\mathfrak{N}$ be an NQBA. Then*

$$y \in \delta\mathfrak{N} \quad \text{and} \quad xy = 0 \;\Rightarrow\; x = 0.$$

Proof. Let $y \in \delta(\mathfrak{N})$ and $xy = 0$. Then $x' = x' \cup xy = (x \cup x')(x' \cup y)$. Now $x \cup x' \in \delta\mathfrak{N}$ and $x' \cup y \geqslant y$. But by Corollary 7.4, $\delta\mathfrak{N}$ is a dual ideal. Hence $x' \in \delta\mathfrak{N}$. Thus $x \in \partial\mathfrak{N}$. By (7.4), $y \geqslant x$. Hence $x = xy = 0$. $\blacksquare$

Quasi-Boolean algebras in relation to Boolean algebras 235

DEFINITION 7.3. An element y of a QBA $\mathfrak{N}$ is *dense* if $xy \neq 0$ for all non-zero elements x of $\mathfrak{N}$ (cf. Theorem T1, Appendix, §T).

THEOREM 7.7. *Let $\mathfrak{N}$ be an NQBA. Then every coboundary element of $\mathfrak{N}$ is dense.*

Proof. By Lemma 7.6 and Definition 7.3. ∎

The hypothesis of normality is necessary in Theorem 7.7 as there exist non-normal QBA's with non-dense coboundary elements. Consider, for instance, Q_4. As $n = n'$ and $m = m'$, n, m are both coboundary elements. But $nm = 0$, $n \neq 0$, and $m \neq 0$.

7.5 Quasi-Boolean algebras in relation to Boolean algebras

Associated with a QBA $\mathfrak{N}$ is a certain homomorphic image which is obtained by 'cancelling' the boundary elements. This system $Z(\mathfrak{N})$, the *nucleus* of $\mathfrak{N}$, is a BA. In the application of this result to logic, the classical propositional calculus is the nucleus of the three-valued logic. Some properties of $Z(\mathfrak{N})$ are developed below.

7.5.1 *Congruences and ideals in a quasi-Boolean algebra*

From an ideal B in a QBA $\mathfrak{N}$ there can be defined a congruence which then yields the reduced algebra $\mathfrak{N}$ mod B.

DEFINITION 7.4. Let B be an ideal of a QBA $\mathfrak{N}$.

(i) The equivalence B^e is defined on $\mathfrak{N}$ by

$$s = t \; mod(B^e) \quad \Leftrightarrow_{def} \quad (Ea, b \in B)(sa' \cup b = ta' \cup b).$$

(ii) $B^\cup$ is the function which assigns to each element x of $\mathfrak{N}$ its $mod(B^e)$ equivalence class $B^+(x)$.

Lemma 7.8 follows easily from Definition 7.4 and the QBA identities.

LEMMA 7.8. (Moisil 1935). *B^e is a congruence.*

The reduced algebra $\mathfrak{N}/B^e$ can be constructed using this lemma. $B^\cup$ is the canonical projection. The elements of $\mathfrak{N}/B^e$ are the congruence classes $B^+(x)$ for $x \in \mathfrak{N}$; B^+ maps $\mathfrak{N}$ homomorphically onto $\mathfrak{N}/B^e$. $\mathfrak{N}/B^e$ is, of course, a QBA which is normal if $\mathfrak{N}$ is normal.

When B is a principal ideal of $\mathfrak{N}$ generated by a boundary element, $\mathfrak{N}/B^e$ is included as a kind of subsystem in $\mathfrak{N}$ itself.

DEFINITION 7.5. Let $\mathfrak{N}$ be a QBA.
(i) For $a, b \in \mathfrak{N}$ if $a \leqslant b$ then $[a, b] =_{\text{def}} \{x \in \mathfrak{N}; a \leqslant x \leqslant b\}$.
(ii) For $a \in \mathfrak{N}$, $H_a(x) =_{\text{def}} xa' \cup a$ for all $x \in \mathfrak{N}$.
(iii) For $a \in \mathfrak{N}$ (a) denotes the ideal generated by a, i.e. $[0, a]$ (Definition 5.12(ii)).

LEMMA 7.9. Let $\mathfrak{N}$ be a QBA and $a \in \partial\mathfrak{N}$. Then for all $x, y \in \mathfrak{N}$:
(i) $x = y \, mod((a)^e) \Leftrightarrow H_a(x) = H_a(y)$;
(ii) $H_a(x) \in [a, a']$;
(iii) $x \in [a, a'] \Rightarrow x = H_a(x)$;
(iv) $[a, a']$ is closed under the QBA operations.

Proof.
(i) Suppose $x = y \, mod((a)^e)$. Then there exist $p, q \in (a)$ such that $xp' \cup q = yp' \cup q$. But $p, q \leqslant a$. Hence $a' \leqslant p'$. And as $a \in \partial\mathfrak{N}$, $a \leqslant a'$. Hence $xa' \cup a = (xp' \cup q)a' \cup a = ya' \cup a$, i.e. $H_a(x) = H_a(y)$.
  Conversely, if $H_a(x) = H_a(y)$, then by Definitions 7.4 and 7.5, $x = y \, mod(a)$.
(ii) As $a \in \partial\mathfrak{N}$, $a \leqslant a'$. Then, for every $x \in \mathfrak{N}$, $a \leqslant xa' \cup a \leqslant a'$ i.e. $H_a(x) \in [a, a']$.
(iii) Suppose $a \leqslant x \leqslant a'$. Then $a'x = x = x \cup a$. Hence $H_a(x) = x$.
(iv) Trivial.

It follows from Lemma 7.9 that $\mathfrak{N}(a) =_{\text{def}} \langle [a, a'], a, a', \cap, \cup, ' \rangle$ is a QBA (a subsystem of $\mathfrak{N}$).

Theorem 7.10 follows easily from Lemma 7.9.

THEOREM 7.10. Let $\mathfrak{N}$ be a QBA and $a \in \partial\mathfrak{N}$. Then H_a is a homomorphism of $\mathfrak{N}$ onto $\mathfrak{N}(a)$ and $(a)^+ \restriction_{\mathfrak{N}(a)}$ is an isomorphism of $\mathfrak{N}(a)$ onto $\mathfrak{N}(a)^e$ with $(a)^+ \restriction_{\mathfrak{N}(a)} \circ H_a = (a)^+$.

7.5.2 The nucleus of a quasi-Boolean algebra

The most important ideal in a QBA $\mathfrak{N}$ is the ideal $(\partial\mathfrak{N})$ generated by the boundary elements (Definition 5.12(ii)).

Quasi-Boolean algebras in relation to Boolean algebras 237

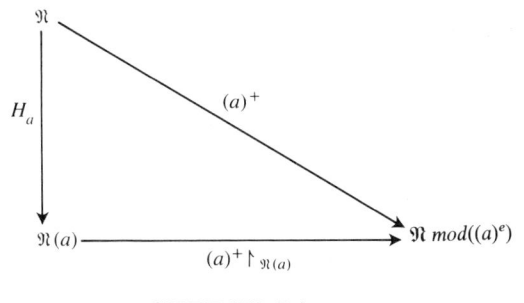

```
              𝔑
             /|\
            / | \
           /  |  \(a)⁺
       Hₐ |   \
          |    \
          ↓     ↘
        𝔑(a) ─────────────→ 𝔑 mod((a)ᵉ)
              (a)⁺ ↾ 𝔑(a)
```

SKETCH 7.1

DEFINITION 7.6. Let $\mathfrak{N}$ be a QBA. The *nucleus* $Z(\mathfrak{N})$ of $\mathfrak{N}$ is the QBA $\mathfrak{N}/\partial\mathfrak{N}^e$. The elements of $Z(\mathfrak{N})$ are the $\partial\mathfrak{N}^e$ congruence classes $[x]$ of the elements $x \in \mathfrak{N}$. Thus $[x] = \{y \in \mathfrak{N}; y = x \, mod(\partial\mathfrak{N}^e)\}$. $(\partial\mathfrak{N})^+$ is the *canonical homomorphism* of $\mathfrak{N}$ onto $Z(\mathfrak{N})$.

The importance of $(\partial\mathfrak{N})$ derives from theorem 7.11, corollary 7.12 and the applications to logic in Chapter 8.

THEOREM 7.11. (Moisil 1935). $Z(\mathfrak{N})$ is a BA.

Proof. As $Z(\mathfrak{N})$ is a homomorphic image of $\mathfrak{N}$, it is a QBA. It remains to show that $Z(\mathfrak{N})$ satisfies **BA** (§7.2).
Let $x \in Z(\mathfrak{N})$. Then $x = [a]$ for some $a \in \mathfrak{N}$. Since $(\partial\mathfrak{N})^+$ is a congruence, $xx' = [aa']$. But $aa' \in (\partial\mathfrak{N})$. Thus, $aa' = 0 \, mod(\partial\mathfrak{N}^e)$. Hence $xx' = [0]$. Likewise $x \cup x' = [1]$. Hence $Z(\mathfrak{N})$ is a BA. ∎

In fact $Z(\mathfrak{N})$ occupies a privileged position amongst all Boolean homomorphic images of $\mathfrak{N}$. For a homomorphism h of $\mathfrak{N}$ onto a BA h maps all elements xx' of $\mathfrak{N}$ to the minimum element of $0_\mathfrak{N}$. Thus $h: (\partial\mathfrak{N}) \to 0_\mathfrak{N}$. Hence the following corollary is obtained.

COROLLARY 7.12. Let $\mathfrak{N}$ be a QBA. Any homomorphism θ of $\mathfrak{N}$ onto a BA $\mathfrak{L}$ can be factored by the canonical homomorphism $(\partial\mathfrak{N})^+$ of $\mathfrak{N}$ onto its nucleus $Z(\mathfrak{N})$.

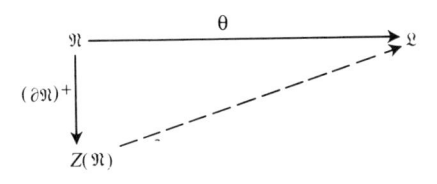

```
              θ
        𝔑ーーーーーーー→ 𝔏
        |          ↗
  (∂𝔑)⁺ |       ↗
        ↓    ↗
       Z(𝔑)
```

SKETCH 7.2

Under some circumstance the nucleus of a QBA is isomorphic to a subsystem.

COROLLARY 7.13. *Let $\mathfrak{N}$ be an NQBA such that $n = lub(\partial\mathfrak{N})$ exists in $\mathfrak{N}$. Then $n \in \partial\mathfrak{N}$, $(n)^+$ is the canonical homomorphism, and $(n)^+ \restriction_{\mathfrak{N}(n)}$ is an isomorphism of $\mathfrak{N}(n)$ onto $Z(\mathfrak{N})$ such that $(n)^+ \restriction_{\mathfrak{N}(n)} \circ H_n = (n)^+$.*

Proof. By Corollary 7.5, $n \leq n'$, i.e. $n \in \partial\mathfrak{N}$. By Theorem 7.3, $\partial\mathfrak{N} = (n)$. Hence the canonical homomorphism is $(n)^+$. The statement then follows by Theorem 7.10. ∎

7.6 The structure of quasi-Boolean algebras

The structure theorem and its proof in this section are due to Kalman (1958). By Birkhoff's theorem (Appendix, §A), every QBA is the subdirect product of subdirectly irreducible QBA's. We therefore identify the subdirectly irreducible QBA's. This task is accomplished by examining certain congruences on QBA's.

Let $\mathfrak{N}$ be a QBA. For each $p \in \mathfrak{N}$ there is an equivalence relation $C(p)$ on $\mathfrak{N}$ defined by

$$(x, y) \in C(p) \Leftrightarrow_{\text{def}} xp = yp \ \& \ x'p = y'p.$$

It is easily seem that $C(p)$ is in fact an equivalence relation and, further, that it is a congruence, i.e. for all $x, y, z \in \mathfrak{N}$, if (x, y) is in $C(p)$ then (x', y'), (xz, yz), $(x \cup z, y \cup z)$ are also in $C(p)$.

LEMMA 7.14.

(i) $C(1) = id_\mathfrak{N}$
(ii) $C(p) \cdot C(q) = C(p \cup q)$
(iii) $C(p) \cdot C(p') = id_\mathfrak{N}$
(iv) $C(p) = id_\mathfrak{N} \Leftrightarrow p' \leq p$

Proof.

(i) This is trivial.
(ii) Suppose that $(x, y) \in C(p) \cdot C(q)$. Then $xp = yp$, $x'p = y'p$, $xq = yp$, and $x'q = y'q$. Hence $x(p \cup q) = xp \cup xq = yp \cup yq = y(p \cup q)$. Likewise

$$x'(p \cup q) = y'(p \cup q).$$

Thus $(x, y) \in C(p \cup q)$.

The structure of quasi-Boolean algebras   239

Conversely, suppose $(x, y) \in C(p \cup q)$. Then $x(p \cup q) = y(p \cup q)$ and $x'(p \cup q) = y'(p \cup q)$. By **K3**, $xp = xp(p \cup q) = yp(p \cup q) = yp$. Likewise $x'p = y'p$. Thus $(x, y) \in C(p)$. Similarly, $(x, y) \in C(q)$. Hence $(x, y) \in C(p) \cdot C(q)$.

This completes the proof of (ii).

(iii) Suppose that $(x, y) \in C(p) \cdot C(p')$. Then $xp = yp$, $x'p = y'p$, $xp' = yp'$, and $x'p' = y'p'$. Hence $x \cup p = y \cup p$ so that

$$x = x(x \cup p) = x(y \cup p) = xy \cup xp$$
$$= xy \cup yp = y(x \cup p) = y(y \cup p) = y.$$

Thus $C(p) \cdot C(p') = id_{\mathfrak{N}}$.

(iv) By **K3**, $pp = (p \cup p')p$. Further, $p'p = (p'p)p = (p \cup p')p$. Hence

$$(p, p \cup p') \in C(p). \tag{7.5}$$

Now suppose that $C(p) = id_{\mathfrak{N}}$. Then by (7.5), $p = p \cup p'$. Hence $p' \leq p$.

Conversely, suppose that $p' \leq p$. Then $p = p \cup p'$. By (ii) and (iii), $C(p) = C(p \cup p') = C(p) \cdot C(p') = id_{\mathfrak{N}}$. ∎

The congruences $C(p)$ shed some light on the structure of subdirectly irreducible QBA's via Birkhoff's theorem.

LEMMA 7.15. *Let $\mathfrak{N}$ be a subdirectly irreducible QBA. Then for all $x, y, z \in \mathfrak{N}$:*

(i) $x \leq x'$ or $x' \leq x$;

(ii) $x > x'$ & $y > y'$ $\Rightarrow$ $xy > (xy)'$;

(iii) $y > y'$ & $z \leq z'$ $\Rightarrow$ $y > z$.

Proof. Since $\mathfrak{N}$ is subdirectly irreducible,

$$C(p) \cdot C(q) = id_{\mathfrak{N}} \Rightarrow C(p) = id_{\mathfrak{N}} \quad \text{or } C(q) = id_{\mathfrak{N}}. \tag{7.6}$$

(i) By Lemma 7.14(iii), $C(x) = id_{\mathfrak{N}}$ or $C(x') = id_{\mathfrak{N}}$. By Lemma 7.14(iv), $x \leq x'$ or $x' \leq x$.

(ii) Suppose that $x > x'$ and $y > y'$. Suppose, contrary to (ii), that $xy > (xy)'$. By (i), $xy \leq (xy)'$. By Lemma 7.14(iv), $C(x' \cup y') = C((xy)') = id_{\mathfrak{N}}$. By Lemma 7.14(ii), $C(x') \cdot C(y') = id_{\mathfrak{N}}$. By (7.6), $C(x') = id_{\mathfrak{N}}$ or $C(y') = id_{\mathfrak{N}}$. By (i), $x \leq x'$ or $y \leq y'$. But by hypothesis, $x > x'$ and $y > y'$. Hence $x' < x'$ or $y' < y'$. This is a contradiction. Hence (ii) holds.

(iii) Suppose $y > y'$ and $z \leq z'$. Then

$$(yz')' \leq yz'. \tag{7.7}$$

For suppose not. Then by (i), $yz' < (yz')'$. By (ii),

$$y(y' \cup z) > y' \cup yz' = (y' \cup y)(y' \cup z')$$
$$= y(y' \cup z') \qquad \text{as } y > y'$$
$$\geq y(y' \cup z) \qquad \text{as } z' \geq z.$$

This is a contradiction. Hence (7.7) holds.

From (7.7), $y \geq yz' \geq y' \cup z \geq z$. Thus $y \geq z$. But since $y > y'$ and $z \leq z'$, $y \neq z$. Hence $y > z$. ∎

COROLLARY 7.16. Let $\mathfrak{N}$ be a subdirectly irreducible QBA. Then for all $x, q \in \mathfrak{N}$:

(i) $q > q'$ & $x > x'$ $\Rightarrow$ $\delta(xq) = q\delta x$;

(ii) $q > q'$ & $x \geq x'$ $\Rightarrow$ $\delta(xq) = \delta x$;

(iii) $q \leq q'$ & $x > x'$ $\Rightarrow$ $\delta(xq) = \delta q$;

(iv) $q \leq q'$ & $x \geq x'$ $\Rightarrow$ $\delta(xq) = \delta x \cup q'$;

(v) $q' > q$ & $x' > x$ $\Rightarrow$ $\delta(x \cup q) = q'\delta x$;

(vi) $q' > q$ & $x \geq x'$ $\Rightarrow$ $\delta(x \cup q) = \delta x$;

(vii) $q' \leq q$ & $x' > x$ $\Rightarrow$ $\delta(x \cup q) = \delta q$;

(viii) $q' \leq q$ & $x \geq x'$ $\Rightarrow$ $\delta(x \cup q) = q' \cup \delta x$.

Proof. (i)–(iv) follow directly from the above lemma. (v)–(viii) follow from (i)–(iv) respectively by interchanging q' with q and x' with x and then using Lemma 7.2(ii). ∎

COROLLARY 7.17. Let $\mathfrak{N}$ be a QBA. Define a relation $\sim$ on $\mathfrak{N}$ by

$$x \sim y \Leftrightarrow$$

    (i) $x > x'$ & $y > y'$
        or
   (ii) $x' > x$ & $y' > y$
        or
   (iii) $x = x'$ & $y = y'$.

Then $\sim$ is a congruence on $\mathfrak{N}$.

Proof. Trivially, $\sim$ is an equivalence relation; to show that it is a congruence we prove that if $x \sim y$ then (i) $x' \sim y'$ and, for all $p \in \mathfrak{N}$, (ii) $xp \sim yp$ and (iii) $x \cup p \sim y \cup p$.

The structure of quasi-Boolean algebras

Suppose that $x \sim y$. Then one of the three conditions (a) $x > x'$ & $y > y'$, (b) $x' > x$ & $y' > y$, (c) $x = x' = y = y'$ holds.

ad (i) This is trivial.

ad (ii) Suppose (a) holds. By Lemma 7.15(i) either (a)(α) $p > p'$ or (a)(β) $p' \geqslant p$. If (a)(α) then by (i) and Lemma 7.15(ii), $xp > (xp)'$ or $yp > (yp)'$, i.e. $xp \sim yp$. If (a)(β) then by (i) and Lemma 7.15(iii), $x, y > p$. Hence $xp = yp = p$. Thus $xp = yp = p$. Hence $xp = yp = (xp)' = (yp)'$ if $p = p'$ and $(xp)' > xp$ & $(yp)' > yp$ if $p' > p$. In either case $xp \sim yp$. Thus, if (a) then $xp \sim yp$.

Suppose (b) holds. By Lemma 7.15(i) either (b)(α) $p \geqslant p'$ or (b)(β) $p' > p$. If (b)(α) then by (ii) and Lemma 7.15(iii), $x', y' > p'$. Hence $px = x$ and $py = y$. By (ii), $(px)' > px$ and $(py)' > py$. Thus, $px \sim py$. If (b)(β) then $x' \cup p' \geqslant p' > p \geqslant xp$. Hence $(xp)' > xp$. Likewise, $(yp)' > yp$. Thus $xp \sim yp$.

Suppose (c) holds. By Lemma 7.15(i) either (c)(α) $p > p'$ or (c)(β) $p = p'$ or (c)(γ) $p' > p$. If (c)(α) then by Lemma 7.15(iii), $p > x, y$. Hence $x = px$, $y = py$. Then, clearly, $px = (px)' = py = (py)'$, i.e. $px \sim py$. If (c)(β) then by (iii) $xp = yp$. By Lemma 7.15(i) one of the cases (δ1) $xp > (xp)'$, (δ2) $xp = (xp)'$, (δ3) $xp < (xp)'$ holds.

If (δ1) then as $xp = yp$, we also have $yp > (yp)'$.
If (δ2) then we also have $yp = (yp)'$.
If (δ3) then $yp < (yp)'$.

Thus if (c)(β) then $xp \sim yp$. If (c)(γ) then by Lemma 7.15(iii), $p' > x', y'$ so that $xp = x$ and $yp = y$. Hence $xp \sim yp$.

This concludes the derivation of $xp \sim yp$ from $x \sim y$.

Likewise $x \cup p \sim y \cup p$ is derivable from $x \sim y'$. ∎

LEMMA 7.18. *Let $\mathfrak{N}$ be a subdirectly irreducible QBA. For each $p \in \mathfrak{N}$ define the relations $D(p)$ and $E(p)$ on $\mathfrak{N}$ by*

$$(x, y) \in D(p) \Leftrightarrow x \sim y \ \& \ p\delta x = p\delta y$$

$$(x, y) \in E(p) \Leftrightarrow x \sim y \ \& \ p \cup \delta x = p \cup \delta y.$$

Then $D(p)$ and $E(p)$ are congruences, and $D(p) \cdot E(p) = id_{\mathfrak{N}}$.

Proof. Let $(x, y) \in D(p)$. We prove that for all $q \in \mathfrak{N}$ $(xq, yp) \in D(p)$ and $(x \cup q, y \cup q) \in D(p)$.

Let $q \in \mathfrak{N}$. By Lemma 7.15(i) either $q \geqslant q'$ or $q < q'$. Consider the first case. Now $x \sim y$. So by Corollary 7.17 one of the three cases (i), (ii), or (iii) holds. Consider case (i). Then we have $x > x'$, $y > y'$. Then, by Corollary 7.16 (viii), $\delta(x \cup q) = \delta x \cup q'$ and $\delta(y \cup q) = \delta y \cup q'$. Hence, as $(x, y) \in D(p)$, $p\delta(x \cup q) = p\delta(x \cup q)$, i.e. since $\sim$ is a congruence (Corollary 7.18), $(x \cup q, y \cup q) \in D(p)$. Cases (ii) and (iii) are similarly proved by appeal to Corollary 7.16.

242   *Quasi-Boolean algebras and empirical continuity*

Likewise, if $q < q'$ then $(x \cup q, y \cup q) \in D(p)$. Thus, for all q, $(x \cup q, y \cup q) \in D(p)$. Similarly, $(xq, yq) \in D(p)$. Thus, $D(p)$ is a congruence. Similarly, $E(p)$ is a congruence.

Next, suppose that $(x, y) \in D(p) \cdot E(p)$. Then

$$x \sim y \tag{7.8}$$

$$p\delta x = p\delta y \qquad p \cup \delta x = p \cup \delta y. \tag{7.9}$$

From (7.9)

$$\delta x = \delta x \cup p\delta x = \delta x \cup p\delta y = (\delta x \cup p)(\delta x \cup \delta y)$$

$$= (\delta y \cup p)(\delta x \cup \delta y) = \delta y.$$

Then by (7.8) and the definition of $\sim$ (Corollary 7.17), $x = y$. Thus, $D(p) \cdot E(p) = id_{\mathfrak{N}}$. ∎

THEOREM 7.19. B_2, N_3, and Q_4 are subdirectly irreducible up to isomorphism and are the only subdirectly irreducible QBA's.

Proof. Each of the given QBA's are subdirectly irreducible. We prove the converse.

By Lemma 7.18, since $\mathfrak{N}$ is subdirectly irreducible

$$D(p) = id_{\mathfrak{N}} \quad \text{or} \quad E(p) = id_{\mathfrak{N}} \qquad (p \in \mathfrak{N}). \tag{7.10}$$

If $1 = 0$ then $\mathfrak{N}$ is trivial. So suppose $1 > 0$. Then for all $x \in \mathfrak{N}$

$$x \neq 0 \ \& \ x \neq 1 \ \Rightarrow \ x = x'. \tag{7.11}$$

For suppose $x \neq 0 \ \& \ x \neq 1$. As $\mathfrak{N}$ is subdirectly irreducible, by Lemma 7.15(i)

$$x > x' \qquad \text{or } x' > x \qquad \text{or } x' = x. \tag{7.12}$$

If $x > x'$ then from the definition of $\sim$, $1 \sim x \sim 0$, Further, $\delta x = x = x\delta 1 = x\delta 0$. Hence $(1, x) \in D(x)$ and $(0, x) \in E(x)$. This contradicts (7.10) as $0 \neq 1$, $x \neq 0$, and $x \neq 1$. If $x > x'$ then similarly $(1, x') \in D(x')$ and $(0, x') \in E(x')$, contradicting (7.11). Thus, by (7.12), $x = x'$. This proves (7.23).

Thus every element x of $\mathfrak{N}$ which is distinct from 0 and 1 satisfies $x = x'$. Next

$$x, y \notin \{0, 1\} \ \& \ x \neq y \ \Rightarrow \ xy = 0. \tag{7.13}$$

For suppose that x and y are distinct elements, neither of which is 0 or 1. Then $xy \neq 1$. For otherwise $x = y = 1$. Suppose that $xy \neq 0$. By (7.11), $x = x'$, $y = y'$, and $xy = (xy)'$. Hence $x' \cup y' = xy = x \cup y$. Hence $x = x \cup xy = x \cup y$ so that $x \geq y$. Likewise, $y \geq x$. Hence $x = y$, which contradicts the hypothesis that $x \neq y$. Thus $xy = 0$.

Finally, suppose that $\mathfrak{N}$ has three distinct elements x, y, z, all of which are distinct from 0 and 1. By (7.13), $xz = yz = 0$. Hence $x' \cup z' = y' \cup z' = 1$. But by (7.11), $x = x'$ and $y = y'$. Thus $x \cup z = y \cup z = 1$. Hence $x = x(x \cup z) = x(y \cup z) = xy \cup xz = 0$. This contradicts the hypothesis that $x \neq 0$.

Thus $\mathfrak{N}$ has at most two elements which are neither 0 nor 1 and each of which satisfies $x = x'$. Hence $\mathfrak{N}$ is isomorphic to B_2, N_3, or Q_4. ∎

Observe that B_2 and N_3 are NQBA's but Q_4 is not. Thus the only non-trivial subdirectly irreducible NQBA's are B_2 and N_3. Further, B_2 is a subsystem of N_3 and N_3 is a subsystem of Q_4. Hence we have the following theorems.

THEOREM 7.20.

(i) Every non-trivial NQBA is a subdirect power of Q_3.
(ii) Every non-trivial QBA and every QBA$^+$ is a subdirect power of Q_4.

7.7 Inexact classes and predicates

A notion of inexact predicate has been developed by Körner (1966) and used to study the relation between empirical and theoretical entities. From this work there arises an algebra of inexact classes and an associated three-valued logic.

A fundamental part is played in empirical discourse by the *empirical* notions of individual, class or relations, and continua. These notions are directly applicable to experience. They must allow that some individuals, at least, are indefinite in the sense of not being sharply separable from their background, that some classes or relations are inexact in the sense of admitting borderline cases, that empirical continua are relative in the sense that what is a continuum under specifiable conditions is a discrete aggregate under others. The idealization of inexact empirical concepts by exact non-empirical predicates is the condition for the applicability of a scientific theory. The analysis of empirical classes etc. requires the transition from inexact concepts to exact ones.

Inexactness inevitably arises with empirical concepts. Borderline cases must always be allowed for in the classification stage of an empirical science. For instance, in economics,

> There is a substantial difference between the sum of economic activity and the non-economic, although we can always find intermediate types which cannot and should not be relegated by definition into one category or the other.
>
> Šik 1976, p. 54

This state of affairs can be idealized as follows. Let A be a given domain of individuals. Consider a definition D of a property P of these individuals which stipulates conditions under which P shall be ascribed to, or denied of, a given individual. Assume that D is consistent, i.e. D does not both affirm and deny P of any individual. In contrast with the customary usage in mathematics in particular, and classical logic in general, we envisage the possibility that there exist individuals in A to which the conditions of D neither ascribe nor deny P. We call such objects the *neutral cases* of D. But still we demand that the conditions be *definite*, i.e. for each individual a of A exactly one of the *three* conditions shall prevail: D ascribes P to a (a is a *positive case* of D), D denies P of a (a is a *negative case* of D), or D neither denies nor affirms P of a (a is a *neutral case* of D). Finally we reach an extensional viewpoint (cf. Körner 1966, pp. 41–2) by declaring two definitions *equal* if they have exactly the same positive, negative, and neutral cases. The consistent definitions (in the above sense) under this relation of equality are called *inexact classes*. There will be instances where a definition D has no neutral cases—every individual of A is either a positive case or a negative case of D. Such a definition gives rise to an *exact class*; such classes are what we usually understand by 'class'.

The above conception of three-valuedness was precisely the mainspring of Körner's development of the theory of inexact classes and the logic of inexactness. It is also basic to Hájek's study (Hájek et al. 1971) of automated research where the problem is to determine automatically (i.e. by computer program) hypotheses that are verifiable on the basis of some finite experimentally determined data. One must allow that for some object a and property P no information is available on whether a has or has not P—and in this case the formula $P(a)$ is assigned a third truth-value, 'undetermined', so that a is in the neutral domain of P.

There are two ways of regarding an inexact class D. D can be seen as either

(i) defining an algorithm for computing, of any x, whether x has the property P, in which interpretation the neutral cases are those undecided by D, and the definiteness of D is a classical condition but not intuitionistic (Kleene 1962, p. 333),

or

(ii) representing a provisional, but decided, stage in classifying individuals into P's and non-P's (i.e. a snapshot) so that at a later stage the neutral cases of D can be freely declared P's or non-P's.

It is clear that inexact classes can be treated by normal mathematical methods if we represent inexact classes by their characteristic functions. Thus, let X be an inexact class of individuals of A. The *characteristic function* of X is the function $\varphi: A \to \{1, n, 0\}$ defined by putting $\varphi(a) = 1, n, 0$ according as a is a positive cases, a neutral case, or a negative case respectively of X,

Inexact classes and predicates

where the n is an object distinct from 0 and 1. By the assumption, $\varphi(a)$ is defined for each a in A. There should be no confusion when, in later work, we identify inexact classes with their characteristic functions (see Appendix, §R).

A valuable heuristic aid in the classical algebra of classes is the Venn diagram. Here a set X is represented by a certain area (shaded) within the given region A (Fig. 7.3). In the algebra of subclasses of A, X cannot be

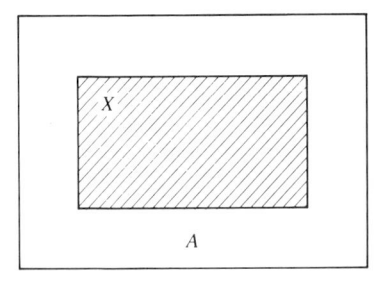

Fig. 7.3.

considered in isolation. It is associated with its complement X', the region which is not shaded. Thus a more complete picture of what X is comes by treating with the odered pair (X, X'). Alternatively, X can be represented by its characteristic function.

Analogously, a picture of an inexact set X divides A into three exclusive zones representing the positive, neutral, and negative cases of X (Fig. 7.4). The complete picture of X is also given by the ordered pair (U, V) where U is the set of non-negative cases and V is the set of non-positive cases. In these representations of an inexact class X it is tempting to regard the neutral cases as being a boundary between the positive cases and the negative cases, i.e. a *class* in the usual classical sense. Thus the boundary cases of X would form an entity of a different logical type from X itself. It is more useful, however, in the algebra of inexact classes to associate with the set of boundary points of X the *inexact class* ∂X whose neutral cases are precisely these boundary points and whose negative cases are the positive cases and the negative cases of X. Thus ∂X has no positive cases and it has the same logical type as X itself, namely a function from A into $\{1, n, 0\}$.

Having decided that neutrality shall be treated provisionally as a truth-value, we now face the task of constructing a suitable logic and a calculus of inexact classes (which will be an extension of the usual calculus of classes because exact classes are a special kind of inexact class). To do this the set $\{1, n, 0\}$ must be equipped with operations. Here we shall choose to make this set into an NQBA, in fact N_3, and justify this choice in Chapter 8. Then the inexact classes (strictly, inexact subsets of A) are exactly the members of N_3^A, i.e. the N_3-valued sets (Chapter 6, §6.8.2). By Theorem A1 (Appendix,

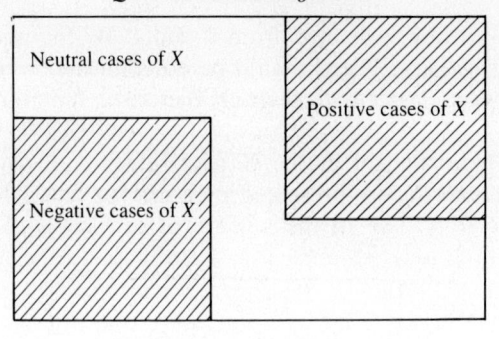

Fig. 7.4.

§A), the N_3-valued sets form an NQBA. Since B_2 is a subsystem of N_3, the BA of the exact classes, B_2^A is a subsystem of the algebra N_3^A of inexact classes. It easily follows from this definition of the algebra of inexact sets that $\partial X = XX'$ in accordance with Definition 7.2. The operations on N_3^A are thus interpreted as join, meet, and involution of the lattice N_3; we shall refer to these as *union*, *intersection*, and *complementation* respectively. By analogy with Definition 5.9, a *field* of inexact classes is a collection of inexact classes which is closed under union, intersection, and complementation.

EXAMPLE 7.5. Let L be the BA of subsets of a set X (Example 5.28(i)). Then X is an ideal in L (Definition 5.12). The elements of (L, X) (see Example 7.3) are pairs (U, V) of sets $U, V \in L$ such that $U + V = X$. For each such pair (U, V) let $\theta[U, V]: X \to \{1, n, 0\}$ be defined by

$$\theta[U, V](x) = \begin{cases} 1 & \text{if } x \in U - V \\ 0 & \text{if } x \in V - U \\ n & \text{if } x \in U \cdot V. \end{cases}$$

Then θ is an isomorphism of (L, X) onto N_3^x. ◆

THEOREM 7.21. Every NQBA is isomorphic to a field of inexact classes.

The proof of this theorem follows from Theorem 7.19. It is analogous to that of the corresponding theorem for BA's, i.e. Theorem 5.61. A structure theorem for finite QBA's, analogous to Theorem 5.35, was given by Muškardin (1978).

The more general notion of *inexact relation* is reached from the notion of *inexact* set in a similar manner. An inexact n-ary relation will be identified with the corresponding inexact set of n-tuples. Thus inexact n-ary relations on the domain A are the functions $\rho: A^n \to \{1, n, 0\}$. Hence we define an *inexact structure* to be a structure $\langle A, R_1, R_2, \ldots \rangle$, where A is a non-empty domain of individuals and, for each i, R_i is an inexact relation on A (see Chapter 6, §6.8.2).

7.8 Empirical continuity

Empirical continuity is a property of a finite set of attributes or classes (and those derived from them by purely logical means) based upon their continuous connection. Körner's method of investigating this notion is based on the theory of inexact classes, the fundamental idea being that

> an immediate transition from one class to another is discontinuous unless it is a kind of 'merging' of the two or a 'shading into each other' which presupposes that the two classes not only have neutral candidates, but also that some of these are common to the two classes.
>
> Körner 1966, p. 50

The procedure we follow here is somewhat more general, though motivated by this idea. Since the above idea depends on the notion of inexact class, more generality is achieved by stating the theory in terms of QBA's. This has the advantage of relating some topological notions (Examples 7.2 and 7.3, and §7.2).

It is assumed that the given finite set of attributes forms a QBA K. The operations of K can be regarded as 'logical' operations. The theory is then about the connection between logical 'parts'. Our notion of 'parts of an attribute' is a purely external concept, depending entirely upon the QBA under consideration and the 'logical' operations. It contributes little to the subtle whole-part question (cf. Russell 1903, Chapter 16). Yet even at this primitive level a species of continuity appears.

Three notions of 'part' will be distinguished and denoted by special terms. By a *factor* (by analogy with Example 7.1) of a given attribute x (i.e. member of K) we mean a formally identifiable subattribute—it is not required that the factor can be 'detached' in any sense from x. The attribute could be analysable into non-trivial factors a, b, i.e. $x = a + b$ where $x \neq a$ and $x \neq b$; we call such factors *formal parts*. Yet if a, b are formal parts of x they could be related in the sense of possessing a common non-zero factor. On the other hand the parts might be quite distinct, with no overlap ($ab = 0$). In this case x is merely the conjunction of two unrelated formal parts—we call such parts *components*. It is terminologically convenient to allow x to be a factor, a formal part, and a component of itself. Three notions of 'part' are given in the following definition, each of which gives rise to a corresponding notion of 'indivisible'.

DEFINITION 7.7. Let x be an element of a QBA K. For $a \in K$, the following hold.

(i) a is a *factor* of x if $a \leqslant x$. An element having no proper factors is an atom.
(ii) a is a *formal part* of x if $a = x$ or there exists $b \in K$ such that $x \neq a$, $x \neq b$, and $x = a \cup b$. An element having no proper formal part is said to be *join-irreducible* (cf. Definition 5.13).
(iii) a is a *component* of x if $a = x$ or there exists $b \in K$ such that $x \neq a$, $x \neq b$, $ab = 0$, and $x = a \cup b$.

The above definitions of 'part' do not at first sight commit us to considering only finite QBA's. It is, of course, desirable to examine QBA's where each element can be analysed into a finite number of parts. Now it is known that in a distributive lattice satisfying the DCC (Definition 6.1) every element can be expressed as the join of join-irreducible elements. But in a QBA, the operation ' is an involution. Hence, if the QBA satisfies the DCC it also satisfies the ACC and hence is finite (Cohn 1965, p. 79, Examples 16, 17). We therefore restrict attention to finite QBA's.

We shall treat the notions of 'component', which relate empirical continuity to the mathematical notion of 'connected topological space', and 'formal part' in turn. Atoms play no part in our theory of continuity.

7.8.1 Connected quasi-Boolean algebras

The notion of connectedness (Appendix, §T) is imported into the theory of QBA's by considering the principal NQBA $\mathfrak{N}(X)$ of a topological space X. We reach a workable definition as a result of the following theorem.

THEOREM 7.22. Let $(c, d) \in \mathfrak{N}(X)$. ($c, d$ are closed sets in X.) Then c is connected (in the topological sense) if and only if, for all $x, y \in \mathfrak{N}(X)$,

$$(c, d) = x \cup y \ \& \ xy = 0 \ \Rightarrow \ x = 0 \quad \text{or} \quad y = 0.$$

Proof. Suppose that c is connected and $(c, d) = x \cup y$, where $x = (a, b)$, $y = (p, q)$, and $xy = 0$. Then by the construction of $\mathfrak{N}(X)$, $c = a \cup p$, $\varnothing = ap$. As c is connected (topologically), $a = \varnothing$ or $p = \varnothing$. Hence $x = 0$ or $y = 0$.

Conversely, suppose that c is not connected. Then there exist closed sets a, p such that $c = a \cup p$ and $ap = \varnothing$. Let $x = (a, p \cup d)$, $y = (p, a \cup d)$. But $a(p \cup d) \subseteq cd$. As cd is a closed boundary set, so also is $p \cup d$. Hence $x \in \mathfrak{N}(X)$. Likewise $y \in \mathfrak{N}(X)$ and, by construction of $\mathfrak{N}(X)$, $x \cup y = (c, d)$ and $xy = 0$. ∎

Empirical continuity

COROLLARY 7.23. The topological space X is connected if and only if, for all $x, y \in \mathfrak{N}(X)$,

$$x \cup y \in \delta\mathfrak{N}(X) \ \& \ xy = 0 \ \Rightarrow \ x = 0 \quad \text{or} \quad y = 0. \tag{7.14}$$

Proof. Suppose that X is connected. If $x \cup y \in \delta\mathfrak{N}(X)$, then $x \cup y = (X, b)$ for some closed boundary set b. Suppose that $xy = 0$. By Theorem 7.22, $x = 0$ or $y = 0$. This proves (7.14).

Conversely, suppose that (7.14) holds for all $x, y \in \mathfrak{N}(X)$. Choose a closed boundary set b. So $(X, b) \in \delta\mathfrak{N}(X)$. By (7.14), for all $x, y \in \mathfrak{N}(X)$,

$$(X, b) = x \cup y \ \& \ xy = 0 \ \Rightarrow \ x = 0 \quad \text{or} \quad y = 0.$$

By Theorem 7.22, X is connected. ∎

Theorem 7.22 and Corollary 7.23 motivate the following definition.

DEFINITION 7.8. Let $\mathfrak{N}$ be a QBA.

(i) An element z of $\mathfrak{N}$ is *connected* if for all $x, y \in \mathfrak{N}$

$$x \cup y = z \ \& \ xy = 0 \ \Rightarrow \ x = 0 \quad \text{or} \quad y = 0.$$

(ii) $\mathfrak{N}$ is *connected* if every element of $\delta\mathfrak{N}$ is connected.

EXAMPLE 7.6.

(i) If x, y are connected elements of a QBA and $xy \neq 0$ then $x \cup y$ is connected.
(ii) An element of a QBA is connected iff it has no proper components.
(iii) Every join-irreducible element of a QBA is connected. Every element x of a finite QBA can be expressed as $x = a_1 \cup \cdots \cup a_n$ where each a_i is connected and $a_i a_j = 0$ for $1 \leq i < j \leq n$. Moreover, if $x = b_1 \cup \cdots \cup b_k$, where each b_j is connected and $b_i b_j = 0$ if $i \neq j$, then $k = n$ and for each i, $1 \leq i \leq n$, $a_i = b_j$ for some j with $1 \leq j \leq n$ (cf. Lemma 5.37). Because of this unique decomposition of x we call $a_1, \ldots, a_n$ the *principal parts* of x. ◆

A convenient test for connectivity for finite QBA's is provided by the following theorem.

THEOREM 7.24. If $n = lub(\partial\mathfrak{N})$, where $\mathfrak{N}$ is a finite NQBA, then $\mathfrak{N}$ is connected iff n' is connected.

Proof. Suppose $\mathfrak{N}$ is connected. By Corollary 7.5, $n' \in \delta\mathfrak{N}$. By Definition 7.8(ii), n' is connected.

Conversely, suppose that $\mathfrak{N}$ is not connected. Then by Definition 7.8(ii) there exists $c \in \delta\mathfrak{N}$ such that c is not connected. By Definition 7.8(i), for some $x, y \in \mathfrak{N}$, $c = x \cup y$, $xy = 0$, $x \neq 0$, and $y \neq 0$. But $n' = lub(\delta\mathfrak{N})$. Hence

$n' = n'c = n'x \cup n'y$. By Theorem 7.7 n' is dense. As $x, y \neq 0$, we have $n'x \neq 0$ and $n'y \neq 0$. But $(n'x)(n'y) = n'xy = 0$. Hence n' is not connected. ∎

7.8.2 Continuity and the connection between parts

The weakest sort of intersection relation between elements of a QBA is given in the following definition.

DEFINITION 7.9. Let $\mathfrak{N}$ be a QBA.

(i) Elements a, b of $\mathfrak{N}$ *overlap* (written $\omega(\alpha, \beta)$) if $ab \neq 0$. (ω is a reflexive and symmetric relation.)
(ii) Let $\Omega \subseteq K$. Ω is *chain-connected* if $\omega^*(a, b)$ (Definition 5.1(iii)) for all $a, b \in \Omega$, i.e. there exist $c_0, \ldots, c_n \in \Omega$ such that $a = c_0$, $b = c_n$, and, for all $i < n$, c_i and c_{i+1} overlap.

We immediately have the following theorem.

THEOREM 7.25. *The set of principal parts of a coboundary element of a finite connected QBA is chain-connected.*

Proof. Let $\mathfrak{N}$ be a finite connected QBA, and $x \in \delta\mathfrak{N}$. Then x is connected (Definition 7.8(i)). Let $y_1, \ldots, y_k$ be the principal parts of x. Now as ω is symmetric and reflexive, ω^* is an equivalence relation: let $\Omega_0, \Omega_1, \ldots, \Omega_r$ be the ω^*-classes of the principal parts of x. Clearly, each equivalence class Ω_j is chain-connected and so $lub(\Omega_j)$ is connected (Definition 7.7). Clearly $lub(\Omega_j) \cap lub(\Omega_{j'}) = 0$ if $j \neq j'$. But

$$x = y_1 \cup \cdots \cup y_k = lub(\Omega_0) \cup \cdots \cup lub(\Omega_r).$$

As x is connected, only one of the elements $lub(\Omega_j)$ is non-zero. Hence there is only one ω^*-class and so $\{y_1, \ldots, y_k\}$ is chain-connected. ∎

The principal parts of some coboundary elements of a connected QBA, however, have a finer intersection relation than the overlap of Definition 7.9. It is best described by means of a relation whose definition is derived from the topological analogy.

Two intersecting closed sets x, y of a topological space X are said to be 'in contact' if every common point is a common boundary point, i.e. $\emptyset \neq x \cdot y \subseteq bd(x) \cdot bd(y)$ ('... those things are "in contact" whose extremities are together' (Aristotle 1936, 226^{b}18, p. 396)). This condition can be expressed in terms of $\mathfrak{N}(X)$. Put $\hat{x} = (x, C(X - x))$ and $\hat{y} = (y, C(X - y))$. Then x and y are in contact iff $0 \neq \hat{x}\hat{y} \leqslant \partial\hat{x}\partial\hat{y}$. Hence we have the following definition.

Empirical continuity 251

DEFINITION 7.10. Let a, b be elements of a QBA $\mathfrak{N}$.

(i) a, b are *in contact* if $0 \neq ab \leqslant \partial a \partial b$
(ii) Let Ω be a set of connected elements of $\mathfrak{N}$ such that for all $a, b \in \Omega$ either $a = b$ or $ab \leqslant \partial a \partial b$. Ω is *continuous* if it is chain-connected.

EXAMPLE 7.7.

(i) For any two elements a, b of a continuous set Ω if $ab \neq 0$ then either $a = b$ or a, b are in contact.
(ii) Any attribute with non-void boundary 'merges' into its negation, i.e. for any $a \in \mathfrak{N}$, if $\partial a \neq 0$ then a, a' are in contact. For by Lemma 7.1(ii), $0 \neq aa' = \partial a = \partial a \partial a'$ ('... two things are continuous when the limits of each become identical and are held together ... continuity then exists in things out of which in virtue of their mutual contact a unity is produced' (Aristotle 1936, 227^{a}6, p. 397)). ◆

THEOREM 7.26. Let $\mathfrak{N}$ be a finite NQBA and let $n = lub(\partial \mathfrak{N})$. If c, d are distinct principal parts of n' then $cd = \partial c \partial d$.

Proof. By Theorem 7.10 and Corollaries 7.12 and 7.13, H_n is a homomorphism onto the BA $\mathfrak{N}(n)$.

If $\mathfrak{N}(n)$ is trivial, the maximum and minimum elements of $\mathfrak{N}(n)$ are identical so $n = n'$. Hence the principal parts of n' are boundary elements. The theorem then holds trivially.

If $\mathfrak{N}(n)$ is non-trivial let $a_1, \ldots, a_k$ be the atoms of $\mathfrak{N}(n)$. Then if B is any join-irreducible element $\leqslant n'$,

$$b \text{ is an atom of } \mathfrak{N}(n) \quad \text{or} \quad b \leqslant n. \quad (7.15)$$

For as $b \leqslant n'$, $H_n(b) = b \cup n \in \mathfrak{N}(n)$. Hence if $b \notin n$, $H_n(b)$ is a join of atoms of $\mathfrak{N}(n)$, say

$$b \cup n = a_{\alpha 1} \cup \cdots \cup a_{\alpha m}. \quad (7.16)$$

Hence $b = ba_{\alpha 1} \cup \cdots \cup ba_{\alpha m}$. As b is join-irreducible $b = a_j$, say. Hence $b \leqslant a_j$ and, by (7.16), $b \cup n = a_j$. This proves (7.15). Now suppose that c, d are distinct principal parts of n'. Then

$$c \leqslant d'. \quad (7.17)$$

For by (7.15) there are three cases.

(a) $c, d \leqslant n$. Hence $d' \geqslant n$. As $\mathfrak{N}$ is normal, by Corollary 7.5, $c \leqslant n \leqslant n' \leqslant d'$.
(b) $d \cup n = a_j$, say, $c \leqslant n$. So $d' \geqslant a_j'$. As $a_j \in \mathfrak{N}(n)$, $a_j' \in \mathfrak{N}(n)$. Hence $a_j' \geqslant n$. Hence $d' \geqslant a_j' \geqslant n \geqslant c$.
(c) $c \geqslant n = a_s$, $d \geqslant n = a_t$, say. Then $a_s \neq a_t$. For suppose otherwise. Then $c \cup n = d \cup n$. Hence $c = cd \geqslant cn$. As c is join-irreducible, $c = cd$ or

$c = cn$. But as $a_s \neq n$, $c \notin n$. So $c \neq cn$. Hence $c = cd$. Likewise $d = cd$. So $c = d$, contrary to the assumption that $c \neq d$. As $a_1, \ldots, a_k$ are atoms of $\mathfrak{N}(n)$, $a'_t = lub(\{a_v; v \neq t\}) > a_s$ as $a_s \neq a_t$. Hence $d' \geqslant a'_t \geqslant a_s \geqslant c$.

This completes the proof of (7.17). It follows from this that $cd \leqslant c'd'$. Hence $acd = cdc'd' = \partial c \partial d$. ∎

As a result of this theorem there follows the main result on the existence of continuous sets of elements in connected QBA's.

THEOREM 7.27. *The set of principal parts of n', where $n = lub(\mathfrak{N})$ and $\mathfrak{N}$ is a finite connected NQBA, is continuous.*

Proof. The principal parts of n' are connected. By Theorem 7.26, for all principal parts a, b of n', either $a = b$ or $ab < \partial a \partial b$. By Theorem 7.25, these principal parts are chain-connected and hence continuous. ∎

8

Three-valued logic

8.1 Introduction

The main aim of this chapter is to study the logic of inexactness arising from Körner's theory of inexact classes and predicates (Chapter 1, §1.3.2 and Chapter 7, §7.7). We first study logics whose truth-values are the QBA's B_2, N_3, and Q_4 (Chapter 7, Fig. 7.1). Though the centre of attention is N_3, a completeness proof is given which is applicable in all three cases (§8.8).

The logic with truth-values in B_2, i.e. $\mathfrak{L}[B_2]$ (Chapter 6, §6.5), is, of course, the classical logic (Chapter 2). We justify the three-valued logic based on N_3, i.e. $\mathfrak{L}[N_3]$, in some detail in Chapter 8, §8.2. There is a deep relation between these two logics: the Lindenbaum algebra $\mathfrak{L}[B_2]\P$ (Example 3.9(ii)) of the classical logic is the nucleus (Chapter 7, §7.5.2) of $\mathfrak{L}[N_3]\P$.

The philosophical ideas do not suffice in themselves to determine the logic completely: the final choice of method of construction of the three-valued logic is based on a 'continuity principle' of truth-values (§8.5) (see Cleave 1974c, 1980). Other choices will give rise to other three-valued logics. The logic based on Q_4, as treated in this chapter, must remain a mere technical artifice until it is justified in Chapter 9 where it arises naturally in multi-subject semantics in connection with relevance logic.

8.2 The classical imperative

It has long been felt that two-valued logic has fundamental limits. Aristotle had doubts about the universal applicability of the principle of the excluded middle (Bocheński 1961, p. 63), particularly to future contingent events. Łukasiewicz justified his three-valued logic on precisely such grounds. Peirce also suggested a third truth-value as a basis for an extension of classical two-valued logic. In a letter to William James in 1909 he wrote:

> I have long felt that it is a serious defect of the existing logic that it takes no heed of the *limit* between two realms. I do not say that the Principle of the Excluded Middle is downright *false*; but I *do* say that in every field of thought whatsoever there is an intermediate ground between *positive assertion* and *positive negation* which is just as real as they. Mathematicians always recognize this, and seek for that limit as the presumable lair of powerful concepts; while metaphysicians and

oldfashioned logicians—the sheep and goat separators—never recognize this. The recognition does not involve any denial of existing logic but it involves a great addition to it.

Fisch and Turquette 1966

More positive steps were later taken for various technical and philosophical considerations by Vasil'ev (1924) (Chapter 4, §4.2.2) (see also Arruda 1977), Łukasiewicz, Post, and Tarski to develop formal logical systems based on three or more truth-values (Chapter 6, §6.2).

The motive for the development of a three-valued logic in this chapter is derived from the role of decision-making in ascertaining truth and its connection with vagueness and inexactness at the interface between empirical observation and scientific theory (Chapter 1, §1.3.2).

But decision-making is familiar enough in practical experience. For instance, as group secretary I have to record in the minutes of our meetings the number of people at last night's meeting. I think that there were 20 people there, counting the babe-in-arms. It was not clear to me at first whether the baby should be counted as a person for the purpose of this count—standing orders gave no help. The officers of the group *decided* to count the babe—it would make the record look better. Joe was there, though he had not paid his subscription—we *decided* to count him as a visitor. I included the person who left after 15 minutes because he had signed the attendance register. I am sure Jane was there, though her name was not on the register: I will phone her later. All these trivial decisions just to answer a question about the number of people at a meeting!

The connection between decision-making and two-valued logic was made by Waismann (1955):

> In jurisprudence there are various guiding principles: one lays down that the judge has to decide every case; according to another he is free to leave a case undecided. (The praetor could dismiss a case by saying 'non liquet'). If we attempt to carry through the analogy into language, we notice that Aristotelian 2-valued logic corresponds to the principle of *deciding* every case. This is not the only possible policy. There is also the possibility of saying 'I want no decision'. And this tendency would find its fulfilment in a language in which it would make sense to say of a given sentence that it is in certain circumstances neither true nor false. This would in no way mean that the statement is 'in itself' true or false and that we are only unable to decide the issue: rather it would mean that we use the sentence according to different rules, in another logical system.

We add that the necessity to make *decisions* is a feature of scientific observation and measurement as well as of jurisprudence.

Waismann's remarks on juridical decision-making shows a difference between scientific thinking and legal–practical thinking. The exact rules for the application of a scientific concept determine once and for all and for every object whether or not it falls under that concept. Legal and practical concepts, however, admit borderline cases and their meaning is partly determined in the course of application in the sense that a legal decision turns a borderline (neutral) case (Chapter 7, §7.7) into a positive or negative case which can set a precedent for all future applications (see Körner 1980). This fact leads to a concept of *continuity* of truth-functions (§8.5).

More than one way of incorporating neutrality into a logic is therefore possible:

> ... the process which leads to the crystallising of a definite logic is not uniquely determined by the actual usage ... Aristotelian logic is the logic that meets our desire for decisiveness better than other types of 2-valued logic which does not include the law of the excluded middle. Behind ordinary logic there is something like a desire for decisiveness. One might go even further and suggest that this attitude is itself imposed on us by the conditions of our life—e.g. by the necessity of making ourselves understood by other people, of social cooperation, and so forth. On the other hand we could think of circumstances in which it would seem preferable to leave things in suspense.
>
> Waismann 1955

8.3 Sources of the third truth-value

In Chapter 7, §7.4, we discussed a third 'truth-value' in connection with the theory of inexactness. There are other motives for introducing a third truth-value—but having truth-values is one thing, constructing a logic from them is another issue. Below we list a few sources of three truth-values.

8.3.1 *Paradoxes of logic and set theory*

There is a long history of attempts to use three-valued logics (with various interpretations and diverse results) to solve the paradoxes of logic (Grelling etc.) and set theory (Russel etc.) Łukasiewicz (1941) applied his three-valued logic to such foundational problems, as did Wang (1961). Bochvar (1939) also used a third truth-value to avoid these paradoxes by declaring crucial sentences involving them to be 'meaningless'.

8.3.2 *Partial function logic*

There are various sources and concepts of *undefinedness* and numerous methods of dealing with it formally, e.g. Hoogewijs (1979). Blamey (1986) gave a penetrating discussion of this topic.

Partial functions arise naturally in the study of the theory of computation because a computation procedure may yield a result for some arguments, and for others may never stop. This situation, in the context of partial recursive functions, was described by Kleene (1962, p. 332). The truth-tables used to construct a logic depend upon whether series or parallel computation is under consideration (Manna and McCarthy, 1969).

Suppose that we are given a *partial* function $f: D^n \to D$ defined on the non-empty subset Ω of D^n. In normal mathematical discourse about a *total* function f (i.e. f is defined on the whole of the domain) one would make such statements as $f(v_1, \ldots, v_n) = a$, or $f(f(v_1, \ldots, v_n), a_2, \ldots, a_n) = b$, where $f(v_1, \ldots, v_n), a_2, \ldots, a_m, a$ and b are assumed to be in D. There is no problem here when f is defined for all its arguments. But consider the case of a partial function f. What are we doing when we assert $f(v_1, \ldots, v_n) = b$ or $f(f(v_1, \ldots, v_n), a_2, \ldots, a_n) = c$? How can discourse about partial functions be regulated? We can acknowledge the fact that f is undefined for arguments $(v_1, \ldots, v_n)$ of $D^n \to \Omega$ by saying that the expression $f(v_1, \ldots, v_v)$ is neutral. Then a logic can be constructed to deal with neutral expressions *as though* neutrality were a truth-value.

8.3.3 *Theory of descriptions*

The failure of denotation in the theory of descriptions gives another source of undefinedness. Consider the description $\iota x F(x)$ (read as 'the x such that $F(x)$'). Quine, simplifying Russell's theory, uses the contextual definition

$$P(\iota x F(x)) = (Ex)((y)(F(y) \Leftrightarrow y = x) \enspace \& \enspace P(x))$$

where $P(x)$ is an elementary open formula. However, numerous statements which hold for all individual denotations A fail to hold in classical logic for descriptions if the existental condition $(Ex)(y)(F(y) \Leftrightarrow y = x)$ is false. For example,

$$A = A$$

$$(x)G(x) \supset G(A)$$

$$G(A) \supset (Ex)G(x).$$

Carnap decides *arbitrarily* that $P(\iota x F(x))$ is to be counted as having the same truth-value as $P(a)$, where a is an arbitrarily chosen individual, if the existential condition fails:

$$P(\iota x F(x)) = (Ex)((y)(F(y) \Leftrightarrow y = x) \enspace \& \enspace P(x))$$
$$\vee \enspace .P(a) \enspace \& \enspace \sim (Ex)((y)(F(y) \Leftrightarrow y = x)).$$

This has the *technical* advantage that theorems which hold for individual constants and other individual denotations are now also valid for descriptions.

For Hilbert, only referential descriptions (i.e. those satisfying the existential condition) are allowed *syntactically*, for then the same axioms and rules hold as for the remaining individual denotations. The axiom $F(\iota x F(x))$ also holds. The peculiarity of Hilbert's procedure is that syntactical admissibility is then undecidable, for it depends on the facts which are formalized by $P(\iota x F(x))$.

The logic of truth-values

Some of these difficulties can be met if we acknowledge the dilemmas concerning the truth-value of $P(\iota x F(x))$, where F fails to satisfy the existential assumptions, by declaring $P(\iota x F(x))$ to be *neutral*. The problem is then to incorporate this judgement into the logic by pretending that neutrality is a kind of truth-value. This matter was treated extensively by Blau (1978).

The semantic rules for these concepts, whether inherited by tradition or conventionally agreed, do not suffice for us to classify all phenomena sharply. In order to apply formal logic to the process of reasoning in natural language we must assume that sentences are truth-definite, that they are interpreted in a linguistic 'snapshot' of the social process. The 'snapshot' assumption permits a sharp classification—not the classical binary division into P and non-P, but a ternary division into positive, negative, and neutral cases. We concede that this is itself an idealization, but it implies that, at any given instant, semantic rules for these predicates give three possibilities of reaction. In practice the situation is fluid. The practical function of language, linguistic decisions, is primarily a preparation for action. The compulsion to decide is the classical imperative. One is called upon, in practical affairs, not merely to record the occasions when semantic conventions fail to give a yes or a no, but to register a decision, to take the neutral case x of P and reclassify it as a P or a non-P. Once the decision has been made, a new fact has been created—we have entered another 'snapshot'. Thus the classical imperative, the enforcement of a yes/no decision, effects the evolution from one snapshot to the next. We recognize that this is a rather schematic account, but it suggests that in some sense classical logic is an asymptotic form of a three-valued logic.

8.3.4 Quantum mechanics

Reichenbach (1975, p. 54) remarked that in quantum mechanics simultaneous values of a measured entity cannot be effected. If q has been measured, we do not know the value of p.

> This lack of knowledge is considered in the Bohr–Heisenberg interpretation as making a statement about p *meaningless*.

Reichenbach accordingly introduced a third truth-value which he called *indeterminacy* and constructed an appropriate logic.

8.4 The logic of truth-values

The set of truth-values is $T = \{0, n, 1\}$ in which 0, 1 are identified with the classical *falsehood* and *truth* respectively. To construct a logic based on these truth-values we follow the method of Chapter 6 by first equipping T with a logical consequence function c_3, to make T into a simple logic $\mathfrak{N} = \langle T, c_3 \rangle$.

In anticipation of the next stage of development let us choose c_3 so that this logic has conjunction. By Theorem 5.14, T then acquires a quasi-ordering. Since the logic is simple the ordering is a partial ordering $\leqslant$. Since the logic is finite it has a maximum element 1 and a minimum element 0. Then $0 \leqslant n \leqslant 1$. This ordering is connected so that the ordering coincides with the converse of the implication relation. Thus, in this logic (cf. Chapter 6, §6.4)

$$x \in c_3(\Gamma) \quad \Leftrightarrow \quad glb(\Gamma) \leqslant x. \tag{8.1}$$

With this ordering conjunction coincides with glb:

$$ab = glb(\{a, b\}). \tag{8.2}$$

There are, of course, other ways of making T into a logic. A frequently used choice comes from taking 1 as a designated truth-value (Chapter 6, §6.2) so that a is a consequence of b if $a = 1$ when $b = 1$. This consequence function is denoted by c_2. By (6.2), for $\Gamma \subseteq T$,

$$x \in c_2(\Gamma) \quad \Leftrightarrow \quad (\Gamma \subseteq \{1\} \Rightarrow x = 1). \tag{8.3}$$

Both these logical consequence functions coincide with the classical one for classical values. c_3 is finer than c_2, i.e.

$$c_3(\Gamma) \subseteq c_2(\Gamma) \quad (\Gamma \subseteq T). \tag{8.4}$$

Further, the logic $\langle T, c_2 \rangle$ is not simple because $0 = n \, mod(c_2^e)$ (Definition 3.10).

Although we cannot yet justify it, we shall frame subsequent developments in this chapter in such a way as to be applicable to truth-values in Q_4 which is made into a logic with the consequence function c_4:

$$x \in c_4(\Gamma) \quad \Leftrightarrow \quad glb(\Gamma) \leqslant x. \tag{8.5}$$

The next task is to enquire into the truth-functions which $\mathfrak{N}$ should have.

8.5 The continuity principle for truth-values

Körner's theory of inexact classes and the logic of inexactness rests on the three 'truth-values' 0, 1, n (0 = falsehood, 1 = truth—the 'classical' values— and n = neutrality, the 'third' value), and the principle that the truth-value of a proposition is developed by stages in such a way that if at some stage the proposition receives a classical value (true or false) then it retains that classical value at all succeeding stages. This leads to the following definition.

The continuity principle for truth-values 259

DEFINITION 8.1. Let $T = \{0, n, 1\}$.

(i) Define the binary relation ▶ on T by

$$\tau \blacktriangleright \sigma \Leftrightarrow_{def} \tau = \sigma \vee (\sigma \notin \{0, 1\} \,\&\, \tau \in \{0, 1\}).$$

Extend ▶ to T^m as follows. Let $\bar{x}, \bar{y} \in T^m$. Then

$$\bar{x} \blacktriangleright \bar{y} \Leftrightarrow_{def} x_i \blacktriangleright y_i \quad \text{for } i = 1, 2, \ldots, m.$$

If $\bar{y} \blacktriangleright \bar{x}$ then $\bar{y}$ is a *refinement* of $\bar{x}$. The classical truth-values, 0 and 1, are preserved by refinement: if $\bar{x} \neq \bar{y}$ then for some i the neutral value of x_i is replaced in $\bar{y}$ by a classical value—the neutral case of $\bar{x}$ becomes a decided (i.e. classical) case in $\bar{y}$.

(ii) A function $\psi: T^m \to T$ is *regular* if

Reg1: $\psi(n, n, \ldots, n) = n$

Reg2: $\bar{x} \in \{0, 1\}^m \Rightarrow \psi(\bar{x}) \in \{0, 1\}$

Reg3: $\bar{x}, \bar{y} \in T^m \,\&\, \bar{x} \blacktriangleright \bar{y} \,\&\, \psi(\bar{y}) \in \{0, 1\} \Rightarrow \psi(\bar{x}) = \psi(\bar{y})$.

(iii) For $a, b \in T$

$$a * b =_{def} \begin{cases} b & \text{if } a = 1 \\ b' & \text{if } a = 0 \\ bb' & \text{if } a = n. \end{cases}$$

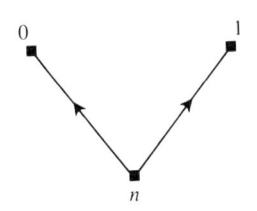

Fig. 8.1. Refinement in N_3.

EXAMPLE 8.1.

(i) Regular functions take classical values on classical arguments (**Reg2**) and have the continuity property (**Reg3**).
(ii) Regular functions are closed under composition.
(iii) Let $\bar{x}, \bar{y} \in T^m$. If $x_r * y_r = 1$ for $1 \leq r \leq m$ then $\bar{x} = \bar{y}$. If $x_r * y_r \in \{1, n\}$ for $1 \leq r \leq m$ then $\bar{x} \blacktriangleright \bar{y}$.
(iv) Consider the partial ordering $\geq$ on $\{0, n, 1\}$ in which $1 \geq n \geq 0$. Let $p, q: T^m \to T$ be regular functions. Let $\bar{x}, \bar{y} \in T^m$ with $\bar{x}$ a refinement of $\bar{y}$. If $p(\bar{y}), q(\bar{y}) \in \{0, 1\}$ and $p(\bar{y}) \geq q(\bar{y})$ then $p(\bar{x}) \geq q(\bar{x})$. ◆

Regularity of connectives has a topological interpretation which gives the classical truth-values of formulas a certain stability or continuity property. Let $\mathbb{O} = \{\emptyset, \{1\}, \{0\}, \{1,0\}, \{1,n,0\}\}$. Then $\mathbb{O}$ is a topology on T and a truth-function is regular if and only if it is continuous with respect to $\mathbb{O}$.

Kleene (1962) used truth functions defined by the following 'strong truth-tables' to construct a logic suitable for application to partial recursive function and predicates (Table 8.1). These truth-functions are extensions of

Table 8.1 Strong truth-tables

x	x'
1	0
0	1
n	n

$\cup$	1	0	n
1	1	1	1
0	1	0	n
n	1	n	n

$\cap$	1	0	n
1	1	0	n
0	0	0	0
n	n	0	n

the classical negation, disjunction, and conjunction, respectively (cf. Table 2.1). They are clearly regular functions in the sense of Definition 8.1. Moreover, the algebra $\langle T, 0, 1, \cap, \cup, ' \rangle$ is the NQBA N_3. Since the lattice operations of N_3 are regular, every QBA polynomial function is also regular. The converse is also true. Kleene also considered 'weak' truth-tables which are obtained from the classical ones by making n the value when at least one of the arguments is n (Table 8.2). However, weak conjunction and

Table 8.2 Weak conjunction and disjunction

$\wedge$	1	0	n
1	1	0	n
0	0	0	n
n	n	n	n

$\vee$	1	0	n
1	1	1	n
0	1	0	n
n	n	n	n

disjunction can be compounded from $\cap$, $\cup$, and $'$, and so are regular:

$$x \wedge y = xy(x \cup x')(y \cup y')$$

$$x \vee y = (x \cup y)(x \cup x')(y \cup y').$$

In the following work in this chapter the notion $\bigwedge, \bigvee$, which in Chapter 2, §2.2.2, was used for arbitrary conjunctions and disjunctions respectively, will be extended to arbitrary meets and joints respectively of lattice polynomials. Thus, if x_{ij}, for $i = 1, 2, 3$ and $j = 1, 2$, are variables then the lattice polynomial $(x_{11} \cap x_{12}) \cup (x_{21} \cap x_{22}) \cup (x_{31} \cap x_{32})$ can also be written as $\bigvee_{i=1,2,3} \bigwedge_{j=1,2} x_{ij}$ or as $\bigvee \{\bigwedge \{x_{ij}; j = 1, 2\}; i = 1, 2, 3\}$.

The continuity principle for truth-values 261

THEOREM 8.1. *Every regular function on T is an N_3 polynomial function.*

Proof. Let $\psi: T^m \to T$ be regular. A lattice polynomial $P(x)$ in the m variables $x_1, \ldots, x_m$ will be constructed for ψ.

Preliminary definitions
 (i) For $t \in T$, put $D(t) = \psi^{-1}(t)$. Then the sets $D(0), D(1), D(n)$ are pairwise disjoint and $D(0) \cup D(1) \cup D(n) = T^m$. By the definition of regularity, for all $\bar{x}, \bar{y}, \bar{z} \in T^m$ and for $t \in \{0, 1\}$

$$\bar{y} \in D(t) \ \& \ \bar{z} \blacktriangleright \bar{y} \ \Rightarrow \ \bar{z} \in D(t) \tag{8.6}$$

$$\bar{z} \blacktriangleright \bar{x}, \bar{y} \ \& \ \bar{x} \in D(t) \ \Rightarrow \ \bar{y} \in D(t). \tag{8.7}$$

 (ii) For any $\bar{z} \in T^m$, $z^{(c1)} =_{\text{def}} \{r; 1 \leqslant r \leqslant m \ \& \ z_r \in \{0, 1\}\}$ and $z^{(n)} =_{\text{def}} \{r; 1 \leqslant r \leqslant m \ \& \ z_r = n\}$.
 (iii) For the purpose of constructing the monomials in $P(\bar{x})$ the following notation will be used in conformity with Definition 8.1(iii): for $r = 1, \ldots, m$ and $a \in T$, $a * x_r =_{\text{def}} x_r', x_r, x_r x_r'$ according as $a = 0, 1, n$ respectively.
 (iv) For any $\bar{z} \in T^m$, we have the following.
  (a) If $z^{(c1)} \neq \emptyset$, then $\mathbb{1}[\bar{z}](\bar{x})$ is defined to be the monomial $\bigwedge \{z_r * x_r; r \in z^{(c1)}\}$. Thus $\mathbb{1}[\bar{z}](\bar{x})$ is defined in $\bar{z} \neq (n, n, \ldots, n)$.
  (b) $\mathbb{N}[\bar{z}](\bar{x})$ is the monomial $\bigwedge \{z_r * x_r; 1 \leqslant r \leqslant m\}$.
 (v) We define the polynomial $P(\bar{x})$ as follows. But first note that $(n, n, \ldots, n) \in D(n)$. Thus, though $D(1)$ or $D(0)$ may be empty, $D(n)$ is always non-empty. Now let $D-$ be the set of elements of $D(1)$ which are minimal under the ordering $\blacktriangleright$. Then $D(1) = \emptyset$ if and only if $D- = \emptyset$. Further, by **Reg1**, if $\bar{z} \in D-$ then $z^{(c1)} \neq \emptyset$ so that $\mathbb{1}[\bar{z}](\bar{x})$ is defined. Then

$$P(\bar{x}) =_{\text{def}} \begin{cases} \bigvee_{\bar{z} \in D(n)} \mathbb{N}[\bar{z}](\bar{x}) \cup \bigvee_{\bar{z} \in D-} \mathbb{1}[\bar{z}](\bar{x}) & \text{if } D(1) \neq \emptyset \\ \bigvee_{\bar{z} \in D(n)} \mathbb{N}[\bar{z}](\bar{x}) & \text{if } D(1) = \emptyset. \end{cases}$$

The action of P as a *function* is determined by the following four propositions:

$$\bar{y} \in D(1) \ \Rightarrow \ P(\bar{y}) = 1 \tag{8.8}$$

$$P(\bar{y}) = 1 \ \Rightarrow \ \bar{y} \in D(1) \tag{8.9}$$

$$\bar{y} \in D(n) \ \Rightarrow \ P(\bar{y}) = n \tag{8.10}$$

$$\bar{y} \in D(0) \ \Rightarrow \ P(\bar{y}) = 0. \tag{8.11}$$

ad (8.8) Suppose that $\bar{y} \in D(1)$. Then there exists $\bar{z} \in D-$ such that $\bar{y} \blacktriangleright \bar{z}$. Hence $y_r = z_r$ for $r \in z^{(c1)}$ by Definition 8.1(i). Thus $z_r * y_r = 1$ for $r \in z^{(c1)}$. Hence $\mathbb{1}[\bar{z}](\bar{y}) = 1$ and so $P(\bar{y}) = 1$.

ad (8.9) Suppose that $P(\bar{y}) = 1$. Then from the definition of $P(\bar{y})$ either

(a) $1[\bar{z}](\bar{y}) = 1$ for some $\bar{z} \in D-$

or

(b) $\mathbb{N}[\bar{z}](\bar{y}) = 1$ for some $\bar{z} \in D(n)$.

First consider case (b). $\mathbb{N}[\bar{z}](\bar{y}) = 1$ where $\bar{z} \in D(n)$. But $\psi(\bar{z}) = n$. By **Reg2**, for some s, $z_s = n$. Then by the definition of $\mathbb{N}$,

$$\mathbb{N}[\bar{z}](y) = y_s y'_s \wedge \{z_r * y_r; r \neq s\}.$$

But $y_s y'_s \in \{0, n\}$. Hence $\mathbb{N}[\bar{z}](\bar{y}) \neq 1$. This is a contradiction and so case (a) must hold. In case (a) $1[\bar{z}](\bar{y}) = 1$ where $\bar{z} \in D-$. Then $z_r * y_r = 1$ for $r \in z^{(c1)}$. By Example 8.1(iii) $\bar{y} \blacktriangleright \bar{z}$. By **Reg3**, $\psi(\bar{y}) = \psi(\bar{z}) = 1$. Thus, $\bar{y} \in D(1)$. This proves (8.9).

ad (8.10) Suppose $\bar{y} \in D(n)$. By (8.9), $P(\bar{y}) \in \{0, n\}$. But

$$\mathbb{N}[\bar{y}](\bar{y}) = \wedge \{y_r * y_r; r \in y^{(n)}\} \cap \wedge \{y_r * y_r; r \in y^{(c1)}\}.$$

By Definition 8.1(iii), $y_r y_r = n$ for $r \in y^{(n)}$, and $y_r y_r = 1$ for $r \in y^{(c1)}$. Hence $\mathbb{N}[\bar{y}](\bar{y}) = n$ and so $P(\bar{y}) = n$.

ad (8.11) Suppose $\bar{y} \in D(0)$. To prove that $P(\bar{y}) = 0$, it suffices to prove that $P(\bar{y}) \neq 1$ and $P(\bar{y}) \neq n$. Suppose $P(\bar{y}) = 1$. By (8.9), $(\bar{y}) \in D(1)$. This contradicts $\bar{y} \in D(0)$. Hence $P(\bar{y}) \neq 1$. Suppose $P(\bar{y}) = n$. Then either

(c) $1[\bar{z}](\bar{y}) = n$, for some $\bar{z} \in D-$,

or

(d) $\mathbb{N}[\bar{z}](\bar{y}) = n$, for some $\bar{z} \in D(n)$.

(c) is false. For suppose (c) holds. Then $z_r * y_r \in \{1, n\}$ for all $r \in z^{(cl)}$ and $z_r * y_r = n$ for some $s \in z^{(c1)}$. By Definition 8.1(iii), $y_r = z_r$ or $y_r = n$ for all $e \in z^{(c1)}$ znd $y_s = n$. Define $q_r =_{\text{def}} z_r$ if $r \in z^{(c1)}$, and $q_r = y_r$ if $r \in z^{(n)}$. Put $\bar{q} = (q_1, \ldots, q_m)$. Then $\bar{q} \blacktriangleright \bar{z}, \bar{y}$. But as $y \in D(0)$, $\psi(\bar{y}) = 0$ and as $\bar{z} \in D-$, $\psi(\bar{z}) = 1$. By **Reg3**, $\psi(\bar{q}) = 1$ and $\psi(\bar{q}) = 0$. This is a contradiction.

(d) is false. For suppose (d). Then $z^{(n)} \neq \emptyset$. Hence, by the definition of $\mathbb{N}$, $z_r * y_r \in \{1, n\}$ for $1 \leqslant r \leqslant m$. By Example 8.1(iii), $\bar{z} \blacktriangleright \bar{y}$. But $\bar{y} \in D(0)$. By **Reg3**, $\bar{z} \in D(0)$. This contradicts $\bar{z} \in D(n)$.

Hence $P(\bar{y}) \neq n$.

The theorem now following immediately from (8.3), (8.5), and (8.6). ■

The concept of regular truth-functions leads, via the above theorem, to our choice of organization of the truth-values as the NQBA N_3 and to the choice of the lattice operations as the primitives. We can now draw on the theory of Chapter 6 for the construction of a suitable logic.

8.6 N_3 propositional calculus

The construction of the propositional logic given here follows the general principles laid down for lattice-valued logics in Chapter 6 (cf. §6.6). A development of the N_3-propositional logic is also contained in Hájek et al. (1971). By Theorem 6.7 the logic inherits the SDPC laws from its lattice of truth-values.

The alphabet of the formal language (Chapter 2, §2.2.1) includes the pairwise disjoint sets P of proposition letters, the set $J = \{t, f, \neg, \wedge, \vee\}$ of junctors: t, f are propositional constants, $\neg$ is a one-place junctor, and $\wedge, \vee$ are two-place junctors. The set $\mathbb{F}(J, P)$ of formulas is defined in the usual way:

(F1) $\{t, f\} + P \subseteq \mathbb{F}(J, P)$

(F2) If $A, B \in \mathbb{F}(J, P)$ then $\neg A, [A \wedge B], [A \vee B] \in \mathbb{F}(J, P)$.

However, we shall use the abbreviations and bracketing conventions as defined in Chapter 2, §2.2.2, namely AB and A' stand for $[A \wedge B]$ and $\neg A$ respectively, etc., and $vb(A)$ denotes the set of propositional letters which occur in the formula A.

A formula of particular importance where $Q \in \mathbb{P}_\omega(P)$ is

$$N(Q) =_{\text{def}} \bigvee \{pp'; p \in Q\}. \tag{8.12}$$

The *algebra of formulas* is $\mathbb{F} =_{\text{def}} \langle \mathbb{F}(J, P), \wedge, \vee, \neg, t, f \rangle$. The valuations of the formulas are the homomorphism of $\mathbb{F}$ into N_3. Since B_2 is a subsystem of N_3, the valuations include homomorphisms which assign only the two values 0 or 1 to the proposition letters. Accordingly, $\mathbb{V}_2, \mathbb{V}_3$, and $\mathbb{V}_4$ denote the sets of homomorphisms of $\mathbb{F}$ into B_2, N_3, and Q_4, respectively. These are two-valued (classical), three-valued, and four-valued valuations respectively. Since B_2 is a subsystem of N_3 and N_3 is a subsystem of Q_3, we have

$$\mathbb{V}_2 \subseteq \mathbb{V}_3 \subseteq \mathbb{V}_4$$

Thus, for each $v \in \mathbb{V}_4$,

$$v(f) = 0 \quad v(t) = 1$$
$$v(\neg A) = v(A)'$$
$$v([A \vee B]) = v(A) \cup v(B)$$
$$v([A \wedge B]) = v(A) \cap v(B).$$

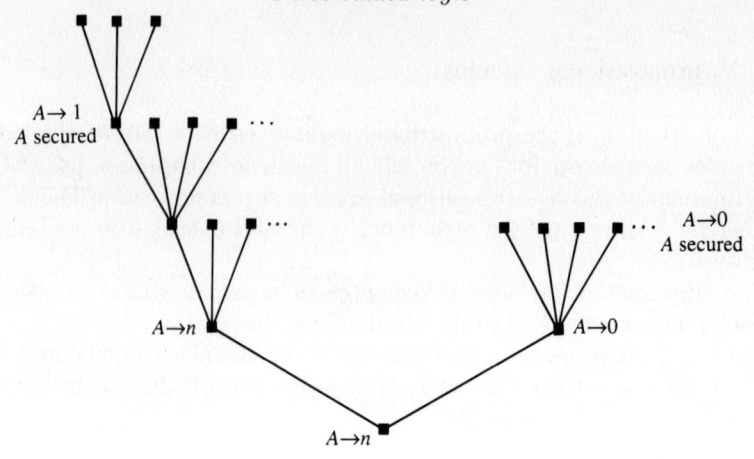

Fig. 8.2. Refinement tree.

The notion of *refinement* (Definition 8.1(i)) can be extended to valuations of formulas in the obvious way: for $u, v \in \mathbb{V}$,

$$u \blacktriangleright v \Leftrightarrow_{\text{def}} u(p) \blacktriangleright v(p) \qquad (p \in P).$$

This orders $\mathbb{V}_3$ as a tree in which the root (minimum element) is the valuation which assigns n to all proposition letters. Then, since the connectives are founded on regular truth-tables, we have the following theorem.

THEOREM 8.2. For every $A \in \mathbb{F}(J, Q)$

Reg1: $v(A) = n$ if there are no occurrences of t, f in A

Reg2: $v \in V_2 \Rightarrow v(A) \in \{0, 1\}$

Reg3: (continuity)

$$u, v \in V_3, \, \& \, u \blacktriangleright v \, \& \, v(A) \in \{0, 1\} \Rightarrow u(A) = v(A).$$

The continuity property (**Reg3**) means that a sequence of refinements $\cdots \blacktriangleright u_n \blacktriangleright \cdots \blacktriangleright u_2 \blacktriangleright u_1$ can be considered as a sequence of 'snapshots' such that, once a valuation, u_n, say, *decides* a formula A, i.e. gives A a classical value, that decision remains in force for all subsequent valuations u_m for $m \geq n$, i.e. $u_m(A) = u_n(A) \in \{0, 1\}$: A is *secured* at u_n (Fig. 8.2).

The notion of refinement of valuations can be extended in an obvious way to Q_4 where there are two classical values 0, 1 and two neutral values n, m.

N_3 propositional calculus

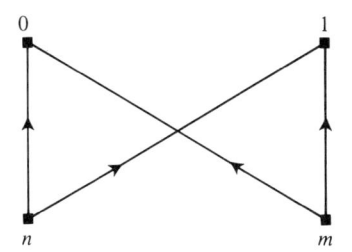

Fig. 8.3. Refinement in Q_4.

Define the refinement relation on Q_4 as in Definition 8.1 (Fig. 8.3). But now the continuity property fails. For instance, consider the formula $p \vee q$ where p and q are distinct proposition letters. Let $u \in \mathbb{V}_4$ be a valuation such that $u(p) = n$ and $u(q) = m$ so that $u(p \vee q) = 1$. Define $v \in \mathbb{V}_4$ by

$$v(r) = \begin{cases} u(r) & \text{if } r \notin \{p, q\} \\ 0 & \text{if } r \in \{p, q\}. \end{cases}$$

Then $v \blacktriangleright u$ but $u(p \vee q) = 0$.

As in Chapter 4 and Chapter 6, §6.6 (particularly (6.18)), the logical consequence functions Cn_i for $i = 2, 3, 4$ are derived from the lattices (considered as logics) B_2, N_3, Q_4. Thus, for any set Δ of formulas (cf. Chapter 6, §6.6),

$$Cn_2 = Cn[B_2] \qquad Cn_3 = Cn[N_3] \qquad Cn_4 = Cn[Q_4]. \qquad (8.13)$$

Since N_3 is finite glbs exist, so that, by (6.18), (8.13) is equivalent to

$$A \in Cn_i(\Delta) \iff (v \in \mathbb{V}_i)(glb(v(\Delta)) \leqslant v(A)) \qquad (i = 2, 3, 4) \qquad (8.14)$$

The case $i = 2$ is also equivalent to

$$A \in Cn_2(\Delta) \iff (v \in \mathbb{V}_3)(v(\Delta) \subseteq \{1\} \Rightarrow v(A) = 1).$$

We shall compare the three-valued logic whose logical consequence function is Cn_3 with the classical two-valued logic. We shall write $A \to_i B$ when $B \in Cn_i(A)$ and $A \leftrightarrow_i B$ when $A = B \, mod(Cn_i^e)$ (Definition 3.10). $[A]_i$ denotes the logical equivalence class of A modulo Cn_i^e. By Theorem 6.7, $\mathfrak{L}[B_2]\P = \mathbb{F}/Cn_2^e$ is a Boolean algebra, $\mathfrak{L}[N_3]\P = \mathbb{F}/Cn_3^e$ is an NQBA, and $\mathfrak{L}[Q_4]\P = \mathbb{F}/Cn_4^e$ is a QBA.

EXAMPLE 8.2.

(i) In BA's and QBA's

$$x \cup y \leqslant wx \quad \text{iff } x \leqslant w,z \text{ and } y \leqslant wz.$$

Hence, if $\rightarrow$ stands for $\rightarrow_2$, $\rightarrow_3$, or $\rightarrow_4$, then for any formulas A, B, C, D, by (8.14),

$$A \vee B \rightarrow CD \quad \text{iff } A \rightarrow C, A \rightarrow D, B \rightarrow C, \text{ and } B \rightarrow D.$$

(ii) By Theorems 6.4 and 6.7, as QBA's are distributive lattices, $\wedge$, $\vee$ are strongly logical operations in $\mathfrak{L}[Q]$ for any QBA Q. Similarly $\neg$ is a logical operation. But it is not a strongly logical operation in $\mathfrak{L}[N_3]$, for $p_1 \in Cn_3(p_2, p_1 p_2)$ and $p_2 \in Cn_3(p_1, p_1 p_2)$ but $\neg p_1 \notin Cn_3(\neg p_2, p_1 p_2)$.

(iii) There are numerous examples of logical laws which hold in classical logic but not in $\mathfrak{L}[N_3]$.

  (a) In classical logic, $\Delta, A \rightarrow B$ implies $\Delta, \neg B \rightarrow \neg A$ (Example 2.1(iv)). This implication only holds in $\mathfrak{L}[N_3]$ for $\Delta = \emptyset$ (see (ii) above).
  (b) Also in classical logic, $A, B \rightarrow C$ and $A, \neg B \rightarrow C$ implies $A \rightarrow C$ (Example 2.1(v)). This implication fails in $\mathfrak{L}[N_3]$, for $p_2 \vee \neg p_2 \in Cn_3(p_1, p_2)$ and $p_2 \vee \neg p_2 \in Cn_3(p_1, \neg p_2)$ but $p_2 \vee \neg p_2 \notin Cn_3(p_1)$.
  (c) The disjunctive syllogism (Example 4.9(ii)) fails in $\mathfrak{L}[N_3]$. Thus, let p, q be distinct proposition letters. Consider a valuation $v \in \mathbb{V}_3$ such that $v(p) = n$, $v(q) = 0$. Then $glb\{v(\neg p), v(p \vee q)\} = n$ but $v(q) = 0$. Hence $\neg p, p \vee q \nvdash_3 q$.

(iv) In contrast with classical logic, no formula in which neither t nor f occurs is a theorem of $\mathfrak{L}[N_3]$ or $\mathfrak{L}[Q_4]$. ◆

Clearly,

$$Cn_4(\Delta) \subseteq Cn_3(\Delta) \subseteq Cn_2(\Delta). \tag{8.15}$$

Thus

$$A \rightarrow_4 B \quad \Rightarrow \quad A \rightarrow_3 B \quad \Rightarrow \quad A \rightarrow_2 B. \tag{8.16}$$

Hence the corresponding notions of logical equivalence are similarly related. From (8.15), $[A]_4 \subseteq [A]_3 \subseteq [A]_2$, i.e.

$$A \leftrightarrow_4 B \quad \Rightarrow \quad A \leftrightarrow_3 B \quad \Rightarrow \quad A \leftrightarrow_2 B. \tag{8.17}$$

Thus Cn_3^e is a finer congruence than Cn_2^e. Hence $\mathcal{F}/Cn_2^e$ is a homomorphic image of $\mathbb{F}/Cn_3^e$.

EXAMPLE 8.3.

(i) By (8.12), $N(Q)' \leftrightarrow_3 \bigwedge \{p \vee p', p \in Q\}$. Thus for all $v \in \mathbb{V}_3$,

$$v(N(Q)) \leqslant n \leqslant v(N(Q)').$$

(ii) Let p, q be distinct proposition letters. Then

$$pp' \leftrightarrow_2 qq' \quad \text{but not} \quad pp' \leftrightarrow_3 qq'$$
$$p \vee p' \leftrightarrow_2 q \vee q' \quad \text{but not} \quad p \vee p' \leftrightarrow_3 q \vee q'$$
$$pp'(p \vee p')(q \vee q') \rightarrow_3 qq' \vee pp'$$
$$pp'(p \vee p')(q \vee q') \nvdash_3 qq'.$$

For the last case, consider a valuation v such that $v(p) = n$ and $v(p) = 1$. Then $v(pp'(p \vee p')(q \vee q')) = n$ and $v(qq') = 0$. ◆

LEMMA 8.3.

(i) $AB \to_3 f \Rightarrow A \leftrightarrow_3 f$ or $B \leftrightarrow_3 f$.

(ii) $t \to_3 A \vee B \Rightarrow A \leftrightarrow_3 t$ or $B \leftrightarrow_3 t$.

Proof.

(i) Let v be the $\mathbb{V}_3$ valuation defined by $v(p) = n$, all $p \in P$. Suppose that $AB \to_3 f$. Then $v(A)v(B) = v(AB) = 0$. Hence $v(A) = 0$ or $v(B) = 0$. If $v(A) = 0$ then, for all valuations u, $u(A) = 0$, i.e. $A \leftrightarrow_3 f$. Similarly, if $v(B) = 0$ then $B \leftrightarrow_3 f$.

(ii) This is similar to (i). ∎

THEOREM 8.4. Let $A, B \in \mathbb{F}(J, P)$ and let $Q = vb(AB)$. Then

$$AN(Q)' \to_3 B \vee N(Q) \Leftrightarrow A \to_2 B.$$

Proof. First, for any $v \in \mathbb{V}_3$, $v(N(Q)') = 1$ iff $v \in \mathbb{V}_2$. It is obvious that $v(N(Q)') = 1$ if $v \in \mathbb{V}_2$. For suppose that $v(N(Q)') = 1$ and $v \notin \mathbb{V}_2$. Then $v(p) = n$ for some variable p. But then $v(N(Q)') = n$. This contradicts the hypothesis and so $v \in \mathbb{V}_2$.

Now suppose that $A \to_2 B$. By Example 8.3(i), for any $v \in \mathbb{V}_3$, either (a) $v(N(Q)') = 1$ or (b) $v(N(Q)') = n$. Suppose (a). Then $v \in V_2$. As $A \to_2 B$, $v(A) \leqslant v(B)$. But $v(AN(Q)') = v(A)$ and $v(B \vee N(Q)) = v(B)$. Thus $v(AN(Q)') = 1 = v(B \vee N(Q))$. Suppose (b). Then

$$v(AN(Q)') \leqslant v(N(Q)') = v(N(Q)) \leqslant v(B \vee N(Q)).$$

Thus in both cases $AN(Q)' \to_3 B \vee N(Q)$.

Next, suppose that $AN(Q)' \to_3 B \vee N(Q)$. Let $v \in V_2$. Then $v(N(Q)') = 1$. Suppose that $v(A) = 1$. Then $v(AN(Q)') = 1$. Hence $v(B) = v(B \vee N(Q)) = 1$. Thus $A \to_2 B$. ∎

COROLLARY 8.5. $B \in \mathbb{F}(J, P)$, $Q = vb(B)$. B is a tautology of classical propositional calculus if, and only if, $N(Q)' \to_3 B \vee N(Q)$.

Example 8.3(ii) shows that the phrase $\vee N(Q)$ cannot be dropped from the statement of Theorem 8.4.

8.7 Normal forms and the analysis of implications

The construction of conjunctive and disjunctive normal forms derives from the fact that the Lindenbaum algebra $\mathfrak{L}[N_3]\P$ is an NQBA so that the algebraic identities **K1–K9** of Chapter 7, §7.1, can be used to establish equivalences.

DEFINITION 8.2. A conjunction C (disjunction D) of literals is *degenerate* if for some $p \in P$ both p and $\neg p$ are conjuncts of C (disjuncts of D). A formula is *pure* if it has no occurrences of t or f.

The construction of conjunctive and disjunctive normal forms is achieved by regarding the NQBA identities as transformation rules. We shall use the notation of Schütte (1961): let $F[P]$ denote a formula F with a distinguished occurrence of a subformula P. Then for any other formula Q, $F[Q]$ denotes the result of substituting Q for that occurrence of P. Each of the NQBA identities Ki has the form $t_1 = t_2$ and so gives rise to the two (reversible) transformation rules

$$\frac{F[t_1]}{F[t_2]} \qquad \frac{F[t_2]}{F[t_1]}$$

We refer to either of these rules as 'rule Ki'. These rules transmit logical equivalence: if A can be derived from B by application of a sequence of the rules then $A \leftrightarrow_3 B$.

By using the QBA laws, particularly the de Morgan laws, every formula can be seen to be logically equivalent to a disjunction $C_1 \vee \cdots \vee C_n$ where each C_i is a conjunction of literals. Thus, if $p, q, r \in P$ then

$$p(qq' \vee r')' \leftrightarrow_3 p((qq')'r'') \leftrightarrow_3 p((q' \vee q'')r'' \leftrightarrow_3 prq' \vee prq.$$

Then each conjunct C_i is equivalent to f, t, or a pure conjunction. Note that in the classical case degenerate conjuncts are equivalent to f—this is not the case with the three-valued logic. A dual result also holds. Thus we have the following theorem.

THEOREM 8.6. For every formula $A \in \mathbb{F}(J, P)$, exactly one of the three cases below hold:

(i) $A \leftrightarrow_3 f$;
(ii) $A \leftrightarrow_3 t$;
(iii) there exist pure formulas $D_1, \ldots, D_k, C_1, \ldots, C_n$ such that each D_i is a disjunction and each C_i is a conjunction of literals, and

$$C_1 \vee \cdots \vee C_n \leftrightarrow_3 A \leftrightarrow_3 D_1 \cdots D_k.$$

Lemma 8.7.

(i) If A is pure then $A \not\leftrightarrow_3 f$ and $A \not\leftrightarrow_3 t$.
(ii) If $A \leftrightarrow_3 t$ and $A \to_3 B$ then $B \leftrightarrow_3 t$.
(iii) If $B \leftrightarrow_3 f$ and $A \to_3 B$ then $A \leftrightarrow_3 f$.

Proof.

(i) Consider the valuation v defined by $v(p) = n$ for all $p \in P$. Suppose A is pure. By **Reg1**, $v(A) = n$. Hence $A \not\leftrightarrow_3 f$ and $A \not\leftrightarrow_3 t$.
(ii) and (iii) These follow directly from the definition of logical consequences. ∎

LEMMA 8.8. Let $\to$ denote either $\to_2$ or $\to_3$. Then

$$C_1 \vee \cdots \vee C_k \to D_1 \cdots D_n$$

iff $C_i \to D_j$ for $1 \leq i \leq k$ and $1 \leq j \leq n$.

Proof. Since conjunction and disjunction are normal in classical and three-valued logic,

$$C_1 \vee \cdots \vee C_k \to D_1 \cdots D_n \quad \text{iff} \quad C_1 \vee \cdots \vee C_k \to D_j \text{ for } 1 \leq j \leq n$$

and

$$C_1 \vee \cdots \vee C_k \to D_j \quad \text{iff} \quad C_i \to D_j \text{ for } 1 \leq 1 \leq k. \quad \blacksquare$$

Theorem 8.6 and Lemma 8.7 enable implications to be analysed into implications $C \to D$ where C and D are respectively conjunctions and disjunctions of literals. The remaining problem is to define the syntactical structure of such conjunctions C and disjunctions D when $C \to D$.

THEOREM 8.9. Let C (D) be a conjunction (disjunction) of literals.

(i) If $C \to_3 D$ then either
    (a) C and D are both degenerate
    or
    (b) some conjunct of C is also a disjunct of D.

(ii) If $C \to_2 D$ then either
    (c) C or D is degenerate
    or
    (d) some conjunct of C is also a disjunct of D.

Proof.

(i) Suppose $C \to_3 D$. Suppose (a) were false. Then either C or D is non-degenerate. Consider the case where D is degenerate. Then each propositional letter occurs at most once in D. Hence there exists a valuation $v \in \mathbb{V}_3$ such that

$$v(p) = \begin{cases} n & \text{if } p \text{ does not occur in } D \\ 1 & \text{if } p' \text{ is a disjunct of } D \\ 0 & \text{if } p \text{ is a disjunct of } D. \end{cases} \quad \begin{array}{c} (\alpha) \\ (\beta) \\ (\gamma) \end{array}$$

Clearly, $v(D) = 0$. But $v(C) \leqslant v(D)$ as $C \to_3 D$. Hence, $v(C) = 0$. But then v assigns 0 to one of the conjuncts P or p', say, of C. By clauses (β) and (γ) of the definition of v that conjunct is a disjunct of D. Thus (b) holds. Thus, either (a) or (b) holds.

(ii) Suppose $C \to_2 D$. Suppose (c) is false. Then C, D are both non-degenerate. Hence there exists $v \in \mathbb{V}_2$ such that

$$v(p) = \begin{cases} 1 & \text{if } p \text{ is a conjunct of } C, \text{ or (neither } p \text{ not } p' \text{ is a} \\ & \text{conjunct of } C, \text{ and } p \text{ is a disjunct of } D) \\ 0 & \text{if } p' \text{ is a conjunct of } C, \text{ or (neither } p \text{ nor } p' \text{ is a} \\ & \text{conjunct of } C, \text{ and } p \text{ is a disjunct of } D). \end{cases}$$

Clearly, $v(C) = 1$. Hence $v(D) = 1$. Hence v assigns 1 to some disjunct of D. By the definition of v, this disjunct is also a conjunct of C. Thus (d) holds. Hence (c) or (d) holds. ∎

For the completeness theorem of the predicate logics based on N_3 and Q_4 we need to consider non-empty sets Γ, Δ of literals which satisfy all or some of the following conditions:

C1: for all $p, q \in P$, $p \notin \Delta$ or $p' \notin \Delta$ or $q \notin \Gamma$ or $q' \notin \Gamma$

C2: $\Delta \cdot \Gamma = \emptyset$

C3: $f, t' \notin \Gamma$ and $f', t \notin \Delta$.

LEMMA 8.10. *Let Γ, Δ be non-empty sets of literals.*

(i) *If **C1** & **C2** & **C3** holds then there exist $v \in V_3$ and $x, y \in N_3$ such that $\sim(x \leqslant y)$, $glb(v(\Gamma)) \geqslant x$, and $lub(v(\Delta)) \leqslant y$.*
(ii) *If **C2** & **C3** holds then there exist $v \in V_4$ and $x, y \in Q_4$ such that $\sim(x \leqslant y)$, $glb(v(\Gamma)) \geqslant x$, and $lub(v(\Delta)) \leqslant y$.*

Proof. For each literal A, exactly one of the three conditions $\mathbf{B}i(A)$ holds, with $i = 1, 2, 3$.

$\mathbf{B1}(A)$: $A \notin \Gamma$ and $A \notin \Delta$

$\mathbf{B2}(A)$: $A \in \Gamma$ and $A \notin \Delta$

$\mathbf{B3}(A)$: $A \notin \Gamma$ and $A \in \Delta$

Let

$$V = \begin{bmatrix} n & 0 & 1 \\ 1 & n & 1 \\ 0 & 0 & n \end{bmatrix}$$

V_{ij} denotes the element in row i, column j of V. Then for each propositional letter p there are unique numbers i, j such that $\mathbf{B}i(p)$ and $\mathbf{B}j(p')$. Define the valuation v by $v(p) = V_{ij}$ and $v(f) = v(t') = 0$, $v(t) = v(f') = 1$. Put $\gamma = glb(v(\Gamma))$ and $\delta = lub(v(\Delta))$.

(i) Let $A \in \Gamma$ and $B \in \Delta$. Then $\mathbf{B2}(A)$ and $\mathbf{B3}(B)$. If A is a propositional letter p then $v(A) = v(p) \geq n$ (row 2 of V). If A is a constant then, by $\mathbf{C3}$, $A = t$ or $A = f'$. Hence $v(A) = 1 \geq n$. If A is p' for some propositional letter p then $v(p) \leq n$ (column 2 of V). Hence $v(A) = v(p') \geq n$. Similarly, $v(B) \leq n$. Thus, $v(A) \geq n \geq v(B)$. Hence

$$\gamma \geq n \geq \delta. \qquad (8.18)$$

Now suppose that $\gamma = n$. Since the elements of N_3 are finite in number and linearly ordered, there exists $A \in \Gamma$ such that $v(A) = n$. By $\mathbf{C3}$, A cannot be a constant. Hence for some propositional letter p, either (a) $A = p$, or (b) $A = p'$.

Suppose (a). Then from V, $\mathbf{B2}(p')$ holds. Hence $p, p' \in \Gamma$. By $\mathbf{C1}$, $q \notin \Delta$ or $q' \notin \Delta$ for all propositional letters q. Hence if $q \in \Delta$, $\mathbf{B3}(q)$ and ($\mathbf{B1}(q')$ or $\mathbf{B2}(q')$). From row 3, columns 1, 2 of V we then have $v(q) = 0$. If $q' \in \Delta$ then $\mathbf{B3}(q')$ and ($\mathbf{B1}(q)$ or $\mathbf{B2}(q)$) which gives $v(q) = 1$ and hence $v(q') = 0$. Thus, $v(B) = 0$ for all $B \in \Delta$.

Suppose (b). A similar argument to that of case (a) shows again that $v(B) = 0$ for all $B \in \Delta$.

Thus, in either case $\delta = 0$. Hence

$$\gamma = n \;\; \Rightarrow \;\; \delta = 0 \qquad (8.19)$$

From (8.18) and (8.19) it follows that

$$\gamma = 1 \;\&\; \delta \leq n \quad \text{or} \quad \gamma = n \;\&\; \delta = 0.$$

(ii) Suppose that $\mathbf{C2} \;\&\; \mathbf{C3}$ holds. Then there exist Γ_0, Γ_1, Δ_0, and Δ_1 such that $\Gamma = \Gamma_0 + \Gamma_1$, $\Delta = \Delta_0 + \Delta_1$, $\Gamma_0 \cdot \Gamma_1 = \Delta_0 \cdot \Delta_1 = \emptyset$, and the following conditions (c) and (d) hold.

(c) Γ_0 and Δ_0 consist of literals only (no constants) and no propositional variable occurs in both of Γ_0 and Δ_0, i.e. by **C2**, for all variables p,

$$p \in \Gamma_0 \;\Rightarrow\; p' \notin \Delta_0$$
$$p \in \Delta_0 \;\Rightarrow\; p' \notin \Gamma_0.$$

(d) Γ_1, Δ^1 consist of constants and literals, and every propositional variable which occurs in one also occurs in the other. Thus by **C2**. for every variable p,

$$p \in \Gamma_1 \;\Leftrightarrow\; p' \in \Delta_1$$
$$p' \in \Gamma_1 \;\Leftrightarrow\; p \in \Delta_1.$$

Now $|Q_4| = \{0, 1, n, m\}$ where $0, 1$ are the minimum and maximum elements respectively. Then from conditions (c) and (d) there can be constructed a $v \in \mathbb{V}_4$ such that

$$v(A) = \begin{cases} 1 & \text{if } A \in \Gamma_1 \\ 0 & \text{if } A \in \Delta_1 \\ n & \text{if } A \in \Gamma_0 \\ m & \text{if } A \in \Delta_0. \end{cases}$$

Hence $glb(v(\Gamma)) \geq n$ and $lub(v(\Delta)) \leq m$. As not-$(n \leq m)$, the lemma is proved. ∎

8.8 The nucleus of $\mathfrak{L}[N_3]\P$

The Lindenbaum algebra $\mathfrak{L}[N_3]\P$ of $\mathfrak{L}[N_3]$ is an NQBA. By Theorem 7.3, the set $\partial\mathfrak{L}[N_3]\P$ of boundary elements of this NQBA is an ideal (Definition 7.2). From Lemma 7.2(i) it follows that every element of $\mathfrak{L}[N_3]\P$ is logically equivalent in $\mathfrak{L}[N_3]$ (i.e. $= mod(Cn_3^e)$) to a formula of the form AA'. It is easily established that there exist formulas $E(p)$ such that, for $p \in vb(A)$, $vb(E(p)) \subseteq vb(A)$ and $AA' = \bigvee \{pp'E(p); p \in vb(A)\}$. Note that if C is a degenerate conjunction then $[C]_3 \in \partial\mathfrak{L}[N_3]\P$.

THEOREM 8.11. *The nucleus (Definition 7.6) $Z(\mathfrak{L}[N_3]\P)$ of $\mathfrak{L}[N_3]\P$ is isomorphic to $\mathfrak{L}[B_2]\P$.*

Proof. Let $\equiv$ denote the equivalence relation $\partial\mathfrak{L}_3\P^e$ (Definition 7.4) determined by the ideal $\partial\mathfrak{L}[N_3]\P$. By Definition 7.6 it is sufficient to prove that, for all $C, D \in \mathbb{F}(J, P)$,

$$C = D \; mod(C_2^e) \;\Leftrightarrow\; [C]_3 \equiv [D]_3. \tag{8.20}$$

First,

$$[C]_3 \equiv [D]_3 \;\Rightarrow\; C \leftrightarrow_2 D.$$

For suppose that $[C]_3 \equiv [D]_3$. By Definition 7.4 and (8.20) there exist $A, B \in \mathbb{F}(J, P)$ such that $C(A \vee A') \vee BB' \leftrightarrow_3 D(A \vee A') \vee BB'$. By (8.17) $C(A \vee A') \vee BB' \leftrightarrow_2 D(A \vee A') \vee BB'$. But $A \vee A' \leftrightarrow_2 t$ and $BB' \leftrightarrow_2 f$. Hence $C \leftrightarrow_2 D$.

Conversely,
$$C \leftrightarrow_2 D \Rightarrow [C]_3 \equiv [D]_3. \tag{8.21}$$

This follows from

(i) $C \rightarrow_2 D \Rightarrow [C \vee D]_3 \equiv [D]_3$

and

(ii) $D \rightarrow_2 C \Rightarrow [D \vee C]_3 \equiv [C]_3$.

ad (i) Suppose that $C \rightarrow_2 D$. Now by Theorem 8.5 there are three exclusive cases in the analysis of C:

(a) $C \leftrightarrow_3 t$;

(b) $C \leftrightarrow_3 f$;

(c) $C \leftrightarrow_3 C_1 \vee \cdots \vee C_k$ where $C_1, \ldots, C_k$ are pure conjunctions

Likewise the three cases in the analysis of D are:

(a') $D \leftrightarrow_3 t$;

(b') $D \leftrightarrow_3 f$;

(c') $D = D_1 \vee \cdots \vee D_k$ where $D_1, \ldots, D_k$ are pure disjunctions.

Consider these cases in pairs.

(a, a') Then $C \vee D \leftrightarrow_3 D$. Hence $[C \vee D]_3 \equiv [D]_3$.
(a, b') This case cannot arise because, by (8.17), $C \leftrightarrow_2 t$ and $D \leftrightarrow_2 f$ so that $C \not\leftrightarrow_2 D$.
(a, c') $t \rightarrow_2 D_1 \ldots D_n$. By Lemma 8.7, $t \rightarrow_2 D_i$ for $1 \leq i \leq n$. As the D_i are pure they are degenerate. Fix D_i. Then $D_i \leftrightarrow_3 p \vee p' \vee D_i^*$, say, for some $p \in P$. Hence

$$D_i(p \vee p') \leftrightarrow_3 p \vee p' \leftrightarrow_3 t(p \vee p') \vee pp'.$$

Thus $[D_i]_3 \equiv [t]_3$.
(c, a') and all cases (b, .) If $C \leftrightarrow_3 f$ or $D \leftrightarrow_3 t$ then $C \vee D \leftrightarrow_3 D$. Hence $[C \vee D]_3 \equiv [D]_3$.
(c, b') Similar to case (a, c).

(c, c') By Lemma 8.7, $C_i \to_2 D_j$ for $1 \leq i \leq n$ and $1 \leq j \leq k$. Fix i, j. By Theorem 8.8(ii), either

(iii) one of C_i, D_j is degenerate

or

(iv) some conjunct of C_i is also a disjunct of D_j.

Suppose (iii). If C_i is degenerate then $C_i = pp'C$ for some formula C and some formula C and some $p \in P$. Thus, $C_i(p \vee p') \vee pp' \leftrightarrow_3 pp'$. Hence

$$(C_i \vee D_j)(p \vee p') \vee pp' \leftrightarrow_3 D_j(p \vee p') \vee pp'.$$

Thus

$$[C_i \vee D_j]_3 \equiv [D_j]_3. \tag{8.22}$$

Similarly for D_j degenerate. Suppose (iv). Then $C_i \leftrightarrow_3 FC$ and $D_j \leftrightarrow_3 F \vee D$ for some formulas F, C, D. But clearly, $FC \to_3 F \vee D$. Hence $C_i \vee D_j \leftrightarrow_3 D_j$ so that (8.22) again holds.

Thus (8.22) holds for all i, j and so $[C \vee D]_3 \equiv [D]_3$. This concludes the proof of (i).

ad (ii) The proof of (ii) is similar to that of (i). ■

8.9 Connectivity and continuity in $\mathfrak{L}[N_3]\P$

The connectivity properties (Chapter 7, §7.8.1) of the three-valued and two-valued logics differ markedly.

THEOREM 8.12. Every element of $\mathfrak{L}[N_3]\P$ is connected. Hence $\mathfrak{L}[N_3]\P$ is connected.

Proof. Let A, B be formulas and suppose $[A]_3[B]_3 = [f]$. Then $AB \to_3 f$. By Lemma 8.2, $[A]_3 = [f]$ or $[B]_3 = [f]$. Thus, by Definition 7.7(iii), every element of $\mathfrak{L}[N_3]\P$ is connected. ■

By Theorem 7.27 we then have the following corollary.

COROLLARY 8.13. Let P be finite and $\mathfrak{L}[N_3]$ the N_3-propositional logic constructed from P. Put $m = lub(\partial\mathfrak{L}[N_3]\P)$. Then the set of principal parts of m' is connected.

In contrast with Theorem 8.12, $\mathfrak{L}[B_2]\P$ is not connected; the only connected elements are the zero element and the atoms.

Corollary 8.13 can be proved directly. Without loss of generality it can be assumed that $P = \{p_0, \ldots, p_n\}$. Then, by Example 8.3(i),

$$m' = [N(P)']_3 = [\bigwedge \{p \vee p'; p \in P\}]_3$$
$$= [\bigvee_\varepsilon \{p_0^{\varepsilon(0)} \cdots p_n^{\varepsilon(n)}\}]_3$$

where ε ranges over all sequences $(\varepsilon(0), \ldots, \varepsilon(n))$ of 0's and 1's and, for any $p \in P$,

$$p^v = \begin{cases} p & \text{if } v = 1 \\ p' & \text{if } v = 0. \end{cases}$$

It can easily be verified that the classes $[p_0^{\varepsilon(0)} \cdots p_n^{\varepsilon(n)}]_3$ are the principal parts of m'. It is then trivial to verify that these principal parts form a continuous set.

8.10 Three-valued predicate calculus

Lattice-valued structures have already been defined generally in Definition 6.4. We take over the semantic notions such as *valuation* and *canonical realization* developed in Chapter 6, §6.8.2. The lattices under consideration are N_3 and Q_4. The relations of the structure are accordingly functions on the domain of the structure with values in $\{0, n, 1\}$ (or $\{0, n, m, 1\}$ for Q_4) and, accordingly, we call such structures *inexact*. The method of constructing three- or four-valued logics of inexact structures has already been defined in Chapter 6, §6.8.2, in the section on the general method of defining lattice-valued logics. The logics $\mathbb{T}$ of truth-values are now $N_3^\dagger$ and $Q_3^\dagger$ (§8.2). The connectives of the predicate logics $Pred(N_3^\dagger)$ and $Pred(Q_4^\dagger)$ so constructed are conjunction, disjunction, and 'negation' as defined for the corresponding propositional logic.

A realization $\mathfrak{A}$ is, of course, an inexact structure. *Exact structures* are those inexact structures whose relations have values in $\{0, 1\}$. Given a realization $\mathfrak{A}$ and an assignment α of terms to $\mathfrak{A}$ there is a function $[\mathfrak{A}, \alpha]$ which assigns truth-values (in $\{0, n, 1\}$ or $\{0, n, m, 1\}$) to formulas: in particular, valuations of quantifiers are defined in Chapter 6, §6.9.2(iii)(c). Logical consequence is again defined by (8.13). We note that, by this definition, the quantifiers exhibit a certain duality similar to that displayed by the de Morgan laws: for $i = 2, 3, 4$

$$\forall a A(a/x) \leftrightarrow_i \neg \exists a \neg A(a/x)$$
$$\exists a A(a/x) \leftrightarrow_i \neg \forall a \neg A(a/x). \tag{8.23}$$

$Pred(N_3^\dagger)$ and $Pred(Q_4^\dagger)$ are compact by Theorem 6.18.

276 Three-valued logic

The notion of *refinement* used in the N_3-propositional logic extends to N_3-valued structures.

DEFINITION 8.3. Let
$$\mathfrak{A} = \langle A, g_1, g_2, \ldots, R_1, R_2, \ldots \rangle$$
and
$$\mathfrak{A}' = \langle A, g_1, g_2, \ldots, R'_1, R'_2, \ldots \rangle$$
be N_3-valued structures of type τ which differ only in their relations. A refinement relation is first defined for the relations in terms of the refinement relation for the truth-values in N_3 (Definition 8.1): for all n, if R_n is an m-place relation,

$$R'_n \blacktriangleright R_n \Leftrightarrow_{\text{def}} R'_n(x_1, \ldots, x_m) \blacktriangleright R_n(x_1, \ldots, x_m) \qquad (x_1, \ldots, x_m \in A).$$

Then
$$\mathfrak{A}' \blacktriangleright \mathfrak{A} \Leftrightarrow_{\text{def}} R'_n \blacktriangleright R_n \qquad (n = 1, 2, 3, \ldots).$$

Continuity is easily established.

THEOREM 8.14. Let $\mathfrak{A}'$, $\mathfrak{A}$ be N_3-valued structures such that $\mathfrak{A}' \blacktriangleright \mathfrak{A}$. Let α be an assignment to $|\mathfrak{A}|$. Then for all formulas F of $Pred(N_3^+)$

$$[\mathfrak{A}', \alpha](F) \blacktriangleright [\mathfrak{A}, \alpha](F).$$

Proof. By induction of $|F|$.

Basis:
$|F| = 0$. Then F is an atomic formula and the result follows immediately from the definition.

Inductive step:
The inductive hypothesis is

$I(n)$: for all formulas F with $|F| \leq n$, and all assignments α to $|\mathfrak{A}|$, $[\mathfrak{A}', \alpha](F) \blacktriangleright [\mathfrak{A}, \alpha](F)$.

Let G be a formula with $|G| = n + 1$. Then G has one of the following forms: (a) $\neg A$, (b) $A \vee B$, (c) $A \wedge B$, (d) $\forall a A(a/x)$, or (e) $\exists a A(a/x)$. Here $|A|, |B| \leq n$. The propositional cases (a)–(c) are proved as in Theorem 8.2, so consider now the case of quantified formulas.

Case (d). Here $G = \exists a A(a/x)$. If $[\mathfrak{A}, \alpha](F) = n$ then clearly $[\mathfrak{A}, \alpha](F) \blacktriangleright n$. If $[\mathfrak{A}, \alpha](F) = 1$, then $[\mathfrak{A}, \alpha[s/x]](A) = 1$ for some $s \in |\mathfrak{A}|$. By $I(n)$, $[\mathfrak{A}', \alpha[s/x]](A) = 1$. Hence $[\mathfrak{A}', \alpha](F) = 1$. Thus $[\mathfrak{A}', \alpha](F) \blacktriangleright [\mathfrak{A}, \alpha](F)$.

If $[\mathfrak{A}, \alpha](F) = 0$ then $[\mathfrak{A}, \alpha[s/x]](A) = 0$ for all $x \in |\mathfrak{A}|$. By $I(n)$, $[\mathfrak{A}', \alpha[s/x]](A) = 0$ for all $s \in |\mathfrak{A}| = |\mathfrak{A}'|$. Thus $[\mathfrak{A}', \alpha](\exists a A(a/x)) = 0$. Hence $[\mathfrak{A}', \alpha](F) \blacktriangleright [\mathfrak{A}, \alpha](F)$.
Case (e). Similar to case (d).

Thus, $I(n+1)$ holds. The stated result now follows by induction on n. ∎

The usual systems of classical predicate logic are complete, i.e. a formula is derivable in the formal system if, and only if, it is universally valid, i.e. a tautology. We cannot attempt to produce a meaningful direct analogue of this for the logic of inexact predicates as there are no pure tautologies. But the notion of logical consequence can be used to construct a Gentzen type of formal system (Chapter 2, §2.2.6(iii), and Chapter 3, §3.10) which is complete in the sense that a sequent $(A_1 \wedge \cdots \wedge A_n : B_1 \vee \cdots \vee B_m)$ is derivable in that system if, and only if, $A_1 \wedge \cdots \wedge A_n \rightarrow_3 B_1 \vee \cdots \vee B_m$. The point of this exercise is effectively to characterize the notion of logical consequence by a finite number of axiom schemas and a finite number of rules. If we are only concerned with propositional calculus the task is relatively trivial. But for the predicate calculus, the quantifiers are infinitary operations. The completeness problem therefore centres on the quantifiers.

The formal systems $\mathfrak{G}_3$ and $\mathfrak{G}_4$, corresponding respectively to $Pred(N_3^{\ddagger})$ and $Pred(Q_4^{\ddagger})$ will be calculi of sequents. By a sequent we mean an expression of the form $(\Gamma : \Delta)$ where Γ and Δ are non-empty sequences of formulas. Usually, Γ and Δ are finite, but at one point in the completeness proof it is expedient to admit the possibility that Γ and Δ are infinite. We write $A \in \Gamma$ if the formula A is a member of the sequence Γ. $\bigwedge \Gamma$ ($\bigvee \Delta$) denotes the conjunction (disjunction) in these logics of the formulas in Γ (Δ).

DEFINITION 8.4.
(i) The sequent $(\Gamma : \Delta)$ is i-valid if $\bigwedge \Gamma \rightarrow_i \bigvee \Delta$.
(ii) Let $\mathfrak{A}$ be an inexact structure and α an assignment to $\mathfrak{A}$. Then $(\mathfrak{A}, \alpha)$ *refutes* $(\Gamma : \Delta)$ if

$$glb\{[\mathfrak{A}, \alpha](A); A \in \Gamma\} \not\leq lub\{[\mathfrak{A}, \alpha](B); B \in \Delta\}.$$

DEFINITION 8.5.
(a) For atomic formulas F, $\check{F} = \hat{F} = F$.
(b) For formulas A, B:
   if F is $[A \vee B]$ then $\check{F} = [\check{A} \vee \check{B}]$, $\hat{F} = \neg[\neg \hat{A} \wedge \neg \hat{B}]$;
   if F is $[A \wedge B]$ then $\check{F} = \neg[\neg \check{A} \vee \neg \check{B}]$, $\hat{F} = [\hat{A} \wedge \hat{B}]$;
   if F is $\neg A$ then $\check{F} = \neg \check{A}$, $\hat{F} = \neg \check{A}$;
   if F is $\exists a A(a/x)$ then $\check{F} = \exists a \check{A}(a/x)$, $\hat{F} = \neg \forall a \neg \hat{A}(a/x)$;
   if F is $\forall a A(a/x)$ then $\check{F} = \neg \exists a \neg \check{A}(a/x)$, $\hat{F} = \forall a \hat{A}(a/x)$.

For a set (sequence) Δ of formulas, $\check{\Delta}$ is the set (corresponding sequence) of formulas $\check{F}$ where $F \in \Delta$. $\hat{\Delta}$ is defined analogously.

EXAMPLE 8.4. For each formula F, $\hat{F}$ is a $\wedge$-formula and $\check{F}$ is a $\vee$-formula. Using the de Morgan laws for connectives and (8.23) for quantifiers it can be shown that, for each formula A, $\hat{A} \leftrightarrow_i \check{A} \leftrightarrow_i A$ for $i = 2, 3, 4$. ◆

As in Schütte (1961, Chapter III) the rules of inference of the formula system attain a certain simplicity by allowing an infinite but decidable set of axioms (Definition 8.6 below) and also by restricting the type of logical symbols that can occur in the antecedent and succedent of a sequent. Thus, as by Example 8.4, $A \to_3 B$ if, and only if, $\hat{A} \to_3 \check{B}$, it will henceforth be assumed that in any sequent $(\Gamma : \Delta)$ the members of Γ are all $\wedge$-formulas and the members of Δ are all $\vee$-formulas. The axioms and rules of inference of $\mathfrak{G}_3$ and $\mathfrak{G}_4$ are defined as follows.

DEFINITION 8.6.

(i) The $\mathfrak{G}_3$-axioms are all finite sequents $(\Gamma : \Delta)$ satisfying **A1**, **A2**, or **A3**.
  A1: There exist atomic formulas p, q such that
  $$p, p' \in \Gamma \quad \& \quad q, q' \in \Delta$$
  A2: There exists an atomic formula p such that
  $$p \in \Gamma \;\&\; p \in \Delta \quad \vee \quad p' \in \Gamma, p' \in \Delta$$
  A3: $f \in \Gamma \quad \vee \quad t' \in \Gamma \quad \vee \quad f' \in \Delta \quad \vee \quad t \in \Delta$

(ii) The $\mathfrak{G}_4$-axioms are all finite sequents $(\Gamma : \Delta)$ satisfying **A2** or **A3**.

It is important in the following theorem to observe that if the sequent $(\Gamma : \Delta)$ consists only of literals then it is a $\mathfrak{G}_3$-axiom iff the pair (Γ, Δ) does not satisfy **C1** & **C2** & **C3** of Lemma 8.10; similarly it is a $\mathfrak{G}_4$-axiom iff it does not satisfy **C2** & **C3**.

DEFINITION 8.7. In the following rules of inference, A and B are formulas, Γ and Δ are finite sequences of zero and more formulas, and Θ is a non-empty finite sequence of formulas.

R1a: $\dfrac{(\Gamma, \Delta, A' : \Theta) \quad (\Gamma, \Delta, B' : \Theta)}{(\Gamma, (A \wedge B)', \Delta : \Theta)}$

R1s: $\dfrac{(\Theta : \Gamma, \Delta, A') \quad (\Theta : \Gamma, \Delta, B')}{(\Theta : \Gamma, (A \vee B)', \Delta)}$

R2a: $\dfrac{(\Gamma, \Delta, A, B : \Theta)}{(\Gamma, AB, \Delta : \Theta)}$

R2s: $\dfrac{(\Theta : \Gamma, \Delta, A, B)}{(\Theta : \Gamma, A \vee B, \Delta)}$

R3a: $\dfrac{(\Gamma, \Delta, A : \Theta)}{(\Gamma, A'', \Delta : \Theta)}$

R3s: $\dfrac{(\Theta : \Gamma, \Delta, A)}{(\Theta : \Gamma, A'', \Delta)}$

R4a: $\dfrac{(\Gamma, \Delta, A' : \Theta)}{(\Gamma, (\forall a A(a/x))', \Delta : \Theta)}$ for $x \in Fv - Fv(\Theta, \Gamma, \Delta)$

R4s: $\dfrac{(\Theta : \Gamma, \Delta, A'(t/x))}{(\Theta : \Gamma, \Delta, (\exists a A(a/x))')}$ for $x \in Fv - Fv(\Theta, \Gamma, \Delta)$

R5a: $\dfrac{(\Gamma, \forall a A(a/x), \Delta, A(t/x) : \Theta)}{(\Gamma, \Delta, \forall a A(a/x) : \Theta)}$

R5s: $\dfrac{(\Theta : \Gamma, \exists a A(a/x), \Delta, A(t/x))}{(\Theta : \Gamma, \exists a A(a/x), \Delta)}$

(The quantifier rules **R4a**–**R5s** should be compared with the axioms ∀1, ∀2, ∃1, ∃2, with Theorem 6.16, and with Example 4.7. Our principal task in this section is to prove the following theorem.

THEOREM 8.15. *For formulas A, B, $A \to_3 B$ ($A \to_4 B$) iff $\hat{A} : \check{B}$ is derivable from the $\mathfrak{G}_3$-axioms ($\mathfrak{G}_4$-axioms).*

COROLLARY 8.16. *For formulas A, B, if $A \not\to_3 B$ (or $A \not\to_4 B$) there exists a canonical realization such that $[\mathfrak{A}, \iota](A) \not\subseteq [\mathfrak{A}, \iota](B)$.*

It is easy to show that (i) every axiom is valid and (ii) for each rule of inference, if the premisses are valid then so is the conclusion. Hence every derivable sequent is valid. Thus if $\hat{A} : \check{B}$ is derivable from the $\mathfrak{G}_3$-axioms ($\mathfrak{G}_4$-axioms) then $\hat{A} \to_3 \check{B}$ and so, by Example 8.4, $A \to_3 B$ ($A \to_4 B$). The proof given here of the converse closely follows the corresponding argument given by Schütte for the completeness of the classical predicate calculus. Thus we define an algorithm for associating with each sequent $(\Gamma : \Delta)$ a 'proof tree' T such that T yields a derivation of $(\Gamma : \Delta)$ (in which case $(\Gamma : \Delta)$ is valid) or a canonical realization (obtained from a certain valuation of atomic formulas as in Theorem 2.18) which refutes $(\Gamma : \Delta)$.

The definition of 'proof tree' is facilitated by the following classification of formulas.

DEFINITION 8.7.

(i) A formula in Γ or Δ of a sequent ($\Gamma : \Delta$) is said to be:
  (a) *analysable* if it has one of the forms A'', $(AB)'$, AB, $(\forall a F(a/x))'$, $(A \vee B)'$, $A \vee B$, $(\exists a F(a/x))'$;
  (b) *critical* if it has one of the forms $\forall a F(a/x)$, $\exists a F(a/x)$;
  (c) a *literal* if it is atomic or the negation of an atomic formula.
  (Each formula in Γ and Δ is analysable, critical, or a literal.)
(ii) The sequent ($\Gamma : \Delta$) is said to be:
  (a) *analysable* if at least one of the formulas in Γ or Δ is analysable;
  (b) *critical* if it is not analysable but contains a critical formula;
  (c) *literal* if all of the formulas in Γ and Δ are literals.

We also need a standard enumeration $t_1, t_2, t_2, \ldots$ of terms. Thus, if $\mathfrak{A}$ is a canonical realization then $|\mathfrak{A}| = \{t_1, t_2, t_3, \ldots\}$.

The construction of proof trees is defined inductively as follows.

DEFINITION 8.8.

(P1) ($\Gamma : \Delta$) is a *sequent of height* 1 in the proof tree of ($\Gamma : \Delta$).
(P2) For a sequent ($\Pi : \Lambda$) of height n in the proof tree of ($\Gamma : \Delta$) the *upper sequents* are defined by cases as follows.
  (i) If ($\Pi : \Lambda$) is analysable let C be the first (reading from the left) analysable formula in the sequent ($\Pi : \Lambda$). Then we have the following.
    (i)(1) If $C \in \Pi$ then $\Pi = (\Theta, C, \Psi)$, say, where $C \notin \Theta$ and C has one of the forms (a) A'', (b) $(AB)'$, (c) AB, (d) $(\forall a A(a/x))'$. The upper sequent(s) are then
      $(\Theta, \Psi, A : \Lambda)$ in case (a)
      $(\Theta, \Psi, A' : \Lambda)$ and $\Theta, \Psi, B' : \Lambda$ in case (b)
      $(\Theta, \Psi, A, B : \Lambda)$ in case (c)
      $(\Theta, \Psi, A'(y/x) : \Lambda)$ in case (d)
      where y is the first free variable (in the standard enumeration of free variables) which does not occur in ($\Pi : \Lambda$).
    (i)(2) If $C \notin \Pi$ then $C \in \Lambda$ and $\Lambda = (\Theta, C, \Psi)$, where $C \notin \Theta$ and C has one of the forms (a) A'', (b) $(A \vee B)'$, (c) $A \vee B$, (d) $(\exists X(A))'$. The upper sequents are then
      $(\Pi : \Theta, \Psi, A)$ in case (a)
      $(\Pi : \Theta, \Psi, A')$ and $(\Pi : \Theta, \Psi, B')$ in case (b)
      $(\Pi : \Theta, \Psi, A, B)$ in case (c)
      $(\Pi : \Theta, \Psi, A'(y/x))$ in case (d)

where y is the first free variable (in the standard enumeration of free variables) which does not occur in $(\Pi : \Lambda)$.

(ii) If $(\Pi : \Lambda)$ is a critical sequent but not an axiom, suppose that the critical formulas are $\forall b_i A_i(b_i/y_i)$, for $1 \leqslant i \leqslant r$ in Π, and $\exists c_j B_j(c_j/z_j)$, for $1 \leqslant j \leqslant s$ in Λ, where $b_1, \ldots, b_r, c_1, \ldots, c_m \in Bv$ and $y_1, \ldots, y_r, z_1, \ldots, z_m \in Fv$. Then the upper sequent of $(\Pi : \Lambda)$ is

$(\Pi, A_1(t_1/y_1), \ldots, A_1(t_n/y_1), A_2(t_1/y_1), \ldots,$

$A_r(t_1/y_1), \ldots, A_r(t_n/y_1):$

$\Lambda, B_1(t_1/z_1), \ldots, B_1(t_n/z_1), B_2(t_1/y_1), \ldots,$

$B_m(t_1/z_1), \ldots, B_m(t_n/z_1)).$

(iii) If $(\Pi : \Lambda)$ is literal or an axiom then $(\Pi : \Lambda)$ has no upper sequent and is called a *vertex sequent*.

The upper sequent(s) of $(\Pi : \Lambda)$ are sequents of height $n + 1$ in the proof tree of $(\Gamma : \Delta)$. A sequent will only belong to the proof tree if it does so by means of the rules (P1) and (P2).

The upper sequent(s) S_1 (and S_2) of a sequent S have been so defined that S is derivable from S_1 (and S_2). Hence if the proof tree of a sequent S is finite and each vertex sequent is an axiom then S is derivable. Our aim now is to examine the situation where the proof tree is either infinite or has a vertex sequent which is not an axiom. To this end we define a *path* Φ in the proof tree of the sequent S to be a sequence $S_1, S_2, \ldots$ of sequents in the proof tree of S such that S_1 is S, S_{n+1} is an upper sequent of S_n, and if Φ is finite then the last sequent is a vertex sequent. S_n therefore has height n. The analysable formulas in a sequent of Φ are broken down into subformulas by rule (P2)(i) and passed to the right-hand end of the antecedent or succedent; those that are non-analysable—critical or literal—are propagated through each succeeding sequent in the following manner.

LEMMA 8.17.

(i) If p is a literal in Γ_n (Δ_n) then $p \in \Gamma_m$ $(p \in \Delta_m)$ for all $m \geqslant n$.
(ii) Suppose that Φ does not contain an axiom. If $\forall a A(a/x) \in \Gamma_n$ $(\exists a A(a/x) \in \Delta_n)$ then Φ is infinite and, for each m, $A(t_m/x) \in \Gamma_j$ $(A(t_m/x) \in \Delta_j)$ for some $j \geqslant m$.

Proof.

(i) This is obvious from the definition of upper sequent.
(ii) If $\forall a A(a/x) \in \Gamma_m$ then $(\Gamma_m : \Delta_m)$ is neither literal nor an axiom. By Definition 8.8(P2) $(\Gamma_{m+1} : \Delta_{m+1})$ is defined and $\forall a A(a/x) \in \Gamma_{m+1}$. Hence if $\forall a A(a/x) \in \Gamma_n$ then Φ is infinite and $\forall a A(a/x) \in \Gamma_m$ for all m.

Now $(\Gamma_m : \Delta_m)$ is critical for infinitely many m. For otherwise $(\Gamma_m : \Delta_m)$ would be analysable for cofinitely many m, and this is impossible as, by Definition 8.8(P2)(i), each upper sequent of an analysable sequent $(\Pi : \Lambda)$ contains fewer symbols that $(\Pi : \Lambda)$. Thus if $\forall aA(a/x) \in \Gamma_n$ then for each m there exists $i > m$ such that $\Gamma_i : \Delta_i$ is critical. By Definition 8.8(P2)(ii), $A(t_m/x) \in \Gamma_{i+1}$.

Similarly for $\exists aA(a/x)$. ∎

Lemma 8.17 enables there to be associated with each path Φ in the proof tree of $\Gamma : \Delta$ which does not contain an axiom, a canonical realization which refutes $(\Gamma : \Delta)$. To effect the construction we extract the constant literals from the sequents in Φ. Let $(^0\Gamma_n : {}^0\Delta_n)$ be the sequent obtained from $(\Gamma_n : \Delta_n)$ by deleting all non-literal formulas. It can easily be seen from Definition 8.8 that

$$^0\Gamma_{n+1} \text{ extends } {}^0\Gamma_n \text{ and } {}^0\Delta_{n+1} \text{ extends } {}^0\Delta_n. \tag{8.24}$$

Now define Γ_ω (Δ_ω) to be the shortest sequence of formulas which extends $^0\Gamma_n$ ($^0\Delta_n$) for all n. By (8.24), Γ_ω and Δ_ω are well defined. Note that Γ_ω or Δ_ω could be infinite.

LEMMA 8.18. *If $(\Gamma_\omega : \Delta_\omega)$ satisfies the conditions* **A1**, **A2**, *and* **A3** (**A2** *and* **A3**) *in the definition of 'axiom' then Φ contains a $\mathfrak{G}_3$-axiom ($\mathfrak{G}_4$-axiom).*

Proof. First consider $\mathfrak{G}_3$. Suppose that $\Gamma_\omega : \Delta_\omega$ satisfies **A1**. Then there exist atomic formulas p, q such that $p, p' \in \Gamma_\omega$ and $q, q' \in \Delta_\omega$. By the definition of Γ_ω and Δ_ω there exists an N such that $p, p' \in {}^0\Gamma_n$ and $q, q' \in {}^0\Delta_n$. Hence $p, p' \in \Gamma_n$ and $q, q' \in \Delta_n$. Thus $(\Gamma_n : \Delta_n)$ is an axiom.

Similarly, if $(\Gamma_\omega : \Delta_\omega)$ satisfies **A2** (or **A3**) then there exists n such that $(\Gamma_n : \Delta_n)$ satisfies **A2** (or **A3**).

The proof for $\mathfrak{G}_4$ is similar to that for $\mathfrak{G}_3$. ∎

LEMMA 8.19. *If Φ does not contain an $\mathfrak{G}_3$-axiom ($\mathfrak{G}_4$-axiom) then there exists a canonical realization $\mathfrak{A}$ and elements $\gamma, \delta \in |N_3|$ ($\gamma, \delta \in |Q_4|$) such that p, q*

$$p \in \Gamma_\omega \Rightarrow v \geq \gamma$$
$$q \in \Delta_\omega \Rightarrow \delta \geq v \tag{8.25}$$

where $v =_{\text{def}} [\mathfrak{A}, \iota]$. (Thus v refutes $(\Gamma_m : \Delta_m)$.)

Proof. First consider the $\mathfrak{G}_3$ case. Suppose that Φ does not contain an axiom. By Lemma 8.18, $(\Gamma_\omega : \Delta_\omega)$ satisfies neither **A1** nor **A2** nor **A3**, and hence satisfies one of **C1**, **C2**, **C3** of Lemma 8.10. By that lemma there exists a (propositional) valuation u in N_3 of atomic formulas and $\gamma, \delta \in N_3$ such that

$\gamma \notin \delta$, $p \in \Gamma_\omega \Rightarrow u(p) \geq \gamma$, and $q \in \Delta_\omega \Rightarrow \delta \geq u(q)$. From u there can be constructed a canonical realization $\mathfrak{A}$ such that $[\mathfrak{A}, \iota](p) = u(p)$ for literals. Hence (8.25) holds.

The proof for $\mathfrak{G}_4$ is similar. ∎

The final step is to show that if the proof tree of $(\Gamma : \Delta)$ contains a path Φ without an axiom then (8.25) holds for *every* sequent in Φ and so v refutes $(\Gamma : \Delta)$. To this end we define, for $1 \leq n \leq \omega$,

$$m(n) = glb\{v(A); A \in \Gamma_n\}$$
$$M(n) = lub\{v(B); B \in \Gamma_n\}.$$

(8.25) is then equivalent to

$$m(\omega) \geq \gamma \qquad \delta \geq M(\omega) \qquad (8.26)$$

In Lemma 8.20 below we deduce from this that, for all i,

$$m(i) \geq \gamma \qquad \delta \geq M(i). \qquad (8.27)$$

In particular, $m(1) > M(1)$ so that v refutes $(\Gamma_1 : \Delta_1)$; thus $(\Gamma : \Delta)$ is not valid. This proves completeness.

Suppose that Φ is a path in the proof tree of $(\Gamma : \Delta)$ which does not contain an axiom. Let $\mathfrak{A}$ be a canonical realization associated with Φ, as in Lemma 8.18, and put $v = [\mathfrak{A}, \iota]$.

LEMMA 8.20. *For all i, $m(i) \geq \gamma$, $\delta \geq M(i)$.*

Proof. We prove by induction on n that $I(n)$ holds for all $n \geq 1$, where $I(n) \Leftrightarrow_{\text{def}} (i)$ (for all $A \in \Gamma_i$ & $B \in \Delta_i$,

$$|A| + |B| \leq n \quad \Rightarrow \quad v(A) \geq \gamma \ \& \ \delta \geq v(B)).$$

Basis:
$I(1)$. This follows directly from (8.25).

Inductive step:
Suppose $I(n)$ for $n \geq 1$. Let $A \in \Gamma_i$ and $B \in \Delta_i$ with $|A| + |B| = n + 1$. We prove that

$$v(A) \geq \gamma \quad \& \quad \delta \geq v(B). \qquad (8.28)$$

Now as $|A| + |B| = n + 1$, there are three cases: (a) $|A|, |B| \leq 1$; (b) $|A| \geq 2$; (c) $|B| \geq 2$.

Case (a) Then (8.28) follows from $I(1)$.

Case (b) By Definition 8.8(P2) there are two subcases: (b1) A is analysable, or (b2) A is critical.

Subcase (b1) As $A \in \Delta_i$, A can assume one of the following forms: (α) F'', (β) $(FG)'$, (γ) FG, or (δ) $(\forall x F(a/x))'$.

ad (α) $A = F''$. By Definition 8.8(P2)(i)(1)(a), for some $j > i$, $F \in \Gamma_j$. But $|F| \leqslant n$. By $I(n)$,
$$v(A) = v(F'') = v(F) \geqslant \gamma.$$

ad (β) $A = (FG)'$. By Definition 8.8(P2)(i)(1)(b), for some $j > i$ either $F' \in \Gamma_j$ or $G' \in \Gamma_j$. Without loss of generality it can be assumed that $F' \in \Gamma_j$. But $|F'| \leqslant n$. As $|(FG)'| > |F'|$, by $I(n)$,
$$v(A) = v((FG)') \geqslant v(F') \geqslant \gamma.$$

ad (γ) $A = FG$. By Definition 8.8(P2)(i)(1)(c), for some $j > 1$, $F, G \in \Gamma_j$. But $|F|, |G| \leqslant n$. By $I(n)$
$$v(A) = v(F) \cdot v(G) > \gamma \cap \gamma = \gamma.$$

ad (δ) $A = (\forall a F(a/x))'$. By (P2)(i)(1)(d) there exists $j > 1$ and a free variable y not occurring in F such that $F'(y/x) \in \Gamma_j$. But $|F'(y/x)| \leqslant n$. By $I(n)$
$$v(F'(y/x)) \geqslant \gamma.$$

But as the domain of $\mathfrak{A}$ is the set of terms, the definition of the truth-value of $\forall a F(a/x)$ in the canonical realization $\mathfrak{A}$ with the identity assignment ι (i.e. clause (iii) of Definition 2.13 modified for lattice-valued structures in Chapter 6, §6.8.2) gives
$$[\mathfrak{A}, \iota]((\forall a F(a/x)) \leqslant [\mathfrak{A}, \iota(y/x)](F) = [\mathfrak{A}, \iota](F(y/x)).$$

Hence $v((\forall a F(a/x))') \geqslant v(F'(y/x))$. But $|(\forall a F(a/x))'| > |F'(y/x)|$. Thus, by $I(n)$, $v((\forall X(F')) \geqslant v(F'(y/x)) \geqslant \gamma$.

Subcase (b2) A is critical. As $A \in \Gamma_i$, A has the form $\forall a F(a/x)$. By Lemma 8.15(ii), for each m there exists $j > m$ such that $A(t_m/x) \in \Gamma_j$. But $|A(t_m/x)| \leqslant n$. By $I(n)$, $v(A\{t_m/x\}) \geqslant \gamma$. Then, as in (d) above, $glb\{v(A(t/x)); t \in |\mathfrak{A}|\} \geqslant \gamma$. Hence
$$v(A) = v(\forall a F(a/x)) \geqslant v(B).$$

Thus in both subcases (b1) and (b2) of case (b), $v(A) \geqslant \gamma$.

Case (c) Then B is either analysable or critical. The case analyses here are the 'duals' of cases considered above and $\delta \geqslant v(B)$ is proved similarly.

This completes the proof of (8.28). Thus, $(n)I(n)$ follows by induction in n. For any given i choose an n which exceeds the sum of the logical complexities of all formulas occurring in Γ_i and Δ_i. Then as $I(i)$ holds, $v(A) \geq \gamma$ and $\delta \geq v(B)$ for all $A \in \Gamma_i$ and $B \in \Delta_i$. Hence $m(i) \geq \gamma$, $\delta \geq M(i)$. ∎

9

Relevance

9.1 Introduction

In many of the standard formulations of the classical propositional calculus there are to found the theorems $A \supset [B \supset A]$ and $\neg A \supset [\neg A \supset B]$. The sense of the first is frequently expressed by saying that if a proposition is true then any proposition whatever implies it. The sense of the second is expressed by asserting that if a proposition is false then it implies any proposition. These theorems are often called the 'paradoxes of (material) implication'. Certain related kinds of implication such as $AA' \rightarrow B$ and $B \rightarrow A \vee A'$ are sometimes regarded as fallacious or paradoxical, even though classically correct, on the grounds that AA' (or $A \vee A'$) is irrelevant to the arbitrary proposition B. Much of the criticism of such implications has focused on the nature of disjunction. Thus the universality of the 'disjunctive syllogism' $\neg p, p \vee q \rightarrow q$ has been called into question because it allows p and q to be totally unrelated. A 'valid' inference, it is claimed, would depend upon p and q being *relevant* to one another (Anderson and Belnap 1975). This criticism of disjunction was made earlier by some nineteenth century idealist philosophers (Bloor 1983, §6.4). Bradley (1922, p. 130), with Bosanquet (1911, p. 323), developing Hegel's remarks on Disjunctive Judgement, said that disjunction needs a 'ground'—it must be formulated on the basis of an 'assumption' about the sphere in which it is affirmed. This idea suggests that formal logic should be restricted to domains possessing an essential unity, i.e. in which all propositions are mutually relevant. The possibility of accomplishing this without difficulty is already present in the relativization of the classical predicate logic by Skølem and Löwenheim (Chapter 2, §2.1). This matter will be taken up later in this chapter in the section on multi-subject semantics (§9.4).

The actual development of the idea of 'relevance', however, has concentrated on the formal and axiomatic aspects, rather than on the semantics. Like many-valued logics, relevance logics seem to lack concrete applications—they certainly have no relevance to mathematics.

Various attempts have been made to characterize a notion of 'valid' implication by constructing formal systems based on postulated properties of 'entailment'. An admirable well-documented example is in the paper by Anderson and Belnap (1962). This is the axiom system of 'tautological entailments' of Anderson and Belnap (1975, Chapter III). They introduced

Introduction

this notion as an outcome of their search for 'first-degree entailments' (i.e. entailments between formulas not themselves containing an entailment connective) which do not commit 'paradoxes of relevance'. It is an axiom system whose theorems are pairs (A, B) such that A entails B where A and B are propositional formulas whose connectives are negation, conjunction, and disjunction only. It is an axiom system whose theorems are pairs (A, B) such that A entails B. We shall write such a pair as $A \vdash B$.

The following axiom schemes (**LA**) and rules of inference (**LR**) of the calculus L comprise the *basic system of logical entailment*.

LA1: $\qquad A \vdash A''$

LA2: $\qquad A'' \vdash A$

LA3: $\qquad AB \vdash A$

LA4: $\qquad AB \vdash B$

LA5: $\qquad A \vdash A \vee B$

LA6: $\qquad B \vdash A \vee B$

LA7: $A(B \vee C) \vdash AB \vee AC$

LR1: $\dfrac{A \vdash B, \quad B \vdash C}{A \vdash C}$

LR2: $\dfrac{A \vdash B}{B' \vdash A'}$

LR3: $\dfrac{A \vdash B, \quad \vdash C}{A \vdash BC}$

LR4: $\dfrac{A \vdash C, \quad B \vdash C}{A \vee B \vdash C}$

It is a simple but tedious exercise to prove by induction on length of proof that this system is equivalent to the propositional substructure $\mathfrak{G}'_4$ of the Q_4-valued system $\mathfrak{G}_4$ in Chapter 8 in the sense that:

$(\bigwedge \Gamma \vdash \bigvee \Delta)$ is provable in L iff $(\hat{\Gamma}: \check{\Delta})$ is provable in $\mathfrak{G}'_4$.

The completeness theorem for $\mathfrak{G}_4$ (Theorem 8.13) then shows the *technical* connection between first-degree entailments and Q_4 semantics. Q_4 coincides with the characteristic matrix of Smiley (Anderson and Belnap 1975, §15.3). This matrix appeared as a technical device: a sound justification in terms of multi-subject semantics has been given by Muškardin (1978) (see §9.4).

Our concern here is to avoid the postulational approach. We consider two strategies here, neither of which is without postulatory critics. In the first we seek to characterize the source of the paradoxes of material implication entirely in terms of traditional syntax and semantics of the classical propositional calculus, as expounded by Körner:

> ... if one does not like the so-called paradoxes of classical deducibility, one can characterize them fully in terms of classical logic. For example, if somebody tells me that from the axioms of Euclidean geometry he can not only deduce Pythagoras' Theorem but also the alternation of Pythagoras' statement and the statement that Pythagoras' mother was Viennese and then by contraposition, antilogism etc., some other odd deducibility statements, I will tell him not to worry because he can separate the odd from the other deducibility statements by a logical criterion rather than a Gricean psychological one. The logical criterion would be my notion of relevance or something near it.
>
> Letter to Professor Anderson, 29 January 1973

This we do, following Körner, by defining *irrelevant* subformulas of a formula, a notion shortly to be made precise and further developed in §§9.2 and 9.3 (see also Schurz 1983). The results in this section are developed from Cleave (1974a).

The second approach (§9.4), due to Muškardin (1978), recognizes the source of paradox as the occurrence within the same sentence of expressions having entirely different meanings. This is also acknowledged in the quotation (above) from Körner's letter. The idea was developed by Muškardin to provide a semantic foundation for a logic of entailment.

9.2 Rigid propositions

Formulas with an irrelevant subformula are called *slack*, and non-slack formulas are termed *rigid*. The classical Bolzano–Tarski notion of logical consequence (Definition 2.4(iii)) will be retained. We note that $A \to B$ iff $A \supset B$ is a tautology (Example 2.1(x)). In view of the absence of agreement on the nature of the paradoxical implications we regard a tautology $A \supset B$ as expressing a valid entailment if the formula $A \supset B$ is rigid. This notion of entailment has the following properties:

 (i) most, if not all, of the commonly agreed paradoxes of implication arise from slack tautologies;
 (ii) the relation of entailment is not transitive, thus confirming the opinion of Lewy, Geach, Smiley, and others (Anderson and Belnap 1962, p. 11).
(iii) there exist valid entailments which, at least in the usual types of Gentzen systems, cannot be proved without use of 'invalid' entailments, i.e. slack tautologies.

It would seem that (i), (ii), and (iii) cast some doubt on the feasibility of axiomatizing entailment.

The basic principle of relating entailments to rigid propositions can be traced to a notion of 'pure entailment' due to Körner (1946–7). It occurs more explicitly in Körner (1955) where an 'essential' logically necessary proposition is defined as one not containing an 'inessential component', i.e. a subformula whose replacement by its negation leaves the logical necessity of the whole proposition intact. A similar concept of 'inessential component' occurs in Körner (1946–7), where an occurrence of a subformula is inessential if its *deletion* leaves the necessity of the whole proposition intact. An essential logically necessary proposition is therefore, in our terminology, a rigid tautology.

Our purpose now is to give some definitions of 'rigid' for the classical propositional calculus. As an alternative to the postulational approach, paradox will be related to the notion of rigidity and the closure properties of the corresponding classes of rigid formulas will be examined.

To fix ideas, consider the formulation of the classical propositional calculus in Chapter 2, §2.2.1, but without the implication junctor. In the abbreviated informal presentation of formulas AB denotes the conjunction of A and B, $A \vee B$ denotes the disjunction of them, and A' denotes the negation of A. $A \supset B$ is an abbreviation of $[\neg A \vee B]$. The notation used by Schütte (1961) (see Chapter 2, §2.2.2) is useful in this work. In actual examples of a formula F with a distinguished occurrence of a subformula A the intended occurrence will be enclosed in the brackets $\{, \}$. Thus if p, q, r are proposition letters, F is $rp \vee q(r \vee (rp)')$ and A is the first occurrence of the subformula rp, then $F[A]$ is written as $\{rp\} \vee q(r \vee (rp)')$; if A is the second occurrence of rp, then $F[A]$ is written as $rp \vee q(r \vee \{rp\}')$.

9.2.1 *d-Rigidity and minimal formulas*

Given a formula F with a distinguished occurrence of a subformula A, $F[\cdot]$ denotes the sequence of signs arising from $F[A]$ when the following *deletion rules* are applied to the *formal* expression (i.e. not the informal abbreviated notation).

1. A in $F[A]$ is deleted.
2. If in $F[A]$ a sign is removed which immediately follows $\neg$ then the negation sign is removed.
3. If $F[A]$ contains a component $[B \wedge C]$, $[C \wedge B]$, $[B \vee C]$, or $[C \vee B]$ in which B must be removed, then this component is replaced by C.

Thus if $F[A]$ is $\{rp\} \vee q(r \vee (rp)')$ then $F[\cdot]$ is $q(r \vee (rp)')$, and if $F[A]$ is $rp \vee q(r \vee \{rp\}')$ then $F[\cdot]$ is $rp \vee qr$.

The concept of *slack formula* can be made precise in more than one way: we choose the following methods.

DEFINITION 9.1.

(i) An occurrence of a subformula X in a formula x is *irrelevant* if the logical equivalence class of F remains unchanged when that occurrence of X, and only that occurrence, is replaced by its negation, i.e. $F[X] \leftrightarrow F[X']$; it is *d-irrelevant* if the logical equivalence class of F remains unchanged when that occurrence of X, and only that occurrence, is deleted (by the above deletion rules), i.e. $F[X] \leftrightarrow F[\cdot]$.
(ii) A formula is *slack* (*d-slack*) if it contains an irrelevant (d-irrelevant) subformula, and is rigid (d-rigid) otherwise.
(iii) $\mathbb{R}$, $\mathbb{S}$, d-$\mathbb{R}$, d-$\mathbb{S}$ denote the classes of rigid, slack, d-rigid, and d-slack formulas respectively. Thus $\mathbb{S}$ is the complement of $\mathbb{R}$ and d-$\mathbb{S}$ is the complement of d-$\mathbb{R}$ with respect to the set of formulas. $\mathbb{T}$ denotes the class of tautologies.

EXAMPLE 9.1. To test a given formula F for rigidity (d-rigidity) each occurrence of a subformula must be replaced by its negation (deleted) and the resulting formula G tested for logical equivalence to F, as in the following example.

(i) The propositional constants t, f are both rigid and d-rigid as also are propositional letters $p, q, r, \ldots$.
(ii) $(q \vee q')p$ is in $\mathbb{R}$ but not in d-$\mathbb{R}$.
(iii) $p \supset p(q \vee q')$ and $p \vee pp'. \supset p$ are rigid tautologies but are not d-rigid since $p \supset p\{q \vee q'\} \leftrightarrow p \supset p$ and $p \vee \{pp'\}. \supset p \leftrightarrow p \supset p$.
(iv) $pq \vee pr \supset p(q \vee r)$ and $p(q \vee r) \supset pq \vee r$ are both rigid tautologies. The first of these is d-slack since $pq \vee pr \supset P$ is also a tautology. Further, $pq \vee pr \supset pq \vee r$ is a slack tautology because

$$pq \vee \{p\}r \supset pq \vee r \leftrightarrow pq \vee p'r. \supset pq \vee r.$$

◆

LEMMA 9.1. For any formula $F[X]$,

$$F[X] \leftrightarrow F[X'] \quad \Leftrightarrow \quad F[X] \leftrightarrow F[t] \leftrightarrow F[f].$$

Proof. By Example 2.2(i), for any valuation $v \in \mathbb{V}$

$$v(X) = 1 \implies v(F[X]) = v(F[t]) \;\&\; v(F[X']) = v(F[f])$$
$$v(X) = 0 \implies v(F[X]) = v(F[f]) \;\&\; v(F[X']) = v(F[t]). \quad (9.1)$$

Suppose now that $F[X] \leftrightarrow F[X']$. Let $v \in \mathbb{V}$. Then $v(F[X]) = v(F[X'])$. Again by Example 2.2(i),

$$v(X) = 1 \implies v(F[X]) = v(F[t]) = v(F[f])$$
$$v(X) = 0 \implies v(F[X]) = v(F[f]) = v(F[t]).$$

Hence for all $v \in \mathbb{V}$ $v(F[X]) = v(F[t]) = v(F[f])$, i.e. $F[X] \leftrightarrow F[t] \leftrightarrow F[f]$.

Rigid propositions 291

Conversely, suppose that $F[X] \leftrightarrow F[t] \leftrightarrow F[f]$. Let $v \in \mathbb{V}$. Then
$$v(F[X]) = v(F[t]) = v(F[f]).$$
By (9.1)
$$v(X) = 1 \Rightarrow v(F[X]) = v(F[X'])$$
$$v(X) = 0 \Rightarrow v(F[X]) = v(F[X']).$$
Hence $v(F[X]) = v(F[X'])$ for all $v \in \mathbb{V}$, i.e. $F[X] \leftrightarrow F[X']$. ∎

This lemma enables the relation between the two notions of slackness and rigidity to be established.

THEOREM 9.2.

(i) $\mathbb{S} \subset \text{d-}\mathbb{S}$

(ii) $\text{d-}\mathbb{R} \subset \mathbb{R}$.

Proof.

(i) Let $F \in \mathbb{S}$. Then there is an occurrence in F of a subformula X such that $F[X] \leftrightarrow F[X']$. By Lemma 9.1, $F[X] \leftrightarrow F[t] \leftrightarrow F[f]$. It may be assumed that X is a subformula of maximum length with this property. Now $X \neq F$, for otherwise $t \leftrightarrow f$. Thus there occurs a subformula Z of F containing that occurrence of X such that Z has one of the forms X', $X \vee Y$, XY. So $F = F[Z]$ and that occurrence of X is the X in one of the three cases: (a) $F = F[X']$, (b) $F = F[X \vee Y]$, (c) $F = F[XY]$.
 (a) This case cannot occur. For if $F = F[X']$ we would have $F[X'] \leftrightarrow F[t'] \leftrightarrow F[f']$, i.e. $F[X'] \leftrightarrow F[f] \leftrightarrow F[t]$. But this contradicts the assumption that X was a formula of maximum length with this property.
 (b) and (c) If (b) then $F[X \vee Y] \leftrightarrow F[f \vee Y]$. Hence $F[X \vee Y] \leftrightarrow F[Y]$. If (c) we likewise have $F[XY] \leftrightarrow F[Y]$. In either case, it can be seen that $F[Y]$, the result of deleting the distinguished occurrence of X from F, is logically equivalent to F itself. Hence F is d-slack. Thus $\mathbb{S} \subseteq \text{d-}\mathbb{S}$. By Example 9.1, this inclusion is strict.

(ii) This follows directly from (i). ∎

DEFINITION 9.2. A formula F is *minimal* if every formula which is logically equivalent to F has length at least that of F. $\mathbb{M}$ is the class of minimal formulas.

EXAMPLE 9.2. t is the only minimal tautology. ◆

THEOREM 9.3. $\text{d-}\mathbb{R}$ properly includes $\mathbb{M}$.

Proof. First, $\mathbb{M} \subseteq$ d-$\mathbb{R}$. For suppose that F is d-slack. Then there exists a subformula X of F such that $F[X] \leftrightarrow F[\cdot]$. Clearly, $|F| = |F[X]| > |F[\cdot]|$. Thus F is not minimal. Hence if F is minimal then it is not d-slack and so is d-rigid. Second, if p is a propositional variable $p \vee p'$ is a d-rigid tautology but it is not minimal (Example 9.3). ∎

Since every formula is logically equivalent to a minimal formula we have the following corollary.

COROLLARY 9.4. *Every formula is logically equivalent to a d-rigid formula.*

9.2.2 *Closure with respect to Boolean operations*

The subject of this subsection is the closure properties of the classes d-$\mathbb{R}$ and d-$\mathbb{S}$ with respect to the Boolean operations.

LEMMA 9.5. *Let $F[X, Y]$ be a formula with distinguished occurrences of subformulas X and Y. Let $Z \leftrightarrow Y$. If that occurrence of X is an irrelevant (d-irrelevant) subformula for $F[X, Y]$ then it is an irrelevant (d-irrelevant) subformula of $F[X, Z]$*

Proof.

(i) Relevance: as X is irrelevant.

$$F \leftrightarrow F[X, Y] \leftrightarrow F[X, Z] \quad \text{as } Z \leftrightarrow Y$$
$$\leftrightarrow F[X', Y] \quad \text{as } X \text{ is irrelevant}$$
$$\leftrightarrow F[X', Z] \quad \text{as } Z \leftrightarrow Y.$$

Thus $F[X, Z] \leftrightarrow F[X', Z]$, i.e. X is an irrelevant subformula of $F[X, Z]$.

(ii) d-Irrelevance: as X is d-irrelevant

$$F \leftrightarrow F[X, Y] \leftrightarrow F[X, Z] \quad \text{as } Z \leftrightarrow Y$$
$$\leftrightarrow F[\cdot, Y] \quad \text{as } X \text{ is d-irrelevant}$$
$$\leftrightarrow F[\cdot, Z] \quad \text{as } Z \leftrightarrow Y.$$

Thus $F[X, Z] \leftrightarrow F[\cdot, Z]$, i.e. X is a d-irrelevant subformula of $F[Z]$. ∎

LEMMA 9.6. *Let X be a formula with a distinguished subformula Y, $X = X[Y]$. Let $Y \leftrightarrow Z$. If $X[Y]$ is an irrelevant (d-irrelevant) subformula of $F[X[Y]]$ then $X[Z]$ is an irrelevant (d-irrelevant) subformula for $F[X[Z]]$.*

Proof.

(i) Relevance: as $X[Z]$ is irrelevant

$$F \leftrightarrow F[X[Y]] \leftrightarrow F[(X[Y])'] \quad \text{as } X \text{ is irrelevant}$$
$$\leftrightarrow F[X[Z]] \quad \text{as } X[Y] \leftrightarrow X[Z]$$
$$\leftrightarrow F[(X[Z])'] \quad \text{as } Y \leftrightarrow Z.$$

Thus $F[X[Z]] \leftrightarrow F[(X[Z])']$, i.e. $X[Z]$ is an irrelevant subformula of F.

(ii) d-Relevance: as $X[Z]$ is d-irrelevant

$$F \leftrightarrow F[X[Y]] \leftrightarrow F[\,\cdot\,] \quad \text{as } X \text{ is d-irrelevant}$$
$$\leftrightarrow F[X[Z]] \quad \text{as } X[Y] \leftrightarrow X[Z].$$

Thus $F[X[Z]] \leftrightarrow F[\,\cdot\,]$, i.e. $X[Z]$ is a d-irrelevant subformula of F. ∎

THEOREM 9.7.

(i) $\qquad\qquad\qquad A \in \text{d-}\mathbb{R} \quad\to\quad A' \in \text{d-}\mathbb{R}.$

(ii) If $F \in \text{d-}\mathbb{R}$ and G is a subformula of F then $G \in \text{d-}\mathbb{R}$.

(iii) (a) $\qquad\qquad F[AB] \in \text{d-}\mathbb{R} \;\Leftrightarrow\; F[BA] \in \text{d-}\mathbb{R}.$

(b) $\qquad\qquad F[A \vee B] \in \text{d-}\mathbb{R} \;\Leftrightarrow\; F[B \vee A] \in \text{d-}\mathbb{R}.$

(iv) (a) $\qquad\qquad F[(AB)C] \in \text{d-}\mathbb{R} \;\Leftrightarrow\; F[A(BC)] \in \text{d-}\mathbb{R}.$

(b) $\qquad F[(A \vee B) \vee C] \in \text{d-}\mathbb{R} \;\Leftrightarrow\; F[A \vee (B \vee C)] \in \text{d-}\mathbb{R}.$

(v) $\qquad\qquad\qquad F[A''] \in \text{d-}\mathbb{R} \;\Leftrightarrow\; F[A] \in \text{d-}\mathbb{R}.$

(vi) (a) $\qquad\qquad F[(AB)'] \in \text{d-}\mathbb{R} \;\Leftrightarrow\; F[A' \vee B'] \in \text{d-}\mathbb{R}.$

(b) $\qquad\qquad F[(A \vee B)'] \in \text{d-}\mathbb{R} \;\Leftrightarrow\; F[A'B'] \in \text{d-}\mathbb{R}.$

(vii) (a) $F[(A \vee B)(A \vee C)] \in \text{d-}\mathbb{R} \;\Rightarrow\; F[A \vee BC] \in \text{d-}\mathbb{R}.$

(b) $\qquad F[AB \vee AC] \in \text{d-}\mathbb{R} \;\Rightarrow\; F[A(B \vee C)] \in \text{d-}\mathbb{R}.$

Proof. Although the issues here are the closure properties of d-$\mathbb{R}$ it is easier to conduct the proofs in terms of the complementary notion of d-slackness. For instance, with regard to (i) it suffices to prove

$$A' \in \text{d-}\mathbb{S} \to A \in \text{d-}\mathbb{S}.$$

(i) Suppose $A' \in \text{d-}\mathbb{S}$. Then A' contains an occurrence of a subformula X whose deletion gives a logically equivalent formula. X cannot be A' or A for the deletion of A or A' annihilates the whole formula. Hence X is a proper subformula of A, i.e. $A = A[X]$. Then $(A[X])' \leftrightarrow (A[\cdot])'$. Hence $A[X] \leftrightarrow A[\cdot]$, i.e. $A \in$ d-S.

(ii) Suppose $G \in$ d-S. Then there occurs a subformula X such that $G[X] \leftrightarrow G[\cdot]$. Hence $F = F[G[X]] \leftrightarrow F[G[\cdot]]$, i.e. $F \in$ d-S.

(iii) (a) Suppose $F[BA] \in$ d-S. Then F has a d-irrelevant subformula, X say. By example 2.2(ii), three cases can occur: (α) X does not overlap BA, (β) BA is a subformula of X, and (γ) X is a subformula of BA.

Case (α) By Lemma 9.5, as $BA \leftrightarrow AB$, X is a d-irrelevant subformula of $F[AB]$.

Case (β) By Lemma 9.6, $X[BA]$ is a d-irrelevant subformula for $F[X[BA]]$.

Case (γ) It can be supposed from (β) that X is a proper subformula of AB so X is a subformula of A or of B. Without loss of generality it can be assumed that $B = B[X]$. So $BA = B[X]A$. But X is a slack subformula of $F[B[X]A]$. Thus $F[B[X]A] \leftrightarrow F[B[\cdot]A]$. Further, $B[X]A \leftrightarrow AB[X]$ and $B[\cdot]A \leftrightarrow AB[\cdot]$. Hence $F[AB[X]] \leftrightarrow F[AB[\cdot]]$, i.e. $F[AB] \in$ d-S. Thus $F[BA] \in$ d-S $\Rightarrow F[AB] \in$ d-S.

The converse follows by interchange of A and B. This proves (iii) (a)

(iii) (b) This case is similar to (iii) (a).

(iv) (a) Suppose that $F[A(AC)] \in$ d-S. By Lemma 9.5 we need only consider the cases that arise when a d-irrelevant subformula of F occurs as a proper subformula of $A(BC)$. There are four cases: (α) X is a subformula of A; (β) X is a subformula of B; X is a subformula of C; (δ) $X = BC$.

Case (α) $A = A[X]$. Then $F[A[X](BC)] \leftrightarrow F[A[\cdot](BC)]$, $A[X](BC) \leftrightarrow (A[X]B)C$, and $A[\cdot](BC) \leftrightarrow (A[\cdot]B)C$. Hence

$$F[(A[X]B)C \leftrightarrow F[A[X](BC)] \leftrightarrow F[A[\cdot](BC)] \leftrightarrow F[(A[\cdot]B)C].$$

Thus $F[(AB)C] \in$ d-S.

Cases (β), (γ) Similar to case (α).

Case (δ) Then $F[A(BC)] \leftrightarrow F[A]$ by the deletion rules. Let $v \in \mathbb{V}$. Then consider the following alternatives: ($\delta 1$) $v(B) = 0$ or $v(C) = 1$; ($\delta 2$) $v(B) = 1$ & $v(C) = 0$. In case ($\delta 1$) $v((AB)C) = v(AB)$. Hence

$$v(F[(AB)C]) = v(F[AB]). \tag{9.2}$$

In case ($\delta 2$), $v(AB) = v(A)$. Hence $v(F[A(BC)]) = v(F[AB])$. As $A(BC) \leftrightarrow (AB)C$. $F[A(BC)] \leftrightarrow F[(AB)C]$. Hence (9.2) also holds in this case. Hence

$F[(AB)C] \leftrightarrow F[AB]$. Thus C is a d-irrelevant subformula of $F[(AB)C]$, i.e. $F[(AB)C] \in $ d-S. This proves that

$$F[A(BC)] \in \text{d-S} \Rightarrow F[(AB)C] \in \text{d-S}.$$

The converse of this result has a similar proof.

(iv) (b) Similar to (iv) (a) but using the cases (α) $v(B) = 1$ or $v(C) = 0$, (β) $v(B) = 0$ & $v(C) = 1$.

(v) Suppose that $F[A] \in $ d-S. By Lemma 9.5 we need only consider the case where a d-irrelevant formula X occurs as a subformula of A. So $A = A[X]$. Then $A[X] \leftrightarrow (A[X])''$ and $A[\cdot] \leftrightarrow (A[\cdot])''$. Now $F[A[X]] \leftrightarrow F[A[\cdot]]$. Hence

$$F[(A[X])''] \leftrightarrow F[A[X]] \leftrightarrow F[A[\cdot]] \leftrightarrow F[(A[\cdot])''].$$

Thus $F[A''] \in $ d-S.

Conversely, suppose that $F[A''] \in $ d-S. Again by Lemma 9.5 we need only consider the case where the d-irrelevant subformula X is a proper subformula of A''. Consider the following two cases (Example 2.2(ii)): (α) $X = A'$ or $X = A$; (β) X a proper subformula of A.

Case (α) Then $F[A''] \leftrightarrow F[\cdot]$ by the deletion rules. As $A'' \leftrightarrow A$, $F[A] \leftrightarrow F[\cdot]$. Thus $F[A] \in $ d-S.

Case (β) Then $F[A[X]''] \in F[A[\cdot]'']$. But $A[X]'' \leftrightarrow A[X]$ and $A[\cdot]'' \leftrightarrow A[\cdot]$. Hence $F[A[X]] \leftrightarrow F[A[X]''] \leftrightarrow F[A[\cdot]''] \leftrightarrow F[A[\cdot]]$. Thus $F[A] \in $ d-S.

(vi) (a) Suppose that $F[A' \vee B'] \in $ d-S. Then without loss of generality it can be assumed that a d-irrelevant subformula X occurs as a proper subformula of $A' \vee B'$. There are therefore two cases to consider: (α) X a subformula of A'; (β) X a subformula of B'.

Case (α) There are three subcases: (α1) $X = A'$; (α2) $X = A$; (α3) X a proper subformula of A.

Subcase (α1) Then by the deletion rules,

$$F[A' \vee B'] \leftrightarrow F[B']. \tag{9.3}$$

Let $G[A] = (AB)'$. Then $G[\cdot] = B'$. But $G[A] \leftrightarrow A' \vee B'$. Hence by (9.3), $F[[G[A]] \leftrightarrow F[(AB)'] \leftrightarrow F[A' \vee B'] \leftrightarrow F[B'] \leftrightarrow F[G[\cdot]]$. Then A is a d-irrelevant subformula of $F[(AB)']$ and so $F[(AB)'] \in $ d-S.

Subcase (α2) Deletion of A from $A' \vee B'$ and from $(AB)'$ gives B' in both cases. Hence as A is an irrelevant subformula of $F[A' \vee B']$, it is also an irrelevant subformula of $F[(AB)']$. Thus $F[(AB)'] \in $ d-S.

Subcase ($\alpha 3$) $A = A[X]$ and $F(A[X]' \vee B') \leftrightarrow F[A[\cdot]' \vee B']$. But $A[X]' \vee B' \leftrightarrow (A[X]B)'$ and $A[\cdot]' \vee B' \leftrightarrow (A[\cdot]B)'$. It easily follows that X is an irrelevant subformula of $F[(AB)']$. Thus $F[(AB)'] \in$ d-S. Hence $F[A' \vee B'] \in$ d-S $\Rightarrow F[(AB)'] \in$ d-S. The proof of the converse is similar.

Case (β) This is similar to case (α).

(vi) (b) Similar to (vi) (a).

(vii) (a) Suppose $F[A \vee BC] \in$ d-S. As above we need only consider where there occurs an irrelevant subformula X in F which occurs as proper subformula of $A \vee BC$. There are two cases: (α) X is a subformula of A: (β) X is a subformula of BC.

Case (α) There are two subcases: ($\alpha 1$) $X = A$; ($\alpha 2$) X is a proper subformula of A.

Subcase ($\alpha 1$) Then $F[A \vee BC] \leftrightarrow F[BC]$. We prove that

$$F[(A \vee B)(A \vee C)] \leftrightarrow F[(A \vee B)C]. \tag{9.4}$$

Let $v \in \mathbb{V}$. There are three possible cases: ($\alpha 1.1$) $v(A \vee BC) = 0 = v(BC)$; ($\alpha 1.2$) $v(A \vee BC) = 1 = v(BC)$; ($\alpha 1.3$) $v(A \vee BC) = 1$, $v(BC) = 0$. We show that in each case

$$v(F[(A \vee B)(A \vee C)]) = v(F[(A \vee B)C]). \tag{9.5}$$

Sub-subcase ($\alpha 1.1$). Then

$$v((A \vee B)(A \vee B)) = v(A \vee BC) = v(AC \vee BC) = v(BC) = 0.$$

Hence (9.5).

Sub-subcase ($\alpha 1.2$) Then $v(A) = v(B) = 1$. Hence $v((A \vee B)(A \vee C)) = 1$. (9.5) follows from this.

Sub-subcase ($\alpha 1.3$) Then $v(A) = 1$ so that $v((A \vee B)(A \vee C)) = 1$. Consider now the two possible truth-values of C. If $v(C) = 1$, then $v((A \vee B)C) = 1 = v((A \vee B)(A \vee C))$. Hence (9.5). If $v(C) = 0$, then $v((A \vee B)C) = 0 = v(BC)$. As $A \vee BC \leftrightarrow (A \vee B)(A \vee C)$ we have $v((A \vee B)(A \vee C)) = v((A \vee B)C) = 0$. Hence (9.5). So (9.5) holds for all $v \in \mathbb{V}$.

Subcase ($\alpha 2$) Then $A = A[X]$. We show that for all $v \in \mathbb{V}$

$$v(F[(A \vee B)(A[X] \vee C)]) = v(F[(A \vee B)(A[\cdot] \vee C)]). \tag{9.6}$$

Now as X is a d-irrelevant subformula

$$F[A[X] \vee BC] \leftrightarrow F[A[\cdot] \vee BC]. \tag{9.7}$$

If $v((A \vee B)(A[X] \vee C)) = v((A \vee B)(A[\cdot] \vee C))$ then (9.6) holds. But if $v((A \vee B)(A[X] \vee C)) \neq v((A \vee B)(A[\cdot] \vee C))$ then $v(A \vee B) = 1$. There are then two cases: (α2.1) $v(A \vee C) = 1$ and $v(A[\cdot] \vee C) = 0$; (α2.2) $v(A \vee C) = 0$ and $v(A[\cdot] \vee C) = 1$.

Sub-subcase (α2.1) Then $v(A[\cdot]) = v(C) = 0$ and $v(A) = 1$. Hence

$$v((A \vee B)(A[\cdot] \vee C)) = v(A[\cdot] \vee BC)$$

$$v(A \vee BC) = v((A \vee B)(A \vee C)) = 1.$$

(9.6) now follows from (9.7).

Sub-subcase (α2.2) Then $v(A) = v(C) = 0$, $v(A[\cdot]) = v(B) = 1$. Hence

$$v((A \vee B)(A[\cdot] \vee C)) = v(A[\cdot] \vee BC) = 1$$

$$v(A \vee BC) = v((A \vee B)(A \vee C)) = 0.$$

(9.6) now follows from (9.7). Thus (9.6) holds in both cases, and, as it holds for all $v \in \mathbb{V}$,

$$F[(A \vee B)(A[X] \vee C)] \leftrightarrow F[(A \vee B)(A[\cdot] \vee C)].$$

So X is a d-irrelevant subformula of $F[(A \vee B)(A \vee C)]$, i.e. $F[(A \vee B)(A \vee C)] \in \text{d-}\mathbb{S}$.

(vii) (b) This is similar to (vii) (i). ∎

A similar theorem holds for relevance.

THEOREM 9.8.

(i) $\qquad\qquad\qquad A \in \mathbb{R} \;\Rightarrow\; A' \in \mathbb{R}.$

(ii) If $F \in \mathbb{R}$ and G is a subformula of F then $G \in \mathbb{R}$.

(iii) (a) $\qquad\qquad F[AB] \in \mathbb{R} \;\Leftrightarrow\; F[BA] \in \mathbb{R}.$

(b) $\qquad\qquad F[A \vee B] \in \mathbb{R} \;\Leftrightarrow\; F[B \vee A] \in \mathbb{R}.$

(iv) (a) $\qquad\qquad F[(AB)C] \in \mathbb{R} \;\Leftrightarrow\; F[A(BC)] \in \mathbb{R}.$

(b) $\qquad F[(A \vee B) \vee C] \in \mathbb{R} \;\Leftrightarrow\; F[A \vee (B \vee C)] \in \mathbb{R}.$

(v) $\qquad\qquad\qquad F[A''] \in \mathbb{R} \;\Leftrightarrow\; F[A] \in \mathbb{R}.$

(vi) (a) $\qquad\qquad F[(AB)'] \in \mathbb{R} \;\Leftrightarrow\; F[A' \vee B'] \in \mathbb{R}.$

(b) $\qquad\qquad F[(A \vee B)'] \in \mathbb{R} \;\Leftrightarrow\; F[A'B'] \in \mathbb{R}.$

(vii) (a) $\quad F[(A \vee B)(A \vee C)] \in \mathbb{R} \;\Rightarrow\; F[A \vee BC] \in \mathbb{R}.$

(b) $\qquad\qquad F[AB \vee AC] \in \mathbb{R} \;\Rightarrow\; F[A(B \vee C)] \in \mathbb{R}.$

(viii) (a) $\qquad\qquad F[AA] \in \mathbb{R} \;\Rightarrow\; F[A] \in \mathbb{R}.$

(b) $\qquad\qquad F[A \vee A] \in \mathbb{R} \;\Rightarrow\; F[A] \in \mathbb{R}.$

Proof. The proofs are similar in principle to the corresponding parts of the preceding theorem. The only real difference appears in (viii) (a) and (viii) (b): the d-relevance analogues of these are false. (Note that $F[AA] \leftrightarrow F[A]$ so that $F[AA]$ is d-slack—but it might not be slack.)

(viii) (a) Suppose that X is an irrelevant subformula of F. By Lemma 9.5 we need only consider the case where X occurs as a proper subformula of A, i.e. $A = A[X]$. We show that $F[AA[X]] \leftrightarrow F[AA[X']]$ which shows that $F[AA]$ is slack. Let $v \in \mathbb{V}$. Consider the cases (α) $v(A[X]) = v(A[X'])$ and (β) $v(A[X]) \neq v(A[X'])$.

Case (α) Then $v(AA[X]) = v(AA[X'])$. Thus

$$v(F[AA[X]]) = v(F[AA[X']]). \tag{9.8}$$

Case (β) Then there are two subcases: (α1) $v(AA[X]) = v(AA[X'])$; (α2) $v(AA[X]) \neq v(AA[X'])$.

Subcase (α1) Then (9.8) holds.
Subcase (α2) As $v(A[X]) \neq v(A[X'])$ we have $v(A[X]) = v(AA[X])$ and $v(A[X']) = v(AA[X'])$ so that $v(F[A[X]]) = v(F[AA[X]])$ and $v(F[A[X']]) = v(F[AA[X']])$. Now X is a slack subformula so that $v(F[A[X]]) = v(F[A[X']])$. Hence (9.8) holds.

Thus (9.8) holds for both cases (α) and (β) and so for all $v \in \mathbb{V}$. Hence $F[AA[X]] \leftrightarrow F[AA[X']]$. ∎

EXAMPLE 9.3. Let p, q, r be distinct propositional letters.

(i) $pp' \in$ d-$\mathbb{R}$ (and so $pp' \in \mathbb{R}$ by Theorem 9.2(ii)) but $p'pp'p \notin \mathbb{R}$ as $pp' \leftrightarrow p'pp'p \leftrightarrow p'pp'p'$. Thus d-$\mathbb{R}$ and $\mathbb{R}$ are not closed under conjunction and the converse of Theorem 9.8(viii)(a) does not hold.
(ii) $(p' \vee p)(p' \vee p) \in \mathbb{R} -$ d-$\mathbb{R}$. Application of the distributive laws does not preserve rigidity since $(p' \vee p)p' \vee (p' \vee p)p \leftrightarrow (p' \vee p')p' \vee (p' \vee p)p$. Hence the converse of Theorem 9.8(vii)(a) does not hold.
(iii) $pq \vee qr \in$ d-$\mathbb{R}$ but $(\{pq\} \vee q)(pq \vee r) \in$ d-$\mathbb{S}$ since $pq \vee q \leftrightarrow q$. Thus the application of the distributive laws does not preserve d-rigidity either. Hence the converse of Theorem 9.9(vii)(a) does not hold.
(iv) $(pq)' \vee pq \in \mathbb{R} \cdot \mathbb{T}$ but $(pq)' \vee pq \in \mathbb{T} \cdot$ d-$\mathbb{S}$ since $(pq)' \vee p\{q\} \leftrightarrow (pq)' \vee p$. But $((pq)' \vee p)((pq)' \vee q) \in \mathbb{S} \cdot \mathbb{T}$ as $((pq')' \vee p)((pq)' \vee p) \in \mathbb{T}$. ◆

9.3 Minimal implications

Redundancy in implications can occur between formulas of premiss and conclusion and within formulas in the guise of irrelevant subformulas. Implications which are devoid of redundancy are *minimal*. There are several

Minimal implications

ways of defining such redundancy, each giving rise to a notion of *entailment*. No single method is *intrinsically* better than the others. We give two such methods below.

9.3.1 Rigid implications

In the spirit of Definition 9.1, a pair (Δ, A) may be called a *slack* implication if $\Delta \to A$ and there exists $B \in \Delta$ such that when B is replaced in Δ by its negation—to give Δ' say (so that $\Delta' = \Delta - \{B\} + \{\neg B\}$)—then $\Delta' \to A$. Classical logic already provides a mechanism for the elimination of the offending formula $\neg B$, for by Example 2.1(v) under the above circumstances $\Delta - \{B\} \to A$. Thus the pair (Δ, A) contains some redundancy. This idea is sharpened in the following definition of rigid implication which extends the notion of rigidity from formulas (Definition 9.1) to relations between formulas. It requires rigidity of the individual formulas concerned and a form of non-redundancy in the logical consequence relation between them.

DEFINITION 9.3. Let Δ be a set of formulas and let A be a formula.

(i) Δ is *minimal with respect to A* (or *A-minimal*) if A is a logical consequence of Δ but A is not a logical consequence of any proper subset of Δ.
(ii) The pair (Δ, A) is a *rigid (d-rigid)* if all members of $\Delta + \{A\}$ are rigid (d-rigid), $\Delta \to A$, and Δ is A-minimal.
(iii) Write $\Delta \to_r A$ ($\Delta \to_{d\text{-}r} A$) when (Δ, A) is a rigid (d-rigid) implication. Following our usual convention we write $B \to_r A$ instead of $\{B\} \to_r A$ etc.
(iv) Define

$$Imp =_{def} \{(A, B); \{A\} \to_r B\}$$
$$\text{d-}Imp =_{def} \{(A, B); \{A\} \to_{d\text{-}r} B\}.$$

EXAMPLE 9.4.

(i) By Theorem 9.2, d-$Imp \subseteq Imp$.
(ii) By Example 9.1(ii), if p, q are distinct proposition letters then $(p, p(q \lor q')) \in Imp -$ d-Imp. ♦

LEMMA 9.9. *Let $B \to A$. Then A is not a tautology if, and only if, $\{B\}$ is A-minimal. If, in addition, A, B are both rigid (d-rigid) then $B \to_r A$ ($B \to_{d\text{-}r} A$).*

Proof. Suppose A is not a tautology. If $\{B\}$ is not A-minimal then there exists a proper subset of $\{B\}$ of which A is a consequence. Thus $A \in Cn(\emptyset)$, i.e. $A \in \mathbb{T}$. This is a contradiction. Hence $\{B\}$ is A-minimal.

Conversely, if A is a tautology then $A \in Cn(\emptyset)$ so that $\{B\}$ is not minimal. If, A, B are also rigid then, by Definition 9.3, $(\{B\}, A)$ is a rigid implication. Similarly for d-rigidity. ∎

We now examine some closure properties of rigid implications. The most important property is transitivity. A limited form of it is exhibited.

THEOREM 9.10. Let $\vdash$ stand for either $\to_r$ or $\to_{d\text{-}r}$. Then $A \vdash B$ and $B \vdash C$ implies $A \vdash C$.

Proof. Suppose $A \to_r B$ and $B \to_r C$. Then A, B, and C are rigid, C is a consequence of A, and B is C-minimal. It remains to prove that A is C-minimal. Suppose that A is not C-minimal. By Lemma 9.9, $\emptyset \to_r C$. This contradicts the C-minimality of B. Hence A is C-minimal.

The proof of the d-rigidity case is similar. ∎

Some relations between d-rigid implications and the connectives are given in the following theorem.

THEOREM 9.11.

(i) $F \lor G \to_{d\text{-}r} H \Rightarrow F \to_{d\text{-}r} H \ \& \ G \to_{d\text{-}r} H$.

(ii) $H \to_{d\text{-}r} FG \Rightarrow H \to_{d\text{-}r} F \ \& \ H \to_{d\text{-}r} G$.

(iii) If $F' \notin \mathbb{T}$ then $F \to_{d\text{-}r} G \Rightarrow G' \to_{d\text{-}r} F'$.

Proof.

(i) Suppose that $F \lor G \to_{d\text{-}r} H$. As disjunction is normal in classical logic, by **Dj1**⁰ (Chapter 4, §4.4), $H \in Cn(F) \cdot C(G)$. Suppose that $\{F\}$ is not H-minimal. Then $H \in Cn(\emptyset)$ by Lemma 9.9. But the $F \lor G$ is not H-minimal, contrary to hypothesis. Hence $\{F\}$ is H-minimal. Now H and $F \lor G$ are d-rigid. By Theorem 9.7(ii), F is d-rigid. Hence $F \to_{d\text{-}r} H$. Likewise $G \to_{d\text{-}r} H$.

(ii) Suppose $H \to_{d\text{-}r} FG$. Then $FG \in Cn(H)$. As conjunction in classical logic is normal, by **Cj2**⁰ $F \in Cn(H)$ and $G \in Cn(H)$. H and FG are d-rigid and by Theorem 9.7(ii) so are F and G. Suppose that H is not F-minimal. By Lemma 9.9, $F \in \mathbb{T}$. But then $FG \leftrightarrow G$ so that FG is not d-rigid. This is a contradiction. Hence H is F-minimal so that $H \to_{d\text{-}r} F$. Likewise $H \to_{d\text{-}r} G$.

(iii) Suppose $F \to_{d\text{-}r} G$ where $F' \notin \mathbb{T}$. Then $G' \to F'$. By Theorem 9.7(i) G', F' are d-rigid. Suppose that $\{F'\}$ is not G'-minimal. By Theorem 9.9, $F' \in \mathbb{T}$. This is a contradiction. Hence $\{F'\}$ is G'-minimal and so $G' \to_{d\text{-}r} F'$. ∎

EXAMPLE 9.5. Further, (iii) does not hold if the clause $F' \notin \mathbb{T}$ is omitted, for $pp' \to_{\text{d-r}} q$ but not $q' \to_{\text{d-r}} (pp')'$. Since $(pp')' \in \mathbb{T}$ we then have a contradiction by Lemma 9.9. ◆

THEOREM 9.12.

(i) $F \vee G \to_r H \Rightarrow F \to_r H \mathbin{\&} G \to_r H$.
(ii) If $F' \notin \mathbb{T}$ then $F \to_r \Rightarrow G' \to_r H'$.

Proof. As for (i) and (iii) respectively of the previous theorem but appealing to the corresponding parts of Theorem 9.8. ∎

EXAMPLE 9.6. The analogue of Theorem 9.11(ii) does not hold, for if p, q are distinct propositional letters, by Example 9.4(ii) $(p, p(q \vee q'))$ is a rigid implication but $(p, q \vee q')$ is not.

Implications which are most obviously paradoxical are those in which either premiss or conclusion are tautologous or contradictory—in classical logic, if B is a tautology then for all formulas A, $A \to B$. But we have the following theorem.

THEOREM 9.13.

(i) If $A \to_r B$ or $A \to_{\text{d-r}} B$ then $A, B \notin \mathbb{T}$.
(ii) If, in addition, $A' \notin \mathbb{T}$ then $B' \notin \mathbb{T}$.

Proof.

(i) Suppose $A \to_r B$ or $A \to_{\text{d-r}} B$. Suppose also $A \in \mathbb{T}$ or $B \in \mathbb{T}$. If $A \in \mathbb{T}$ then $B \in \mathbb{T}$. If $B \in \mathbb{T}$ then $\varnothing \to B$. Thus $\{A\}$ is not B-minimal. This contradicts $A \to_r B$ or $A \to_{\text{d-r}} B$. Hence $A, B \notin \mathbb{T}$.
(ii) Suppose $A' \notin \mathbb{T}$. Then by Theorems 9.11(iii) and 9.12(ii), $B' \to A'$ or $B' \to_{\text{d-r}} A'$ and so (i), $B' \notin \mathbb{T}$. ∎

COROLLARY 9.14. If $A \in \mathbb{T}$ there does not exist a formula B such that $A \to_r B$, $A \to_{\text{d-r}} B$, $B \to_r A$, or $B \to_{\text{d-r}} A$.

(This is somewhat stronger than a criterion of Smiley referred to by Anderson and Belnap (1962, p. 10).)

THEOREM 9.15.

(i) If $F' \notin \mathbb{T}$ and either $F \to_r G$ or $F \to_{\text{d-r}} G$ then F and G have a propositional variable in common.
(ii) If F, G are rigid (d-rigid) and F is a tautology then $F' \to_r G$ ($F \to_{\text{d-r}} G$).

Proof.

(i) Suppose that $F' \notin \mathbb{T}$ and either $F \rightarrow_r G$ or $F \rightarrow_{d-r} G$. Then $F \rightarrow G$. Also, by Theorem 9.13, $G \notin \mathbb{T}$. Then there exist $u, v \in \mathbb{V}$ such that $u(F) = 1$ and $v(G) = 0$. If F and G had no propositional variable in common, then from u and v there could be constructed $w \in \mathbb{V}$ such that $w(F) = u(F)$ and $w(G) = v(G)$, i.e. $w(F) = 1$ and $w(G) = 0$. But this contradicts $F \rightarrow G$. Hence F and G have a common propositional variable.

(ii) This follows directly from Lemma 9.9. ∎

9.3.2 *Rigid tautologies*

The notions of rigid implication given above do not impose a very strong connection between premiss and conclusion—they allow cases where premiss and conclusion have no variable in common. This is often regarded as a sin in the relevance trade. There are many ways of regaining virtue. We could enter a monster-barring mode and append to Definition 9.3(i) a clause which excludes inconsistent premisses. Another strategy is to base the non-paradoxical implications on rigid tautologies since then the notion of rigidity applies indiscriminately to premiss and conclusion which are united in the tautology.

DEFINITION 9.4. The relations $\vdash$ and $\vdash_d$ are defined by

$$A \vdash B \Leftrightarrow_{def} A \supset B \in \mathbb{R} \cdot \mathbb{T}$$
$$A \vdash_d B \Leftrightarrow_{def} A \supset B \in \mathbb{T} \cdot \text{d-}\mathbb{R}.$$

If $A \vdash B$ then A entails B, and if $A \vdash_d B$ then A d-entails B;

$$\text{d-}Ent =_{def} \{(A, B); A \vdash_d B\}$$

$$Ent =_{def} \{(A, B); A \vdash B\}.$$

THEOREM 9.16.

(i) Neither $Ent \subseteq \text{d-}Imp$ nor $\text{d-}Imp \subseteq Ent$.
(ii) $\text{d-}Ent \subset \text{d-}Imp$, $Ent \subset Imp$.

Proof.

(i) Let p, q be distinct propositional letters. Then $p \supset p(q \vee q') \in \mathbb{R} \cdot \mathbb{T}$ so that $(p, p(q \vee q')) \in Ent$. But $p(q \vee q') \notin \text{d-}\mathbb{R}$ so that $(p, p(q \vee q')) \notin \text{d-}Imp$. Also, $pp' \rightarrow_d q$ but $pp' \supset q \in \mathbb{S} \cdot \mathbb{T}$. Then $(pp', q) \in \text{d-}Imp$ and $(pp', q) \notin Ent$.

(ii) By Theorem 9.2(ii), $\mathbb{T} \cdot \text{d-}\mathbb{R} \subseteq \mathbb{R} \cdot \mathbb{T}$. Hence $\text{d-}Ent \subseteq Ent$. It follow immediately from the definitions that $\text{d-}Ent \subseteq \text{d-}Imp$ and $Ent \subseteq Imp$. Further, $\text{d-}Imp \subseteq Imp$ (Example 9.4). From (i), $\text{d-}Ent \neq \text{d-}Imp$ and $\text{d-}Ent \neq Ent$. Likewise $\text{d-}Imp \neq Imp$ and $Ent \neq Imp$. ∎

Minimal implications 303

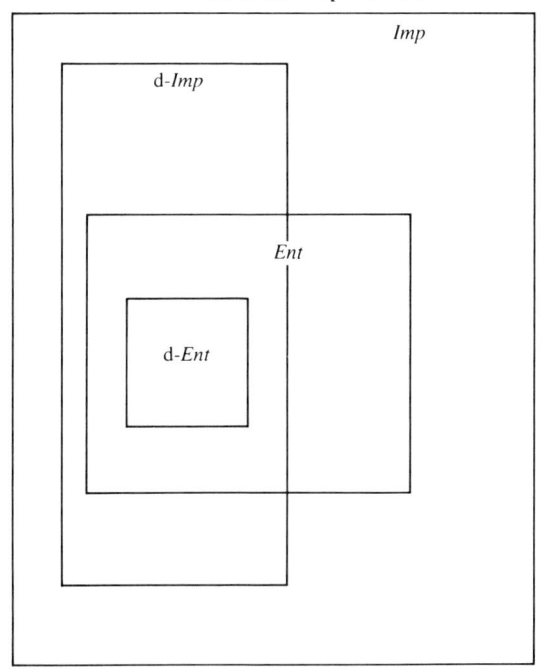

Fig. 9.1.

For entailments a slightly stronger version of Theorem 9.13 holds.

THEOREM 9.17. *If $A \vdash B$ or $A \vdash_d B$ then $A, B, A', B' \notin \mathbb{T}$.*

Proof. By Theorem 9.2 we need only consider $A \vdash B$. If $A \vdash B$ then $A \supset B \in \mathbb{T}$. Suppose $A' \in \mathbb{T}$ or $B \in \mathbb{T}$. Then $A \supset B \leftrightarrow A' \supset B$. Thus $A \supset B$ is slack. This contradicts the hypothesis. Hence if $A \vdash B$ then $A', B \notin \mathbb{T}$. Similarly, if $A \in \mathbb{T}$ or $B' \in \mathbb{T}$ then $A \supset B \leftrightarrow A \supset B'$ so that again $A \supset B$ is slack. Hence $A, B' \notin \mathbb{T}$. ∎

The class of rigid (d-rigid) tautologies is, of course, closed under the operations defined in Theorem 9.8 (Theorem 9.7). In addition we have the following theorem.

THEOREM 9.18.

(i) $AB \in \mathbb{R} \cdot \mathbb{T} \Leftrightarrow A, B \in \mathbb{R} \cdot \mathbb{T}$.

(ii) $AB \in \text{d-}\mathbb{R} \cdot \mathbb{T} \Rightarrow A, B \in \text{d-}\mathbb{R} \cdot \mathbb{T}$

   $A \in \text{d-}\mathbb{R} \cdot \mathbb{T} \Rightarrow AA \in \text{d-}\mathbb{S} \cdot \mathbb{T}$.

Proof.

(i) Suppose $AB \in \mathbb{R} \cdot \mathbb{T}$. Then $A, B \in \mathbb{T}$. By Theorem 9.8(ii), $A, B \in \mathbb{R}$. So $A, B \in \mathbb{R} \cdot \mathbb{T}$.

Conversely, suppose AB is a tautology but is not rigid. Then it is a slack tautology, i.e. $F = AB \in \mathbb{S} \cdot \mathbb{T}$. Then F contains a subformula $F[X] \equiv F[X']$. Clearly, X is not AB itself. Hence it is a subformula of A or B. Suppose $A = A[X]$. Then $F[X] = A[X]B \equiv A[X']B = F[X']$. But $AB \in \mathbb{T}$. Hence $B \in \mathbb{T}$. Thus $A[X] \equiv A[X']$ so that A is slack. Similarly, if X is a subformula of B then B is slack. Hence, if AB is a slack tautology then one of A, B is also a slack tautology, i.e. one of them is not rigid.

(ii) Suppose $AB \in \mathbb{T} \cdot \text{d-}\mathbb{R}$. Then $A, B \in \mathbb{T}$. By Theorem 9.7(ii), $A, B \in \text{d-}\mathbb{R}$. So $A, B \in \mathbb{T} \cdot \text{d-}\mathbb{R}$.

Suppose $A \in \mathbb{T} \cdot \text{d-}\mathbb{R}$. Then $AA \in \mathbb{T}$. But $AA \in \text{d-}\mathbb{S}$. Thus $AA \in \text{d-}\mathbb{S} \cdot \mathbb{T}$. ∎

Some closure properties of entailments are given in the following theorem.

THEOREM 9.19.

(i) $A \vdash B \ \& \ A \vdash C \Rightarrow A \vdash BC$.

(ii) $A \vdash C \ \& \ B \vdash C \Rightarrow A \vee B \vdash C$.

(iii) $A \vdash B \Leftrightarrow A' \vdash B'$.

(iv) $A \vdash_d B \Leftrightarrow A' \vdash_d B'$.

Proof.

(i) Suppose $A \vdash B \ \& \ A \vdash C$. Then $A' \vee B$, $A' \vee C \in \mathbb{R} \cdot \mathbb{T}$. By Theorem 9.8(vii)(a), $A' \vee BC \in \mathbb{R} \cdot \mathbb{T}$. Hence $A \vdash BC$.

(ii) Similar but using Theorem 9.8(vii)(b).

(iii) Suppose $A \vdash B$. Then $A' \vee B \in \mathbb{R} \cdot \mathbb{T}$. By Theorem 9.8(v), $B'' \vee A' \in \mathbb{R} \cdot \mathbb{T}$. Hence $B' \vdash A'$.

(iv) Similar to (iii). ∎

EXAMPLE 9.7.

(i) The analogues of Theorems 9.11 and 9.12, parts (i) and (ii), do not hold for entailments, e.g. $p \vdash p(q \vee q')$ but not $p \vdash q \vee q'$.

(ii) The versions of (i) and (ii) in the above theorem for d-entailments do not hold, for $p \vdash_d p$ but not $p \vdash_d pp$, and not $p \vee p \vdash_d p$.

(iii) $A \supset A$ is not always a d-rigid tautology. $pq \supset \{p\}q$ is a slack tautology. ◆

It is quite easy to find 'desirable' properties of logical consequence that fail to hold for entailment. Thus, for no formulas A, B is it the case that AB implies A or A implies $A \vee B$ for any of the forms of rigid implication or entailment. Further, Example 9.1(iv) shows that entailment is not transitive, in contrast with rigid, or d-rigid, implication (Theorem 9.10). The converses of Theorem 9.19, parts (i) and (ii), fail (see also Example 9.5). Finally, it seems unlikely that paradoxical formulas can be excluded from any of the usual formulation of formal calculi; at least, in Gentzen type systems, there are rigid tautologies which cannot be derived without the use of slack tautologies. Consider, for example, the system defined by Schütte (1961, p. 36). $pq \vee pr \supset p(q \vee r)$ is a rigid tautology (Example 9.1(iv)). It is derivable by inference rule **S1b** only from $pq \vee pr \supset p$ and $pq \vee pr \supset q \vee r$: there are no other premisses from which it can be derived. But $pq \supset p(q \vee r)$, $pq \supset p(q \vee r)$, $pq \vee pr \supset p$, and $pq \vee pr \supset q \vee r$ are all slack tautologies.

9.4 A semantic method

Current practice in mathematics is, in effect, to regard the universe of mathematical discourse as a 'unified' sphere to which formal logic—classical or intuitionist—can be applied. For instance, whilst working on Euclidean geometry one might apply some results from the theory of real closed fields (i.e. from algebra). At first sight this part of algebra might seem irrelevant to the geometry. But a deep connection exists between the two: algebra and geometry are already a unified system. This does not mean to say that further mathematical ideas cannot be imported into the area to make a larger unity. However, no one has demonstrated a unifying connection between Pythagoras' grandmother and Euclidean geometry. To attempt to introduce such an alien idea would, accordingly, be viewed as indicative of a grasshopper mind—as eccentric, to say the least. Thus a judgement of relevance of subjects occurs *before* the application of formal logic.

There is no evil in the grasshopper syndrome if it is kept away from serious thinking. Yet there might be occasions for putting two disparate subjects into the same discourse, perhaps in a theory of multiple data-bases. This raises the problem of regulating the grasshopper mind, i.e. how to build a logic applicable to two or more subjects without arbitrarily obliterating their difference. Below we give a brief introduction to Muškardin's (1978) multi-subject semantics and the related concept of *entailment* which accomplishes this aim.

The central idea is that the paradoxes mentioned above arise from mixing statements about totally unrelated subjects. The procedure adopted is to develop the semantics of a formal language in the standard way (e.g. Chapter 2), but simply to refrain from identifying truth in one subject with truth in another.

When dealing with a propositional logic, the definition of truth (Definition 2.4) was framed in terms of valuations (Definition 2.3) which mapped the algebra of formulas homomorphically into the algebra $\mathbb{T}$ of truth-values. From the present point of view $\mathbb{T}$, which here will be assumed to be finite, is regarded as the algebra of truth-values in a unified subject. If there are now n different subjects, let us assume that the truth-values in the different subjects are isomorphic but not identical: there should be a positive reason for making such an identification. The truth-values can therefore be considered as $\mathbb{T}$ coupled with some notional indices $1, 2, \ldots, n$ to indicate the different subject matters. Thus we begin with a set $\mathbb{F}(J, P)$ where $J = \{\neg, \vee, \wedge\}$. (Following our usual practice we shall contract $A \wedge B$ to AB and write A' instead of $\neg A$.) For greater generality it will be assumed that $\mathbb{T}$ is a QBA (which includes the usual classical two-valued case in Chapter 2). Thus the new sets of 'truth-values' is $S^{(n)} \times T$, which we denote by $T^{(n)}$, where $S^{(n)} = \{1, 2, \ldots, n\}$. A logic is obtained by extending valuations of the set P of propositional variables in $T^{(n)}$ to $\mathbb{F}(J, P)$ in such a way that, if $n = 1$ (i.e. only one subject), the logic based on $T^{(1)}$ coincides with $\mathfrak{L}[T]$ (Chapter 6, §6.6).

Consider now the definition of valuations in $T^{(n)}$. Each mapping $v: P \to T^{(n)}$ can be split into two functions s, t such that $s: P \to S^{(n)}$, $t: P \to T$, and $v(p) = (s(p), t(p))$ for all $p \in P$. v must now be extended to a valuation of all formulas. Let us here choose to extend v componentwise, i.e.

$$v(A \vee B) = (s(A \vee B), t(A \vee B)) =_{\text{def}} (s(A) + s(B), t(A) + t(B))$$

$$v(AB) = (s(AB), t(AB)) =_{\text{def}} (s(A) \cdot s(B), t(A) \cdot t(B))$$

$$v(A') = (s(A'), t(A')) =_{\text{def}} (s(A)', t(A)').$$

(This is a procedural decision—there must be other ways of doing this, perhaps producing different results.) This construction is viable provided that a QBA $G(n)$ is constructed from $S^{(n)}$. The elements of $G(n)$ are the *contexts*. Eventually the logic will be based on both context and truth-value. The method adopted here is to take $G(n)$ to be the free product $\bar{1} * \bar{2} * \cdots * \bar{n}$ of the one-element algebras $\bar{1}, \bar{2}, \ldots, \bar{n}$ (Chapter 7, §7.3). This algebra is called the *context algebra*. Then $G(1)$ is a one-element QBA and the algebra of truth-values, $G(1) \times T$, is isomorphic to T. Thus $\mathbb{F}(J, P)$ with valuations in the finite QBA $G(n) \times T$ is $\mathfrak{L}(G(n) \times T)$ (Chapter 6, §6.6). The notion of logical consequence is the usual one, for lattice-valued logics, i.e. (6.22); let $Cn^{(n)}$ denote this logical consequence function. This notion of logical consequence depends on both truth-value and context. The auxiliary logical consequence function $Cnt^{(n)}$, based on the context algebra $G(n)$, depends only on the contexts and not the truth-values. The main results, due to Muškardin (1978), are that only two contexts are necessary and that the logic with two contexts, i.e. $\mathfrak{L}[(G(2) \times T)]$, is isomorphic to $\mathfrak{L}[Q_4]$.

DEFINITION 9.5. Let $\Delta, A, B \subseteq \mathbb{F}(J, P)$.

(i) Δ *logically n-entails* A when $A \in Cn^{(n)}(\Delta)$. A is *logically n-equivalent* to B (written $A[n] \equiv B$) when $A = B \, mod(Cn^{(n)e}$ (Definition 3.2(i)).
(ii) Δ contextually n-entails A when $A \in Cnt^{(n)}(\Delta)$. A is *contextually n-equivalent* to B (written $A \, c(n) \equiv B$ when $A = B \, mod(Cnt^{(n)e})$.

EXAMPLE 9.8. (cf. Chapter 7, §7.3).

(i) $G(2)$ is isomorphic to Q_4.
(ii) For $n \geq 2$, $G(2)$ is a subsystem of $G(n)$ which itself is a subsystem of $G(n) \times T$. $G(2)$ is also a subsystem of $G(2) \times T$ which is also a subsystem of $G(n) \times T$. Hence

$$C^{(n)} \subseteq Cnt^{(n)} \qquad Cn^{(2)} \subseteq Cnt^{(2)}. \tag{9.9}$$

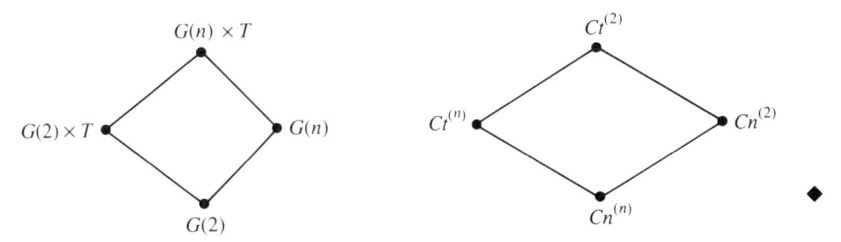

THEOREM 9.20. For any $A, B \in \mathbb{F}(J, P)$,

$$A[n] \equiv B \iff A \, c(2) \equiv B$$

Proof. $G(2)$ is isomorphic to Q_4. By Theorems 6.14 and 7.20(ii), $G(2)$ is functionally free over QBA's. Hence $Ac(2) \equiv B \Rightarrow A[n] \equiv B$. Conversely, by (9.9), $A[n] \equiv B \Rightarrow A \, c(2) \equiv B$. ∎

COROLLARY 9.21. $Cn^{(n)} = Cnt^{(n)} = Cn^{(2)} = Cnt^{(2)}$.

Proof. By (9.9) we need only prove that $Cnt^{(2)} \subseteq Cn^{(n)}$. Let $\Delta, A \subseteq \mathbb{F}(J, P)$. Suppose $A \in Cnt^{(2)}(\Delta)$. Since $G(n) \times T$ is finite, by Theorem 6.11, $Cnt^{(2)}$ is compact. Hence there exists a finite subset $\Gamma \subseteq \Delta$ such that $A \in Cnt^{(2)}(\Gamma)$. Let B be the conjunction of all members of Γ. By Theorem 6.7 conjunction is normal. Hence $B \, cn(2) \equiv AB$. By Theorem 9.20, $B[n] \equiv AB$. Hence $A \in Cn^{(n)}(B)$. Hence $A \in Cn^{(n)}(\Gamma) \subseteq Cn^{(n)}(\Delta)$. ∎

It follows from Corollary 9.21 that $Cn^{(n)} = Cn[Q_4]$ and so $\mathfrak{L}[(G(n) \times T)] = \mathfrak{L}[Q_4]$. Thus, as a formal system, the *n*-subject logic with $n \geq 2$ is identical with the basic system of logical entailment. Observe also that this result depends only on the algebra of truth-values being a QBA.

The extension of multi-subject semantics to predicate logic offers no difficulty. Consider a predicate language of type τ. Assume that the structures in the different subjects are isomorphic, but not necessarily identical. Then these structures can be taken as ordered pairs $(i, \mathfrak{A})$, where i is a subject index and $\mathfrak{A}$ is a structure of type τ. Definition 2.11 can easily be extended, via the principles used above, to give the set of valuations in $G(n) \times T$.

10

The calculus of logics: effective logic

10.1 Introduction

The semantically based logics of previous chapters can be considered as the logics of particular sciences or formal systems. The task we now undertake is to regularize the discourse about logics: can a science of logics (Chapter 1, §1.2) be built?

The construction of effective logic given in this chapter is guided by the initial aim of formulating logic as *rules of correct reasoning* rather than as a machine for the generation of universal truths.

> ... effective logic (explicitly in its operative interpretation), must be looked at as a system of 'universal' rules which are accepted whenever a system of rules of action e.g. rules for producing proofs or rules for producing arbitrary strings of signs has been laid down. In this case the field of application for the rules of logic is not the world as a totality of facts but rather the world as seen in terms of specific kinds of human activities. Within mathematics, for example, the rules of effective logic may be used without restriction. And this obtains because mathematics is not viewed as a system of truths, even less, logical ones, but is treated as an independent scientific activity, which, together with its intrinsic rules, may use the rules of logic as additional 'admissible' ones. Effective logic is the material content of an scientific investigation, ... the 'empirical' core even within mathematics ...
>
> Lorenz 1973

But effective logic is itself a logic and so is simultaneously subject and object. It unites in itself two competing aspects of logic:

(1) logic as a set of rules for correct thinking;
(2) logic as a set of theorems on the general behaviour of (correct!) thought—in particular, formalizations of sciences, is arbitrarily logics.

This phenomenon appeared even in Aristotle's logic where syllogisms are used for both arts and sciences.

> ... syllogisms are treated neither as theorems nor as rules because in a way they are *both*, depending on the level of argumentation. They can, any one of them, be considered as rules of inference, but as soon as syllogisms are not considered with respect to their producing something out of something, but as entities *sui generis*, those rules of inference may be transferred in (logical) implications i.e. three-place (meta)propositions, and thus theorems. (This is in accord with the characterizations of arts and sciences in A. Post. 100 a 88f. Arts are concerned with the world of coming-to-be and passing away, sciences are concerned with the world of being).
>
> Lorenz 1973

With the early Greek logicians, the function of logic ('dialectic') was to provide the means of distinguishing between valid and invalid arguments. In pre-Aristotelian dialectic:

> ... it is always a matter of rules, now laws; they are principles stating how one should proceed, not laws which describe an objective state of affair....
>
> Bocheński 1961, p. 32

Kapp (1942) showed that the origin of Aristotle's Syllogistic was in the actual practice of public discussion of civic affairs in the city states. Its dialogical nature is evident in Aristotle's development of logic:

> If we attend carefully to Aristotle's explanations in the early chapters of the *Topics* we learn that he simply presupposes, as a matter of course and as a device commonly used, a curious kind of mental gymnastics. It consists of either arguing about a proposed problem—and debatable problem—from probable premises, or, if one is attacked in argument, in avoiding self-contradiction. For this kind of philosophical exercise there are always two persons required, plus a problem; one person has the part of the questioner, the other person the part of the respondent or opponent ...
>
> Kapp 1942, p. 12

The discursive nature of Aristotle's logic appears in the first sentence of the *Topics* (100a 18ff) on the purpose of the treatise as:

> ... finding a method by which we shall be able to argue on any problem set before us starting from accepted premises such that when sustaining an argument we shall avoid saying anything self-contradictory.

The effective logic $\mathfrak{E}(\mathfrak{L})$ of a given logic $\mathfrak{L}$ is well-founded rational discourse about $\mathfrak{L}$. It is rational because the rules of debate are formal rules which *introduce* linguistic elements without important philosophical presuppositions from $\mathfrak{L}$; it is well-founded because the justification of a compound statement is nothing other than the discharge of the introduced linguistic elements (junctors) by application of the formal rules which govern the introduction of the logical particles. Discourse in $\mathfrak{E}(\mathfrak{L})$ ultimately reduces to the uncontested primitive facts of $\mathfrak{L}$. It is *about* $\mathfrak{L}$ in the literal sense of being an extension of $\mathfrak{L}$: $\mathfrak{E}(\mathfrak{L})$ is not a metalogic of $\mathfrak{L}$ with symbols *denoting* the elements of $\mathfrak{L}$—these elements occur autonomously in $\mathfrak{E}(\mathfrak{L})$. But whilst $\mathfrak{E}(\mathfrak{L})$ is particularly about $\mathfrak{L}$ the construction method $\mathfrak{E}$ applies to all logics of arbitrary type and therefore treats the elements of each $\mathfrak{L}$ as unanalysed (i.e. atomic) elements of (cf. $Qf_2(\lambda)$). The only information which $\mathfrak{E}(\mathfrak{L})$ inherits from $\mathfrak{L}$ comes from the positive diagram of the logical consequence function.

Our approach is implicit in Kolmogoroff's (1932) proof-theoretical interpretation of intuitionistic logic and is similar in spirit to Hertz (1922, 1923) and the operative logic of Lorenzen (1955). The philosophical basis for Lorenzen's later dialogical construction, in which the junctors correspond to the finite number of possible forms of attack and defence, is best argued

Introduction

in Kamlah and Lorenzen (1967). Lorenzen (1962) employs a dialogical form of logic, but for the investigation of formal arithmetic it is converted to a type of sequent calculus. Lorenz (1968) gave a systematic definition of the dialogue games in which, by suitable but not very clearly motivated modification of the rules for the 'effective logic', a dialogue game for the classical logic is obtained. The modifications allow the proponent P to retract his defences against the opponent's attack: if, in the course of the dialogue, P becomes convinced that a different form of defence is preferable, then retraction and a new start is allowed. The simple tree structure of effective logic is lost by this modification. Stegmüller (1964) gave a related, but more plausible, account of classical logic. Dialogically, the difference between effective logic and classical logic is very small (see also Curry 1965, §4).

Effective propositional and first-order predicate logic (§§10.2–10.6) is formally very closely connected with formal intuitionistic first-order logic. Intuitionist logic and philosophy—the principal opposition movement to the classical establishment—are briefly treated in §10.10. Some other varieties of constructivism in mathematics are touched on in §§10.11 (Bishop) and 10.12.

Our construction of effective propositional and predicate logic (§§10.2–10.6) proceeds via an inductive definition of the extended class of formulas and the extension of logical consequence. The classical propositional logic (Chapter 2) and the many-valued logics (Chapter 4) were based on the assumption that sentences concerned are truth-definite. There are, however, quite simple sentences of number theory which are undecided, e.g. the 'Fermat conjecture'—for all integers $n > 2$, the equation $x^n = y^n + z^n$ has no solution in the positive integers. Why should it be assumed that this sentence has one of the truth-valued true, false when no one knows which is the case? Such an assumption can only follow from an ancient Greek dogma (though even Aristotle had some reservations about its universality; see Bocheński (1961, p. 63)). The hypothesis of truth-definiteness must therefore be abandoned and the junctors redefined. This raises the question as to which junctors should be used. The clauses of the inductive definitions formally introduce junctors into formulas and into the consequence relation. The dialogical sense of inductive definition was explained in Chapter 3.) The 'meaning' of each junctor is just its set of permissible operations as defined by these rules—there is no definition by truth-tables.

The inductive definition of the logical consequence function of effective logic is most easily effected in terms of sequent calculi (Chapter 3, §3.11), thus bringing our development of effective logic into a close relation with Curry's (1963, pp. 184–244) systems of propositional calculus $LA(\mathfrak{S})$ based on a formal system $\mathfrak{S}$. Two related problems occur here. Which junctors should be used? Does the resulting structure yield a logic in the sense of Definition 3.7? The *regular junctors* (§10.5) afford a positive answer: they are exactly those junctors which can be defined in terms of the traditional intuitionistic positive junctors of conjunction, disjunction, and implication.

Negation in effective logic is treated in §10.6. Effective quantifier logics can be built on similar principles (§10.7).

The *pure effective quantifier logic* is formally identical with the intuitionistic predicate logic (§10.8). The philosophical motives behind intuitionistic logic are not usually mentioned. Therefore a brief summary of Brouwer's philosophy is presented in §10.9. The material for this section was mostly taken from the penetrating study by Van Stigt (1971). The remaining section of the chapter deal with alternative approaches to constructive logics.

10.2 Effective extension of a logic

The present purpose of constructing effective logic is to systematize theoretical discourse about a given denumerable logic $\mathfrak{L} = \langle \Sigma, Cn \rangle$ by extending Σ to a larger domain of 'formulas', i.e. compound sentences which are ultimately 'about' the consequence function Cn. Since the method of construction of effective logic is to apply to logics of arbitrary type, the elements in Σ are treated as unanalysed entities, there is no loss of generality in assuming that the logics $\mathfrak{L}$ have a fixed standard set $P = \{p_1, p_2, p_3, \ldots\}$ of elements which can be referred to as *proposition letters*. Thus we assume that $\Sigma = P$ so that Cn is a logical consequence function on P.

To say that the new compound sentences of $\mathfrak{E}(\mathfrak{L})$ are 'ultimately' about Cn means that disputation about formulas is so regulated that objections to a compound formula are rebutted by reference to subformulas. Disputation therefore rests ultimately on the positive 'facts' of $\mathfrak{L}$. The rules governing this disputation are best given in the form of a Gentzen type system of sequents $\mathfrak{S}(\mathfrak{L})$.

The construction of the effective propositional logic begins with the definition of the class $F_{\text{eff}}(\mathfrak{L}) =_{\text{def}} \mathbb{F}(J, P)$ of formulas which are constructed from P by means of symbols which are assumed to be not in P, namely the set $J = \{c_1, c_2, \ldots\}$ of junctors (logical symbols) and the list G of left bracket [, right bracket], and comma , (cf. Chapter 6, §6.6). For the moment there is no commitment to having a finite number of junctors nor to having junctors which are recognizably related to the traditional ones (Chapter 2, §2.2) or those with which $\mathfrak{L}$ is equipped. With each junctor $c \in J$ there is associated the number $n(c)$ of its 'arguments' and non-empty finite sets $R(c), L(c)$ of rules which govern the introduction of c. The junctors will not be interpreted as truth-functions of any kind (cf. Chapter 6, §6.6); whatever 'meaning' they have derives solely from these rules.

The set $F_{\text{eff}}(\mathfrak{L})$ of formulas of $\mathfrak{E}(\mathfrak{L})$ *generated by* P is given by the inductive definition:

(F1) $P \subseteq F_{\text{eff}}(\mathfrak{L})$
(F2) if $A_1, \ldots, A_n \in F_{\text{eff}}(\mathfrak{L})$ and $c \in J$ with $n(c) = n$ then $c[A_1, \ldots, A_n] \in F_{\text{eff}}(\mathfrak{L})$.

Effective extension of a logic

As usual, $|A|$ is the number of occurrences of logical symbols (here, junctors in J) in the formula A. The formulas are logical combinations (cf. Definition 2.1(iv)) of elements of P, i.e. $F_{\text{eff}}(\mathfrak{L}) = \Lambda(P)$. The elements of the original logic $\mathfrak{L}$ are logically unanalysed so that, by definition, $|A| = 0$ for $A \in P$. Further, the syntactical variables $A_0, A_1, A_2, \ldots$ for formulas in F_{eff} and $\Delta, \Gamma, \Theta, \Xi, \ldots$ for finite subsets of $F_{\text{eff}}(\mathfrak{L})$ will be used when the rules are discussed.

We conceive discourse about $\mathfrak{L}$ as a dispute over whether one formula $A_0 \in F_{\text{eff}}(\mathfrak{L})$ is a logical consequence of formulas $A_1, \ldots, A_k \in F_{\text{eff}}(\mathfrak{L})$. Thus, dispute concerns ordered pairs $\langle \{A_1, \ldots, A_k\}, A_0 \rangle$, i.e. sequents (cf. Chapter 3, §3.10) which, as in §3.11, we shall write as $(\{A_1, \ldots, A_k\} : A_0)$ or $(A_1, \ldots, A_k : A_0)$. Any challenge to the claim that A is a logical consequence of $A_1, \ldots, A_k$ can only be met by reference to the rules by which the compound formulas $A_0, A_1, \ldots, A_k$ were introduced into the succedent and the antecedent of the sequent $(A_1, \ldots, A_k : A_0)$. Such an argument should be well-founded, leading ultimately to the simplest claims $(A_1, \ldots, A_k : A_0)$ where now $A_1, \ldots, A_k, A_0 \in P$, i.e. claims which are justified by reference to the data provided by the original logic $\mathfrak{L}$ (in general, as a result of the undecidability of various logics, e.g. Church (1936) and Chapter 2, §2.3.7(iii), negative information is not effectively ascertainable).

Thus the set of *axioms of* $\mathfrak{S}(\mathfrak{L})$ is defined to be $Ax_0 + Ax_{\mathfrak{L}}$.

(i) $Ax_0 =_{\text{def}} \{(\Theta : A_0); A_0 \in \Theta \in \mathbb{P}_\omega(\mathbb{F}(J, P))\}$. These are the *general axioms*—they contain no information about the logical structure of $\mathfrak{L}$ beyond the mere identity of its members.

(ii) $Ax_{\mathfrak{L}} =_{\text{def}} \{(\Theta : A_0); \Theta, A_0 \in \mathbb{P}_\omega(P) \ \& \ A_0 \in Cn(\Theta)\}$. These are the *special axioms*—they contain the assumed positive finite data from the given logic $\mathfrak{L}$.

In the following discussion of rules, if I is a set of natural numbers, $A_I =_{\text{def}} \{A_i; i \in I\}$, the finite set of formula variables whose indices are in I. (Thus if $I = \emptyset$ then $A_I = \emptyset$.)

The *rules of inference* of $\mathfrak{S}(\mathfrak{L})$ include the thinning rule

$$\frac{(\Gamma : A_0)}{(\Gamma, \Delta : A_0)}.$$

The remaining rules concern the introduction of junctors into succedent and antecedent. Associated with each $c \in J$ are the sets $L(c)$ and $R(c)$ of *left-rules* and *right-rules* respectively.

A *right-rule* $R \in R(c)$ for the introduction of the junctor c *into the succedent* is determined by a non-empty *index set* $indR$ which itself is a set of sequents $(I : i)$, where $I, i \subseteq \{1, \ldots, n(c)\}$ (see Chapter 3, §3.11). (In accordance with our conventions concerning sequents, I is allowed to be the empty set.) Further, if I is the finite set $\{n, m, \ldots\}$, $(I : i)$ will also be expressed as

$(n, m, \ldots : i)$. Thus $(\{1\} : 3)$ will also be written as $(1: 3)$. The right-rule R is conveniently expressed as

$$\frac{\{(\Gamma, A_I : A_i);\ (I:i) \in indR\}}{(\Gamma : c(A_1, \ldots, A_{n(c)}))}\ R$$

Thus any instance of the conclusion $(\Gamma : c(A_1, \ldots, A_{n(c)}))$ is directly derivable from the set $\{(\Gamma, A_I : A_i); (I:i) \in indR)$ of premises. For instance, if c is a two-place junctor (i.e. $n(c) = 2$) and $R(c)$ contains a rule R with $indR = \{1: 2)\}$ then R is

$$\frac{(\Gamma, A_1 : A_2)}{(\Gamma : c(A_1, A_2))}.$$

A *left-rule* $L \in L(c)$ *for the introduction of c into the antecedent* is likewise determined by a non-empty *index* set *indL* of sequents $(I:i)$, where $I, i \subseteq \{0, 1, \ldots, n(c)\}$ and so is expressed as

$$\frac{\{(\Gamma, A_I : A_i);\ (I:i) \in indL\}}{(\Gamma, c(A_1, \ldots, A_{n(c)}) : A_0)}\ L$$

Thus for $n(c) = 2$, if $L(c)$ contains a rule L with $indL = \{(\emptyset: 1), (2: 0)\}$, then L is the two-premiss rule

$$\frac{(\Gamma, A_2 : A_0),\ (\Gamma : A_1)}{(\Gamma, c(A_1, A_2) : A_0)}.$$

To avoid circularity we do not allow support for the claim that the formula A follows from the hypothesis Θ to be drawn from premises in which A itself occurs as a hypothesis. Hence we declare that

for all $L \in L(c)$ and all $(I: 0) \in indL$, $0 \notin I$.

It is technically convenient to divide the index set *indL* of a rule $L \in L(c)$ into two exclusive subsets:

$$ind^+(L) =_{def} \{(I: i) \in indL;\ i \neq 0\}$$
$$ind^0(L) =_{def} \{(I: i) \in indL;\ i = 0\}.$$

Clearly, for each n there are finitely many possible rules of the above type for introducing an n-place junctor into the antecedent or succedent. For each junctor c there is at least one (possibly more than one) rule in $R(c) + L(c)$.

EXAMPLE 10.1. Here are three important examples of two-place junctors. In accordance with our usual convention, we use the expressions $A_1 A_2$, $A_1 \vee A_2$, and $A_1 \supset A_2$ as abbreviations of $\wedge(A_1, A_2)$, $\vee(A_1, A_2)$, and $\supset(A_1, A_2)$ respectively.

(i) ∧. $L(\wedge) = \{L_1, L_2\}$, $R(\wedge) = \{R\}$

Succedent:

R. $indR = \{(\emptyset: 1), (\emptyset: 2)\}$

$$\frac{(\Gamma : A_1), \quad (\Gamma : A_2)}{(\Gamma : A_1 A_2)}$$

Antecedent:

L_1. $ind^+(L_1) = \emptyset$, $ind^0(L_1) = \{(1:0)\}$

$$\frac{(\Gamma, A_1 : A_0)}{(\Gamma, A_1 A_2 : A_0)}$$

L_2. $ind^+(L_2) = \emptyset$, $ind^0(L_2) = \{(2:0)\}$

$$\frac{(\Gamma, A_2 : A_0)}{(\Gamma, A_1 A_2 : A_0)}$$

(ii) ∨. $L(\vee) = \{L\}$. $R(\vee) = \{R_1, R_2\}$

Succedent:

R_1. $indR_1 = \{(\emptyset: 1)\}$

$$\frac{(\Gamma : A_1)}{(\Gamma : A_1 \vee A_2)}$$

R_2. $indR_2 = \{(\emptyset: 2)\}$

$$\frac{(\Gamma : A_2)}{(\Gamma : A \vee A_2)}$$

Antecedent:

L. $ind^+(L) = \emptyset$, $ind^0(L) = \{(1:0), (2:0)\}$

$$\frac{(\Gamma, A_1 : A_0), \quad (\Gamma, A_2 : A_0)}{(\Gamma, A_1 \vee A_2 : A_0)}$$

(iii) ⊃. $L(\supset) = \{L\}$, $R(\supset) = \{R\}$

Antecedent:

L. $ind^+(L) = \{(\emptyset: 0)\}$, $N(L) = \{(2:0)\}$

$$\frac{(\Gamma : A_1), \quad (\Gamma, A_2 : A_0)}{(\Gamma, A_1 \supset A_2 : A_0)}$$

Succedent:

R. $indR = \{(1:2)]$

$$\frac{(\Gamma, A_1 : A_2)}{(\Gamma : A_1 \supset A_2)}$$

These rules govern conjunction, disjunction, and implication in the classical and intuitionistic propositional calculus. ◆

The set of sequents together with the set of such right- and left-rules constitute a formal system $\mathfrak{S}(\mathfrak{L})$ in the sense of Definition 3.15. A sequent will be called *valid* (in $\mathfrak{S}(\mathfrak{L})$) if it is derivable by means of the chosen set of rules from the axioms: let $Vd(\mathfrak{L})$ denote the set of valid sequents. $Vd(\mathfrak{L})$ is determined by the initial logic $\mathfrak{L}$ and the set J of junctors together with their associated sets of rules.

$Vd(\mathfrak{L})$ was defined inductively and so has a constructive and dialogical sense (Chapter 2, §2.2.5, and Lorenzen 1961a, 1962). The premisses of a rule are sequents whose elements are subformulas of the conclusions. Thus in defence of the claim 'A follows from Δ' the proponent can only make further claims 'B follows from Γ' where the formulas in $\Gamma + \{B\}$ are subformulas of those in $\Delta + \{A\}$. There are finitely many such possible pairs (Γ, B)—the precise number can easily be computed from (Δ, A)—and so *by barring repetitions* (i.e. avoiding circular argument) the attempt to justify 'A follows from Δ' terminates in a predictable number of steps. $\mathfrak{S}(\mathfrak{L})$ depends on *arbitrarily* chosen rules of inference, and there is no reason, apart from failure to communicate, why one should not conduct the discourse about $\mathfrak{L}$ in such arbitrary terms. However, by placing certain restrictions on the rules the debate takes on a more familiar aspect which entitles us to claim that we have a logic. Now $Vd(\mathfrak{L})$ is the graph of a finitary set-to-set function $Cn_{Vd(\mathfrak{L})}$: can this be extended to the logical consequence function so that we actually have a logic? By Definition 3.18(i), this is the question of whether the structure $J(\mathfrak{L}) =_{\text{def}} \langle F_{\text{eff}}(\mathfrak{L}), Vd(\mathfrak{L}) \rangle$ is a G-structure, i.e. whether **D1**$^\omega$ & **D2**$^\omega$ & **D3**$^\omega$ (or, by Example 3.19, **H1** and **H2**) holds. Our task, then, is to frame some conditions under which $J(\mathfrak{L})$ is a G-structure for all logics $\mathfrak{L}$. Now **D1**$^\omega$ holds by the definition of the set of axioms of $\mathfrak{S}(\mathfrak{L})$ and **D2**$^\omega$ holds because $\mathfrak{S}(\mathfrak{L})$ is equipped with the thinning rule. There remains the following.

D3$^\omega$ (cut):   For all $\Delta, \Gamma \in \mathbb{P}_\omega(F_{\text{eff}}(\mathfrak{L}))$ and $A, B \in F_{\text{eff}}(\mathfrak{L})$

$$(\Delta : A) \in Vd \ \& \ (\Gamma, A : B) \in Vd \ \Rightarrow \ (\Delta, \Gamma : B) \in Vd.$$

(10.1)

10.3 The cut condition

The purpose of this section is to impose such conditions on the rules that $J(\mathfrak{L})$ satisfies (10.1). So far no conditions have been specified which relate a left-rule of a junctor to a right-rule. Such a condition is given in the definition below. It is framed in terms of the index sets of rules. They are finite sets of sequents whose elements are natural numbers and accordingly can be transformed by certain rules of inference ('transformation rules'—Carnap), in this case, cut and thinning.

DEFINITION 10.1. A junctor c satisfies the *cut condition* if for each $R \in R(c)$ and each $L \in L(c)$ the sequent $(\emptyset: 0)$ is derivable by cut and thinning from $ind R + ind L$.

EXAMPLE 10.2. The junctors $\wedge$, $\vee$, $\supset$ of Example 10.1 each satisfy their corresponding cut conditions. For instance, in (i) $ind R = \{(\emptyset: 1), (\emptyset: 2)\}$, $ind L_1 = \{(1: \emptyset)\}$, $ind L_2 = \{(2: 0)\}$. Then

$$ind R + ind L_1 = \{(\emptyset: 1), (\emptyset: 2), (1: 0)\}$$
$$ind R + ind L_2 = \{(\emptyset: 1), (\emptyset: 2), (2: 0)\}.$$

$(\emptyset, 0)$ is derivable from each of these by cut. ◆

Throughout the rest of this chapter for $\Delta, A \in \mathbb{P}_\omega(F)$,

$$\Delta/A =_{\text{def}} \begin{cases} \Delta & \text{if } A \notin \Delta \\ \Delta - \{A\} & \text{if } A \in \Delta. \end{cases}$$

LEMMA 10.1. *The axioms of $\mathfrak{F}$ are closed under applications of cut.*

Proof. Suppose $(\Gamma : A)$ and $(\Delta : B)$ are axioms of $\mathfrak{F}$. Then $A \in \Gamma$ and $B \in \Delta$. If $A = B$ then $B \in \Gamma$ so that $(\Gamma, \Delta/A : B)$ is derivable by thinning from the axiom $(B : B)$. But if $A \neq B$ then $B \in \Delta/A$, so again $(\Gamma, \Delta/A : B)$ is derivable from $(B : B)$. ∎

THEOREM 10.2 (cut admissibility). *If each junctor $c \in J$ satisfies its cut condition then (10.1) holds.*

Proof. The proof of this theorem is modelled on the proof of Gentzen's *Hauptsatz* or 'normal form theorem' given by Kleene (1962, Chapter 15). Certain places in our proof are marked with a dagger (†). These signals should be ignored at first reading—they indicate an expansion port to which various add-on devices will later be connected to handle quantifiers and modalities (Chapter 11).

The statement of the theorem concerns the admissibility of the cut rule in $\mathfrak{S}(\mathfrak{L})$. It can be conceived as a statement about proofs in the following way. Given two proofs Π_1, Π_2, assumed to be arranged in tree form, there can be extracted the unique level 0 sequents (i.e. the *end-sequents*) $(\Gamma : A), (\Delta : B)$ respectively. From these there can be found the sequent $(\Gamma, \Delta/A : B)$ which we shall call $M(\Pi_1, \Pi_2)$. Then (10.1) can be rephrased as follows.

S: for all $\mathfrak{S}$-proofs $\Pi_1, \Pi_2, M(\Pi_1, \Pi_2) \in Vd.$ (10.2)

The proof of S is based on a double induction on certain parameters computed from Π_1 and Π_2. In fact, the proof defines a procedure which constructs a proof of $M(\Pi_1, \Pi_2)$ from Π_1 and Π_2.

Let Π_1, Π_2 be two proofs arranged in tree form whose *end-sequents* (level 0) are $(\Gamma : A)$ and $(\Delta : B)$ respectively. The pair (Π_1, Π_2) will be called *special* if $A \in \Gamma$ or $A \notin \Delta$. If (Π_1, Π_2) is *special* then, if $A \in \Gamma$, $\Gamma + \Delta/A = \Gamma + \Delta$ so $M(\Pi_1, \Pi_2) = (\Gamma, \Delta/A : B)$ is derivable by thinning from $(\Delta : B)$ and therefore $(\Gamma, \Delta/A : B) \in Vd$; if $A \notin \Delta$ then $\Delta/A = \Delta$ so that $(\Gamma, \Delta/A : B) = (\Gamma, \Delta : B)$ and can therefore be derived from $(\Delta : B)$ by thinning—again $(\Gamma, \Delta/A : B) \in Vd$. Thus if (Π_1, Π_2) is *special* then $M(\Pi_1, \Pi_2) \in Vd$. So S is trivial for special pairs of proofs.

Let (Π_1, Π_2) be a non-special pair of proofs arranged in tree form whose end-sequents (level 0) are $(\Gamma : A)$ and $(\Delta : B)$ respectively. Let l be the greatest number such that there exists a branch in Π_1 whose sequents of levels $\leqslant l$ each contain A as succedent: l is the *left-rank*. Let r be the greatest number such that there exists a branch in Π_2 whose sequents of levels $\leqslant r$ each contain A in the antecedent: r is the *right-rank*. As (Π_1, Π_2) is non-special, $A \in \Delta$ so that r is well defined.

level $l+1$	$\ldots(\Theta : C)$		
level l	$(\Gamma_l : A)$		
	$\vdots$	$(\Delta_r + 1 : B)$	level $r+1$
	$\vdots$	$(\Delta_r, A : B)$	level r
level 2	$\vdots$		$\vdots$
level 1	$(\Gamma_1 : A)$	$(\Delta_1, A : B)$	
level 0	$(\Gamma : A)$	$(\Delta : B)$	(note: $A \in \Delta$)
	Π_1	Π_2	

The two parameters ρ (*rank* of (Π_1, Π_2)) and γ (*grade* of (Π_1, Π_2)) required for induction are defined by

$$\rho(\Pi_1, \Pi_2) =_{\text{def}} l + r \qquad \gamma(\Pi_1, \Pi_2) =_{\text{def}} |A|. \qquad (10.3)$$

The theorem is established by proving the following statement $S(n, m)$ by induction on rank n and grade m:

$S(n, m)$: for all proof pairs (Π_1, Π_2) in $\mathfrak{S}$ such that $\rho(\Pi_1, \Pi_2) \leqslant n$ and $\gamma(\Pi_1, \Pi_2) \leqslant m$,

$$M(\Pi_1, \Pi_2) \in Vd.$$

The cut condition   319

Then S (i.e. (10.2)) is equivalent to

$$S(n, m) \quad \text{for all } n \geq 0 \text{ and } m \geq 0. \tag{10.4}$$

Clearly, in the proof of $S(n, m)$ we can ignore special pairs of proofs.

The proof of (10.4) is established by double induction on rank and grade. The four steps in the proof require the proof of

$$S(0, 0) \tag{10.5}$$

$$(r \leq n)S(r, 0) \Rightarrow S(n + 1, 0) \tag{10.6}$$

$$(m \leq g)(r)S(r, m) \Rightarrow S(0, g + 1) \tag{10.7}$$

$$(r \leq n)S(r, g + 1) \Rightarrow S(n + 1, g + 1). \tag{10.8}$$

The theorem is derived from these statements as follows. From (10.5) and (10.6), by complete induction n, there follows

$$(n)S(n, 0). \tag{10.9}$$

From (10.7) and (10.8), by complete induction on n, there follows

$$(m \leq g)((r)S(r, m)) \Rightarrow (r)S(r, g + 1) \tag{10.10}$$

(10.4) follows from (10.9) and (10.10), by complete induction on g.

Proof of (10.5). Let (Π_1, Π_2) be a non-special pair of proofs with lowest rank and grade, i.e. $\rho = 0$ and $\gamma = 0$. Suppose their end-sequents are $(\Gamma : A_0)$ and $(\Delta : C)$ respectively. Then

$$A_0 \notin \Gamma \quad \text{and} \quad A_0 \in \Delta \tag{10.11}$$

The conditions $\rho = 0$ and $\gamma = 0$ impose certain restrictions on the form of the end-sequents.

ad $(\Gamma : A_0)$. This sequent cannot be the conclusion of a right-rule as $\gamma = |A_0| = 0$. Since $\rho = 0$ it cannot be the conclusion of a thinning or a left-rule, one of whose premisses contains A_0 in the succedent. By (10.11) $(\Gamma : A_0)$ is not an axiom. Hence

(†1) $\Gamma : A_0$ is the conclusion of a left-rule from premisses none of which contain A_0 in the succedent.

ad $(\Delta : C)$. By (10.11) $A_0 \in \Delta$. As $\rho = 0$, $(\Delta : C)$ is either an axiom or the conclusion of a rule by which A_0 is introduced into the antecedent—thus either thinning or a left-rule. As $\gamma = |A_0| = 0$ the latter case is ruled out. Hence

$(A : C)$ is an axiom or the conclusion of a thinning.

(†1) Since $(\Gamma : A_0)$ is the conclusion of a left-rule $L \in L(c)$, the premisses of the application of L are $(\Gamma', A_I : A_i)$, say, with $(I: i) \in indL$, $i \neq 0$, and $\Gamma = \Gamma' + \{c[A_1, \ldots, A_n]\}$, where $n = n(c)$:

$$\left. \frac{\{(\Gamma', A_I : A_i); \quad (I: i) \in indL\}}{\underbrace{(\Gamma', c[A_1, \ldots, A_n] : A_0)}_{\Gamma}} \; L \; \right\} \Pi_1$$

But then $(\Gamma', c[A_1, \ldots, A_n] : C)$ is directly derivable from these premisses via L for *any* formula C. Hence $(\Gamma : C) \in Vd$. By thinning $(\Gamma, \Delta/A_0 : C) \in Vd$. This completes the proof of (10.5).

Proof of (10.6). We deduce that $S(n + 1, 0)$ from the supposition

$$(r \leqslant n)S(r, 0) \tag{10.12}$$

Let (Π_1, Π_2) be a non-special pair of proofs with rank $n + 1$ and grade 0, i.e. $\rho = n + 1, \gamma = 0$. Suppose their end-sequents are $(\Gamma : A_0)$ and $(\Delta : C)$ respectively. Then $\rho \geqslant 1$ so, by (10.3), there are two cases to consider: (i) $l \geqslant 1$, and (ii) $r \geqslant 1$.

Case (i) $l \geqslant 1$. Then in the proof Π_1, $(\Gamma : A_0)$ is derived from premisses which include a sequent in which A_0 occurs in the antecedent. Thus $(\Gamma : A_0)$ is the conclusion of (i)(a) a thinning rule, or (i)(b) a left-rule $L \in L(c)$. It cannot be the conclusion of a right-rule as such rules make the logical length of the succedent >0; here we are assuming $\gamma = 0$. Now consider these two subcases.

Subcase (i)(a)

$$\text{level 1} \quad \left. \frac{(\Theta : A_0)}{(\Gamma : A_0)} \quad \text{thinning} \right\} \Pi_1$$

with Π' above $(\Theta : A_0)$.

In Π_1, $(\Gamma : A_0)$ is derived from $(\Theta : A_0)$, say, by thinning. Then in the proof tree Π_1 there can be found a proof-tree Π' whose end-sequent is $(\Theta : A_0)$ such that $\rho(\Pi', \Pi_2) < \rho(\Pi_1, \Pi_2)$. By (10.12), $(\Theta, \Delta/A_0 : C) \in Vd$. But $\Gamma \supseteq \Theta$. Hence by thinning, $(\Gamma, \Delta/A_0 : C) \in Vd$.

(†2) Subcase (i)(b) In the proof Π_1, $(\Gamma : A_0)$ is derived by a rule $L \in L(c)$. Put $n = n(c)$. Thus $\Gamma = \Theta + \{c[A_1, \ldots, A_n]\}$ and the premisses of this application of L are $\{(\Theta, A_I : A_i); (I: i) \in indL\}$. For each $(I: i) \in indL$,

$(\Theta, A_I : A_i)$ has a subproof $\Pi_{1,8(I,i)}$ in Π_1.

$$L \frac{\{(\Theta, A_K: A_0); (K:0) \in Ind^0 L\} \quad \{(\Theta, A_I: A_i); (I:i) \in Ind^+ L\}}{\underbrace{(\Theta, c[A_1, \ldots, A_n] : A_0)}_{\Gamma}} \left.\begin{array}{c} \Pi_{1,(K,0)} \quad\quad \Pi_{1,(I,i)} \\ \\ \\ \end{array}\right\} \Pi_1$$

Consider the premisses in this application of rule L. For $(K:0) \in ind^0 L$ the subproof $\Pi_{1,(K,0)}$ in Π_1 of $(\Theta, A_K : A_0)$, where $(I, i) \in ind L$, has lower left-rank. Thus $\rho(\Pi_{1,(K,0)}, \Pi_2) < \rho(\Pi_1, \Pi_2)$. By (10.12).

$$(\Theta, A_K, \Delta/A_0 : C) \in Vd \qquad ((K:0) \in ind^0 L). \qquad (10.13)$$

For $(I:i) \in ind^+ L$, $(\Theta, A_I : A_i) \in Vd$. By thinning

$$(\Theta, A_I, \Delta/A_0 : A_i) \in Vd \qquad ((I:i) \in ind^+ L) \qquad (10.14)$$

From (10.13) and (10.14) an application of L gives

$$(\Theta, \Delta/A_0, c[A_1, \ldots, A_n] : C) \in Vd,$$

i.e. $(\Gamma, \Delta/A_0 : C) \in Vd$.

Case (ii) $r \geq 1$ Then the end-sequent $(A : C)$ in the proof Π_2 is the conclusion of (a) thinning, or (†3)(b) a left-rule, or (†4)(c) a right-rule
Subcase (ii)(a)

$$\begin{array}{rl} & \Pi'_2 \\ \text{level 1} & (\Theta : C) \\ \text{level 0} & (A : C) \end{array} \left.\begin{array}{c} \\ \\ \end{array}\right\} \Pi_2$$

$(A : C)$ is directly derived in Π_2 from the premiss $(\Theta : C)$ by thinning, where $\Delta \supseteq \Theta$ and, since $r \geq 1$, $A_0 \in \Theta$. Let Π'_2 be the subproof in Π_2 whose end-sequent is $(\Theta : C)$. Then $\rho(\Pi_1, \Pi'_2) < \rho(\Pi_1, \Pi_2)$. By (10.12) $(\Gamma, \Theta/A_0 : C) \in Vd$. But $\Delta/A_0 \supseteq \Theta/A_0$. By thinning, $(\Gamma, \Delta/A_0 : C) \in Vd$.
(†3) Subcase (ii)(b)

$$\begin{array}{rl} & \Pi'_2 \\ \text{level 1} & (\Theta, B_I : B_i) \quad (\Theta, B_K : C) \\ \text{level 0} & \underbrace{(\Theta, c[B_1, \ldots, B_n] : C)}_{\Delta} \end{array} \left.\begin{array}{c} L \\ \end{array}\right\} \Pi_2$$

$(A : C)$ is directly derived in Π_2 by an application of a rule $L \in L(c)$. Thus $\Delta = \Theta + \{c[B_1, \ldots, B_n]\}$. As $\gamma = 0$, $|A_0| = 0$, so $A_0 \neq c[B_1, \ldots, B_n]$. Hence $A \in \Theta$. The premises of the application of L are the two sets of sequents

$$\{(\Theta, B_I : B_i) \; ; \; (I: i) \in ind^+ L\}$$
$$\{(\Theta, B_K : C) \; ; \; (K: 0) \in ind^\circ L\}.$$

Each member of the first set is the end-sequent of a subproof Π'_2, in Π_2 say; likewise each member of the second set has a subproof in Π_2. But $\rho(\Pi_1, \Pi'_2) < \rho(\Pi_1, \Pi_2)$. So by (10.12), from $(\Gamma : A) \in Vd$ and $(\Theta, B_I : B_i) \in Vd$ there follows $(\Gamma, (\Theta, B_I)/A_0 : B_i) \in Vd$ and by thinning, if necessary, $(\Gamma, \Theta/A_0, B_I : B_i) \in Vd$. As $(\Theta, B_K : C) \in Vd$, by L,

$$(\Gamma, \Theta/A_0, c[B_1, \ldots, B_n] : C) \in Vd,$$

i.e. $(\Gamma, \Delta/A_0 : C) \in Vd$.
(†4) Subcase (ii)(c).

$$\left. \begin{array}{ll} & \Pi'_2 \\ \text{level 1} & \dfrac{(\Delta, B_J : B_j)}{(\Delta : c[B_1, \ldots, B_n])} \; R \\ \text{level 2} & \underbrace{}_{C} \end{array} \right\} \Pi_2$$

$(A : C)$ is directly derived in Π_2 by an application of a rule $R \in R(c)$ so that $C = c[B_1, \ldots, B_n]$ and the premises of the application of R are the sequents $(\Delta, B_J : B_j)$ for $(J: i) \in indR$. Each such sequent has a subproof Π'_2 of Π such that $\rho(\Pi_1, \Pi'_2) < \rho(\Pi_1, \Pi_2)$. Then, by (10.12), $(\Gamma, \Delta/A_0, B_J : B_j) \in Vd$. By R, $(\Gamma, \Delta/A_0 : c[B_1, \ldots, B_n])$.

Hence $S(n + 1, 0)$ holds. This completes the proof of (10.6).

Proof of (10.7). Suppose

$$(m \leqslant g)(r)S(r, m). \tag{10.15}$$

We deduce that $S(0, g + 1)$.

Let (Π_1, Π_2) be a non-trivial proof-pair of rank 0 and grade $g + 1$. $\gamma(\Pi_1, \Pi_2) = g + 1$ and $\rho(\Pi_1, \Pi_2) = 0$. Then Π_1, Π_2 are proofs of (Γ, A_0) and $(A : C)$, say, respectively, where $|A_0| = g + 1$, $A_0 \notin \Gamma$, and $A_0 \in \Delta$. Then A_0 is a compound formula: $A_0 = c[B_1, \ldots, B_n]$ where $n = n(c)$, say, and $|B_1|, \ldots, |B_n| \leqslant g$. Since $\rho(\Pi_1, \Pi_2) = 0$ and $|A_0| = g + 1$, $(\Gamma : A_0)$ is directly derived in Π_1 from premises which do not contain A_0 in the succedent. Thus $(\Gamma : A_0)$ is the conclusion in Π_1 of (†5)(i) a right-rule or (ii) a left-rule.

Case (i) Here $A_0 = c[B_1, \ldots, B_n]$ and $(\Gamma : A_0)$ is the conclusion of an application of the right-rule $R \in R(c)$ from the set of premises $\{(\Gamma, B_J : B_j); (J:j) \in indR\}$, each of which has a subproof in Π_1. Further, since $\rho(\Pi_1, \Pi_2) = 0$ and $|A_0| = g + 1$, $(A : C)$ is directly derived in Π_2 from premises which do not contain A_0 in the antecedent. Thus $(A : C)$ is the conclusion in Π_2 of a left-rule $L \in L(c)$. i.e. of premises $\{(\Psi, B_I : B_i); (I:i) \in indL\}$ where $\Delta = \Psi, c[B_1, \ldots, B_n]$; each of these premises has a subproof of Π_2.

$$\Pi_1 \begin{cases} \text{level 1} \\ \text{level 0} \end{cases} R \frac{(\Gamma, B_J : B_j) \quad ((J:j) \in indR)}{\underbrace{(\Gamma : c[B_1, \ldots])}_{A_0}} \quad \frac{(\Psi, B_I : B_i) \quad ((I:i) \in indL)}{\underbrace{(\Psi, c[B_1, \ldots] : C)}_{A_0}} L \Bigg\} \Pi_2$$

Now the junctor c satisfies the cut condition (Definition 10.1). Thus $(\emptyset : 0)$ has a derivation Ω, say, from $IndR + indL$, which uses only cut and thinning. In Ω substitute B_i for i with $i > 0$ and C for 0. This yields a derivation Ω' in $\mathfrak{S}$, again using only cut and thinning of $(\emptyset : C)$ from $\{(B_J : B_j); (J:j) \in indR\}$ and $\{(B_I : B_i); (I:i) \in indL\}$. ($\Omega'$ is not a derivation in $\mathfrak{S}(\Omega)$ since cut is used). The formulas eliminated by the cut rule here are amongst the B_i, i.e. formulas of logical length $\leq g$. Then by adding Γ and Ψ to the appropriate antecedents of sequents in Ω' and interpolating some applications of thinning if necessary, we obtain a derivation $S_1, S_2, \ldots, S_n$ of $(\Psi, \Gamma : C)$ from the set

$$\Xi = \{(\Gamma, B_J : B_j); (J:j) \, indR\} + \{(\Psi, B_I : B_i); (I:i) \in indL\}$$

of sequents using only cut and thinning and in which again the formulas eliminated by cut are amongst the B_i, i.e. have logical length $\leq g$. Here $S_n = (\Psi, \Gamma : C)$ and for each r with $1 \leq r \leq n$ one of the following three conditions holds:

(a) $S_r \in \Xi$;
(b) S_r is derived by thinning from S_s for some $s < r$;
(c) S_r is derived by cut from S_u, S_v for some $u, v < r$ and the formula so eliminated is one of the B_i.

It easily follows from (10.15) by induction on r that $S_r \in Vd$ for $1 \leq r \leq n$. Hence $(\Psi, \Gamma : C) \in Vd$, i.e. $(\Gamma, \Delta/A_0 : C) \in Vd$.

Case (ii) $(\Gamma : A_0)$ is the conclusion in Π_1 of a left-rule $L \in L(c)$. Thus $\Gamma = \Psi, c[A_1, \ldots, A_n]$ and $(\Gamma : A_0)$ is directly derived in Π_1 from $\{(\Psi, A_I : A_i); (I:i) \in indL\}$. Since the rank is zero A_0 does not appear in the succedent of any of the premises. Hence $i \neq 0$ for all $(I:i) \in indL$. But then an application of this rule to these premises taking $A_0 = C$ gives a proof of $(\Gamma : C)$. By thinning, $(\Gamma, \Delta/A_0 : C) \in Vd$.

Hence $S(0, g + 1)$ holds. This completes the proof of (10.7).

Proof of (10.8). The deduction of $S(n+1, g+1)$ from the supposition $(r \leq n)S(r, g+1)$ is similar to the proof of (10.6) but with appeal to (10.15) instead of (10.11). This completes the proof of (10.8).

The proof of Theorem 10.2 is now complete. ∎

COROLLARY 10.3. *If all the junctors in J of $\mathfrak{S}(\mathfrak{L})$ satisfy the cut condition then $J(\mathfrak{L}) =_{\text{def}} \langle F_{\text{eff}}(\mathfrak{L}), Vd(\mathfrak{L}) \rangle$ is a G-structure.*

From the G-structure there can be constructed the logic $\langle F, Cn^{\clubsuit}_{Vd} \rangle$ as in Definition 3.18(ii). By Definition 3.1(iv), for all $\Delta, A \subseteq F$,

$$A \in Cn^{\clubsuit}_{Vd}(\Delta) \quad \text{iff} \quad (\Delta' : A) \in Vd \quad \text{for some } \Delta' \in \mathbb{P}_\omega(\Delta). \quad (10.16)$$

10.4 The triviality condition

Another condition on a junctor c, which is independent of the cut condition, is the *triviality condition*, which requires that c be a connective in the sense of Definition 3.11.

DEFINITION 10.2. A junctor c is said to *satisfy the c-triviality condition* if it satisfies at least one of the conditions $T_1(c), T_2(c)$.

$T_1(c)$: there exists a left-rule $L \in L(c)$ such that $\Phi_1(L)$ holds, where $\Phi_1(L)$ is equivalent to (i) and (ii):
    (i) $\text{ind}^+(L) = \emptyset$;
    (ii) for each $(K:0) \in \text{ind}^\circ(L)$ there exists right-rule $R \in R(c)$ such that $i \in K$ for all $(I:i) \in \text{ind}R$.

$T_2(c)$: there exists a right-rule $R \in R(c)$ such that $\Phi_2(R)$ holds, where $\Phi_2(R)$ is equivalent to: for each $(I:i) \in \text{ind}R$ there exist a left-rule $L \in L(c)$ such that $j \in I$ and $i \in K$ for all $(K:0) \in \text{ind}^\circ L$ and all $(Q:j) \in \text{ind}^+ L$.

EXAMPLE 10.3.

(i) The junctors $\wedge, \vee, \supset$ of Example 10.1 satisfy the cut condition (Example 10.2) and the triviality conditions. $\wedge, \vee$ satisfy T_1 and $\supset$ satisfies T_2 but not T_1.
(ii) $T_1(c) \Rightarrow T_2(c)$. Thus c fails the triviality condition iff $-T_2(c)$. Hence a junctor c with one left-rule L and one right-rule R fails the triviality condition iff there exist $(I:i) \in \text{ind}R, (K:0) \in \text{ind}^\circ L$, and $(Q:j) \in \text{ind}^+ L$ such that $j \notin I$ or $i \notin K$. Thus the two-place junctor c defined by the rules L and R

$$\frac{(\Gamma, A_2 : A_1) \quad (\Gamma : A_2)}{(\Gamma : c[A_1, A_2])} R$$

$$\frac{(\Gamma, A_2 : A_0) \quad (\Gamma, A_2 : A_1)}{(\Gamma, c[A_1, A_2] : A_0)} L$$

satisfies the cut condition but not the triviality condition ($I = K = Q = \{2\}$, $i = j = 1$). ◆

LEMMA 10.4. Let c be an n-place junctor which satisfies the triviality condition. Then, for all $A_1, \ldots, A_n, B_1, \ldots, B_n \in F_{\text{eff}}(\mathfrak{L})$,

$$(c[A_1, \ldots, A_n] : c[B_1, \ldots, B_n])$$

is derivable in $\mathfrak{S}(\mathfrak{L})$ from $\Psi = \{(A_i : B_i), (B_i : A_i); 1 \leq i \leq n\}$ without general or special axioms.

Proof. Let $A_1, \ldots, A_n, B_1, \ldots, B_n \in F$. Suppose that $\Phi_1(L)$ for some $L \in L(c)$. Then $(c[A_1, \ldots] : c[B_1, \ldots])$ is the conclusion of an application of L with the set of premises $\{(A_K : c[B_1, \ldots]); (K: 0) \in \text{ind}^0 L\}$. For each of these, by $\Phi_1(L)$ there exists $R_K \in R(c)$ such that $i \in K$ for all $(I: i) \in \text{ind} R$. Now $(A_K : c[B_1, \ldots])$ is directly derivable from $\{(A_K, B_I : B_i); (I: i) \in \text{ind} R_K\}$ by R_K. As $i \in K$, each of these premises is directly derivable by thinning from a member of Ψ. Hence $(c[A_1, \ldots] : c[B_1, \ldots])$ is derivable in $\mathfrak{S}(\mathfrak{L})$ from Ψ.

$$\frac{\begin{array}{c}(A_i : B_i)\\ \hline (A_K, B_I) : B_i)\end{array} \quad ((I: i) \in \text{ind} R_K)}{\{A_K : c[B_1, \ldots]) \quad ((K: 0) \in \text{ind}^0 L)} \qquad \begin{array}{c}\Psi\\ \text{thinning}\\ R_K\end{array}$$
$$\overline{(c[A_1, \ldots] : c[B_1, \ldots])} \qquad L$$

Similarly for the case where $\Phi_2(R)$ holds for some $R \in R(c)$. ■

THEOREM 10.5. Let c be an n-place junctor which satisfies the triviality condition. Let $\Delta, A_1, \ldots, A_n, B_1, \ldots, B_n \in \mathbb{P}_\omega(F_{\text{eff}}(\mathfrak{L}))$. If $(\Delta, A_i : B_i)$ and $(\Delta, B_i : A_i)$ are provable without special axioms for $1 \leq i \leq n$ then $(\Delta, c[A_1, \ldots] : c[B_1, \ldots])$ is also provable without special axioms.

Proof. Suppose that $(\Delta, A_i : B_i)$ and $(\Delta, B_i : A_i)$ are both provable without special axioms for $1 \leq i \leq n$. By Lemma 10.4 there is a derivation D in $\mathfrak{S}(\mathfrak{L})$ of $(c[A_1, \ldots] : c[B_1, \ldots])$ from Ψ (of Lemma 10.4) without special axioms. Append Δ to the antecedents of all sequents in D. This gives a derivation of $(\Delta, c[A_1, \ldots] : c[B_1, \ldots])$ from

$$\{(\Delta, A_i : B_i), (\Delta, B_i : A_i), (\Delta, A_i : A_i), (\Delta, B_i : B_i); 1 \leq i \leq n\}$$

without special axioms. But $(\Delta, A_i : B_i)$ and $(\Delta, B_i : A_i)$ are derivable in $\mathfrak{S}(\mathfrak{L})$ by hypothesis. Further, $(\Delta, A_i : A_i)$ and $(\Delta, B_i : B_i)$ are derivable by thinning from axioms of $\mathfrak{S}$. Hence $(\Delta, c[A_1, \ldots] : c[B_1, \ldots])$ is derivable without special axioms. ■

Corollaries 10.6 and 10.7 follow immediately.

COROLLARY 10.6. If the only junctors which occur in the formula A satisfy the triviality condition then $(\Delta, A : A)$ is provable without the special axioms.

COROLLARY 10.7. If all the junctors in J satisfy the cut and the triviality conditions then the junctors are strong connectives in the logic $\langle \mathbb{F}(J, P), Cn^{\heartsuit}_{Vd(\mathfrak{L})}\rangle$.

10.5 Regular junctors and effective positive logic

In view of Corollary 10.7 we make the following definition.

DEFINITION 10.3. A junctor is *regular* if it satisfies the triviality and cut conditions.

EXAMPLE 10.4. The junctors $\wedge, \vee, \supset$ (Example 10.1) are regular. This set of junctors will be denoted by B and called the *basic* set of junctors. The set of effective formulas built from B together with the set-to-set function $Cn^{\heartsuit}_{Vd}$ also constructed from B is a compact logic in which the basic junctors are strong connectives (Corollary 10.7). ◆

Our aim now is to show that if the junctors in the set J saisfy the cut condition (so that $\langle \mathbb{F}(J, P), Cn^{\heartsuit}_{Vd}\rangle$ is a logic) and if $B \subseteq J$ then every formula constructed from regular junctors is logically equivalent to one constructed from basic junctors only.

Let I be a non-empty set of natural numbers: $I = \{a, b, \ldots, d\}$, say, where $a < b < \cdots d$. If $A_0, A_1, A_2, \ldots$ is a list of formulas in $\mathbb{F}(J, P)$ then

$$\bigwedge \{A_i; i \in I\} =_{\text{def}} \bigwedge A_I =_{\text{def}} [A_a \wedge [A_b \wedge [\cdots \wedge A_d]]]$$
$$\bigvee \{A_i; i \in I\} =_{\text{def}} \bigvee A_I =_{\text{def}} [A_a \vee [A_b \vee [\cdots \vee A_d]]].$$

THEOREM 10.8. Let J be a set of junctors which satisfy the cut condition and suppose that $B \subseteq J$. Let c be an n-place regular junctor in J and let $p_1, \ldots, p_n$ be distinct atomic formulas. Then a formula A can be constructed such that
(i) $A \in \mathbb{F}(B; \{p_1, \ldots, p_n\})$,
(ii) A is logically equivalent to $c[p_1, \ldots, p_n]$.

Proof. c satisfies the triviality condition. Hence either $T_1(c)$ or $T_2(c)$ hold.

If $T_1(c)$ then there is a left-rule $L \in L(c)$ such that $ind^+ L = \emptyset$ and for each $(K: 0) \in ind^\circ L$ there exists $R_K \in R(c)$ such that $i \in K$ for all $(I: i) \in indR_K$. Then since index sets are non-empty, $indR_K \neq \emptyset$, $K \neq \emptyset$ for $(K: 0) \in ind^\circ L$. Further, by applications of thinning and R_K, $(p_K : c[p_1, \ldots]) \in Vd$ for all $(K: 0) \in Ind^\circ L$. Hence, by iteration of the left-rule for $\wedge$, we have

$(\bigwedge p_K : c[p_1,\ldots]) \in Vd$. Note also that since $\text{ind}^+ L = \emptyset$, $\text{ind}^\circ L \neq \emptyset$. Hence, by iteration of the left-rule for $\vee$,

$$(A : c[p_1,\ldots]) \in Vd \tag{10.17}$$

where $A =_{\text{def}} \bigvee \{\bigwedge p_K; (K:0) \in \text{ind}^\circ L\}$. A clearly satisfies (i). Now by iteration of the right-rule of $\wedge$, $(p_K : \bigwedge p_K) \in Vd$. By iteration of the right-rule for $\vee$, $(p_K : A) \in Vd$. Hence by L,

$$(c[p_1,\ldots] : A) \in Vd. \tag{10.18}$$

From (10.17) and (10.18) it follows that A satisfies (ii).

If $T_2(c)$ then there is a right-rule $R \in R(c)$ such that for each $(I:i) \in \text{ind}R$ there exists $L_{I,i} \in L(c)$ with $i \in K$ for all $(K:0) \in \text{ind}^\circ L_{I,i}$ and $j \in I$ for all $(Q:j) \in \text{ind}^+ L_{I,i}$. Now put $P(I:i) =_{\text{def}} (\bigwedge p_I) \supset p_i$ when $I \neq \emptyset$, and $P(I:i) =_{\text{def}} p_i$ when $I = \emptyset$. Then put $A =_{\text{def}} \bigwedge \{P(I:i); (I:i) \in \text{ind}R\}$. A satisfies (i). To prove (ii) we first prove that

$$(P(I:i), p_I : p_i) \in Vd. \tag{10.19}$$

There are cases (a) $I = \emptyset$ and (b) $I \neq \emptyset$.

ad (a). Then $P(I:i) = p_i$ and $p_I = \emptyset$. Thus $(P(I:i), p_I : p_i)$ is an axiom of $\mathfrak{S}(\mathfrak{L})$, so (10.19) holds.

ad (b). For all $i \in I$, $(p_I : p_i)$ is an axiom of $\mathfrak{S}(\mathfrak{L})$. Then, by iteration of the right-rule for $\wedge$, $(p_I : \bigwedge p_I) \in Vd$. Then by thinning and the left-rule for $\supset$, $((\bigwedge p_I) \supset p_i, p_I : p_i) \in Vd$. Again, (10.19) holds.

From (10.19), by the left-rule for $\wedge$, $(A, p_I : p_i) \in Vd$ for $(I:i) \in \text{ind}R$. Then by R,

$$(A : c[p_1,\ldots]) \in Vd. \tag{10.20}$$

(ii) follows from (10.19) and (10.20). ∎

In view of these results, regular junctors other than the basic junctors are superfluous, though there remains the problem of investigating the non-regular junctors. Therefore, where logics with regular junctors are considered, there is no loss of generality in using only the basic junctors. Thus with such logics we consider formal systems $\mathfrak{S}(\mathfrak{L})$ with $\mathbb{F}(B; P)$ as the set of formulas, the corresponding G-structures $\mathbb{B}(\mathfrak{L}) = \langle \mathbb{F}(B, P), Vd(\mathfrak{L}) \rangle$, and the corresponding logics $\mathfrak{E}^p(\mathfrak{L}) =_{\text{def}} \langle \mathbb{F}(B, P), Cn^\vee_{Vd(\mathfrak{L})} \rangle$, the *effective positive logics over* $\mathfrak{L}$. It is easy to see from Example 10.1 and Corollary 10.7 that $\wedge$, $\vee$, $\supset$ are normal conjunction, disjunction, and implication in $\mathfrak{E}^p(\mathfrak{L})$. The *pure effective positive logic* is $\mathfrak{E}^p(\mathfrak{L}_0)$ where $\mathfrak{L}_0$ is the minimum logic of P (Example 3.6(iii)(b)). Note that the special axioms of $\mathfrak{L}_0$ (i.e. derived from the identity function on $\mathbb{P}(P)$) are also general axioms. Thus $Vd(\mathfrak{L}_0) \subseteq Vd(\mathfrak{L})$ for all logics $\mathfrak{L}$ on P.

10.6 Negation in effective logic

In Lorenzen's dialogical interpretation of effective logic, negation, like the other junctors, is just one of a finite number of moves allowed to the protagonists of the dialogue game (cf. Lorenz 1968). The discussion of negation and contradiction in Chapter 4, §4.2.2, indicates, however, that negation is a special case and deserves a different treatment. In our development negation (as a one-place connective) does not occur uniformly within the positive logic, and, in particular, not in the pure logic. This follows from the following lemma.

LEMMA 10.9. *Let p, q be distinct proposition letters in P and let $A \in \mathbb{F}(B; P)$ be a formula in which the propositional letter p is the only one that occurs. Then $(p, A : q) \notin Vd(\mathfrak{L}_0)$.*

Proof. If $B, C \in \mathbb{F}(B; P)$, and $(p, B \oplus C : q) \in Vd(\mathfrak{L}_0)$, where $\oplus$ is one of the basic junctors, then $(p, B : q) \in Vd(\mathfrak{L}_0)$ or $(p, C : q) \in Vd(\mathfrak{L}_0)$. Hence
$$(p, A : q) \in Vd(\mathfrak{L}_0) \Rightarrow (p : q) \in Vd(\mathfrak{L}_0).$$
But as p, q are distinct, $(p : q) \notin Vd(\mathfrak{L}_0)$. Hence $(p, A : q) \notin Vd(\mathfrak{L}_0)$. ∎

COROLLARY 10.10. *There is no formula $A \in \mathbb{F}(B; P)$ such that the propositional letter p is the only one that occurs in A and A satisfies **N1** (Chapter 4, §4.2.2), i.e. $(p, A : C) \in Vd(\mathfrak{L}_0)$ for all $C \in \mathbb{F}(B; P)$.*

Accordingly, we equip $\mathfrak{E}^p(\mathfrak{L})$ with negation by adjoining a 'dummy' (constant) proposition f which we compel to behave as a contradiction (**F**, Chapter 4, §4.2.2)—by *convention* it has every proposition as a consequence. We count f as a logical symbol so that $|f| = 1$.

Let $\mathfrak{L} = \langle P; Cn \rangle$ be a given logic and J a set of junctors, assumed not to be in P, and f a symbol not in $J + P$. Let $\Lambda \subseteq P$. The set $\mathbb{F}^-(J; P)$ of formulas of $\mathfrak{S}^-(\mathfrak{L})$ is defined inductively by

F1: $P + \{f\} \subseteq \mathbb{F}^-(J; P)$
F2: if $F_1, \ldots, F_n \in \mathbb{F}^-(J; P)$ and $c \in J$ with $n(c) = n$ then $c[F_1, \ldots, F_n] \in \mathbb{F}^-(J; P)$.

The rules of inferences of the system $\mathfrak{S}^-(\mathfrak{L})$ are the introduction rules R for the junctors J together with thinning.

The axioms of $\mathfrak{S}^-(\mathfrak{L})$ are special axioms $Ax_\mathfrak{L}$ and general axioms $Ax_0 + Ax_1$ defined as follows.
(i) (a) Ax_0 consists of all sequents $(\Gamma : A)$ where $A \in \Gamma \in \mathbb{P}_\omega(F^-(J, P))$.
   (b) Ax_1 consists of all sequents $(f : A)$, where $A \in P + \{f\}$.
(ii) $Ax_\mathfrak{L}$ consists of all sequents of atomic formulas $(\Delta : A)$ such that $A \in Cn(\Delta)$ and $\Delta \in \mathbb{P}_\omega(P)$.

Effective quantifier logics

The notion of validity is defined as before in terms of derivation from the axioms. Again there is the question of whether $J^-(\mathfrak{L}) =_{\text{def}} \langle F^-(J:P), Vd(\mathfrak{L})\rangle$ is a G-structure.

First note that a simple argument by induction on logical length shows that $(f : A)$ is valid for all $A \in F^-(J, P)$. In order to conclude that f is a contradiction we must first prove that $J^-(\mathfrak{L})$ is a G-structure when the junctors all satisfy the cut condition. This requires the admissibility of the cut rule which depends on Lemma 10.2.

LEMMA 10.11. *The axioms of* $\mathfrak{G}^-(\mathfrak{L})$ *are closed under applications of cut.*

Proof. Let $(\Gamma : A)$ and $(\Delta : C)$ be axioms of $\mathfrak{G}^-(\mathfrak{L})$. If f does not occur in either $(\Gamma : A)$ or $(\Delta : C)$ then, as in Lemma 10.2, $(\Gamma, \Delta/A : C) \in Vd$. But if f occurs in one of them then one of the following four cases arises.

Case (i) $A = f$. Then $(\Gamma : A)$ is a general axiom so that $f \in \Gamma$. Then $(\Gamma : C)$ is derived by thinning from $(f : C)$, an axiom in Ax_1. By thinning, $(\Gamma, \Delta/A : C) \in Vd$.
Case (ii) $C = f$ & $A \neq f$. As $(\Delta : C)$ is an axiom, $f \in \Delta$. As $A \neq f, f \in \Delta/A$. Thus $(\Gamma, \Delta/A : C)$ is valid as it is an axiom in Ax_0.
Case (iii) $f \in \Delta$ & $A \neq f$. Then $f \in \Delta/A$. Hence $(\Gamma, \Delta/A : C) \in Vd$ as $(f : C) \in Vd$.
Case (iv) $f \in \Gamma$. As $(f : C) \in Vd$, by thinning, $(\Gamma, \Delta/A : C) \in Vd$. ∎

Theorem 10.2 now carries over to $\mathfrak{G}^-$, the only change required being an appeal to Lemma 10.11 instead of Lemma 10.1.

Again, when considering logics with regular junctors over a logic $\mathfrak{L}$, there is no loss of generality in confining attention to the basic junctors. The resulting G-structure is $\mathbb{B}^-(\mathfrak{L}) =_{\text{def}} \langle \mathbb{F}(B, P), Vd(\mathfrak{L})\rangle$ and we shall call the corresponding logic $\mathfrak{E}(\mathfrak{L}) =_{\text{def}} \langle \mathbb{F}(B, P), Cn^{\heartsuit}_{Vd(\mathfrak{L})}\rangle$ the *effective logic over* $\mathfrak{L}$. If $\mathfrak{L}_0$ is the minimum logic on the set P of proposition letters (Example 3.4(iii)) then $\mathfrak{E}(\mathfrak{L}_0)$ is the *pure effective logic*. In fact it is formally identical with the formal intuitionistic logic I in the sense that A is a theorem of I iff $(\varnothing : A)$ is provable in $\mathfrak{E}(\mathfrak{L}_0)$.

THEOREM 10.12. *Conjunction, disjunction, and implication are normal in* $\mathfrak{E}(\mathfrak{L})$.

Proof. The sets of rules $L(\wedge), L(\vee)$ of Example 10.1(i),(ii) ensure that $\mathbf{Cj1}^+$, $\mathbf{Cj2}^+$, $\mathbf{Dj1}^+$, $\mathbf{Dj2}^+$ hold in $\mathfrak{E}(\mathfrak{L})$. Likewise, $L(\supset)$ ensures that $B, B \supset C \in Vd(\mathfrak{L})$ so that $\mathbf{MP1}$ holds. $R(\supset)$ ensures that $\mathbf{DT}$ holds. ∎

10.7 Effective quantifier logics

The strategy for regularizing discourse about a logic $\mathfrak{L}$ was to extract from $\mathfrak{L}$ the minimum information by which $\mathfrak{L}$ was recognized as a logic. This was done by representing the formulas of $\mathfrak{L}$ by distinct standard 'proposition letters'. The logical consequence function was therefore represented as a logical consequence function on the proposition letters.

To systematize discourse about a free-variable logic $\mathfrak{L}$ we represent in a standard form the minimum information by which $\mathfrak{L}$ can be so recognized, namely the logical consequence function and the substitution properties of terms (Definition 3.13). This enables discourse about arbitrary free-variable logics with a fixed set $\mathbb{T}$ of terms to be formalized. As with the propositional logic of Chapter 10, §10.2, each formula A of $\mathfrak{L}$ is represented by a distinct letter P, but, in addition, the free variables are also recorded with P, Thus P can be considered as a predicate letter, and A is more fully represented by $P[y_1, \ldots, y_n]$ where $y_1, \ldots, y_n \in Fv(A)$. (We suppose that the free variables are in a fixed order.)

The free-variable logics with terms $\mathbb{T}$ will be assumed to have formulas with unbounded numbers of free variables, i.e. in such logics, for each n there is a formula A such that $Fv(A)$ has at least n members. Since the set of formulas is closed under substitution of terms (Definition 3.13(ii)), and hence of free variables, for each n there is a formula with exactly n free variables. There is therefore no loss of generality in restricting attention to free-variable logics whose formulas are constructed from the standard set $\{P_n^i; i, n = 0, 1, 2, \ldots\}$ of predicate letters, where P_n^i has n argument places, together with the usual marker symbols of brackets and commas. Then the standard set P of formulas consists of all expressions $P_n^i[t_1, \ldots, t_n]$ where $t_1, \ldots, t_n \in \mathbb{T}$. Since P is derived from free-variable logics on $\mathbb{T}$ (Definition 3.13), it, like $\mathbb{T}$ itself, is closed under substitution of terms for free variables:

$$A\sigma \in P \quad (A \in P, \sigma \in Sb(\mathbb{T}, Fv))$$

$$t\sigma \in \mathbb{T} \quad (t \in \mathbb{T}, \sigma \in Sb(\mathbb{T}, Fv)).$$

Thus we consider free-variable logics $\langle P, Cn \rangle$ where Cn is a logical consequence function on P which is derived from the given free-variable logic. Substitution of terms (Definition 3.13) for free variables therefore also preserves logical consequence, i.e.

$$A \in Cn(\Delta) \implies A\sigma \in Cn(\Delta\sigma) \quad (\Delta, A \subseteq P, \sigma \in Sb(\mathbb{T}, Fv)) \quad (10.21)$$

In the same spirit as the previous sections we now systematically extend given free-variable logics $\langle P, Cn \rangle$, including quantifier logics, by adjoining certain junctors to the set of basic junctors. The new basic junctors are the contradiction and quantifier symbols, $\exists$ and $\forall$ respectively, which will

Effective quantifier logics

eventually turn out to be the Aristotelian quantifiers (Chapter 4, §4.4). As before, the extended logic inherits the logical structure of the given logic via 'special axioms' which are analogous to the special axioms in the propositional case (§10.2). The extended set of formulas also inherits the substitution properties relating to terms. The rules of inference governing the quantifier symbols are modelled on the axioms ∀'1, ∀2, ∃1, ∃2' of Chapter 4, §4.11. Unfortunately we do not yet possess a general theory of quantifiers which is comparable in generality with the previous theory of junctors.

Suppose there are available the sets Bv (bound variables), J (basic junctors B), $Q = \{\forall, \exists\}$ (quantifier symbols), $\{f\}$ (propositional constant), and G (left, right brackets and comma) as in Chapter 2, §2.2.1, where these sets are pairwise disjoint and disjoint from the alphabets of P and $\mathbb{T}$.

The class $\mathbb{F}^q(J, P)$ of *effective quantifier formulas* over the given free-variable logic $\mathfrak{L}$ is then defined by induction as before, but with appropriate addition of a clause (F3) for the quantifier symbols as in Definition 2.12(i).

(F1) $\{f\} + P \subseteq \mathbb{F}^q(J, P)$.

(F2) If $F_1, \ldots, F_n \in \mathbb{F}^q(J, P)$ and $c \in J$ with $n(c) = n$ then $c[F_1, \ldots, F_n] \in \mathbb{F}^q(J, P)$.

(F3) If $A \in \mathbb{F}^q(J, P)$, $x \in Fv$ and a is a bound variable not occurring in A then the words $\forall a A(a/x)$ and $\exists a A(a/x)$ are also in $\mathbb{F}^q(J, P)$.

For any formula A, $Bv(A)$, $Fv(A)$ denote the sets of bound variables and free variables respectively which occur in A. Note that

$$Fv(\forall a A(a/x)) = Fv(\exists a A(a/x)) = Fv(A) - \{x\}$$
$$Bv(\forall a A(a/x)) = Bv(\exists a A(a/x)) = Bv(B) - \{a\}.$$

The rules of this system $\mathfrak{G}^q(\mathfrak{L})$ are the introduction rules R for the junctors J together with thinning as in §10.2. But they now include rules for the introduction of the quantifier symbols. In the absence of a general theory of quantifiers comparable with that of connectives, the rules here are modelled on the axioms ∀1, ∀2, ∃1, ∃2 of Chapter 4, §4.3.4. In the following statement of the quantifier rules $x \in Fv$ and a is a bound variable which does not occur in the formula A and $t \in \mathbb{T}$.

$$\forall_l: \quad \frac{(\Delta, A(t/x) \ : \ B)}{(\Delta, \forall a A(a/x) \ : \ B)}$$

$$\forall_r: \quad \frac{(\Delta \ : \ A)}{(\Delta \ : \ \forall a A(a/x))} \quad \text{where } x \notin Fv(\Delta)$$

$$\exists_r: \quad \frac{(\Delta, A\ :\ B)}{(\Delta, \exists x A(a/x)\ :\ B)}$$

where $x \notin Fv(\Delta, B)$

$$\exists_l: \quad \frac{(\Delta\ :\ A(t/x))}{(\Delta\ :\ \exists a A(a/x))}$$

The axioms of the system are $Ax_\varrho + Ax_0 + Ax_1$ which are defined as follows.

(i) (a) Ax_0 consists of all sequents $(\Gamma\ :\ A)$ where $A \in \Gamma \in \mathbb{P}_\omega(\mathscr{F}^q(J, \mathfrak{L}))$.
 (b) Ax_1 consists of all sequents $(f\ :\ A)$, where $A \in \Sigma + \{f\}$.
 $Ax_0 + Ax_1$ are the *general axioms*.
(ii) Ax_ϱ consists of all sequents of atomic formulas $(\Delta\ :\ A)$ such that $A \in Cn(\Delta)$ and $\Delta \in \mathbb{P}_\omega(\Sigma)$ (special axioms).

EXAMPLE 10.5. Let $F, G \in \mathbb{F}^q(J, P)$. If $a \in Bv - Bv(F)$, $b \in Bv - Bv(G)$, and $x \in Fv$, then $(\forall a F(a/x)\ :\ \forall b G(x/b))$ is derivable from $(F\ :\ G)$. ◆

Following the general strategy of the previous sections we show that this system gives rise to a G-system (and so a logic) by extending Theorem 10.2 to the quantifier case. If J is a set of regular junctors the proof goes through provided that the steps where compound formulas are analysed now include the cases of quantified formulas. But there are some small technical points to watch.

(i) The logical length $|F|$ of a formula is now defined by including $f, \forall, \exists$ amongst the logical symbols.
(ii) It can easily be shown by induction on length of formulas that, like P, $\mathbb{F}^q(J, P)$ is closed under substitution of terms for free variables:

$$A\sigma \in \mathbb{F}(J, P) \qquad (A \in \mathbb{F}(J, P), \sigma \in Sb(\mathbb{T}, Fv)).$$

(iii) Clauses relating to quantifiers cannot arbitrarily be inserted into the proof of the cut admissibility theorem—a careful accounting of the occurrence of free variables must be made and this is expedited by the notion of *compatibility of a proof with a finite set of free variables*.

DEFINITION 10.4. Let Π be a proof arranged in tree form and Z a finite set of free variables.

(i) The variable x is the application of the above quantifier rules in the *special variable* in that application. The special variable becomes bound in the conclusion.

(ii) Π is *compatible with* Z if
   (a) for each application of $\forall_r$ in Π and each application of $\exists_l$ the special variable in that application occurs in Π only in the sequents above the conclusion of that application,
   (b) no special variable in Π is in Z.

The success of Theorem 10.2 lies in our being able to assume that each of the proofs Π_1, Π_2 of that theorem is compatible with the free variable of the other. Two simple lemmas are needed to define the conditions under which direct derivation is preserved by substitution.

LEMMA 10.13. *If* $(\Gamma : B)$ *is directly derivable from* $(\Gamma' : B')$ *by a rule* R *then for any free variable* x *and term* t *such that no variable in* $Fv(t)$ *occurs in* $(\Gamma : B)$ *then* $(\Gamma(t/x) : B(t/x))$ *is also directly derivable from* $(\Gamma'(t/x) : B'(t/x))$, *by rule* R.

This is easily established by a case analysis. The next lemma concerns those quantifier rules $\exists_l$ and $\forall_r$ in which a condition is placed on the bound variables.

LEMMA 10.14. *Let* $t_1, \ldots, t_n$ *be terms, let* $z_1, \ldots, z_n$ *be free variables, and* $x \in Fv - \{z_1, \ldots, z_n\} - Fv(t_1, \ldots, t_n)$. *Put* $\sigma = (t_1, \ldots, t_n/z_1, \ldots, z_n)$. *Then*:

(i) *if* $(\Gamma : \forall a B(a/x))$ *is directly derived from* $(\Gamma : B)$ *by* $\forall_r$ *then* $(\Gamma\sigma : \forall a B(a/x)\sigma)$ *is also by* $\forall_r$;
(ii) *if* $(\Gamma, \exists a B(a/x) : C)$ *is directly derived from* $(\Gamma, B : C)$ *by* $\exists_l$ *then* $(\Gamma\sigma, \exists a B(a/x)\sigma : C)$ *is also by* $\exists_l$.

Proof.

(i) Suppose that $(\Gamma : \forall a B(a/x))$ is directly derived from $\Gamma : B$ by $\forall_r$. As $x \notin Fv(t_1, \ldots, t_n)$ we also have $x \notin Fv(\Gamma\sigma)$. Hence by $\forall_r$, $(\Gamma\sigma : \forall a B\sigma(a/x))$. But again as $x \in Fv - \{z_1, \ldots, z_n\} - Fv(t_1, \ldots, t_n)$, $\forall a B(a/x)\sigma = \forall a B\sigma(a/x)$. This proves the stated result.
(ii) Similar to (i). ∎

From this result it can be deduced that $\mathfrak{G}^q(\mathfrak{L})$ has a substitution property analogous to (10.21). The analogy becomes exact when it has been shown that $\mathfrak{G}^q(\mathfrak{L})$ is a logic.

THEOREM 10.15. *Let* $t_1, \ldots, t_n$ *be terms, let* $z_1, \ldots, z_n$ *be free variables, and put* $\sigma = (t_1, \ldots, t_n/z_1, \ldots, z_n)$. *Then*

$$(\Delta : A) \in Vd(\mathfrak{L}) \Rightarrow (\Delta\sigma : A\sigma) \in Vd(\mathfrak{L}).$$

Proof. This result can be proved by induction on length of proof. The basis of induction is correct by (10.21). The inductive step is a case analysis according to the rule of inference used. The only difficulty occurs with $\forall_r$ and $\exists_l$.

Suppose the sequent $(\Delta : \forall a B(a/z)) \in Vd(\mathfrak{L})$ and in its proof Π is directly derived by $\forall_r$ from $(\Delta : B)$ where $z \in Fv - Fv(\Delta)$. Let $Fv(\Pi)$ be the (finite) set of free variables which occur in Π. Choose $x \in Fv - Fv(\Pi) - \{z, z_1, \ldots, z_n\}$. Substitution of x for z throughout Π will give a proof of $(\Delta : B(x/z))$, where again x does not occur in Δ. By inductive hypothesis, $(\Delta\sigma : B(x/z)\sigma) \in Vd(\mathfrak{L})$. But $B(x/z)\sigma = B\sigma(x/z)$ (Appendix, §W) and $x \notin Fv(\Delta\sigma)$. By $\forall_r$, $(\Delta\sigma : \forall a B(x/z)\sigma(a/x)) \in Vd(\mathfrak{L})$. But by choice of x, the substitution operators σ and (a/x) commute. Thus $\forall a B(x/z)\sigma(a/x) = \forall a B(x/z)(a/x)\sigma = \forall a B(a/z)\sigma$. Hence $(\Delta\sigma : \forall a B(a/z)\sigma) \in Vd(\mathfrak{L})$.

The case where $(\Delta, \exists a B(a/z) : C)$ is derived by $\exists_l$ is treated similarly. The stated result now follows by induction. ∎

From Lemmas 10.13 and 10.14 it can now be deduced that $\mathbb{Z}$-compatible proofs can always be constructed:

THEOREM 10.16. *Let $Z \in \mathbb{P}_\omega(Fv)$. Given a proof Π of a sequent $(\Gamma : B)$ and $Z \in \mathbb{P}_\omega(Fv)$ by changes in free variables only, a proof Π' of $(\Gamma : B)$ which is compatible with Z can be found.*

Proof. Let Y be the set of free variables which occur in Π. Consider the tree form of Π. An application of $\forall_r$ or $\exists_l$ in Π is Z-compatible if the subtree Π' of Π above the conclusion of that application, considered as a proof by itself, is Z-compatible.

$$\left.\begin{array}{c} \Pi' \\ \diagdown \diagup \\ \hline (\Gamma : B) \quad \forall_r \end{array}\right\}\Pi$$

If all the applications of $\forall_r$ or $\exists_l$ in Π are Z-compatible then Π itself is compatible with Z. If not, then choose an application of $\forall_r$ or $\exists_l$ in Π which is not Z-compatible but in which every application of $\forall_r$ or $\exists_l$ in the subproof Π' of Π of the conclusion $(\Gamma : B)$ of that application is Z-compatible. Suppose $Y = \{y_1, \ldots, y_n\}$ is the set of special variables of this application. Let $y \in Y$ and let $z_1, \ldots, z_n$ be distinct free variables which do not occur in Π' or in Y or in Z. Substitute z_1 for $y_1, \ldots, z_n$ for y_n throughout Π'. By Lemma 10.12, Π' becomes a proof Π'' of

$$(\Gamma(z_1/y_1) \cdots (z_n/y_n) : B(z_1/y_1) \cdots (z_n/y_n)).$$

Effective quantifier logics 335

But by the restriction of variables in the application, the members of Y do not occur in Γ or B. Hence

$$(\Gamma(z_1/y_1)\cdots(z_n/y_n) : B(z_1/y_1)\cdots(z_n/y_n)) = (\Gamma : B).$$

Thus Π'' is a proof of $(\Gamma : B)$ which is compatible with Z. Now replace Π' in Π with Π'' to give a proof $\tilde{\Pi}$, say. Then $\tilde{\Pi}$ differs from Π by the renaming of certain free variables but contains one less application of $\forall_r$ or $\exists_l$ which is not Z-compatible. Now repeat this procedure on $\tilde{\Pi}$ until there are no applications of $\forall_r$ or $\exists_l$ which are not Z-compatible. ∎

COROLLARY 10.17. *If Π is a Z-compatible proof of $(\Delta : C)$ and x is a special variable in Π and $z \in Z$, then substituting z for x throughout Π yields another proof of $(\Delta : C)$.*

Finally, mutually compatible proofs can always be constructed.

LEMMA 10.18. *Let Π_1, Π_2 be proofs of $(\Gamma : B)$ and $(\Delta : C)$ respectively. Proofs Π'_1 of $(\Gamma : B)$ and Π'_2 of $(\Delta : C)$ can be constructed such that (i) Π'_1 (Π'_2) is compatible with the set of free variables occurring in Π'_2 (Π'_1) and (ii) Π'_1 (Π'_2) differs from Π_1 (Π_2) only by the renaming of free variables.*

Proof. Let Fv_2 be the set of free variables occurring in Π_2. By Theorem 10.16 there exists an Fv_2-compatible proof Π'_1 of $(\Gamma : B)$. Let Fv_1 be the set of free variables occurring in Π'_1. Then there exists an Fv_1-compatible proof Π'_2 of $(\Delta : C)$. ∎

We now return to consider those steps in Theorem 10.2 where consideration of quantified formulas is relevant. These are the places marked by (†i).

ad (†1) (Proof of (10.5)). Add the following text. Since A_0 is not in the succedent, $(\Gamma : A_0)$ is not the conclusion of an application of $\exists_l$ or $\forall_l$.

ad (†2) (Additional subcase (i)(b) in proof of (10.6)). Add the following text. $(\Gamma : A_0)$ is an immediate consequence by (α) $\exists_l$ or (β) $\forall_l$.
Sub-subcase (i)(b)(α) In Π_1, $(\Gamma : A_0)$ is derived by $\exists_l$ from $\Theta, B : A_0$ with

$$\left.\begin{array}{c}\Pi'_1\\ \dfrac{(\Theta, B : A_0)}{\underbrace{(\Theta, \exists aB(a/x) : A_0)}_{\Gamma}}\ \exists_l\end{array}\right\}\Pi$$

Subproof Π_1' $(\Delta : C)$ has a proof Π_2 such that $\rho(\Pi_1', \Pi_2) < \rho(\Pi_1, \Pi_2)$. Choose $z \in Fv - Fv(\Theta, A_0, \Delta, C)$. Let ψ be the permutation of Fv which interchanges x and z. Now $\psi((\Theta, B : A_0)) = (\Theta, \psi(B) : A_0)$ since $x \notin Fv(\Theta, A_0)$. Hence, by Lemma 10.16, $(\Theta, \psi(B) : A_0)$ has a proof Π_1'' isomorphic to Π_1' so that $\rho(\Pi_1'', \Pi_2) = \rho(\Pi_1', \Pi_2) < \rho(\Pi_1, \Pi_2)$. Then, by inductive hypothesis, $(\Theta, \psi(B), \Delta/A_0 : C) \in Vd(\mathfrak{L})$. Now by the construction of ψ, $\psi(x) \notin Fv(\Theta, \Delta/A_0, C)$. By $\exists_l$,

$$(\Theta, \forall a\psi(B)(a/\psi(x))), \Delta/A_0 : C) \in Vd(\mathfrak{L}).$$

But $\forall a\psi(B)(a/\psi(x)) = \forall aB(a/x)$. Hence, $(\Theta, \forall aB(a/x), \Delta/A_0 : C) \in Vd(\mathfrak{L})$, i.e. $(\Gamma, \Delta/A_0 : C) \in Vd(\mathfrak{L})$.

Sub-subcase (i)(b)(β) Similar to (i)(b)(α).

ad (†3) (Additional subcase (ii)(b) in proof of (10.6)). Add the following text. $(\Delta : C)$ is an immediate consequence in Π_2 by (α) $\exists_l$ or (β) $\forall_r$.

Sub-subcase (ii)(b)(α).

$$\Pi_2 \left\{ \begin{array}{c} \Pi_2' \\ \dfrac{(\Theta, B : C)}{(\Theta, \exists aB(a/x) : A_0)} \\ \underbrace{}_{\Delta} \end{array} \;\exists_l \right\} \Pi_2$$

where $x \notin Fv(\Theta, C)$. As $|A_0| = 0$, $A_0 \in \Theta$. $(\Theta, B : C)$ has a subproof Π_2' in Π_2' so that $\rho(\Pi_1, \Pi_2') < \rho(\Pi_1, \Pi_2)$. Choose a permutation $\psi: Fv \to Fv$ which leaves $Bv(\Theta, C)$ fixed and such that $\psi(x) \notin Bv(\Gamma, A_0)$. Then, by Lemma 10.16, $(\Theta, \psi(B) : C)$ has a proof Π_2'' isomorphic to Π_2' so that $\rho(\Pi_1, \Pi_2'') = \rho(\Pi_1, \Pi_2') < \rho(\Pi_1, \Pi_2)$. Then by inductive hypothesis, $(\Gamma, \Theta/A_0, \psi(B) : C) \in Vd(\mathfrak{L})$. By $\exists_l$, $(\Gamma, \Theta/A_0, \exists a\psi(B)(a/\psi(x) : C)$. But $\exists a\psi(B)(a/\psi(x)) = \exists aB(a/x)$. Hence $(\Gamma, \Theta/A_0, \exists aB(a/x) : C) \in Vd(\mathfrak{L})$, i.e. $(\Gamma, \Delta/A_0 : C) \in Vd(\mathfrak{L})$.

Sub-subcase (ii)(b)(β). Similar to (ii)(b)(α).

ad (†4) (Additional subcase (ii)(c) in proof of (10.6)). Add the following text. $(\Delta : C)$ is directly derived in Π_2 by an application of (α) $\exists_r$, (β) $\forall_r$.

Sub-subcase (ii)(c)(α) $C = \exists aB(a/x)$, say, and the premiss is $(\Delta : B(t/x))$

$$\Pi_2 \left\{ \begin{array}{c} \Pi_2' \\ \dfrac{\Delta : B(t/x)}{\Delta : \underbrace{\exists aB(a/x)}_{C}} \;\exists_r \end{array} \right.$$

which has a proof Π_2', a subproof of Π_2. Then $\rho(\Pi_1, \Pi_2') < \rho(\Pi_1, \Pi_2)$. By (10.12), $(\Gamma, \Delta/A_0 : B(t/x)) \in Vd(\mathfrak{L})$. An application of $\exists_r$ then gives $(\Gamma, \Delta/A_0 : C) \in Vd(\mathfrak{L})$.

Sub-subcase (ii)(c)(β) Here $C = \forall aB(a/x)$. By Lemma 10.16 it can be assumed that Π_1, Π_2 are mutually compatible.

$$\left.\begin{array}{c} \Pi_2' \\ \dfrac{(\Delta : B)}{\underbrace{(\Delta : \forall aB(a/x))}_{C}} \; \forall_r \end{array}\right\} \Pi_2$$

The premiss $\Delta : B$ has a proof Π_2', a subproof in Π_2, such that $\rho(\Pi_1, \Pi_2') < \rho(\Pi_1, \Pi_2)$. By (10.12), $(\Gamma, \Delta/A_0 : B) \in Vd(\mathfrak{L})$. But by the mutual compatibility, the special variable x does not occur in Π_2 and so not in Γ or in Δ/A_0. By an application of $\forall_r$,

$$(\Gamma, \Delta/A_0 : \forall aB(a/x)) \in Vd(\mathfrak{L}).$$

(†5) (Additional subcase (i)(a) of case (i) in the proof of (10.7)). Add the following text.

Subcase (i)(a) A_0 is (α) $\exists aB(a/x)$ or (β) $\forall aB(a/x)$.

Sub-subcase (i)(a)(α) $(\Gamma, \exists aB(a/x))$ is the conclusion in Π_1 of an application of $\exists_r$ and $(\Delta : C)$ is the conclusion in Π_2 of an application of $\exists_l$, where $\Delta = \Psi, \exists aB(a/y)$, and $y \notin Fv(\Psi, C)$.

$$(\Pi_1 \left\{ \dfrac{(\Gamma : B(t/x))}{(\Gamma : \exists aB(a/x))} \quad \underbrace{\dfrac{(\Psi, B(y/x) : C)}{(\Psi, \exists aB(a/y) : C)}}_{\Delta} \right\} \Pi_2$$

Then Π_1 contains a subproof Π_1' of $\Gamma : B(y/x)$ where $y \notin Fv(\Psi)$. By Theorem 10.15, $(\Psi, B(y/x)(t/y) : C) \in Vd(\mathfrak{L})$. Thus $(\Psi, B(t/x) : C) \in Vd(\mathfrak{L})$ and has a proof Π_2', say. Then

$$\gamma(\Pi_1', \Pi_2') = |B(t/x)| < |\exists aB(a/y)| = g + 1.$$

By (10.15), $(\Gamma, \Psi/B(t/x) : C) \in Vd(\mathfrak{L})$. Thus, by thinning,

$$(\Gamma, \Psi : C) = (\Gamma, \Delta/A_0) : C) \in Vd(\mathfrak{L}).$$

Sub-subcase (i)(a)(β) Similar to the above case.

(†5) (Additional subcase (ii)(a) of case (ii) in the proof of (10.7)). Add the following text.

or $(\Gamma : A_0)$ is the conclusion in Π_1 of an application of (α) $\exists_l$ or (β) $\forall_l$. Thus we have the following subcase.

Subcase (ii)(a)(α) $\Gamma = \Psi, \exists aB(a/x)$, and $(\Gamma : A_0)$ is directly derived in Π_1 by $\exists_I$ from $(\Psi, B : A_0)$ where the special variable x does not occur in Ψ, A_0.

$$\underbrace{\dfrac{\overset{\Pi'_2}{(\Psi, B : A_0)}}{(\Psi, \exists aB(a/x) : A_0)}}_{\Gamma} \exists_I \Biggr\} \Pi_2$$

$(\Psi, B : A_0)$ has a proof Π'_2 and a subproof of Π_2, and

$$\rho(\Pi_1, \Pi'_2) < \rho(\Pi_1, \Pi_2).$$

By (10.12), $(\Psi, B, \Delta/A_0 : C) \in Vd(\mathfrak{L})$. Now by Lemma 10.16 it can be assumed that Π_1 and Π_2 are compatible. Thus the special variable x does not occur in Δ, C and therefore not in $\Psi, \Delta/A_0, C$. An application of $\exists_I$ then gives $(\Psi, \exists aB(a/x), \Delta/A_0 : C) \in Vd(\mathfrak{L})$.

Subcase (ii)(a)(β) Similar to the above case.

Addition of these clauses to Theorem 10.2 completes the proof of the cut elimination from the system with regular connectives. Theorem 10.5 and Corollary 10.6 also hold so, as before, there is no loss of generality in restricting the junctors to the basic set B. The logic $\mathfrak{E}^q(\mathfrak{L})$ is the *effective quantifier logic over* $\mathfrak{L}$. It readily follows from the quantifier rules and Example 10.4 that $\exists, \forall$ are quantifiers (Definition 3.13) in $\mathfrak{E}^q(\mathfrak{L})$; they also satisfy $\forall 1'$ and $\exists 2'$ (Chapter 4, §4.4). By Lemma 4.23, it also satisfies $\forall 1, \forall 2, \exists 1, \exists 2$. So by Lemma 4.22, $\forall$ and $\exists$ are strong normal quantifiers.

By taking $\mathfrak{L}$ to be the minimum logic $\mathfrak{L}_0$ (Example 3.6(iii)(b) on P we obtain the *pure effective quantifier logic*. This occupies a special position in relation to all effective quantifier logics based on free-variable logics with terms $\mathbb{T}$.

THEOREM 10.19. *A sequent is valid in* $\mathfrak{G}^q(\mathfrak{L}_0)$ *iff it is valid in* $\mathfrak{G}^q(\mathfrak{L})$ *for all free-variable logics* $\mathfrak{L}$ *with terms* $\mathbb{T}$.

Proof. Suppose $(\Delta : A) \in Vd(\mathfrak{L}_0)$. Since the logical consequence function of the minimum logic is the identity on $\mathbb{P}(P)$, the special axioms are included in the general axioms. Hence the proof of this sequent is also a proof in $\mathfrak{G}^q(\mathfrak{L})$ for any $\mathfrak{L}$. Thus $(\Delta : A) \in Vd(\mathfrak{L})$.

Conversely, suppose $(\Delta : A) \in Vd(\mathfrak{L})$ for all free-variable logics with terms $\mathbb{T}$. As $\mathfrak{L}_0$ is such a logic, $(\Delta : A) \in Vd(\mathfrak{L}_0)$. ∎

The *pure effective quantifier logic* is formally, but not philosophically, identical with the intuitionist predicate logic. This logic was first formulated in 1928 by Heyting as a Hilbert-type system and later as a sequent calculus by Gentzen. The latter formulation enabled exact comparison to be made with the classical and other calculi. Some of these results are recorded by Kleene (1962) and Dummett (1977). A few of these, mentioned below, as well as illustrating some metalogical differences from other logics, are of historical importance.

10.8 Effective, intuitionistic, and classical logics

Our formulation of effective logic is a version of formal intuitionistic logic expressed as a sequent calculus. The formulation of the intuitionist calculus given by Dummett (1977, §4.2) differs from ours only in that negation is taken as a one-place connective, thus necessitating sequents $(\Gamma : C)$ where C is either a formula or *the empty set*. This is difficult to sustain if we understand a sequent to mean a claim that C is a consequence of Δ. We define negation in terms of contradiction and the implication connective.

The intuitionist system G1 in Kleene (1962, §77) likewise uses negation, but expresses sequents in terms of *sequences* of formulas as was originally done in Gentzen's system *LJ* (cf. Chapter 3, §3.10; see also Lorenzen (1962, Chapter 1)).

A formalization of classical logic can be obtained from the effective logic by allowing sequents $(\Gamma : \Delta)$, where both Γ and Δ are finite, possibly empty, sets of formulas. The notion of *basic sequent* can be generalized to give an appealing symmetry of antecedent and succedent. The sequents of this generalized type are intended to be derivable if, and only if at least one member of Δ is a logical consequence in the semantic sense—of Γ. But the sequents then lose their sense of defining logical consequence by closure conditions.

The formalization of the intuitionist calculus enables certain unprovability results to be established. The most well known of these is that whereas $A \vee \neg A$ is a classical tautology (PEM) it is intuitionistically unprovable (Kleene 1962, p. 483). However, the double negation of it is provable. This is a consequence of a more general result (Kleene 1962, §81).

THEOREM 10.20. Let B be a formula of the pure effective propositional logic. If B is a classical tautology then $(\varnothing : \neg\neg B)$ is valid.

This theorem was proved by a different method by Glivenko (1929). Note that $\forall a[A \vee \neg A](a/x)$ is a theorem of classical logic, but $\neg\neg\forall a[A \vee \neg A](a/x)$ is not a theorem of the intuitionistic calculus. This result follows from a formal analysis of a Gentzen-type formalization of the intuitionistic calculus

(Kleene 1962, §80). It was first proved by Heyting (1930), then by Kleene (1945) and Nelson (1947) (see Kleene 1962, p. 487). And in contrast with the classical calculus, $A \vee B$ is a theorem iff both A and B are theorems. Further $\exists a A(a/x)$ is a theorem iff there exists a term t such that $A(t/x)$ is a theorem (cf. Dummett 1977, p. 117).

The classical representations of the intuitionistic formal logic—e.g. the topological interpretation (Tarski 1956, Chapter 17), Kripke's model structures (Chapter 6, §6.10), or Cohen's forcing (Fitting 1969) etc.—enable various contrasts with classical logic to be readily established and generally contribute to a deeper understanding of the relation between classical and constructive mathematics. Thus, whereas the classical monadic predicate calculus is decidable, the intuitionistic monadic calculus is not (Kripke 1965).

Intuitionistic number theory (*Heyting arithmetic*) consists of the usual first-order induction schema together with axioms for successor, addition, and multiplication. It is based on intuitionistic predicate logic. This theory has also been interpreted classically by Kleene's notion of *realizability* (Kleene 1962, §82). In this interpretation partial recursive functions (strictly speaking, their Gödel numbers) are systematically associated with formulas of arithmetic in such a way that they are analogous to Skølem functions (Chapter 2, §2.3.6) and so that *provable* formulas are realized by *recursive functions*. (For a criticism of the interpretation of constructivism by recursive function see Heyting (1961).)

An intuitionistic version of Zermelo–Fränkel set theory (IZF) has also been formulated (e.g. Ščedrov 1985) and has found application in the theory of information systems (McCarty 1984). IZF is a first-order theory based on intuitionistic predicate logic with $\in$ and $=$ as its only non-logical predicates. The axioms of IZF are classically equivalent to the traditional ZF axioms. In the 1970's it was discovered that Kleene's concept of realizability could be applied to IZF (e.g. Friedman 1973, Beeson 1979) so that there is a natural cumulative structure of 'realizable' sets which is a model of IZF.

But such representations do not directly engage the kernel of non-classical constructivism, the notions of *existence* and *construction* in mathematics:

> ... Heyting's axioms are very neat. Gentzen's investigation made the systems seem even neater and more natural. Quite remarkable is the isomorphism of the formal system to a topological calculus discovered by Stone and Tarski. The so-called Kripke interpretation and its relation to Cohen's forcing lends even greater apparent depth. But are these connections relevant to intuitionism? Insofar as they provide formal independence proofs (as with Kleene's extensive work on realizability interpretations), the relevance is clear enough. As methods of formal investigation they are excellent. But they cannot fully elucidate the meaning of intuitionism. ... The underlying concept of intuitionism is that of *mathematical construction*. Here construction also includes *proof*. Brouwer regarded proofs as abstract objects in the way we manipulate other mathematical objects. The idea smacks of psychologism in logic. It is not quite the same, however, and it *is* the basis on which the

intuitionists explain themselves. Before rejecting it as non-sense we ought to see whether or not a rigorous theory of mathematical constructions is posible. It is a difficult task, but it does not seem out of the question. The attempts so far have not been quite satisfactory. The whole programme of intuitionism demands a rethinking of mathematical philosophy.

Scott 1972

10.9 Brouwer's intuitionism

Having mentioned Heyting's formalization of intuitionist logic, we give in the following section a brief sketch of the philosophical motives which lead Brouwer to his radical formulation of the foundations of mathematics. It is a distinct philosophical stance and mathematical practice radically different from the established—usually unacknowledged—naive realist view permeating mathematics and science. A deeper and more advanced analysis of intuitionist mathematics can be found in Kleene and Vesley (1965) and van Dalen and Troelstra (1988). The notes below are based on the thesis 'Brouwer's intuitionism' by van Stigt (1971).

One of Brouwer's earliest publications, 'Leven, Kunst en Mystiek' (Brouwer 1905), expresses views which were consistently held throughout his life. Towards the end of his life Brouwer attempted to republish it. It is a collection of mystical–philosophical pronouncements on a wide range of topics, the dominant themes of which are the excellence of the inner vision, the spiritual nature of the individual consciousness, and the distrust of over-rationalization— man's noble animal nature has been betrayed by his intellect. His views on language, logic, and mathematics, which form the basis of intuitionistic philosophy, find their ultimate source in these mystic beliefs.

Brouwer's attack on formalistic trends in mathematics was first published in his dissertation 'On the foundations of mathematics' in 1904. The first version of his thesis, discovered in 1906 (van Stigt 1979), showed that he held extreme solipsistic views even before his intuitionist campaign began.

Brouwer's philosophical motives are evident in all his work. Whereas his disciples tried to eliminate philosophical elements from intuitionism, Brouwer himself saw them as inextricably linked with mathematics and ridiculed 'those who wear philosophy like a Sunday suit'.

10.9.1 *Solipsism*

Brouwer refused to accept the plurality of minds:

> Other individuals, i.e. human bodies, are only iterative complexes of sensations whose elements are permutable in time ... there is no plurality of mind, so much the less is there a science of plural mind.
>
> Brouwer 1949a

There is no room for intersubjectivity in this solipsistic philosophy and his conception of mathematics. In all its aspects intersubjectivity concerns social interaction about which Brouwer had grave doubts. He did not always live up to these beliefs in practice, as is evidenced by his resentment at being misunderstood.

10.9.2 *The three stages of consciousness*

In Brouwer (1905) life is described as the journey of a spiritual soul imprisoned in a body through a veil of tears. Its vision in accordance with its spiritual nature is direct—intuition in its purest form. In this mystic vision all is seen directly, not interconnected through causality. Our bodily senses force their impression on the soul, blur their immediate vision, and impose 'the chain of plurality, separation, time, space ... and causality'. Life in this world, mathematical and perceptual cognition, and social life are seen by Brouwer as consciousness gradually moving farther away from its spiritual nature. Brouwer (1949a) described this transition of consciousness from its 'deepest home' (i.e. the 'self') to the external world as taking place in three phases.

(i) The *naive phase*—the creation of the world of sensation. The 'primordial phenomenon', the move of time, the genesis of mathematical thought, occurs here.
(ii) The *isolated causal phase*—the onset of causal activity. Brouwer identifies sequences with causal systems and sees the origin of science here.
(iii) The *social phase*—involvement with the co-operation with other individuals. At this stage, farthest away from the world of intuitive consciousness, man loses his concentration on his inner soul and acts on the assumption of the existence of other minds. There is total dependence on experience and the physical senses in all aspects of this world of co-operation. Language and logic develop at this stage.

Brouwer's definition of mathematics as occurring in phase (i) is not just the case of a general claim for the subjectivity of all knowledge. He claims a place for mathematics completely outside the main division of the sciences (in the more general sense of any systematized knowledge)—it is an activity more akin to artistic endeavour.

10.9.3 *Connection with Kant*

In his inaugural address (1912) Brouwer states that, with some adjustments, Kant's philosophy of the a priori and synthetic judgements could be accepted as the ultimate basis of mathematics: 'However weak the position of intuitionism seemed to be ... it has recovered by abandoning Kant's a priority of space but adhering the more resolutely to the a priority of time.'

Kant had placed the basis of the validity of mathematics firmly in the structure of knowledge itself. In this respect Brouwer's philosophy can be called Kantian. Apart from this most general basis of agreement and a more detailed account of time as the a priori form of intuition fundamental to mathematics, Brouwer's philosophy is but a consequence of an extreme individualistic interpretation of a fragment of Kant's doctrines.

Kant presupposed the a priori character of mathematics and its synthetic nature, using it to illustrate his more general metaphysical thesis for which he needed an elaborate and complex system of concepts and distinctions. Brouwer avoided the general metaphysical problem of the nature of knowledge; his prime concern was with mathematical objects which, because of the special abstract nature, can be more easily accommodated within the solipsistic isolation of the individual mind. Brouwer's mystical beliefs already inclined him to this restricted conception of human knowledge and ultimately determined the central theme of his thinking.

10.9.4 Sequences and causality

Especially against those who made language and logic the starting point of mathematics, Brouwer (1907) wanted to show the independence of mathematics from logic by placing

> 'mathematical thinking' at the root of all human mental activity. Man has the faculty 'to view his life *mathematically*, to see in the world repetition of sequences, i.e. causal systems'.

'Things' are nothing but 'iterative complexes of sensations whose elements are permutable in point of time' and so are 'individuals, i.e. human bodies, the home of the subject included' (Brouwer 1949a).

Causal attention is the faculty ('originating in the primordial sins of fear and desire') of

> seeing repetition sequences, qualitatively different but supposedly equal'.
> Brouwer 1908

The relation of cause and effect is not one that exists in the external world; it is reduced to a matter of temporal order between elements of sequences, willed by man.

By reducing causality to mere repetition, Brouwer established the dependence of all forms of science on mathematics. By identifying logic with regularity in language, he placed logic in the sphere of social acting (phase (iii)) and removed it altogether from the fundamental area of human thinking, but at the cost of almost complete identification of mathematics with causality. By making the distinction between 'mathematical viewing' and pure mathematics Brouwer could redeem pure mathematics to such an extent that he could attribute wisdom and truth to its beauty, but

> searching for *wisdom* we may find it in knowing that causal thinking and acting is non-beautiful and hard to justify...
>
> Brouwer 1949a

10.9.5 Brouwer's intuition

Despite Brouwer's claim that all intuitionists had to do was strength Kant's a priori intuition of time, Brouwer's intuition of time differs from Kant's. Whereas Kant's a priori form of intuition is a passive highly abstract element in the process of cognition, very much intertwined with the more active schemata and categories of the synthesizing mind, Brouwer's intuition of time is a psychological first step, a simple awareness of time:

> the *primordial intuition* is identical with the awareness (consciousness) of time as nothing but change.
>
> Brouwer 1907

This is the origin of mathematics: 'mathematics is developed from one single a priori intuition which may be called *permanence in change* or *unity in plurality*'. More generally, Brouwer's intuition is the activity of the human mind restricted exclusively to the data of consciousness itself and the tools provided by it. The fundamental and tools

> ...are the elements of construction which can be read from the primordial intuition, concepts such as continuous, unity, again, 'and-so-forth',...
>
> Brouwer 1905

They also include the principle of complete induction.

10.9.6 Mathematics and intuition

Brouwer's pure mathematics is the a priori synthetic constructive activity of the mind independent of experience. Brouwer remained true to this concept, expounded in the *Grondslagen* (1907), for the rest of his life:

> Intuitionist mathematics is an essentially languageless mental structure which comes into being by the self-unfolding of the abstraction of two-ity as the primordial intuition ....

And in 1949 he wrote:

> mathematics comes into being when the two-ity created by a move of time is divested of all quality by the subject, and when the remaining empty form of the common substratum of all two-ities, as the basic intuition of mathematics is left to an unlimited unfolding.
>
> Brouwer 1949a

No objectivity is claimed for mathematics. The objectivity Brouwer is prepared to prescribe to physical sciences is only 'invariance in our image of nature relative to an important group of phenomena'.

Brouwer's solipsistic denial of intersubjectivity of mathematics does not call into doubt the 'exactness' of pure mathematics, but only its social practice on the basis of language:

> The languageless constructions originating in the self-unfolding of the primordial intuition are exact and true as they are present in the human memory.
> Brouwer 1933

The power of human memory which has to be taken in these constructions is limited and fallible. To a human mind with an unlimited power of memory, pure mathematics practised in solitude and without linguistic symbols would be exact (Brouwer 1942). In his subsequent writings Brouwer refers to this idealized mathematician as 'the subject' or the 'creating subject' (Brouwer 1948, 1949a, b). It becomes a mathematical object when he defines choice sequences depending on the experience of the mathematician defining them. Attempts have been made to give an objective interpretation to this concept. Kreisel (1967) used Brouwer's notion in a formal way to derive purely mathematical results, but deviated from Brouwer's usage by allowing more than one creative subject.

10.9.7 *Language*

Rejection of language The dominant themes of Brouwer's work on foundations are (i) mathematics as the activity of the mind, (ii) the essential inadequacy of language as a medium of communication, and (iii) the separation between mathematics and its recording. Concerning (i) see §10.8.2, 10.8.5, and 10.8.6. (ii) follows from Brouwer's solipsism (§10.8.1 and 10.8.2):

> Language, the slave of the illusion of reality, cannot be the instrument of truth.
> Brouwer 1905

And by language

> ... there is no exchange of thought ... by so-called exchange with anoher being the subject only touches the outer walls of an automaton.
> Brouwer 1949a

The radical rejection of language as contributary to mathematics follows from §10.8.6.

Concerning (iii), the absolute distinction between mathematics and its expression was offered by Brouwer as the cure for the crisis in the foundations of mathematics. In 1905 he criticized Poincaré for not going far enough in his criticism of Cantorism:

He does not go to the root of the trouble, which is much deeper, i.e. the confusion of the act of constructing mathematics and the language of mathematics.
Brouwer 1907

Opposition to formalization Unwillingness to accept the possibility of grasping living *mathematical* reality completely and exhaustively in definite pre-defined words or symbols is the fundamental reason for Brouwer's consistent opposition to formalization of mathematics:

> Freedom of mathematics can never be exhausted in any one system, and develops in a self-unfolding guided by a free arbitrariness.
> Brouwer 1907

Brouwer, in retrospect, characterized his activity during his early period—'the first act of intuitionism'—as purging mathematics of the non-mathematical elements of language and logic.

> The first act of intuitionism completely separates mathematics from mathematical language, in particular from the phenomena of language which are described by theoretical logic, and recognises that intuitionistic mathematics is an essentially languageless activity of the mind.
> Brouwer 1952

He accepted that this negative programme

> seems necessarily to lead to destructive and sterilising consequences.
> Brouwer 1953

10.9.8 *Constructions*

Brouwer completely rejected the Platonic or realist existence of mathematical objects independent of human thought (Chapter 2, §2.1), and so excludes the notion of truth as conformity of thought to object. Equally unacceptable to Brouwer was the formalist notion of truth as derivability within a formal system. The truth of a theorem is *created* by a construction. Unknown truths neither given by intuition nor proved by mathematical construction do not exist. As to the truth of a proof, the construction at any of its stages, including the operations of construction originates in the primordial intuitionism (§10.9.5).

Brouwer's conception of construction is nothing else but mathematical activity in the strictest sense, i.e. step-by-step self-development within the a priori limits of the mind. These intuitive constructions are not necessarily simple or immediately clear. Any linguistic recording of them will reflect their coherence, but, being outside the domain of 'first-order' pure mathematics, no active role can be ascribed to it.

The identity of existence and construction is repeatedly asserted by Brouwer and his disciples:

> If 'to exist' does not mean 'to be constructed', it must have some metaphysical meaning. It cannot be the task of mathematics to investigate this meaning or to decide whether it is tenable or not. In the study of mental mathematical constructions 'to exist' must be synonymous with 'to be constructed'.
> Heyting 1955, p. 2

Such assertions frequently assume a characteristic subjectivist form, revealing the addition to solipsism:

> Intuitionist mathematics consists ... in mental constructions; a mathematical theorem expresses a purely empirical fact, namely the success of a certain construction. '2 + 2 = 3 + 1' must be read as an abbreviation for the statement: "I have effected the mental constructions indicated by '2 + 2' and by '3 + 1' and I have found that they lead to the same result".
> Heyting 1956, p. 8

10.9.9 *Logic*

For Brouwer and fellow intuitionists, in contrast with classical mathematics (Chapter 2, §2.1),

> The characteristic of mathematical thought is that it does not convey truth about the external world, but is only concerned with mental constructions.
> Heyting 1956, p. 9

There is no sense in talking about the truth of a proposition independently of a proof. Thus intuitionists explain 'A implies B' as a construction which would transform any proof of A into a proof of B. This conception of mathematical thought gives rise to a logic which is significantly different from the classical logic.

Negation In 1907 Brouwer reproached Poincaré for 'looking very much like his opponent Russell' (Brouwer 1907) in his demand for freedom from contradiction and 'not seeing the confusion of the act of constructing mathematics and the language of mathematics' (cf. §10.9.7). His later, more mature view, recognized Poincaré's merit:

> For these 'separable' parts of mathematics (i.e. natural numbers, principle of complete induction and mathematical entities and theories springing from this source) they [Poincaré, Borel, Lebesgue] postulated an existence and exactness independent of language and logic.
> Brouwer 1952

This change of heart is partly due to Brouwer's different views on non-contradiction after 1923 when he published his 'Calculus of absurdity' (1923, 1925). Non-contradiction, previously dismissed as mathematically irrelevant, is given the constructive meaning of 'absurdity of absurdity'; in 1948 he frequently equated the two.

To Brouwer a contradiction is a 'purely linguistic phenomenon' (Brouwer 1933). (Contradiction was so treated in our construction of effective logic $\mathfrak{G}^-(\mathfrak{L})$ by the introduction of the symbol f together with the axioms Ax_1, in §10.6.) In a verbal description of a mathematical construction a contradiction cannot arise: 'two contradictory theorems cannot be true of a mathematical construction' (Brouwer 1907). Non-contradiction is therefore accepted as a necessary condition for mathematical existence. Contradiction is described as a kind of incompatibility of two mathematical constructions: 'the two systems do not fit into each other' (Brouwer 1908). Verbal contradiction reports an obstruction:

> ... the words of your mathematical demonstration are only the accompaniment of a wordless mathematical *building*, and where you pronounce a contradiction, I simply observe that the construction cannot be further, that in the given construction there is no room for the posited structure.
>
> Brouwer 1907

The word 'posited' here indicates that Brouwer had in mind that the constructive part of a false statement is a verbal hypothesis, a mathematical non-entity which leads to an absurdity like $0 = 1$. Brouwer's insistence on the completely mathematical, i.e. languageless, nature of negation does not seem to be justified.

Principle of the excluded middle Besides his distinctive notion of the continuum, together with the associated concept of *free choice sequence*, the rejection of the PEM is Brouwer's most original contribution to the study of foundations.

The crux of the dispute over PEM is the notion of existence in mathematics. By applying PEM and the principle of contradiction to an existential proposition the existence of a mathematical object is established by showing that the assumption of its non-existence is contradictory. This sort of *indirect proof* is frequently used in mathematics, for instance in Brouwer's fixed-point theorem, which he later rejected.

Most of the components of this rejection of PEM occur in 1907. The full realization that the application of the PEM in mathematics is also a problem of existence and presupposes the solvability of every mathematical problem came in 1908. His publications after the Second World War adamantly reject the PEM as 'an instrument to discover new mathematical truth' (Brouwer 1949a).

Most of Brouwer's work after the First World War consists of reconstructing mathematics expressly without PEM. When, in some of his work, he discusses the PEM itself he gives a historical interpretation. He described 'the dogma of the PEM' as a phenomenon of history of civilization of the same kind as the old belief in the rationality of π or the rotation of the universe on an axis through the earth. First, classical logic was abstracted from the language of the mathematics of 'part sets' of a certain finite system. A priori existence, independent of mathematics, was then attributed to this logic and on the basis of this apriority the rules of logic were then applied to the mathematics of infinite sets.

Brouwer's stand on the PEM in the period 1908-1923 can be summarized as follows.

(i) The question of PEM is to be identified with the possibility of actually carrying out the investigation.
(ii) Within a finite mathematical system such an investigation can be terminated and the PEM is a valid but trivial principle leading to true mathematical statements.
(iii) In an infinite mathematical system the use of PEM is not justified.
(iv) Even the unjustified use of the PEM will never lead to a contradiction.
(v) Freedom from contradiction is not a guarantee of the validity of a mathematical argument.

His rejection of classical double negation in 1923 undermined the PEM even for the finite case. Absurdity and absurdity of absurdity were thereafter applied to relations without any distinction being made between finite and infinite systems. The complete rejection of PEM as a principle can be found from 1928 onwards.

Apart from true mathematical statements, Brouwer distinguished those that are non-contradictory. Such a distinction was hinted at in 1903 (Brouwer 1907):

> ...therefore in mathematics one should, in all the theorems that are usually accepted as having been proved, distinguish between those that are *correct* and those that are *non-contradictory*.

Brouwer's search for a proof that PEM is contradictory in an infinite system was suddenly halted by a publication by Glivenko (1929) which followed a series of controversial articles on Brouwer's 'calculus of absurdity' and the PEM. Glivenko made the first attempt at formalizing intuitionistic logic by stating the axioms acceptable to Brouwer and proving some of Brouwer's results. Glivenko showed that the addition of $p \vee \neg p$ to the axioms gives the classical propositional calculus. But most disturbing to Brouwer must have been Glivenko's first theorem: $\neg\neg(p \vee \neg p)$ (Glivenko 1929). Brouwer's immediate reaction is difficult to assess from his writings.

With the exception of a paper in 1933 he did not mention PEM again until 1948. This almost complete silence starting after Glivenko's publication could itself indicate Brouwer's reaction.

Heyting's formalization Brouwer himself never produced a comprehensive systematic analysis of the language of intuitionist mathematics, even though he claimed (Brouwer 1953) that

> intuitionist mathematics has its general introspective theory of mathematical assertions, which with some right may be called intuitionist logic..

Brouwer did not refer here to Heyting's formalization but to his own fragmentary observations on such topics as PEM and the Brouwer negative.

The first attempt at a formalization of intuitionist mathematics and logic was made by Heyting (1928, unpublished) in response to a challenge by Mannoury. This was the basis for the well-known 1930 publication. Brouwer's reaction was negative. He agreed to its publication because of Heyting's expressed support for Brouwer's position.

While Brouwer insisted on the languageless nature of mathematics and could only bring himself to admit that language in everyday practice is difficult to dispense with, for Heyting the attempt to express the most important parts of mathematics symbolically is justified by precision and conciseness.

Heyting consistently denied any claim that his formalization could completely reflect the whole of intuitionist mathematics. He remained loyal to Brouwer's fundamental maxim that every mathematical or logical theorem must express the result of a mathematical construction. And in 1969:

> I have become more convinced that, at least in the communication of mathematics, formalization has its great advantages. From recent research into the notion of choice sequences it has appeared that formalization is *necessary* for any sufficiently clear representation.

10.9.10 *The continuum*

The systematization of intuitionist set theory and the development of his distinctive theory of the continuum dominates Brouwer's work in the period 1917–1927.

For Brouwer the continuum is given by the primordial intuition of time (§§10.9.5, 10.9.6):

> ... with the primordial intuition of mathematics in which the connected and the separate, the continuous and the discrete are united, the linear continuum is immediately present ...
>
> Brouwer 1907

The continuum is thought of as a whole, not just as a set of points. According to Borel and Poincaré we can speak of *all* the points of the continuum, since this collection as a whole is given, but not of every arbitrary point, since not every point is definable in a finite procedure. The essential unity of the continuum was stressed by Brouwer:

> The *continuum as a whole* is intuitively given: a construction or act which would create 'all' its points individually by means of the mathematical intuition is unthinkable and impossible.
>
> Brouwer 1907

Brouwer's notion of the continuum was probably influenced by Bergson's (1911) work *Creative Evolution*, which was well known at the time and which expounded the 'becoming' of time and motion in contrast to their being constituted by successive states. In his post-1917 work Brouwer tried to extend the becoming to the continuum. A 'point of the continuum' is conceived as an 'indefinitely proceeding sequence' (i.p.s.) which is never completed either as a set or as a method. Free choice sequences (f.c.s.) were used to emphasize the 'becoming', growing aspect of all infinite sequences, the difference between the discrete point and the 'point of the continuum'. In the same way that in Bergson's notion of time there is no place for an atomistic discrete 'now', there is ultimately, in Brouwer's continuum, no room for a discrete point. The fact that in some way sequences determined by an algorithm can be thought of as finished, has obscured the true nature of the infinite mathematical sequence which is the progressive allocation of values by the human mind, according to a predetermined law or otherwise, and this *ad infinitum* is never ending or for ever proceeding. This becoming, growing, unfinished character exemplified by the f.c.s. (but also present in the law-like sequence) distinguishes the point of the continuum from the discrete. The notion of f.c.s., now the central point in current work on intuitionistic analysis, was a genuine innovation by Brouwer. It leads to a radical reinterpretation in this context of logical particles, particularly the universal and existential quantifiers. The first-order intuitionist logic is a subsystem of the classical logic; higher-order logics arise from this conception of the continuum—they introduce essentially novel ideas not included in classical logic.

The increasing prominence of measure theory in mathematics prior to the First World War and the reliance of the 'neo-intuitionists' (Lebesgue, Borel, etc.) on set-theoretical methods highlighted the need for a set theory which could bridge the gap between the continuum and the dense set of definable numbers which has measure zero. This need inspired Weyl to write *Das Kontinuum* (a work which inspired Lorenzen's operative logic) in 1917, an attempt abandoned a few years later in favour of Brouwer's 'continuum of free becoming'. To 'Brouwer we owe the new solution of the problem of the continuum' (Weyl 1921).

10.9.11 Historical

Brouwer's first important work (Brouwer 1907) expounded (i) the language-less constructive development of mathematics from the primordial intuition of time, (ii) wider views (§§10.8.2–10.8.8) on the nature of knowledge, and (iii) rejection of logicism and formalism. Nearly all his critical views originate in this work. *Het wezen der meetkunde* (Brouwer 1909) provided a link between his topology and foundations. It iterates the 1907 conclusion that the only a priori element is the intuition of time, asserts that with the co-ordinatization of space the a priori basis of geometry is no longer a problem, and takes a geometrical (i.e. formula-less) view of topology.

Brouwer's topological work was done in the period 1909–1919. Freudenthal and Heyting blame national isolation during the First World War for Brouwer's loss of interest. The movement of topology away from 'visual–perceptual' simplicity with the growth of algebraic topology, and the impact of set theory on the notion of geometry following Frechet's introduction of abstract spaces, induced Brouwer to concentrate on the notions of set, continuum, and function underlying topology. His continued interest in 'topology proper' is shown in his work with his students (assistants) even after 1925 (Alexandrov, Menger, Vietoris, H. Hopf, Freudenthal, Hurewicz).

Brouwer's opposition to PEM began in 1913 when he was an editor of *Mathematische Annalen*. He rejected submitted papers which applied PEM to undecided propositions. Brouwer was removed from the editorial board in 1921. Hilbert undertook the defence of classical mathematics by initiating the 'formalist' programme which entailed finding a consistency proof. This programme was abandoned in its original form in 1931 after Gödel had proved that every sufficiently rich formal system contains demonstrably unprovable propositions. Modified forms of the programme survived following the work of Gentzen.

The first part of Brouwer's programme, 'the first act of intuitionism', was the liberation of mathematics from logic and language. Most of his earlier work was a negative criticism of classical mathematics, particularly set theory which was dismissed as trivial. Brouwer (1918) inaugurated the 'second act of intuitionism', the construction of analysis on the basis of Brouwer sets and species.

Brouwer (1927) summarized his theory of functions and proved his uniform continuity theorem (which he regarded as the most fundamental theorem of intuitionist function theory) using the bar theorem and the fan theorem. Brouwer's main argument for the uniform continuity of all full functions centres round his definition of function and follows from his notion of mathematical existence, i.e. constructibility.

Brouwer's Vienna lectures (Brouwer 1928a, b) followed a long, bitter, and unedifying controversy with Hilbert in which he finally lost the general support of the mathematical world. The grand visions of these lectures inspired Wittgenstein to resume philosophical activity.

At the time of Hilbert's retirement, Brouwer, in his late forties, abandoned the public mathematical scene, leaving the intuitionist lead in the hands of Heyting who was about to publish his formalization of intuitionist logic.

Brouwer made a temporary return to public mathematical activity after the Second World War. Brouwer (1949a) returned to the mystical tendencies of 1905. His writings after his retirement in 1951 at the age of 70 concentrated on the broader issues of language and logic in mathematics or summarized his views on PEM, sequences, and the continuum.

10.10 Bishop's constructivism

10.10.1 *Bishop's programme*

Most of the work on intuitionism since the Second World War has been concerned with formalization of foundational aspects of Brouwer's ideas, particularly higher-order logic (free choice sequences)—contemplating the intuitionist navel—rather than extending intuitionist mathematics. E. Bishop is the outstanding exception: he carried constructive methods deep into functional analysis (Bishop 1967).

Bishop's programme is to replace non-constructive mathematics (which he characterizes by the pejorative term 'idealist') by constructive 'realist' mathematics. (Strictly, classical mathematics should be called 'naive realist': since Bishop claimed that mathematics is the creation of the mind, his own position might better be called 'pragmatic realist'.) Bishop was sensitive to the common impression that constructive mathematics was merely concerned with finding various constructive substitutes for classical results, thus appearing as a subordinate mathematical discipline:

> Constructive mathematics is in its infancy. According to some, it is doomed to the role of scavenger. These people conceive classical mathematics as establishing the grand design and the imaginative insight, leaving the constructivists to add whatever embellishments their credos demand. Although totally wrong, this viewpoint hints at a truth. The most urgent task of the constructivist is to give predictive embodiment to the ideas and techniques of classical mathematics. Classical mathematics is not totally divorced from reality.... The point rather is to use classical mathematics at least initially as a guide.... The emphasis will be on the discovery of useful and incisive numerical information, not the elegance of the format.
>
> Bishop 1970

Unlike Brouwer, Bishop did not articulate a more general critique of knowledge.

Bishop followed Kronecker's plan of developing mathematics from the positive integers, thus, with Brouwer, rejecting a priori geometrical knowledge (§10.9.3). His constructive intent was stoutly expressed as allows:

> Mathematics belongs to men, not to God. We are not interested in properties of the positive integers that have no descriptive meaning for finite man. When a man proves a positive integer to exist, he should show how to find it. If God has a mathematics of his own that needs to be done, let him do it himself.
>
> Bishop 1967

> ... integers are the only irreducible mathematical construct. This is not an arbitrary restriction but follows from the basic constructivist goal—that mathematics concerns itself with the precise description of finitely performable abstract operations. It is an empirical fact that all such operations reduce to operations with the integers. There is no reason why mathematics should not concern itself with finitely performable operations of other kinds, in the event that such are ever discovered; our insistence on the primacy of the integers is not absolute.
>
> Bishop 1970

10.10.2 *Formalization*

Like Brouwer, Bishop scorned formalization:

> In fairness to Brouwer it should be said that he did not associate himself with their (formalists) efforts to formalize reality; it is the fault of logicians that many mathematicians who think they know something of the constructive point of view have in mind a dinky formal system or, just as bad, confuse constructivism with recursion theory.
>
> Bishop 1967

We shall see that later Bishop (1970) conceded that formalization has some virtues.

10.10.3 *The continuum*

In Bishop (1967), Brouwer's concept of the continuum is rejected as mystical. Bishop simply uses special Cauchy sequences of rational numbers for the construction of real numbers: his notion of 'set' is similar to Brouwer's 'species' except that the constructive content is emphasized. A set is defined by describing what must be done to construct an element of the set and what must be done to show that two elements of the set are equal. Thus 'equality' for Bishop has a special computational meaning relative to each set—it is not absolute as in classical mathematics.

10.10.4 Logic

For Bishop, mathematical existence is a matter of computational evidence, similar to Brouwer's use of 'construction' but without the solipsistic connotations:

> Constructive existence is more restrictive than the ideal existence of classical mathematics. The only way to show that an object exists is to give a finite routine for finding out.

Classical mathematics, however, often deduces existence from the non-constructive 'principle of omniscience': either all elements of a set A have the property P or there exists an element of A with the property not-P. In its simplest form it can be stated as follows: 'If $\{n_k\}$ is a sequence of integers, then either $n_k = 0$ for some k, or $n_k \neq 0$ for all k'. Bishop showed that many deep classical theorems (Brouwer's fixed-point theorem, the ergodic theorem, the Hahn–Banach theorem, etc.) which depend on this principle have constructive substitutes.

The logical connectives have the same immediacy as with Brouwer. To prove $P \wedge Q$ one proves both P and Q as in classical mathematics (i.e. where the classical proof is direct); a proof of $P \vee Q$ is a proof of one of the alternatives. Bishop defines not-P to be P implies $0 = 1$, but in his mathematical practice he replaces negative conditions as far as possible by positive conditions. (It should be observed that Brouwer rejected the attempt by van Dantzig and Griss to eliminate negation entirely from mathematics—cf. Heyting (1956, §8.2).) A proof of $P \supset Q$ is more difficult to express generally:

> The validity of the computational facts implicit in the statement P must ensure the validity of the computational facts implicit in the statement Q, but the way this happens can only be seen by looking at the proof of the statement P implies Q. Statements formed with this connective, for example, statements of the type ((P implies Q) implies R) have a less immediate meaning than the statements from which they are formed, although in practice this does not seem to lead to difficulties in interpretations.
>
> Bishop 1967

Later, Bishop (1970) expressed dissatisfaction with this explanation of implication (see also Bishop and Bridges 1980, p. 13) and said that the problem of the numerical meaning of implication was '... the most urgent foundational problem of constructive mathematics':

> Although the numerical meaning of implication is a priori unclear, in each particular instance the meaning is clear... hopefully there exists a philosophical explanation of the empirical fact that intuitionistic implication in each instance admits a deeper analysis of the content of a theorem of constructive mathematics.
>
> Bishop 1970

Bishop attempted such an analysis by adopting an interpretation due to Gödel (1958) which used recursive functionals of finite type (see also Myhill (1975) for a formalization of Bishop's methods). This led Bishop, contrary to his earlier strictures on formalization, to seek a

> formal system that will efficiently express existing predictive mathematics.

In this system, Bishop claimed, every function mapping sequences of integers to integers, whose existence can be proved, is continuous in a certain sense:

> This can be regarded as the central result of Brouwer's theory of spreads. Thus we can develop Brouwer's ideas as metatheory.

Although this finally conceded some merits in Brouwer's 'mystical' view of the continuum, Bishop declared the theory of spreads to be unimportant as it has no significant mathematical applications.

10.11 Other kinds of constructivism

There are pseudo-constructive tendencies within classical mathematics, mainly based on Church's thesis. Constructive *objects* (usually recursive) are the centre of attention whilst classical reasoning is employed. There is a spectrum of possible developments in this direction (cf. Mostowski 1959). A quite different approach using classical logic and constructing objects by inductive definition is to be found in Lorenzen (1971).

There are several strands of philosophical thought, descending from Kant and Hegel, which deny both the realism on which classical mathematics is founded and Brouwer's mystical solipsism by maintaining that mathematics, logic, and language are *primarily* forms of social (i.e. public) consciousness. The most explicit development of mathematics from such a position is in Kamlah and Lorenzen (1967).

The views expresssed below are probably an inexact and biased reflection of Lorenzen's ideas, but they are sufficiently close to provide a rationale for several directions of development of constructive mathematics, including those of Brouwer, Bishop, and Lorenzen—the philosophical base does not uniquely determine the mathematical and logical superstructure.

10.11.1 *Public language*

Language—conventionalized use of signs—including measuring and counting, is an instrument of social activity. The individual possesses a language only as a member of a community. Language as the product of a solitary individual is senseless; the private meanderings of a pure mathematician must be validated by essentially public criteria—mathematics is an institution.

10.11.2 *Reality*

Like any other instrument, language enables people to do things which they could not do without it; thus it is a creator of facts. There is no way of grasping the 'external' world except by means of language itself: we do not *first* contact reality and *then* express this contact verbally. Rather, what we call 'the world', 'reality', 'the real world', ... is itself a development of language use. Facts concern language in action: the 'objects' of the 'real world' and the linguistic symbols 'referring' to them are derivative aspects of this activity. Language is *in* the world, not an external mirror for it (cf. Kamlah and Lorenzen 1967). In particular, mathematics creates its own reality rather than reflecting a pre-existing world of ideal objects. The 'reality' to which mathematical argumentation and computation refers is the corpus of inductive definitions by which mathematical entities are introduced (Lorenzen 1959) and spatial norms (see §10.11.5).

10.11.3 *Truth*

The realist conception of the relation between thought (i.e. *correct* thought), language, and reality is the 'correspondence' theory in which truth is an abstract correspondence by which a true proposition mirrors a part of the world. This view is overcome when truth is regarded as a certain aspect of the making and justifying of *truth claims*:

> The assumption that truth depends on some pre-existent external fact which has nothing to do with truth claims is a metaphysical delusion.
> Pivčević 1974

('Justify' here means 'establish in accordance with the appropriate (public) canons of validity'. It does not mean 'make plausible' or 'present intuitive arguments for'. In a mathematical context one could use the word 'prove' instead of 'justify', but 'prove' bears a heavy burden of formalist prejudice.) The reality to which the language signs correspond and the mode of such correspondence is unfolded by, and is an inseparable aspect of, the process of justifying truth claims in the appropriate manner. Truth, then, is not passive and abstract, but is *developed* by a concrete active process:

> To make a truth-claim is not just to claim that things are as is said they are: it is also to claim that they can be seen so to be.... Truth does not reduce either to true propositions or true statements it involves *experience* of truth.
> Pivčević 1974

Heyting (1956, p. 19), speaking of mathematical statements, expresses a closely related idea in the characteristic solipsistic manner of intuitionism:

> Every mathematical assertion can be expressed in the form: 'I have effected the construction A in my mind'...

10.11.4 Logic

All non-classical constructive tendencies agree in rejecting truth-values as an adequate foundation for formal logic. Let F denote the statement of the Fermat conjecture. F is assuredly a mathematical statement worthy of attention. Has F a truth-value? To claim that F has a truth-value but that no one can yet justify that claim, is, from the point of view of the preceding section, simply contradictory. The constructivist, therefore, cannot maintain that all mathematical sentences are truth-definite, i.e. have a well-defined truth-value. Thus constructive logic cannot be founded on truth-values. Constructive logical connectives and quantifiers cannot be found on an abstract concept of truth, but must be defined by laying down rules for their use in the practice of formal justification and computation. There are several ways of doing this. For the practical purpose of actually doing constructive mathematics Bishop's explanation of the connectives is sufficiently clear. Kamlah and Lorenzen (1967, Chapter 6) interpret connectives in terms of defending compound propositions against opposing doubts: the connectives then exhaust certain combinatorial possibilities of attack and defence.

A notable feature of constructive hypothetical reasoning is that, in accordance with the active conception of truth, to suppose a mathematical statement A is to suppose that it has been justified or proved. Thus, to suppose $A \vee B$ means to suppose that there is a proof of one of the disjuncts at hand, to suppose that $\exists x A(x)$ means to assume that a particular x has been constructed and that there is an available proof of $A(x)$, to suppose that $\forall x A(x)$ means to suppose that there has been constructed an algorithm which assigns to each x a proof of $A(x)$, etc.

10.11.5 Mathematics

Mathematics is ultimately a normative science, deriving from norms of spatial construction and measurement (geometry) and counting (arithmetic), and so concerns non-empirical truth (Kamlah and Lorenzen 1967, Chapter 6).

Brouwer upheld the reduction of geometry to analysis via co-ordinatization (§10.8.3). Bishop regarded mathematics as a purely numerical science. The work of Lorenzen, based on ideas of Dingler, indicates an alternative evaluation of geometry.

Current opinion regards geometry as having an empirical element—the type of geometry to be used in a particular science should be discovered by empirical means. But following Kant, and particularly Dingler, it is possible to build Euclidean geometry as an a priori science. Geometry as conceived here does not 'model' spatial relationships found in 'nature' or 'reality'—it is not a theory of an externally given space independent of conscious spatial activity. Rather, it concerns relations which are *prescribed* for artefacts: it lays down norms for geometrical figures (points, lines, planes) and relations between them (parallel, perpendicular, etc.). These norms are prescriptions for management of spatial constructions. Like rules, they are neither true nor false. But is not physical space non-Euclidean? If this question is reoriented towards our fundamental starting point of social activity, the question becomes: is not non-Euclidean geometry used in electromagnetic measurements? Yes. And Euclidean geometry *governs* precision engineering practice and design.

The natural numbers for the constructivist have their origin in the practical activity of counting, and this involves systematic operations on the number-signs (numerals). Arithmetic studies the norms for the construction of numerals and for effecting formal operations on them (e.g. Goodstein 1964 etc.).

The simplest way of constructing arithmetic (Lorenzen 1958) begins with the numerals $\mathfrak{N}$ (*Stichzahlen*'),

$$1, 11, 111, \ldots$$

(decimal notation is merely a useful contraction) which can be generated by the inductive definition (cf. Chapter 2, §2.2.5)

1 is an $\mathfrak{N}$

if n is an $\mathfrak{N}$ then so is $n1$.

Concatenation of these numerals is addition. Such numerical results as '11 + 111 gives 11111' and '11 × 111 gives 111111' are ternary relations on the numerals: there are obvious inductive definitions of them. Lorenzen called these rules 'practical arithmetic'. It makes no sense to ask whether they are true or false because they are merely rules.

The *theory* of numbers by constructing expressions $t_1 + t_2 = t_3, t_1 \times t_2 = t_3$, etc., where t_1, t_2, t_3 are numerals. $t_1 + t_2 = t_3$ is said to be *true* if $t_1 + t_2$ gives t_3 is a consequence of the inductive definition of '$\cdots + \cdots$ gives $\cdots$', i.e. is derivable in practical arithmetic. Thus the content of such an expression is the development of practical arithmetic and the passage from rule to statement as just outlined. The classical conception that $t_1 + t_2 = t_3$ expresses a certain abstract relation between the natural numbers denoted by t_1, t_2, t_3 is a distorted, reified view of the facts (cf. Lorenzen 1959).

The principle of induction is obtained by reflection on the principles of proof (cf. Heyting 1956, Chapter II, Lorenzen 1962, Chapter I). Suppose $E(x)$ is a predicate of natural numbers such that $E(1)$ has been proved and for each n, $E(n) \supset E(n + 1)$ can be proved (i.e. there is an algorithm α which for each n gives a proof of $E(n) \supset E(n + 1)$). Then by iteration of *modus ponens* an algorithm β can be constructed from α which for each n gives a proof of $E(n)$. Hence $E(1) \wedge (\forall n)(E(n) \supset E(n + 1))$ implies $(\forall n)E(n)$.

The arithmetic of rational numbers can then be constructed. There are a variety of ways of proceeding to analysis.

11

Modal logics

11.1 Introduction

Questions about modal logic were posed by Aristotle and were thoroughly investigated by the Scholastics, but the subject seems to have been completely neglected until it acquired a new lease of life in mathematical logic, particularly through the work of McColl and Lewis. A recent nodal point in the development of the modal industry was the introduction of 'possible world' semantics (Kripke) which has lead to an unprecedented boom.

Aristotle's theory of modal statements is developed in *De Interpretatione*, Chapters 12 and 13, and *Prior Analytics*, i3 and 13; that of modal syllogisms is developed in *Prior Analytics*, i8–22. His metaphysical views lead to the basing of modality on a notion of contingency and to a particular structure of modal sentences. A distinction was made between belonging and necessarily belonging:

> Since there is a difference between belonging, necessarily belonging, and being able to belong (for many things belong, but not necessarily, others neither necessarily nor at all, but all able to belong), it is clear that there will be syllogisms in each of these cases ...
>
> Aristotle 1949, 8(29^b29–35)

Bocheński says of this passage that

> ... it can be gathered from this text that the modal functor does not determine the sentence as a whole, but part of it. So that for Aristotle a modal sentence is not conceived in such a sense as: 'It is possible that A belongs to B'. The modal functor does not precede the whole sentence but one of its arguments. This distinction becomes still clearer for the distinction is three times made between two possible cases:
> 1. to that which B belongs, A also can belong,
> 2. to that to which B can belong A also can belong.
>
> In the first case the modal functor determines only the consequent, in the second case it determines the antecedent too.
>
> Bocheński 1961, p. 83

Kneale (1962, p. 83) expresses some hesitation in this matter and points to two distinct usages:

> It appears that in his theory of the conversion of modal statements, Aristotle takes what may be called the external view of modality, while in treating of modal

syllogisms he is inclined to the internal interpretation. There is therefore a certain incoherence between the two parts of his theory...

Kneale 1962, p. 91

The logical content of Aristotle's notion of *possibility* is quite clear when he defines it as the non-self-contradictory (*Metaphysics*, Δ 1019^{b}27ff). The notion of contingency is, however, ambiguous:

> The contingent he sometimes treats as if it were an element in the real simply juxtaposed to the necessary and the result is his logical doctrine that one judgement differs modally from another only because it categorically asserts a different type of predicate. Yet at other times Aristotle regards the contingent as merely the defective appearance of the necessary, and chance as a cloak for ignorance.
>
> Mure 1950, p. 134

Aristotle's view was later rejected by Theophrastus, who thought that the modal functor governs the whole sentence. In the middle of the thirteenth century there arose a generally accepted doctrine about the structure of modal propositions. A text of Thomas Aquinas classifies propositions according to the place which the mode has in it. There is also an explicit distinction between the Aristotelian and Theophrastian structures. The modalities *de re* and *de dicto* correspond to the Aristotelian structure in which the modal signs are essentially internal, and the Theophrastian doctrine that it must apply to a proposition as a whole (see Kneale 1962, p. 236). Kneale (1960) maintained that the *de re/de dicto* distinction is one between the different ways in which modal words can occur in ordinary language and not as an important metaphysical distinction: the primary notion is *de dicto*:

> ... If I say 'Every Scotsman must have heard of Burns', I do not mean that the dictum of the sentence 'Every Scotsman has heard of Burns' is intrinsically necessary, which it certainly is not, but rather that there is a fact about present Scotsmen, namely that they have been subject to Scottish national propaganda which makes it inevitable that they have heard of Burns. My examples are in fact elliptical statements of relative modality and just for that reason unsuitable as illustrations for a formal theory of absolute modality in which modal statements are assumed to be fully explicit. But defenders of the theory of modality *de re* have not made that point about it. On the contrary, they have sometimes talked as though it were an independent use of modal words and even perhaps the basic use.
>
> Kneale 1960, p. 626

Modern interest in modal logic begins with the work of Lewis (1918). His theory is commonly called the theory of *strict implication*. It was advanced in opposition to Russell's account of implication (e.g. in *Principia Mathematica*) which, in more modern terms, confuses use with mention. In his definition of implication Lewis followed some suggestions of McColl (1906) on modal logic. Lewis's main interest was to maintain that the deducibility of a conclusion in a valid argument depends on there holding *between the propositions* a relation called 'strict implication' which is not truth-functional. But instead of preserving this distinction between use and mention, Lewis's strict implication propagated the confusion. Lewis presented one version of his system of strict implication in his *Survey of Symbolic Logic* (1918). Further developments were given in his part of Lewis and Langford (1932). At the end of that book he presented five systems S1–S5 in order of increasing strength. He expressed himself as unable to decide which of his systems 'expresses the acceptable principles of deduction'. The first axiomatization of modal logic by augmentation of the propositional calculus with extra rules and axioms appears to have been given by Gödel (1933) (see also Cresswell and Hughes 1968, Kneale 1962, p. 680). The extra apparatus embodied the concept of 'p is provable' (this also gave an interpretation of Heyting's formalization of intuitionistic propositional calculus).

The use versus mention distinction also seems to be blurred in connection with quantification. For in modal predicate logic quantification into a modal context was formally permissible. But

> ... you obviously cannot coherently quantify into a mentioned sentence from outside the mention of it.'
>
> Quine 1966, p. 175

Indeed, Quine was severely critical of the whole exercise of attempting to characterize a notion of 'analytic' axiomatically:

> There are logicians, myself among them, to whom the ideas of modal logic (e.g. Lewis') are not intuitively clear until explained in non-modal terms. But so long as modal logic stops short of quantification theory, it is possible to provide somewhat the type of explanation desired. When modal logic was extended to include quantification theory on the other hand, serious obstacles were encountered.... The notion of analyticity thus appears at present writing to lack a satisfactory foundation. Even so, the notion is clearer to many of us and obscurer surely to none than the notions of modal logic: so we are still obliged to explain the latter in terms of it.
>
> Quine 1947

The modal industry expanded enormously with the advent of 'possible world semantics' which, incidentally, gave a rationale of sorts to quantification into modal contexts. The modal operators are interpreted in terms of model structures (Chapter 6, §6.10) in which the relations are characterized abstractly *but are undefined* (Löb 1966). The problem is exiled to the understanding of the relation between the 'possible worlds'. Doubts about this latest approach to modal logic were expressed by Scott (see also Scott 1972, p. 244):

> *Some* would now claim a coherent understanding of modal contexts in the light of extensive work on possible-world semantics. And we ought to take them seriously; for they have done *something* coherent. There remains a nagging doubt in my mind, however, that the problem solved by the use of the possible-worlds

approach is *not quite* the original problem of modal logic, only an analogous one sharing certain formal similarities.

Scott 1973, p. 788

Nevertheless, realistic applications of 'possible world semantics' are arising in computer science. Pratt (1976) has introduced a logical framework for discussing computer programs based on this approach to modal logic. Such a system has been called *dynamic logic*. The idea is to incorporate programs into the assertion language in the form of modal operators (Fisher and Ladner 1979). For instance, if π is a (possibly non-deterministic) program and A an assertion, then a new assertion, $\langle \pi \rangle A$ can be made which is interpreted as 'π can terminate with A holding on termination'. This compares with 'possibly A'. The dual modal operation $[\pi]$, comparable with 'necessarily A' is defined as $\neg \langle \pi \rangle \neg A$ which means 'whenever π terminates, A holds on termination'. The Hoare assertion $\{A\}\pi\{B\}$ (Chapter 2, §2.3.9) can then be expressed as $A \supset [\pi]B$. The fact that π can terminate can be expressed by the assertion $\langle p \rangle t$, where t is the constant true proposition. The determinacy of a program π can be expressed by $\langle \pi \rangle A \supset [\pi]A$, where A expresses determinacy.

11.2 Necessity and logic

Leibniz called propositions which are true because the world is what it is 'truths of fact'. Other truths which do not depend on the world are truths of logic or 'logically necessary: the latter are true in all possible worlds. To say that a proposition has logical necessity is to say that its denial would involve a contradiction' (Parkinson 1960, p. 121). The related term 'possible', following Aristotle, is defined in terms of non-contradiction:

> That which does not contain a contradictory term, i.e. A not-A is 'possible'. That which is not Y not-Y is possible.
>
> Parkinson 1960, p. 54

In his *Begriffschrift*, Frege (1879) maintained that modal ideas—words like 'must', 'may', 'necessary', and 'possible'—involved a covert reference to human knowledge and so had no place in pure logic. He regarded the effort of Aristotle and his followers to erect a logic of modal propositions as erroneous. Many later philosophers accepted these modalities only as defined in 'metalogical' terms, e.g.

> We must allow that necessity is not a bond between existing things. For logic, what is necessary is nothing beyond a logical consequence.
>
> Bradley 1922, p. 200

Russell regarded the confusion about these ideas as arising from lack of analysis of metalogical types (e.g. the use/mention distinction—Quine):

Another set of notions as to which philosophy has allowed itself to fall into hopeless confusion through not sufficiently separating propositions and propositional functions are the notions of 'modality': *necessary, possible* and *impossible*... In fact, however, there was never any clear account of what was added to truth by the conception of necessity. In the case of propositional functions the three-fold division is obvious. If '$\varphi(x)$' is an undetermined value of a certain propositional function, it will be *necessary* if the function is always true, *possible* if it is sometimes true and *impossible* if it is never true.

Russell 1950, p. 165

Like Bradley, Wittgenstein (of the *Tractatus*) saw 'necessity' as strictly logical:

> 6.37. A necessity for one thing to happen does not exist. There is only *logical* necessity.
>
> 6.375. As there is only *logical* necessity, so there is only *logical* impossibility.

Wittgenstein 1922

Carnap (1949, §§69–71) maintained that sentences containing modalities ('possible', 'impossible', 'necessary', . . .) could be translated into 'syntactical' sentences. Thus, '*A* is possible' is translated as '*A* is not contradictory', though, because of the ambiguity of modal terms, alternative syntactical versions can be found. This makes modal logic redundant:

> ... *syntax already contains the whole logic of modalities* and the construction of a special intensional logic of modalities is not required.

Carnap 1949, p. 256

The first of Carnap's claims can be conceded without rendering the study of modalities entirely useless. Indeed, to define how logics work we cannot easily avoid the use of modal terms:

> ... we cannot consider logic itself without using modal terms. For the formal connections which we consider in logic are naturally described as *necessary*; and when we try to explain what is meant by the validity of a syllogism, we commonly say that it is *impossible* for the premisses to be true while the conclusion is false. Even if the study of these notions were assigned to a different science, called, for instance, metalogic, this would still be a natural extension of logic.

Kneale 1962, p. 96

The aim of this chapter is to begin the development of modal logic as a 'natural extension of logic', i.e. in contrast with the postulational approach or possible world semantics, by extending logic to include its 'syntactical' notions. Rather than trying to axiomatize a notion of necessity, we shall *construct* modal logics on principles which recognize that

(i) necessity is primarily logical and derived from a notion of logical consequence, and consequently

(ii) necessity is relative to given hypotheses and relative to a given logic.

In §11.4 there is a construction of a classical modal logic which first appeared in Cleave (1974b). An effective modal logic, which is constructed in the same spirit as effective logic in Chapter 10, is established in §11.7.

11.3 Relative necessity

A notion of necessity, which was called *hypothetical necessity*, was formulated by Leibniz: a proposition is *hypothetically necessary* if it is necessary, *given* that such and such is the case. This notion is clearly derived from some concept of implication. It was clearly stated by Russell, referring to a paper of Moore (1900):

> The only logical meaning of necessity seems to be derived from implication. A proposition is more or less necessary according as the class of propositions for which it is a premiss is greater or smaller. In this sense the propositions of logic have the greatest necessity and those of geometry have a high degree of necessity.
> Russell 1903, p. 454

This notion is central to the idea of *material necessity*. A distinction is frequently made between empirical truths, the truths of logic, and a third class of truths—the *materially necessary*. In the latter category one might place geometry (considered as a priori exact sciences (Lorenzen 1961b, Kamlah and Lorenzen 1967)) and also some of the propositions of set theory such as the axiom of infinity and the axiom of choice. In the realm of mathematical-physical theories, materially necessary propositions frequently occur as explicitly stated or tacitly assumed restrictions on the class of models to be used. For instance, it is customary to formulate a physical law by a set of ordinary or partial differential equations (Newton's equations of motion for gravitating bodies, Maxwell's equations for the propagation of electromagnetic waves, etc), it's being tacitly assumed that the differential calculus is the 'correct' mathematical method. One could externalize this procedural decision by expressing it as an axiom: 'the universe is a differentiable manifold'. The logical consequences of this proposition are materially necessary propositions of physics. They are not forced on one by logical considerations only—no amount of empirical evidence can refute or deny them. They can be justified on pragmatic grounds and by a priori conceptions of space and time. Similarly, the 'dogma of structural stability' (Abraham 1967, Thom 1967), which requires that models of physical theory have certain qualitative features, finds its justification in some presuppositions about the nature of physical enquiry.

The exact borderlines between the three types of truth is a matter of philosophical dispute. Körner (1973) observed, however, that all the various definitions of material necessity have this feature in common:

> To assert the material necessity of a proposition is to assert that the proposition is true and follows logically from a conjunction of principles which for some reason or other have a privileged status and confer it on their logical consequences.

A related notion of relative necessity was recognized by Carnap (1949, pp. 251ff) in his discussion of the difference between the so-called *logical* and *real* modalities. The real modalities, in this conception, relate to logics which incorporate elements from physical theory and so modify the conception of logical consequence (see the definition of 'L-consequence' and 'P-consequence', (Carnap 1949, p. 180)).

Some general problems emerge at the outset of the enterprise of extending a given logic to embrace a notion of relative necessity. Take a predicate logic L of type τ with its notion logical consequence. Suppose a principle P of material necessity is formulated as a set of sentences of type τ. We suppose that the set $\mathbb{F}_{pred}(\tau)$ of formulas of this language is defined by Definition 2.12, i.e. it is constructed from At, the set of atomic formulas, by the formation of 'logical combinations' to give $\mathbb{F}_{pred}(\tau) = \Lambda(At) =_{def} \mathbb{F}_0$. (Here, $\Lambda(X)$ is the set of all logical combinations of formulas in X (Chapter, §2.2.1, and Definition 2.12(ii)).) For each formula $A \in \mathbb{F}_0$ a new 'formula' is now constructed by prefixing it with a new symbol N, the *necessity operator*; NA will mean that A is a logical consequence of P, i.e. A is a *materially necessary* proposition, or A is *P-analytic*. It is a statement *about* A—it is a different logical type from that of A and so the usual syntactical properties of formulas in $\mathbb{F}_0$ cannot, without special justification, apply to the new formulas. Therefore it must be treated in some respects as a prime formula. In accordance with the type difference—the use/mention distinction—we declare that no occurrence of a variable in A will count as a free or bound variable in NA. Thus

$$Bv(NA) = Fv(NA) = \emptyset. \tag{11.1}$$

Nevertheless, there are circumstances under which it is still useful to count A as a subformula of A.

There is no formal problem in expanding the class of formulas to include 'logical' combinations of modal formulas with $\mathbb{F}_{pred}(\tau)$; we simply put $\mathbb{F}_1 =_{def} \Lambda(\mathbb{F}_0 + \{NA; A \in \mathbb{F}_0\})$. There is then no *formal* reason to refrain from formally constructing a further set of modal formulas and their logical combinations to give $\mathbb{F}_{n+1} =_{def} \Lambda(F_n + \{NA; A \in F_n\})$. This process gives a chain

$$\mathbb{F}_0 \subset \mathbb{F}_1 \subset \mathbb{F}_2 \subset \cdots$$

of classes of formulas, the union of which we call F^N. The problem is how to extend the notion of logical consequence from $\mathbb{F}_0$ to $\mathbb{F}_1$. Supposing this problem to be solved, there then arises a similar problem at the next stage of construction etc. Consider a semantically based logic. The notion of logical consequence is derived from a truth definition (Chapter 6, §6.3) based on a set T of truth-values. The truth-definition must be extended from $\mathbb{F}_0$ to $\mathbb{F}_1$. How shall a truth-value be assigned to NA where $A \in \mathbb{F}_0$? From a classical point of view, NA is either true or false. But are *true* and *false* in this sense adequately represented by members of T? Since there is a difference of logical type between NA and A there is a case for declaring that these NA have a different kind of truth-value from that of A. Hence there should be different sorts of truth values at each level of the nesting of the necessity operator. The truth definition should then proceed as in Muškardin's multi-subject semantics (Chapter 9, §9.4). The procedure is considerably simplified if the set T of truth-values includes a maximum element 1 and a minimum element 0. The truth-value of NA can then be *identified* with 1 if A is a consequence of P and *identified* with 0 otherwise. This is a procedural decision we have to make.

Observe that as NA means 'A is a logical consequence of P' and P is a set of *sentences*, then NA will turn out to be logically equivalent to $N\forall A$, where $\forall A$ is the universal closure of A, i.e. a *sentence* (Definition 2.14). Thus we could restrict N to apply to *sentences* only (cf. von Wright 1951, pp. 26–8): this course was chosen by Cleave (1974b). In the development chosen below we prefer N to apply to any formula—the difference is insignificant. There are some deep problems concerned with combining modalities with quantifiers which are discussed in Quine (1947) and treated in depth by Carnap (1947, Chapter 5) (for a criticism of Carnap's view see Hintikka (1973)). We cut the Gordian knot here by strictly observing the use/mention distinction via §11.1.

11.4 Constructing a classical modal logic

Take a predicate logic L of type τ with notions of truth and logical consequence defined in the usual way (Chapter 2, §2.3.2). Suppose a principle P of material necessity is formulated as a set of sentences of type τ. As above, the language of L is then extended to L_N by adjoining the (new) *necessity junctor* N to the set J of junctors, treating N as a one-place junctor. The truth definition of L_N is then obtained by adjoining to the classical (two-valued) truth definition of L a clause declaring NA to be true in a given structure if A is a logical consequence of P. The set P^N of sentences of L_N which are true in all structures of type τ satisfying P is the *theory of necessity in P*. It is clear that the conception of necessity occurring here is based entirely on the well-known principles of classical semantics, the class of 'possible worlds' being simply an elementary class in the wider sense, to use the standard terminology of model theory (e.g. Tarski and Vaught 1957). Our approach is therefore closely related, in spirit at least, to the work of Montague (1960) and Löb (1966). We present some results on the axiomatization of theories of necessity. The principal result (Theorem 11.3) implies that a theory of necessity P^N is axiomatizable (and decidable) if, and only if, the theory P is decidable. It follows from this that the theory of logical necessity—obtained by taking P to be the set of universally valid sentences of the predicate calculus—is non-axiomatizable. Later it is shown that the *general theory of necessity*—the set of sentences of L_N common to all theories of necessity—can be axiomatized by a S5-like system, as is already suggested by results of Montague (1960), Löb (1966), Körner (1973), and Scott (1973, 1974a). It follows that the general theory of necessity is a proper subsystem of the theory of logical necessity.

The class $\mathbb{F}^N$ can also be defined inductively by adding to Definition 2.12 a clause (F4) relating to the construction of modal formulas. At the same time the *modal depth* $m(A)$ of A—the depth of nesting of the modal operator—is defined.

(F1) If $A \in At$ then $A \in \mathbb{F}^N$ and $m(A) = 0$.

(F2) If $A, B \in \mathbb{F}^N$ then $\neg A, [A \vee B], [A \wedge B], [A \supset B] \in \mathbb{F}^N$ and $m(\neg A) = m(A), m([A \vee B]) = m([A \wedge B]) = m([A \supset B]) = \max\{m(A), m(B)\}$.

(F3) If $A \in \mathbb{F}^N, x \in Fv, a \in Bv - Bv(A)$ then $\forall a A(a/x) \in \mathbb{F}^N$ and $\exists a A(a/x) \in \mathbb{F}^N$ and $m(\forall a A(a/x)) = m(\exists a A(a/x)) = m(A)$.

(F4) If $A \in \mathbb{F}^N$ then $NA \in \mathbb{F}^N$, $m(NA) = m(A) + 1$, and $Bv(NA) = Fv(NA) = \emptyset$.

The logical length of a formula is defined in the usual way but counting N as a logical symbol (Definition 2.1(v)).

By a *modal formula* we mean one whose prime subformulas are within the scope of a necessity operator. Thus $\forall a \forall b [P[a, b] \wedge N[Q[a, b] \vee P[a, x]]]$ is not a model formula because the subformula in bold type is prime but not within the scope of N. But $\forall a \forall b [NP[a, b] \wedge N[Q[a, b] \vee P[a, x]]]$ is a modal formula (with modal depth 1) because its only prime subformulas are $P[a, b]$ and $[Q[a, b] \vee P[a, x]]$, and the first occurs as a subformula of $NP[a, b]$ whilst the second occurs as a subformula of $N[Q[a, b] \wedge P[a, x]]$. For all $A_1, A_2 \in \mathbb{F}^N$, $\neg NA_1 \supset N\neg NA_1$, and $N[A_1 \supset A_2] \supset [NA_1 \supset NA_2]$ are modal formulas.

Some standard notions of first-order semantics will be assumed and we assume that terminology of Definition 2.13, such as *satisfaction*, *model*, etc., once the truth definition has been made. The assignment of truth-values given in Definition 2.13(iii) needs to be extended by a clause (iii)(d) to cover modal formulas in addition to (iii)(a)–(iii)(c):

(iii) (d) $[\mathfrak{A}, \alpha](NA) = glb\{[\mathfrak{B}, \beta](A); (\mathfrak{B}, \beta) \models P\}$.

Thus $(\mathfrak{A}, \alpha) \models NA$ iff every model of P is a model of A. (We have used the symbol $\models$ as in Definition 2.13, even though, in this context, it depends upon P.)

DEFINITION 11.1.

(i) A is a *P-consequence* of the set M of formulas (written $M \to_P A$) if for all $(\mathfrak{A}, \alpha)$, $(\mathfrak{A}, \alpha) \models A$ whenever $(\mathfrak{A}, \alpha) \models M$. A and B are *P-equivalent* (written $A \leftrightarrow_P B$) when $A \to_P B$ and $B \to_P A$.

(ii) A is *P-valid* if $\varnothing \to_P A$, i.e. $(\mathfrak{A}, \alpha) \models A$ for all $(\mathfrak{A}, \alpha)$ of type τ. $P^N =_{\text{def}} \{A \in \mathbb{F}^N : P \to_P A\}$ is the *theory of necessity in P*.

(iii) If P is a set of first-order sentences (of type τ) then P^N is an *elementary theory of necessity* (of type τ). If P is a finite set of first-order sentences (of type τ) then P^N is a *finite theory of necessity* (of type τ).

(iv) $\varnothing^N$ is the *theory of logical necessity* of type τ. N is defined as the intersection of all elementary theories of necessity of type τ and is the *general theory of necessity*.

It is easily established that $\to_P$ is a logical consequence relation and $\leftrightarrow_P$ is logical equivalence. Let Cn_P be the corresponding logical consequence function. Thus $\mathfrak{L}^N(P) =_{\text{def}} \langle \mathbb{F}^N, Cn_P \rangle$ is a logic and $\varnothing^N$ is its set of theorems. Further,

$$A \leftrightarrow_P B \implies NA \leftrightarrow_P NB. \tag{11.2}$$

Hence if $G[A]$ is a formula with a distinguished occurrence of a subformula A (Chapter 2, §2.2.2) then

$$A \leftrightarrow_P B \implies G[A] \leftrightarrow_P G[B]. \tag{11.3}$$

EXAMPLE 11.1. For any formulas $A, B \in \mathbb{F}^N$ the following formulas are *P*-valid for all P and so are in N:

$$NA \supset A$$

$$\neg NA \supset N\neg NA$$

$$N[A \supset B] \supset [NA \supset NB]$$

$$N[A \wedge B] \supset [NA \wedge NB]$$

$$[NA \wedge NB] \supset N[A \wedge B]$$

$$NA \supset [[A \supset NB] \supset NB]. \qquad \blacklozenge$$

Lemma 11.1.

(i) $(\mathfrak{A}, \alpha) \vDash NA$ iff $A \in P^N$.

(ii) Let $A \in \mathbb{F}_{\text{pred}}(\tau)$. Then $P \to A$ iff $A \in P^N$.

(iii) For all $A \in \mathbb{F}^N$, $NA \leftrightarrow_P t$ or $NA \leftrightarrow_P f$.

(iv) $N[A \supset B] \in P^N$ iff $A \to_P B$.

(v) For all $A, B \in \mathbb{F}^N$, $N[A \wedge B] \leftrightarrow_P [NA \wedge NB]$.

(vi) If $A, A \supset B \in P^N$ then $B \in P^N$.

(vii) $A \in P^N$ iff $NA \in P^N$.

(viii) $\neg NA \in P^N$ iff $A \notin P^N$.

(ix) For all $A, A_1, \ldots, A_{n+1} \in \mathbb{F}_{\text{pred}}(\tau)$,
$$[NA \supset NA_1] \vee \cdots \vee [NA \supset A_n] \vee [NA \supset A_{n+1}] \in P^N.$$
$$\Leftrightarrow \quad A \notin P^{**} \quad \text{or } A_1 \in P^{**} \quad \text{or } \cdots A_{n+1} \in P^{**}$$

(P^{**} is the set of logical consequences of P in the predicate logic. See Definition 2.4(i) and Example 2.1(i).)

Proof. (i)–(viii) are easy consequences of the truth-definition and Definition 11.1.

(ix) Suppose that
$$[NA \supset NA_1] \vee \cdots \vee [NA \supset NA_n] \vee [NA \supset A_{N+1}] \in P^N.$$

By (iii) $NA, NA_1, \ldots, NA_n$ are each P-equivalent to t or f. Consider the two cases (a) $NA \leftrightarrow_P t$ and (b) $NA \leftrightarrow_P f$.

(a) Then $NA_1 \vee \cdots \vee NA_n \vee A_{n+1}$ is P-valid. If $NA_i \leftrightarrow_P t$ for some i, $1 \le i \le n$, then $NA_i \in P^N$. By (ii) and (vii), $A_i \in P^{**}$. If $NA_i \leftrightarrow_P f$ for $1 \le i \le n$ then $A_{n+1} \in P^N$. By (ii). $A_{n+1} \in P^{**}$.

(b) Then by (i), (ii), and (viii), $A \notin P^{**}$. ∎

Corollary 11.2 easily follows from (iii) of the above lemma.

Corollary 11.2.
Every modal formula is logically equivalent to t or to f. Every formula of $\mathbb{F}^N$ is logically equivalent to a formula of $\mathbb{F}_0$.

Corollary 11.3.
$\mathfrak{L}^N(P)$ is compact.

THEOREM 11.4.

(i) Let Δ be a set of modal formulas and $\Gamma \in \mathbb{P}_\omega(\mathbb{F}^N)$. Then

$$A \in Cn_P(\Delta, \Gamma) \quad \Rightarrow \quad N[\bigwedge \Gamma \supset A] \in Cn_P(\Delta).$$

(ii) Let $\Delta, \Theta \subseteq \mathbb{F}^N$ and $\Gamma \in \mathbb{P}_\omega(\mathbb{F}^N)$. Then

$$A \in Cn_P(\Delta, \Theta) \quad \Rightarrow \quad A \in Cn_P(\Delta, \Gamma, \{N[\bigwedge \Gamma \supset B]; B \in \Theta\}).$$

Proof.

(i) For each $G \in \Delta$, $G \leftrightarrow_P t$ or $G \leftrightarrow_P f$ by Theorem 11.2. If $G \leftrightarrow_P t$ for all $G \in \Delta$ then $Cn_P(\Delta, \Gamma) = Cn_P(\Gamma)$. Hence $A \in Cn_P(\Gamma)$ so that $\bigwedge \Gamma \supset A \in Cn_P(\Delta)$. But if $G \leftrightarrow_P f$ for some $G \in \Delta$ then $Cn_P(\Delta, \Gamma) = \mathbb{F}^N$ so again $\bigwedge \Gamma \supset A \in Cn_P(\Delta)$.

(ii) If $N[\bigwedge \Gamma \supset B] \leftrightarrow_P f$ for some $B \in \Theta$ then

$$Cn_P(\{N[\bigwedge \Gamma \supset B]; B \in \Theta\}) = \mathbb{F}^N$$

so that the stated relation holds trivially. But if $N[\bigwedge \Gamma \supset B] \leftrightarrow_P t$ for all $B \in \Theta$ then $N[\bigwedge \Gamma \supset B] \in P^N$ for all $B \in \Theta$. By Lemma 11.1(iv), $\Theta \subseteq Cn_P(\Gamma)$. Hence

$$Cn_P(\Delta, \Theta) \subseteq Cn_P(\Delta, \Gamma, \{N[\bigwedge \Gamma \supset B]; B \in \Theta\}). \blacksquare$$

The basic fact on which the analysis of theories of necessity depends is a refinement of Corollary 11.2.

THEOREM 11.5. Let $G[NB]$ be a formula of $\mathbb{F}^N$ with a distinguished occurrence of a subformula NB. Then

$$G[NB] \leftrightarrow_P [[NB \wedge G[t]] \vee [\neg NB \vee G[f]]]. \qquad (11.4)$$

Proof. By Lemma 11.1(iii), $NB \leftrightarrow_P t$ or $NB \leftrightarrow_P f$. If $NB \leftrightarrow_P t$ then, by (11.3), $G[NB] \leftrightarrow_P G[t]$ and

$$[[NB \wedge G[t]] \vee [\neg NB \vee G[f]]] \leftrightarrow_P [[t \wedge G[t]] \vee [f \wedge G[f]]]$$
$$\leftrightarrow_P G[t].$$

Hence (11.4) holds. If $NB \leftrightarrow_P f$ then $G[NB] \leftrightarrow_P G[f]$ and

$$[[NB \wedge G[t]] \vee [\neg NB \vee G[f]]] \leftrightarrow_P [[f \wedge G[t]] \vee [t \wedge G[f]]]$$
$$\leftrightarrow_P G[f].$$

Again, (11.4) holds. $\blacksquare$

It is obvious that Lemma 11.5 can be used to 'unnest' the occurrences of the operator N in any given formula and, moreover, remove them from within the scope of a quantifier to obtain a propositional combination (i.e. with junctors only) of formulas of the form A and NA for $A \in F_1$.

COROLLARY 11.6. *There exists a recursive function* $\beta: \mathbb{F}^N \to \mathbb{F}^N$ *such that, for every* $A \in \mathbb{F}^N$ *and every* $P \subseteq \mathbb{F}_{\text{pred}}(\tau)$, $A \leftrightarrow_P \beta(A)$ *and* $\beta(A)$ *is a conjunction* $A_1 \wedge \cdots \wedge A_n$ *where each conjunct* A_i *is a disjunction of formulas of the form* B, NB, *or* $\neg NB$ *where* $B \in \mathbb{F}_{\text{pred}}(\tau)$.

Once the principles of necessity P have been fixed, the truth-values of formulas NA are, by Lemma 11.1(i), independent of the structure in which the above formulas A_i are evaluated. Hence each $B \in \mathbb{F}^N$ is P-equivalent to a formula B' in $\mathbb{F}_{\text{pred}}(\tau)$. This seems to imply that P^N is *merely* a transcription of P^{**}. However, this conclusion is not quite accurate because it follows from Theorem 11.7 below that if P^{**} is undecidable then the transformation B to B' is non-recursive.

11.5 Axiomatizability of theories of necessity

The possibility of axiomatizing theories of necessity rests upon the following relations of reducibility between P^{**} and P^N.

THEOREM 11.7. *Let P be a set of sentences in* $\mathbb{F}_{\text{pred}}(\tau)$. *Then we have the following.*

(i) P^N *is Turing-equivalent to* P^{**}.
(ii) *If* P^N *is recursively enumerable then* P^{**} *is recursive.*

Proof.

(i) Let $A \in F^N$. $\beta(A) = A_1 \wedge \cdots \wedge A_n$, say, where $A_i = \neg NG_1^i \vee \cdots \vee \neg NG_{m(i)}^i \vee NH_1^i \vee \cdots \vee NH_{n(i)}^i \vee H_{n(i)+1}^i$, and the G^i and H^i are formulas in $\mathbb{F}_{\text{pred}}$ (Corollary 11.6). Then

$$A \in P^N \iff A_i \in P^N \quad \text{for } 1 \leq i \leq n. \qquad (11.5)$$

But

$$A_i \leftrightarrow_P \bigvee \{\bigwedge \{G_3^i; 1 \leq j \leq m(i)\} \supset NH_k^i; 1 \leq k \leq n(i)\} \vee$$
$$\bigwedge \{G_3^i; 1 \leq j \leq m(i)\} \supset H_{n(i)+1}^i\}.$$

By Lemma 11.1(i)

$$A_i \in P^N \iff G_1^i \notin P^{**} \text{ or } \cdots \text{ or } H_1^i \in P^{**} \text{ or } \cdots \text{ or } H_{n(i)+1}^i.$$

$$(11.6)$$

(11.5) and (11.6) define the required reduction of P^N to P^{**}. Conversely, P^{**} is reducible to P^N.

Next, by Lemma 11.1(ii), for any $A \in \mathbb{F}_{\text{pred}}(\tau)$,

$$A \in P^N \quad \Leftrightarrow \quad A \in P^{**} \tag{11.7}$$

and by Lemma 11.1(viii)

$$\neg NA \in P^N \quad \Leftrightarrow \quad A \notin P^{**}. \tag{11.8}$$

(11.7) and (11.8) show that P^{**} is reducible to P^N. Thus, P^{**} is Turing-equivalent to P^N.

(ii) If P^N were recursively enumerable, then by (11.7) and (11.8) both P^{**} and its complement would be recursively enumerable and so P^{**} would be recursive. ∎

COROLLARY 11.8.

(i) If P^{**} is not decidable then P^N is not axiomatizable.
(ii) P^N is either non-axiomatizable or decidable.

Proof.

(i) Suppose P^{**} is not decidable. By Theorem 11.7(ii), P^N is not recursively enumerable and so is not axiomatizable.
(ii) Suppose P^N is axiomatizable. By (i) P^{**} is decidable. By Theorem 11.7(i) so also is P^N. ∎

EXAMPLE 11.2. The theory of logical necessity $\varnothing^N$. Let τ be a similarity class of structures with at least one binary relation. Then $\varnothing^{**}$ is the set of valid formulas of the predicate calculus (of type τ) with at least one two-place predicate letter. By theorems of Church (1936) and Kalmár (1936), $\varnothing^{**}$ is recursively enumerable but non-recursive. By Corollary 11.8(i) $\varnothing^N$ is non-axiomatizable. ◆

EXAMPLE 11.3. The theory of necessity in groups. Let $\mathfrak{G}$ be the (finite) set of group axioms (see Robinson 1963). The elementary theory of groups is undecidable (Tarski et al. 1953). By Corollary 11.8(i), $\mathfrak{G}^N$ is not axiomatizable. (Being non-axiomatizable does not make a theory unusable, as any number theorist can testify.) ◆

EXAMPLE 11.4. (the theory of necessity in real closed fields). Let $\mathfrak{R}$ denote the set of first-order axioms for real closed fields. In contrast with $\mathfrak{S}^N$, $\mathfrak{R}^N$ is not a finite theory of necessity. By a result of Tarski (1948), $\mathfrak{R}^{**}$ is decidable. By Corollary 11.8(i), $\mathfrak{R}^N$ is decidable. ◆

11.6 Axiomatization of the general theory of necessity

Although individual theories of necessity may not be axiomatizable (Examples 11.2 and 11.3) the general theory is. Example 11.1 suggests some version of Lewis's S5. The aim of this section is to define a formal system π and to prove that it gives a complete axiomatization of N (Corollary 11.13).

The well-formed formulas of π are $\mathbb{F}^N$. The axioms of π are all formulas given by the axiom schemas of the propositional logic (Chapter 2, §2.2.5) and the quantifier schemas **A11**, **A12** of the system H of Chapter 2, §2,3,7(i), together with the schemas

A13: $\quad NA \supset A$

A14: $\quad \neg NA \supset N \neg NA$

A15: $\quad N[A \supset B] \supset [NA \supset NB]$.

The rules of inference of π are those of H (i.e. **MP**, **R∀**, **R∃**) together with the modal rule

RN: From A infer NA.

$\vdash_\pi A$ stands for 'A is a theorem of π'. In the remainder of this chapter $A \equiv B$ is an abbreviation of $[[A \supset B] \wedge [B \supset A]]$ (see Example 4.4(ii)).

EXAMPLE 11.5. For all $A, B \in \mathbb{F}^N$ the following formulas are theorems of π: $NNA \supset A$, $N[A \wedge B] \equiv [NA \wedge NB]$, $N[NA \supset B] \supset [NA \supset NB]$, $N[NA \vee B] \equiv [NA \vee NB]$. (See S5 results in Cresswell and Hughes (1968, Chapter 3).) ◆

It is easily established that all axioms of π are in N (see Example 11.1) and that this property is transmitted by the rules of inference. Hence we have the following theorem.

THEOREM 11.9. Let $A \in \mathbb{F}^N$. If $\vdash_\pi A$ then $A \in N$ (the general theory of necessity—Definition 11.1(iv)).

The converse of this theorem is based on a syntactical version of Theorem 11.6.

LEMMA 11.10. Let $G[NB] \in \mathbb{F}^N$. Then

$$\vdash_\pi \beta(G[NB]) \quad \Rightarrow \quad \vdash_\pi G[NB]. \tag{11.9}$$

Proof. By induction on $|G[t]|$.

Basis:
$n = 0$. Then $G[NB]$ is NB. Using the propositional calculus rules and axioms, $\vdash_\pi NB \equiv [NB \wedge t] \vee [\neg NB \wedge f]$. Thus, (11.9) holds for $n = 0$.

Inductive step:
Suppose (11.9) holds for all formulas $G[NB]$ for which $|G[t]| = n + 1$. Then $G[t]$ has one of the following forms: (i) $H[t] \supset J$, (ii) $J \supset H[t]$, (iii) $H[t] \wedge J$, (iv) $J \wedge H[t]$, (v) $H[t] \vee J$, (vi) $J \vee H[t]$, (vii) $\neg H[t]$, (viii) $\forall a H[t](a/x)$, (ix) $\exists a H[t](a/x)$, or (x) $NH[t]$ where by the inductive hypothesis

$$\vdash_\pi \beta(H[NB]) \quad \Rightarrow \quad \vdash_\pi H[NB]. \tag{11.10}$$

Cases (i)–(vii) (11.9) follows from (11.10) by means of the propositional calculus.

Case (viii) By the propositional calculus

$$\vdash_\pi \beta(G[NB]) \equiv [[NB \supset G[t]] \wedge [\neg NB \supset G[f]]]. \tag{11.11}$$

Suppose $\vdash_\pi \beta(G[NB])$. Then for the free variable x, $\vdash_\pi NB \supset \forall a H[t](a/x)$ and $\vdash_\pi \neg NB \supset \forall a H[f](a/x)$. Since no variables occur in NB (11.1), by the predicate calculus rules, $\vdash_\pi \forall a[NB \supset H[t]](a/x)$ and

$$\vdash_\pi \forall a[\neg NB \supset H[f]](a/x).$$

By **A11** (Chapter 2, §2.3.7(i)), $\vdash_\pi NB \supset H[t]$ and $\vdash_\pi \neg NH \supset H[f]$. Then by the propositional calculus $\vdash_\pi \beta(H[NB])$. By (11.10), $\vdash_\pi H[NB]$, so again by the propositional calculus, $\vdash_\pi t \supset H[NB]$. By **R∀** (Chapter 2, §2.3.7(i)), $\vdash_\pi t \supset \forall a H[NB](a/x)$. Hence $\vdash_\pi \forall a H[NB](a/x)$. Thus $\vdash_\pi G[NB]$. This proves (11.9) for case (viii).

Case (ix) This case is similar to case (viii).

Case (x) Suppose $\vdash_\pi \beta(G[NB])$. By (11.11), $\vdash_\pi NB \supset G[t]$ and $\vdash_\pi \neg NB \supset G[f]$. Thus $\vdash_\pi NB \supset NH[t]$ and $\vdash_\pi \neg NB \supset NH[f]$. By **A13** (Chapter 2, §2.3.7(i)) and the propositional calculus, $\vdash_\pi NB \supset H[t]$ and $\vdash_\pi \neg NB \supset H[f]$. By (11.11), $\vdash_\pi \beta(H[NB])$. By inductive hypothesis (11.10), $\vdash_\pi H[NB]$. By **RN**, $\vdash_\pi NH[NB]$, i.e. $\vdash_\pi G[NB]$. This proves (11.9) for case (x).

The case analysis for the inductive step is now complete. The theorem follows by induction. ∎

LEMMA 11.11. *Let* $A, A_1, \ldots, A_n, A_{n+1} \in \mathbb{F}_{\text{pred}}(\tau)$. *Let* G *be the formula* $[NA \supset NA_1] \vee \cdots \vee [NA \supset NA_n] \vee [NA \supset A_{n+1}]$ *and* $P = \{\forall A\}$, *where* $\forall A$ *is the universal closure of* A *(Definition 2.14(iv)). If* $G \in P^N$ *then* $\vdash_\pi G$.

Proof. Suppose $G \in P^N$. As $A \in P^{**}$, by Lemma 11.1(ix) $A_j \in P^{**}$ for some j, i.e. $A \to A_j$. Thus, $\varnothing \to [A \supset A_j]$, i.e. $A \supset A_j$ is universally valid. By the completeness of the predicate calculus, $\vdash_\pi A \supset A_j$. By **RN**, $\vdash_\pi N[A \supset A_j]$ and by **A15**, $\vdash_\pi NA \supset NA_j$. If $j = m + 1$ then by **A13** and the propositional calculus $\vdash_\pi NA \supset A_{m+1}$. Hence $\vdash_\pi G$ by the propositional calculus. ∎

THEOREM 11.12. *Let $G \in \mathbb{F}^N$. If $G \in \prod \{\{S\}^N; S \in \mathbb{F}_{\mathrm{pred}}(\tau)\}$ then $\vdash_\pi G$.*

Proof. By Theorem 11.6, $\beta(G) \in N$. Then $\beta(G) = G_1 \wedge \cdots \wedge G_k$, say, where G_i is $\neg NG_1^i \vee \cdots \vee \neg NG_{n(i)}^i \vee NH_1^i \vee \cdots \vee NH_{m(i)}^i \vee H_{m(i)+1}$. Then $G_i \in N$. But for all P, $G_i \leftrightarrow_P G_i'$ where $G_i' = [NA^i \supset NH_1^i] \vee \cdots \vee [NA^i \supset NH_{m(i)}^i] \vee [NA^i \supset H_{m(i)+1}^i]$ where $A^i = G_1^i \wedge \cdots \wedge G_{n(i)}^i$. By Lemma 11.11 (with $P = \{A^i\}$), $\vdash_\pi G_i'$. By the propositional calculus $\vdash_\pi G$. ∎

COROLLARY 11.13. *Let $G \in \mathbb{F}^N$. Then $G \in N$ iff $\vdash_\pi G$.*

Proof. If $\vdash_\pi G$ then $G \in N$ by Theorem 11.9. Conversely, if $G \in N$ then $G \in \prod \{\{S\}^N; S \in \mathbb{F}_{\mathrm{pred}}(\tau)\}$. By Theorem 11.12, $\vdash_\pi G$. ∎

COROLLARY 11.14. *N is recursively enumerable.*

This result contrasts the general theory of necessity with the theory of logical necessity (Example 11.2).

The obverse of Theorem 11.12 shows that if a formula of $\mathbb{F}^N$ is not provable in π then there exists a finite 'counterexample'.

COROLLARY 11.15. *Let $G \in \mathbb{F}^N$. If $\nvdash_\pi G$ then there exists $A \in \mathbb{F}_{\mathrm{pred}}(\tau)$ such that $G \notin \{A\}^N$.*

The general theory of necessity was defined by quantifying over *all* subsets of $\mathbb{F}_{\mathrm{pred}}(\tau)$, but it actually coincides with the formulas which are common to all finite theories of necessity.

COROLLARY 11.16. *$N = \prod \{\{S\}^N; S \in \mathbb{F}_{\mathrm{pred}}(\tau)\}$.*

Proof. Clearly, $N \subseteq \prod \{\{S\}^N; S \in \mathbb{F}_{\mathrm{pred}}(\tau)\}$. Suppose $G \notin N$. By Corollary 11.13, $\nvdash_\pi G$. By Corollary 11.15 there exists $A \in \mathbb{F}_{\mathrm{pred}}(\tau)$ such that $G \notin \{A\}^N$. Thus $G \notin \prod \{\{S\}^N; S \in \mathbb{F}_{\mathrm{pred}}(\tau)\}$. Hence $\{\{S\}^N; S \in \mathbb{F}_{\mathrm{pred}}(\tau)\} \subseteq N$. ∎

11.7 Effective modal logics

The problem we now turn to is that of regularizing discourse on logical modalities (consequence, possibility, necessity, ...) of a given logic without introducing metaphysical assumptions or non-effective negative information (see also Scott 1974b). The primary modality here is logical consequence: theoremhood, and hence *necessity*, is defined as consequence of the empty set—then *possibility* is defined, as usual, in terms of necessity and negation.

In the spirit of the construction of effective logic in Chapter 10, we extend the given (free-variable) logic by addition of regular junctors, contradiction, and Aristotelian (universal and existential) quantifiers, but now we expand the class of formulas so as to include reference to *logical consequence, the primary modality*. (In the previous section there was oblique reference to logical consequence via the necessity operator.)

As in Chapter 10, §10.7, we consider a free-variable logic $\mathfrak{L} = \langle P, Cn \rangle$ (Definition 3.13) on the fixed set $\mathbb{T}$ of terms. The alphabet with which the formulas are constructed differs from that of $\mathbb{F}^q(J, P)$ only by the presence of a junctor K whose arguments include finite *sets* of formulas. Thus we suppose there are available the sets Bv (bound variables), J (here we shall take $J = B$, the set of basic junctors (Example 10.4)), $Q = \{\forall, \exists\}$ (quantifier symbols), $\{f\}$ (propositional constant), and G (left, right brackets and comma) as in Chapter 2, §2.2.1, together with the *consequence junctor* K where these sets are pairwise disjoint and disjoint from the alphabet of $\mathbb{T}$.

The class $\mathbb{F}^K(B, P)$ of effective formulas over $\mathfrak{L}$ is then defined by induction using the clauses (F1)–(F3) as in the definition of $\mathbb{F}^q(B, P)$ (Chapter 10, §10.7) but with appropriate addition of a clause (F4) for the modal operator K. Reference to logical consequence is achieved by means of K: for a finite set Δ of formulas and formula A, $K[\Delta, A]$ is also counted as a formula and it is intended to mean that A is a logical consequence of the set of formulas which occur in Δ. Thus, for the construction of formulas, K is a kind of two-place junctor in which the first argument is a finite sequence of formulas: in this context two sequences Δ and Δ' are counted as *equal* if every formula which occurs in one also occurs in the other, i.e. Δ represents the same *set* as Δ'.

In the computation of logical length the symbol K is counted as a logical symbol. The *modal depth* $m(A)$ of a formula A is defined inductively.

(F1) If $A \in \{f\} + P$ then $A \in \mathbb{F}^K(B, P)$ and $m(A) = 0$.

(F2) If $A_1, \ldots, A_n \in \mathbb{F}^K(B, P)$ and $c \in B$ with $n(c) = n$ then $c[A_1, \ldots, A_n] \in \mathbb{F}^K(B, P)$ and $m(c[A_1, \ldots, A_n]) = \max\{m(A_i); 1 \leq i \leq n\}$.

(F3) If $C \in \mathbb{F}^K(B, P)$, $x \in Fv$, $a \in Bv - Bv(C)$, then $\forall a C(a/x)$ and $\exists a C(a/x)$ are also in $\mathbb{F}^K(B, P)$ and each has modal depth $m(C)$.

(F4) If $\Delta, D \in \mathbb{P}_\omega(\mathbb{F}^K(B, P))$ then $K[\Delta, D] \in \mathbb{F}^K(B, P)$ and $m(K[\Delta, D]) = 1 + \max\{m(A); A \in \Delta, D\}$.

For any formula B, $Bv(B)$, $Fv(B)$ denote the sets of bound variables and free variables respectively which occur in B. As in the effective quantifier logic

$$Fv(\forall a B(a/x)) = Fv(\exists a B(a/x)) = Fv(B) - \{x\}$$
$$Bv(\forall a B(a/x)) = Bv(\exists a B(a/x)) = Bv(B) - \{a\}$$

and, in addition, in accordance with Quine's strictures (cf. Chapter 11, §11.1)

$$Bv(K[\Delta, D]) = Fv(K[\Delta, D]) = \emptyset. \tag{11.12}$$

The formulas of $\mathbb{F}^K(B, P)$ are not all words on the expanded alphabet A' because of the use the occurrence of a finite set Δ in clause (F4). There are, however, no more difficulties in dealing with such entities than we have already encountered in using sequents composed of finite sets of formulas.

As in §11.4, a *modal formula* is one whose prime subformulas are within the scope of the modal junctor (clearly, if A is a modal formula then $|A| > 0$).

As in the construction of effective logic, the logical particles acquire their meaning through their introduction in sequents. So a Gentzen type of formal system $\mathfrak{G}^K(\mathfrak{L})$ can be defined, the rules of which are thinning, the introduction rules for the junctors and quantifier symbols, and in addition the modal rules K_l and K_r for the introduction of K in the antecedent and succedent respectively (cf. Chapter 10, §10.8). Once more there is considerable choice of method of dealing with arguments containing assertions of modality; the rules we have adopted here are the analogues of Theorem 11.4(ii) and 11.4(i) respectively, reading $K[\Gamma \supset A]$ as $K[\Gamma, A]$. For $\Delta, \Theta, \Psi, A, D \in \mathbb{P}_\omega(\mathbb{F}^K(B, P))$

$$K_l \quad \frac{(\Delta, D : A) \quad \{(\Theta : F); F \in \Psi\}}{(\Delta, \Theta, K[\Psi, D] : A)}$$

$$K_r \quad \frac{(\Delta, \Gamma : A)}{(\Delta : K[\Gamma, A])} \quad \text{where } \Delta \text{ is a finite set of } \textit{modal} \text{ formulas.}$$

where Δ is a finite set of *modal* formulas.

The axioms of the system $\mathfrak{G}^K(\mathfrak{L})$ are $Ax_\mathfrak{L} + Ax_0 + Ax_1$ (cf. Chapter 10, §10.2) where the following hold.

(i) (a) Ax_0 consists of all sequents $(\Gamma : A)$ where $A \in \Gamma \in \mathbb{P}_\omega(\mathbb{F}^K(B, P))$.
 (b) Ax_1 consists of all sequents $(f : A)$, where $A \in P + [f]$.
(ii) $Ax_\mathfrak{L}$ consists of all sequents of atomic formulas $(\Delta : A)$ such that $A \in Cn(\Delta)$ and $\Delta \in \mathbb{P}_\omega(P)$ (special axioms).

As in Chapter 10, we let Vd be the set of sequents which are derivable in this system from the axioms—Vd is the set of *valid* sequents.

Following the general strategy of the previous sections we show that this system gives rise to a G-structure (and so a logic) by establishing the cut admissibility theorem (Theorem 10.2). If B is a set of regular junctors the proof goes through provided that the steps where compound formulas are analysed now include the cases of quantified formulas together with modal formulas. But there are some small technical points to watch concerning the occurrence of variables in proofs, but (11.9) ensures that Lemmas 10.11–10.16 hold for proofs in $\mathfrak{S}^K(\mathfrak{L})$. Theorem 10.2 can now be extended to the effective modal system $\mathfrak{S}^K(\mathfrak{L})$.

THEOREM 11.17. If $(\Gamma : A) \in Vd$ and $(\Delta : C) \in Vd$, then $(\Gamma, \Delta/A : C) \in Vd$.

Proof. The proof is based on Theorem 10.2. We merely insert in the case analysis in the positions marked (†) the appropriate consideration of the modal rules. These are documented below.

ad (†1) Add the clause:
'$(\Gamma : A_0)$ is not the conclusion of K_l since $\rho = 0$ and not the conclusion of K_r since $\gamma = 1$.'

ad (†2) (Additional subcase (i)(b) in the proof of (10.6)). Add the text:
'$(\Gamma : A_0)$ is the conclusion of an application of K_l:

$$\left. \begin{array}{c} \Pi_1^{(1)} \qquad\qquad \Pi_{1F}^{(2)} \\ \diagdown \quad \diagup \quad \diagdown \quad \diagup \\ \underline{(\Phi, D : A_0) \quad \{(\Xi : F); F \in \Psi\}} \\ (\Phi, \Xi, K[\Psi, D] : A_0) \\ \underbrace{}_{\Gamma} \end{array} \right\} K_l \Bigg\} \Pi_1.$$

In Π_1, $(\Phi, \Theta : A_0)$ has a subproof $\Pi_1^{(1)}$ and for each $F \in \Psi$, $(\Xi : F)$ has a subproof $\Pi_{1F}^{(2)}$ where $\rho(\Pi_1^{(1)}, \Pi_2) \langle \rho(\Pi_1, \Pi_2)$. By (10.13) $(\Phi, D, \Delta/A_0 : C) \in Vd$. By K_l a proof of $(\Phi, \Delta/A_0, \Xi, K[\Psi, D] : C)$ can be constructed, i.e. $(\Gamma, \Delta/A_0 : C) \in Vd$.'

ad (†3) (Additional subcase (ii)(b) in the proof of (10.6)) Add the following text:
'$(\Delta : C)$ is an immediate consequence in Π_2 by K_l:

$$\left. \begin{array}{c} \Pi_2^{(1)} \qquad\qquad \Pi_2^{(F)} \\ \diagdown \quad \diagup \quad \diagdown \quad \diagup \\ \underline{(\Delta', D : C) \quad \{(\Theta : F); F \in \Psi\}} \\ (\Delta', \Theta, K[\Psi, D] : C) \\ \underbrace{}_{\Delta} \end{array} \right\} K_l \Bigg\} \Pi_2.$$

Since $A_0 \in \Delta$, $\gamma = 0$, and $|K[\Psi, D]| > 0$, then $A_0 \in \Delta'$ or $A_0 \in \Theta$. Then, by (10.13), proofs can be constructed of $(\Gamma, \Delta'/A_0, D : C)$ (using an additional thinning if $A_0 = D$) and $(\Gamma, \Theta/A_0 : F)$ for each $F \in \Psi$. By K_l, $(\Gamma, \Delta'/A_0, \Theta/A_0, K[\Psi, D] : C) \in Vd$, i.e. $(\Gamma, \Delta/A_0 : C) \in Vd'$.

ad (†4) (Additional subcase (ii)(c) in the proof of (10.6)) Add the following text:

'If $(\Delta : C)$ were an immediate consequence in Π_2 by K_r, then since $A_0 \in \Delta$, A_0 would be a modal formula and so $|A_0| > 0$. This contradicts the assumption of this case that $\gamma = 0$. Hence $(\Delta : C)$ is not an immediate consequence in Π_2 by K_r.'

ad (†5) (Additional subcase of case (i) in the proof of (10.7)) Add the following text:

'Since $\rho = 0$, $(\Delta : C)$ is the conclusion in Π_2 of a rule which introduces A_0 into the conclusion, i.e. K_l:

$$\left. \begin{array}{c} \Pi_2^{(1)} \qquad \Pi_2^{(F)} \\ \diagdown \qquad \diagup \\ \underline{(\Delta' \approx, D : C) \quad \{(\Theta' : F); F \in \Psi\}} \\ \underbrace{(\Delta', \Theta, K[\Psi, D] : C)}_{\Delta} \end{array} \right\} K_l \Bigg\} \Pi_2.$$

Here $A_0 = K[\Theta, A]$ and Γ is a set of modal formulas:

$$\left. \begin{array}{c} \Pi' \\ \diagdown \quad \diagup \\ \underline{(\Gamma, \Theta : A)} \\ (\Gamma : K[\Theta, A]) \end{array} \right\} K_r \Bigg\} \Pi_1.$$

As $\rho = 0$, A_0 does not occur in the premises and so $A_0 \notin \Delta', \Gamma', \Theta'$. As $A_0 \in \Delta$, $A_0 = K[\Psi, D]$. Thus $\Psi = \Theta$ and $D = A$ and $A_0 \notin \Theta$. Consider now the proof in Π_1 of $(\Gamma, \Theta : A)$ and the proofs in Π_2 of $(\Theta' : F)$ for $F \in \Psi$. Now $|K[\Psi, D]| = g + 1$. Hence $|F| \leq g$ for $F \in \Psi$ and $|D| \leq g$. By (10.16) a sequence of cuts—with cut formula each time in Ψ—yields a proof of $(\Gamma, \Theta' : A)$ (as in subcase (i) of the proof of (10.7), by (10.15)). Further, as $|A| = |D| \leq g$, by (10.15), the cut rule applied to $(\Gamma, \Theta' : A)$ and $(\Delta', \Delta : C)$ (recall that $A = D$) yields a proof of $(\Gamma, \Theta', \Delta/AC)$ and hence a proof of $(\Gamma, \Theta', \Delta' : C)$, i.e. $(\Gamma, \Delta/A_0 : C) \in Vd$.'

ad (†5) (Additional subcase of case (ii) in the proof of (10.7)) Add the following text:

'As $\rho = 0$, $(\Gamma : A_0)$ is not the conclusion of K_l—otherwise A_0 would appear in one of the premises.' ∎

The calculus of sequents $\mathfrak{G}^K(B, P)$ therefore yields a G-structure $\langle \mathbb{F}^K(B, P), Vd(\mathfrak{L}) \rangle$. The corresponding logic (Definition 3.17) is $\mathfrak{E}^{\leq}(\mathfrak{L}) = \langle \mathbb{F}^K(B, P), Cn^{\heartsuit}_{Vd(\mathfrak{L})} \rangle$. Here, for a finite set Δ of formulas,

$$A \in Cn^{\heartsuit}_{Vd(\mathfrak{L})} \Leftrightarrow (\Delta : A) \in Vd(\mathfrak{L}).$$

The hierarchical structure of this logic is revealed by defining $\mathbb{F}^K_n$ to be all formulas of $\mathbb{F}^K(B, \mathfrak{L})$ of model depth $\leq n$. Put $Cn_n(\Gamma) =_{\text{def}} \mathbb{F}^K_n \cdot Cn^{\heartsuit}_{Vd(\mathfrak{L})}(\Gamma)$ for $\Gamma \subseteq \mathbb{F}^K_n$. Then $\mathfrak{M}_n =_{\text{def}} \langle F^K_n, Cn^K_n \rangle$ is a logic. Further $\mathfrak{M}_{n+1}$ is an extension of $\mathfrak{M}^n$ and $\mathfrak{M}_0 = \mathfrak{G}^q(\mathfrak{L})$.

EXAMPLE 11.6. Let $A \in \mathbb{F}^K_n$ and $\Delta \in \mathbb{P}_\omega(\mathbb{F}^K_n)$. Then we have the following.

(i) $A \in Cn^K_n(\Delta) \Leftrightarrow K[\Delta, A] \in Cn^K_{n+1}$. (Thus the logic at each level incorporates the notion of logical consequence of the previous level.)
(ii) Let $\leftrightarrow$ denote logical equivalence in $\mathfrak{G}^K(\mathfrak{L})$ and $\leftrightarrow_n$ denote logical equivalence in $\mathfrak{M}_n$. Let $A, B \in \mathbb{F}^K(B, \mathfrak{L})$. Let $m \geq m(A), m(B)$. Then from (i): $A \leftrightarrow B$ iff $A \leftrightarrow_m B$.
(iii) $K[\Delta, A] \leftrightarrow K[\varnothing : \bigwedge \Delta \supset A]$. ◆

By taking $\mathfrak{L}$ to be the minimum logic $\mathfrak{L}_0$ on P, the *pure effective modal logic* is obtained. Then, as in Theorem 10.20, $\mathfrak{E}^K(\mathfrak{L}_0)$ occupies a special position in relation to all effective logics on free-variable logics with terms $\mathbb{T}$.

THEOREM 11.18. *A sequent is valid in $\mathfrak{E}^K(\mathfrak{L}_0)$ iff it is valid in $\mathfrak{E}^K(\mathfrak{L})$ for all free-variable logics with terms $\mathbb{T}$.*

Example 11.6(iii) suggests that effective modal logic could then be constructed on the basis of theoremhood instead of logical consequence. For this purpose we need to consider the special subclass $\mathbb{F}^{SP}$ of $\mathbb{F}^K(B, P)$ consisting of formulas A of $\mathbb{F}^K(B, P)$ such that every occurrence of the modal operator K has $\varnothing$ as its first argument. Each formula $A \in \mathbb{F}^K(B, P)$ can be effectively transformed into a formula $A\$ \in \mathbb{F}^{SP}$. This transformation is defined inductively as follows:

(i) For $F \in P$, $F\$ =_{\text{def}} F$.
(ii) $[A \vee B]\$ =_{\text{def}} [A\$ \vee B\$]$, $\quad [A \wedge B]\$ =_{\text{def}} [A\$ \wedge B\$]$,
   $[A \supset B]\$ =_{\text{def}} [A\$ \supset B\$]$.
(iii) $(\forall a A(a/x))\$ =_{\text{def}} \forall a A\$(a/x)$
   $(\exists a A(a/x))\$ =_{\text{def}} \exists a A\$(a/x)$
(iv) $(K[\Delta, A])\$ =_{\text{def}} K[\varnothing, \bigwedge \{B\$; B \in \Delta\} \supset A\$]$.

Then by induction on the logical length of formulas and using Example 11.6(ii) it can easily be shown that every formula is logically equivalent to a special formula.

LEMMA 11.19. $F \leftrightarrow F\$$ $(F \in \mathbb{F}^K(B, P))$.

Let Cn^{sp} be the restriction of $Cn^{\heartsuit}_{Vd(\mathfrak{L})}$ to $\mathbb{F}^{sp}$, i.e.

$$Cn^{sp}(\Delta) =_{\text{def}} \mathbb{F}^{sp} \cdot Cn^{\heartsuit}_{Vd(\mathfrak{L})} \qquad (\Delta \in \mathbb{P}_\omega(\mathbb{F}^{sp})). \tag{11.13}$$

Then $\mathfrak{S}_{sp}(\mathfrak{L}) = \langle \mathbb{F}^{sp}, Cn^{sp} \rangle$ is a sublogic of the modal logic $\mathfrak{S}^K(\mathfrak{L})$ and, in fact, is the logic derived from the formal system $\mathfrak{S}^{sp}(\mathfrak{L})$ which differs from $\mathfrak{S}^K(\mathfrak{L})$ only by the restriction of the formulas to $\mathbb{F}^{sp}$ and the replacement of the modal rules by special cases of K_l and K_r appropriate for the introduction of formulas $K[\emptyset, A]$. More exactly, the formulas of $\mathfrak{S}^{sp}(\mathfrak{L})$ are $\mathbb{F}^{sp}$. The axioms of $\mathfrak{S}^{sp}(\mathfrak{L})$ are those of $\mathfrak{S}^K(\mathfrak{L})$ restricted to $\mathbb{F}^{sp}$. The rules of $\mathfrak{S}^{sp}(\mathfrak{L})$ are those of $\mathfrak{S}^K(\mathfrak{L})$ except for the modal rules. The modal rules of $\mathfrak{S}^{sp}(\mathfrak{L})$ are $K\$_l$ and $K\$_r$: for $\Delta, A, D \in \mathbb{P}_\omega(\mathbb{F}^{sp})$

$$K\$_l \qquad \frac{(\Delta, D : A)}{(\Delta, K[\emptyset, D] : A)}$$

$$K\$_r \qquad \frac{(\Delta : A)}{(\Delta : K[\emptyset, A])}$$

where Δ is a finite set of modal formulas. ($K\$_l$ is K_l with $\Theta = \Psi = \emptyset$ and $K\$_r$ is K_r with $\Gamma = \emptyset$.)

The symbol $\emptyset$ is redundant in the expressions $K[\emptyset, A]$ of $\mathfrak{S}^{sp}(\mathfrak{L})$. Because of this it is convenient to introduce an abbreviation: write NA instead of $K[\emptyset, A]$. This artifice has the added advantage of underlining the older traditional meaning of 'analytic'. The logic $\mathfrak{S}^{sp}(\mathfrak{L})$ will also be known as $\mathfrak{S}^{\text{mod}}(\mathfrak{L})$.

THEOREM 11.20. Let $\Delta, A \in \mathbb{P}_\omega(\mathbb{F}^{sp})$. Then $(\Delta : A)$ is vaid if $\mathfrak{S}^{\text{mod}}(\mathfrak{L})$ iff it is valid in $\mathfrak{S}^K(\mathfrak{L})$.

Proof. Suppose $(\Delta : A)$ has a derivation $\mathfrak{D}$ in $\mathfrak{S}^{\text{mod}}(\mathfrak{L})$. Since $K\$_l$ and $K\$_r$ are special cases of N_l and N_r respectively, $\mathfrak{D}$ is also a derivation in $\mathfrak{S}^K(\mathfrak{L})$.

Conversely, suppose that $(\Delta : A)$ has a derivation $\mathfrak{D}$ in $\mathfrak{S}^K(\mathfrak{L})$. Consider an application of K_l in $\mathfrak{D}$. Since $\Delta, A \in \mathbb{P}_\omega(F^{so})$, all sequents in the derivation contain only special formulas. Therefore the application of K_l introduces a special modal formula $K[\emptyset, D]$ into the antecedent and so the application takes the form

$$\frac{(\Gamma, D : B)}{(\Gamma, \Theta, K[\emptyset, D] : B)} \quad K_l.$$

Effective modal logics

This can be replaced by an application of $K\$_l$ and thinning:

$$\frac{\dfrac{(\Gamma, D \ : \ B)}{(\Gamma, ND \ : \ B)} \ K\$_l}{(\Gamma, \Theta, ND \ : \ B).} \text{ thinning.}$$

Consider an application of K_r in $\mathfrak{D}$. It likewise introduces a special modal formula $K[\varnothing, D]$ into the succedent, and so the application takes the form

$$\frac{(\Gamma \ : \ B)}{(\Gamma \ : \ N[\varnothing, B])} \ K_r.$$

This is already an application of $K\$_r$.

Thus each application of K_l (K_r in $\mathfrak{D}$ can be replaced by an application of $K\$_l$ ($K\$_r$) to give a derivation of $(\varDelta \ : \ A)$ in $\mathfrak{S}^{\text{mod}}(\mathfrak{L})$. ∎

By Lemma 11.19 and Theorem 11.20 we then have the following Corollary.

COROLLARY 11.21. $(\varDelta \ : \ A)$ is valid in $\mathfrak{S}^K(\mathfrak{L})$ iff $(\{B\$; B \in \varDelta\} \ : \ A\$)$ is valid $\mathfrak{S}^{\text{sp}}(\mathfrak{L})$.

If $\mathfrak{L} = \mathfrak{L}_0$ we have the *pure effective modal logic* which can now be compared with a Hilbert-type system *NI* obtained from the intuitionistic predicate calculus *I* of Chapter 2, §2.3.7, by adjoining certain axioms and modal rules. The formulas of *NI* are $\mathbb{F}^{\text{sp}}(P)$. The axiom schemas of *NI* are those of *I* together with the schema

NO: $NA \supset A$.

The rules of inference are those of *I* together with the rule

RN: $\dfrac{Z \supset B}{Z \supset NB}$

where Z is a modal formula.

THEOREM 11.22. If $\vdash_{NI} A$ then $(\varnothing \ : \ A)$ is valid in $\mathfrak{S}^{\text{sp}}(\mathfrak{L}_0)$.

Proof. By premiss induction on derivations in *NI*. The proof follows exactly the steps required for the predicate calculus only (e.g. Kleene 1962, §77). Accordingly, we deal here only with the cases involving the modal expressions:

Basis:
F is an axiom $NA \supset A$. Now by induction on the length of A it can be proved that $(A \ : \ A)$ is valid in $\mathfrak{S}^{\text{mod}}(\mathfrak{L}_0)$. By $K\$_l$, $(NA \ : \ A)$ is also valid. Hence so is $(\varnothing \ : \ NA \supset A)$.

384   *Modal logics*

Inductive step:
Suppose that $\vdash_{NI} Z \supset B$ and $(\emptyset : Z \supset B)$ is valid in $\mathfrak{S}^{\text{mod}}(\mathfrak{L}_0)$, where Z is a modal formula. Then $\vdash_{NI} Z \supset NB$ by **RN**. Now in the derivation of $(\emptyset : Z \supset B)$ in $\mathfrak{S}^{\text{sp}}(\mathfrak{L}_0)$, $(\emptyset : Z \supset B)$ is not an axiom. It is therefore the conclusion of an application of a right rule for $\supset$. The premiss of this application is therefore $(Z : B)$. Hence $(Z : NB)$ is derivable in $\mathfrak{S}^{\text{sp}}(\mathfrak{L}_0)$ by an application of $K\$_r$. Hence $(\emptyset : Z \supset NB)$ is valid in $\mathfrak{S}^{\text{sp}}(\mathfrak{L}_0)$.

The theorem now follows by induction.   ∎

The converse of this theorem is as follows.

THEOREM 11.23. *If $(\Gamma : A)$ is valid in $\mathfrak{S}^{\text{sp}}(\mathfrak{L}_0)$ then $\vdash_{NI} \Gamma \supset A$.*

Proof. By premiss induction *NH*. As in the previous theorem the proof follows exactly the steps required for the predicate calculus only. Accordingly we deal here only with the cases involving the modal expressions. Since modal expressions do not occur in the axioms of $\mathfrak{S}^{\text{sp}}(\mathfrak{L}_0)$ the basis of the induction is trivial.

Inductive step:
Suppose that $(\Gamma : A)$ is derived by $K\$_l$ from the premiss $(\Gamma\$: A\$)$ where $\vdash_{NI} \Gamma\$ \supset A\$$. Then $(\Gamma : A)$ is $(\Delta, NB : A)$, say. Hence $\Gamma\$ = \Delta + \{B\}$ and $A = A\$$. Put $C = \Delta$. Then

$$\vdash_{NI} [B \wedge C] \supset A. \tag{11.14}$$

Now by the propositional axioms and rules of *NI* (Example 2.6)

$$\vdash_{NI} [NB \supset B] \supset [[[B \wedge C] \supset A] \supset [[NB \wedge C] \supset A]]].$$

But $NB \supset B$ is an axiom (i.e. **NO**). Hence, by **MP**, $\vdash_{NI} [[B \wedge C] \supset A] \supset [[NB \wedge C] \supset A]]$. By (11.14) and **MP**, $\vdash_{NI} [NB \wedge C] \supset A$, i.e. $\vdash_{NI} \Gamma \supset A$. The theorem now follows by induction.   ∎

EXAMPLE 11.7. For all $A, B \in F^{\text{sp}}$ the following formulas are theorems of *NI*: $NA \supset A$, $N[A \supset B] \supset [NA \supset NB]$, $[NA \supset f] \supset N[NA \supset f]$, $[t \supset NA] \supset N[t \supset NA]$, and (if A is modal) $A \supset NA$. These are most easily established first in $\mathfrak{S}^{\text{mod}}(\mathfrak{L}_0)$ and then in *NI* via Theorem 11.23.   ◆

The system *NI* of intuitionistic modal logic could be classes as an S5 system since, by the above example, the S5 axioms **A13**, **A14**, **A15** (§11.6) are theorems of *NI*, though, bearing in mind that the basic logic of *NI* is intuitionistic, some of the traditional S5 theorems are not provable. Thus, letting $\neg A$ stand for $A \supset f$ and PA stand for $\neg N \neg A$, whilst $[PA \vee PB] \supset P[A \vee B]$ is a theorem of *NI*, $P[A \vee B] \supset [PA \vee PB]$ is not. Similarly $NA \supset \neg P \neg A$ is a theorem but $\neg P \neg A \supset NA$ is not.

11.8 Deontic logic

Certain medieval logicians were aware of the possibility of a logic of obligation based on modal logic. But modern deontic logic, the logic of *obligation*, was initiated by von Wright whose 1951 paper and subsequent work has shaped contemporary discussion of the subject: his 1968 book provides perspective on its foundations. Deontic logic is now, in essence, a branch of modal logic with applications in ethics and jurisprudence.

Deontic logic is customarily formulated with an absolute obligation operator O (e.g. Føllesdal and Hilpinen 1971) and attempts are made at laying down axioms for it. An analysis of deontic concepts, using classical concepts of logic and metalogic, was given by Körner (1976, Chapter 2). The analysis of codes of conduct leads to a notion of *obligation relative to* such a code. It is our intention to develop a logic of this notion of relative obligation based, as was Körner's analysis of material necessity, on the classical modal logic of §11.4. For a precise definition and discussion of the formulas $ad(C)$ and $d(C)$ below, the reader is referred to Körner (1976).

Körner used the language L_N (§11.4) to express the formal adequacy of a code of conduct C for operation in a 'world' W. By 'world' is to be understood a class of structures of type τ defined by a set W of formulas of L. Then a sentence $ad(C)$ in the language L can be found such that

$$C \text{ is adequate in } W \iff W \to ad(C).$$

The notion of obligation based on a world W and a code of conduct C can be defined by associating with C another sentence $d(C)$—the *descriptive counterpart* of C—and defining 'A is obligatory on the basis of C' (written $O[d(C), A]$) by $N[d(C) \supset A]$. Unlike Körner's method we allow here that descriptive counterparts of codes of conduct can themselves contain obligations. But by finite nesting of these obligations—guaranteed by the formation rules of formulas—these obligations are ultimately relative to descriptive counterparts without obligations. Sentences describe what it would be like for the obligation to be carried out; law imposes the obligation. This definition enables obligation relative to a code of conduct (with respect to a given 'world' W) to be treated entirely in terms of classical logic. To complete the task two simplifying assumptions can reasonably be made.

(i) Every formula of $\mathbb{F}_{pred}$ is logically equivalent to the descriptive counterpart of some code of conduct. (This justifies us in considering *all* formulas of the form $O[D, A]$ (indicating obligation relative to the code corresponding to D).
(ii) An arbitrary set W of formulas in $\mathbb{F}_{pred}$ defines a 'world'.

We shall allow iterated obligation operators and allow both arguments to be any formulas in $\mathbb{F}^N$.

Thus the class $\mathbb{F}^o$ of *deontic formulas* can be defined by analogy with $\mathbb{F}^K(B, \mathfrak{L})$:

(F1) If $A \in At$ then $A \in \mathbb{F}^o$.
(F2) If $A, B \in \mathbb{F}^o$ then $\neg A, [A \vee B], [A \wedge B], [A \supset B] \in \mathbb{F}^o$.
(F3) If $A \in \mathbb{F}^o$, $x \in Fv$, and $a \in Bv - Bv(A)$, then $\forall aA(a/x), \exists aA(a/x) \in \mathbb{F}^o$.
(F4) If $A, B \in \mathbb{F}^o$ then $O[A, B] \in \mathbb{F}^o$ and $Bv(O[A, B]) = Fv(O[A, B]) = \emptyset$.

It should be noted that there has been some controversy over the meaning of iterated obligation statements. But the above definition shows how to construct and read them systematically. The following definition of truth gives a systematic meaning to them.

The notion of *satisfaction in a structure* is defined as for F^N (§11.4) but instead of (iii)(d) we have the following.

(iii) (d^o) $[\mathfrak{A}, \alpha](O[A, B]) = glb\{[\mathfrak{B}, \beta](A \supset B); (\mathfrak{B}, \beta) \vDash W\}$.

Definition 11.1 can easily be adapted to this notion. In particular, W^o is defined as the set of formulas A of $\mathbb{F}^o$ such that, for all structures $(\mathfrak{A}, \alpha)$ of type τ, $(\mathfrak{A}, \alpha) \vDash A$ if $(\mathfrak{A}, \alpha) \vDash W$. W^o is the *theory of obligation in W*. O is defined as the intersection of all elementary theories of obligation—it is the *general theory of obligation*.

There is clearly a close connection between theories of obligation and theories of necessity. This becomes clearer by the following transformation between $\mathbb{F}^N$ and $\mathbb{F}^o$: for each $A \in \mathbb{F}^o$, $A@$ is the formula of $\mathbb{F}^N$ obtained from A by replacing each subformula of the form $O[A, B]$ by $N[A \supset B]$ and, conversely, for each $B \in \mathbb{F}^N$, $B@$ is the formula of $\mathbb{F}^o$ obtained from B by replacing each subformula of the form $N[A]$ by $O[t, A]$. It easily follows from the truth definitions by induction on the length of formulas that

$$A \in W^o \Leftrightarrow A@ \in W^N \quad (A \in \mathbb{F}^o) \quad (11.15a)$$
$$B \in W^N \Leftrightarrow B@ \in W^o \quad (B \in \mathbb{F}^N). \quad (11.15b)$$

Therefore

$$A \in O \Leftrightarrow A@ \in N \quad (A \in F^o), \quad (11.16a)$$
$$B \in N \Leftrightarrow B@ \in O \quad (B \in F^N). \quad (11.16b)$$

Further,

$$A \equiv F@@ \in O \quad (A \in \mathbb{F}^o) \quad (11.17a)$$
$$B \equiv B@@ \in N \quad (B \in \mathbb{F}^N). \quad (11.17b)$$

EXAMPLE 11.8. Let $\Phi = O[C, A] \supset [[[C \supset A] \supset O[C, B]] \supset B]$ where $A, B, C \in \mathbb{F}^0$. Then $\Phi@ = N[C@ \supset A@] \supset [[[C@ \supset A@] \supset N[C@ \supset B@]] \supset B@]$. By Example 11.1, $\Phi@ \in N$. By (11.13a), $\Phi \in O$. ◆

As with theories of necessity, the general theory of obligation can be axiomatized. We give below a set of axioms for O which are in fact already relativized versions of formulas which are familiar from previous studies of deontic logic. The formulas of the formal system $\pi@$ are $\mathbb{F}^0$. The axiom schemes of $\pi@$ are those of π but instead of **A13**, **A14**, **A15** (§11.6) which concern N there are the following schema **A13⁰**–**A16⁰** which concern O, where $A, B, C, D \in \mathbb{F}^0$:

A13⁰: $O[A, B] \supset [A \supset B]$

A14⁰: $\neg O[A, B] \supset O[A, \neg O[A, B]]$

A15⁰: $O[A, B \supset C] \supset [O[A, B] \supset O[A, C]]$

A16⁰: $O[A, B] \supset O[D, A \supset B]$.

The rules of inference of $\pi@$ are those of π, but instead of **RN** we have

RO: For $B \in \mathbb{F}^0$ and $A \in \mathbb{F}_{\text{pred}}$, from $A \supset B$ infer $O[A, B]$.

EXAMPLE 11.9. For every axiom A of π, $\vdash_{\pi@} A@$. For every axiom B of $\pi@$, $\vdash_\pi B$.

It is a simple exercise to prove that $\pi@$ is sound, i.e. for all sets W of formulas of L, and formulas $A \in \mathbb{F}^0$,

$$\vdash_{\pi@} A \quad \Rightarrow \quad A@ \in W^0.$$

It then follows from (11.15a) that

$$\vdash_{\pi@} A \quad \Rightarrow \quad \vdash_\pi A@ \qquad (A \in \mathbb{F}^0). \tag{11.18a}$$

The related result

$$\vdash_\pi B \quad \Rightarrow \quad \vdash_{\pi@} B@ \qquad (B \in \mathbb{F}^N) \tag{11.18b}$$

can be proved by premiss induction on derivations of π. The basis of the induction is provided by Example 11.9.

LEMMA 11.24. For all $A, B \in \mathbb{F}^0$,

(i) $\vdash_{\pi@} O[A, B] \equiv O[t, A \supset B]$
(ii) $\vdash_{\pi@} A \equiv B \Rightarrow \vdash_{\pi@} O[t, A] \equiv O[t, B]$.

Proof.

(i) By **A16⁰**

$$\vdash_{\pi@} O[A, B] \supset O[t, A \supset B]. \tag{11.19}$$

For the converse, by propositional calculus and **RO**,

$$\vdash_{\pi@} O[A, t] \quad \text{and} \quad \vdash_{\pi@} O(A, A). \tag{11.20}$$

Now by **A16⁰** and **A15⁰** we also have

$$\vdash_{\pi@} O[t, A \supset B] \supset O[A, t \supset [A \supset B]]$$

and

$$\vdash_{\pi@} O[A, t \supset [A \supset B]] \supset [O[A, t] \supset O[A, A \supset B]]$$

respectively. From these we have

$$\vdash_{\pi@} O[t, A \supset B] \supset [O[A, t] \supset O[A, A \supset B]].$$

By propositional calculus,

$$\vdash_{\pi@} O[A, t] \supset [O[t, A \supset B] \supset O[A, A \supset B]].$$

Then, by (11.20),

$$\vdash_{\pi@} O[t, A \supset B] \supset O[A, A \supset B]. \tag{11.21}$$

Next, $A \supset [[A \supset B] \supset B]$ is a propositional tautology. By **RO**, $\vdash_{\pi@} O[A, [A \supset B] \supset B]$. By **A15⁰**,

$$\vdash_{\pi@} O[A, A \supset B] \supset O[A, B]. \tag{11.22}$$

By (11.21) and (11.22),

$$\vdash_{\pi@} O[t, A \supset B] \supset O[A, B]. \tag{11.23}$$

(11.19) and (11.23) give the required equivalence.

(ii) Suppose $\vdash_{\pi@} A \equiv B$. By **RO**, $\vdash_{\pi@} O[t, A \supset B]$. By **A15⁰**,

$$\vdash_{\pi@} O[t, A] \supset O[t, B].$$

Likewise, $\vdash_{\pi@} O[t, B] \supset O[t, A]$. Hence $\vdash_{\pi@} O[t, A] \equiv O[t, B]$. ∎

COROLLARY 11.25. *For all* $A \in \mathbb{F}^0$, $\vdash_{\pi@} A \equiv A@@$.

Proof. By induction on length of formula using Lemma 11.24 and

$$O[A, B]@@ = (N[A@ \supset B@])@ = O[t, A@@ \supset B@@]. \quad ∎$$

A completeness theorem can now be established which is comparable with Corollary 11.13.

THEOREM 11.26. *For all* $A \in \mathbb{F}^{0'}$ $A \in O \Leftrightarrow \vdash_{\pi@} A$.

Proof. Suppose $A \in O$. By (11.13), $A@ \in N$. By Corollary 11.13, $\vdash_\pi A@$. By (11.18b), $\vdash_{\pi@} A@@$. By Corollary 11.25 $\vdash_{\pi@} A$.

Conversely, suppose $\vdash_{\pi@} A$. By (11.18a), $\vdash_\pi A@$. By Corollary 11.13, $A@ \in N$. By (11.16b), $A@@ \in O$. By (11.17a), $A \in O$. ∎

The theorems of $\pi@$ include versions of the axioms of 'standard deontic logic': For each $A, B, C \in \mathbb{F}^0$

$$\vdash_{\pi@} \neg O[A, f] \supset [O[A, B] \supset \neg O[A, \neg B]]$$

$$\vdash_{\pi@} O[C, A \wedge B] \equiv [O[C, A] \wedge O[C, B]]$$

$$\vdash_{\pi@} O[C, A \vee \neg A].$$

Deontic operators are not iterated in von Wright (1951) but they have been permitted in later works. Including amongst the theorems of $\pi@$ are formulas in which iterated operators occur and which have been described as 'plausible candidates for logical truth' (Føllesdal and Hilpinen 1971):

$$\vdash_{\pi@} O[C, A] \supset O[C, O[C, A]]$$

$$\vdash_{\pi@} O[C, O[C, A]] \supset O[C, A]$$

$$\vdash_{\pi@} O[C, [O[C, A] \supset A].$$

According to Prior (1962), the last formula is valid and should be added to von Wright's system. Prior also suggested that $OA \supset [[A \supset OB] \supset OB]$ is valid. Its relativized form, i.e. $O[C, A] \supset [[[C \supset A] \supset O[C, B]] \supset B]$, is in O (Example 11.6) and is therefore a theorem of $\pi@$.

Appendix

A Algebra

The concepts of universal algebra are treated in several excellent books such as Grätzer (1968) to which the reader is referred for a complete and clear exposition.

An *n-ary operation* g on a non-empty set A is a function which assigns to each sequence $a_1, a_2, \ldots, a_n$ of elements of A a *value* in A. Thus $g: A^n \to A$. A *0-ary operation* is a member of A, and so is also called a *constant*.

An *algebra* is a non-empty set A on which operations $g_1, g_2, \ldots$ are defined. Each g_i is an $m(i)$-ary operation on A. The algebra is written as $\langle A, g_1, g_2, \ldots \rangle$. The *domain* of this algebra is A and the function τ is its *type*. Where the sequence of operations is finite, $g_1, \ldots, g_n$, say, the type of the algebra is defined by the sequence $\langle \tau(1), \ldots, \tau(n) \rangle$. Thus, consider the set $\mathbb{Z}$ of integers, together with the binary operations of addition $(+)$ and multiplication $(\times)$ and the constant (O-ary operation) zero (0). This defines the algebra $\langle \mathbb{Z}, +, \times, 0 \rangle$ which has type $\langle 2, 2, 0 \rangle$.

There are several ways of comparing algebras. From the logician's point of view it is convenient to fix attention on algebras of fixed type, τ say, and to consider an appropriate language (of *type* τ) consisting of an infinite list

$$x_1, x_2, \ldots$$

of variables and a (finite or infinite) list F

$$f_1, f_2, \ldots$$

of function letters corresponding to the operations of the algebra, together with brackets [,]. For each n, f_n is an $m(n)$-place function symbol. The function $\tau: F \to \mathbb{N}$ defined by $\tau(f_i) = m(i)$ for $i = 1, 2, 3, \ldots$ is the *type* of the language and of the algebra. Thus in an algebra $\langle A, g_1, g_2, \ldots \rangle$ of type τ, the $\tau(f_n)$-place function symbol f_n corresponds to the $\tau(f_n)$-ary operation g_n.

The class $\mathbb{T}$ of terms of this language is defined inductively (Chapter 2, §2.2.5) as follows.

(i) Every variable is a term.
(ii) If $t_1, \ldots, t_{m(n)}$ are terms then $f_n[t_1, \ldots, t_{m(n)}]$ is a term, where $m(n) = \tau(f_n)$.

$\mathbb{T}_n$ is the set of terms t such that the only variables which occur in t are in the set $\{x_1, \ldots, x_n\}$. $\mathbb{T}_0$ is the set of constant terms. Thus $\mathbb{T}_n \subseteq \mathbb{T}_k$ if $n \leq k$ and $\mathbb{T} = \sum \{\mathbb{T}_n; n \in \mathbb{N}\}$.

Let $\mathfrak{A} = \langle A, g_1, g_2, \ldots \rangle$ be an algebra of type τ. An *assignment* to $\mathfrak{A}$ is a mapping α which assigns elements of A to the variables. This assignment extends to a mapping $\bar{\alpha}: \mathbb{T} \to A$ which is defined inductively as follows.

(i) $\bar{\alpha}(x_i) = \alpha(x_i)$ for $i = 1, 2, \ldots$.
(ii) If $t_1, \ldots, t_m \in \mathbb{T}$, where $m = \tau(f_n)$, then

$$\bar{\alpha}(f_n(t_1, \ldots, t_m)) = g_n(\bar{\alpha}(t_1), \ldots, \bar{\alpha}(t_m)).$$

By a harmless abuse of language we can write α instead of $\bar{\alpha}$.

Let $t \in \mathbb{T}_n$. As the variables which occur in t are in $\{x_1, \ldots, x_n\}$, the value $\alpha(t)$ given by the assignment α to $\mathfrak{A}$ depends only on the values $\alpha(x_1), \ldots, \alpha(x_n)$, i.e. if β is another assignment to $\mathfrak{A}$ and $\alpha(x_i) = \beta(x_i)$ for $1 \leq i \leq n$, then $\alpha(t) = \beta(t)$. Thus, for each $n \in \mathbb{N}$, each term t of $\mathbb{T}_n$ gives rise to an n-ary operation $p[t]$ on A which is defined as follows. Let $a_1, \ldots, a_n \in A$. Define

$$\alpha(x_i) = \begin{cases} a_i & \text{if } 1 \leq i \leq n \\ a_n & \text{if } n \leq i. \end{cases}$$

Then $p[t] =_{\text{def}} \alpha(t)$. $p[t]$ is the n-ary *polynomial associated with* t.

The set $P_n(\mathfrak{A})$ of n-ary polynomials on $\mathfrak{A}$ is defined inductively. First, define the functions e_i^n, where $1 \leq i \leq n$, by

$$e_i^n(a_1, \ldots, a_n) =_{\text{def}} a_i \qquad (a_1, \ldots, a_n \in A).$$

Next, we have the following.

(i) $e_i^n \in P_n(\mathfrak{A})$ for $1 \leq i \leq n$.
(ii) If $h_1, \ldots, h_m \in P_n(\mathfrak{A})$, where $m = \tau(f_n)$, then the composition of functions $g_n(h_1, \ldots, h_m)$ is also in $P_n(\mathfrak{A})$.

Every n-ary polynomial on $\mathfrak{A}$ is the polynomial associated with some term in $\mathbb{T}_n$.

An *equation* or *identity* is an expression $t_1 = t_2$ where $t_1, t_2 \in \mathbb{T}$. This identity *holds in* $\mathfrak{A}$ (or $\mathfrak{A}$ *satisfies* $t_1 = t_2$) if, for all assignments α to A, $\alpha(t_1) = \alpha(t_2)$. An *equational class* of algebras (*variety*) is the class of algebras satisfying a given set of equations.

Derived systems (Cohn 1965, Chapter III, Grätzer 1968, §8). Let $\mathfrak{A} = \langle A, g_1, g_2, \ldots \rangle$ be an algebra of type τ. Let $m: \mathbb{N} \to \mathbb{N}$ and let $t_i \in \mathbb{T}_{m(i)}$ for $i = 1, 2, \ldots$. Consider the algebra $\mathfrak{A}'$ whose domain is A, the domain of $\mathfrak{A}$, and whose operations are polynomials associated with these terms, i.e.

$$\mathfrak{A}' = \langle A, p[t_1], p[t_2], \ldots \rangle.$$

Thus $p[t_i]$ is an $m(i)$-ary operation on A. $\mathfrak{A}'$ is known as a *derived algebra* and the polynomials $p[t_i]$ are said to be *derived operations*.

The language appropriate for $\mathfrak{A}'$ has a list F' of function letters, $k_1, k_2, \ldots,$ say, corresponding to the operations $p[t_1], p[t_2], \ldots$ respectively of $\mathfrak{A}'$. It has the same list $x_1, x_2, \ldots$ of variables. As the function letter k_i corresponds to the $m(i)$-ary operation $p[t_i]$, the type of this language is the function $\mu \colon F' \to \mathbb{N}$ defined by $\mu(k_i) =_{\text{def}} m(i)$ for $i = 1, 2, 3, \ldots$ Let $\mathbb{T}'$ denote the set of terms in this language. The derived structure $\mathfrak{A}'$ is determined by the correspondence $F_i \to t_i$ between the function symbols of the language of type μ and the function symbols of the language of type τ. This correspondence extends to a mapping $d \colon \mathbb{T}' \to \mathbb{T}$ (called the *term derivation corresponding to $\mathfrak{A}'$*) defined by induction on the construction of $\mathbb{T}'$ as follows.

(i) $d(x_i) = x_i$ for $i = 1, 2, \ldots.$
(ii) Consider the term $k_i(s_1, \ldots, s_m)$ in the language of type μ, where $s_1, \ldots, s_m \in \mathbb{T}'$ and $m = \mu(k_i)$. Then $d(k_i(s_1, \ldots, s_m))$ is the result of simultaneously substituting $d(s_1), \ldots, d(s_m)$ for the variables $x_1, \ldots x_m$, respectively in t_i.

A mapping α which assigns members of A to the variables $x_1, x_2, \ldots$ extends to an assignment $\bar{\alpha}$ to $\mathfrak{A}$ and also to an assignment $\hat{\alpha}$ to $\mathfrak{A}'$. Thus, for any term t in the language of type τ and any term s in the language of type μ, $\bar{\alpha}(t)$ and $\hat{\alpha}(s)$ are both elements of A. Then

$$\bar{\alpha}(d(s)) = \hat{\alpha}(s) \qquad (s \in \mathbb{T}').$$

Subsystems Let $\mathfrak{A} = \langle A, g_1, g_2, \ldots \rangle$ and $\mathfrak{B} = \langle B, h_1, h_2, \ldots \rangle$ be two algebras of type τ. $\mathfrak{B}$ is a *subsystem* of $\mathfrak{A}$ if A includes B and, for each n, $h_n(z_1, \ldots, z_m) = g_n(z_1, \ldots, z_m)$, where $m = \tau(f_n)$, for all $z_1, \ldots, z_m \in B$, i.e. each operation h_n on B is the restriction of the corresponding operation g_n to B.

Homomorphisms Let $\mathfrak{A} = \langle A, g_1, g_2, \ldots \rangle$ and $\mathfrak{B} = \langle B, h_1, h_2, \ldots \rangle$ be two algebras of type τ. A *homomorphism* of $\mathfrak{A}$ to $\mathfrak{B}$ is a mapping $\theta \colon A \to B$ such that for each operation g_n, for all $z_1, \ldots, z_m \in A$, where $m = \tau(f_n)$,

$$\theta(g_n(z_1, \ldots, z_m)) = h_n(\theta(z_1), \ldots, \theta(z_m)).$$

If $\theta(A) = B$ (i.e. θ maps A on to B) then $\mathfrak{B}$ is a *homomorphic image* of $\mathfrak{A}$.

Congruences The notion of a *congruence* on an algebra and the notion of a *quotient algebra* is given in Definition 5.3(i). The identity relation $id_\mathfrak{A}$ on an algebra $\mathfrak{A}$ is a congruence. Let K be a family of congruences on $\mathfrak{A}$. Since each member of K is a set of ordered pairs, the intersection $\bigcap K$ of all the congruences is defined: $\prod K$ is also a congruence. K is said to be a *separating family* of congruences if $\prod K = id_\mathfrak{A}$.

Direct products The *direct product* $\mathfrak{A} \times \mathfrak{B}$ of the two algebras of type τ, $\mathfrak{A} = \langle A, g_1, g_2, \ldots \rangle$ and $\mathfrak{B} = \langle B, h_1, h_2, \ldots \rangle$, is the algebra $\langle A \times B, \langle g_1, h_1 \rangle, \langle g_2, h_2 \rangle, \ldots \rangle$ where

$$A \times B = \{\langle s, t \rangle, s \in A \ \& \ t \in B\}$$

and the operations $\langle g_n, h_n \rangle$ act componentwise on $A \times B$, i.e.

$$\langle g_n, h_n \rangle(\langle s_1, t_1 \rangle, \ldots, \langle s_m, t_m \rangle) = \langle g_n(s_1, \ldots, s_m), h_n(t_1, \ldots, t_m) \rangle$$

for all $s_1, \ldots, s_m \in A$ and $t_1, \ldots, t_m \in B$ where $\tau(n) = m$. The notion can be extended to products of families of algebras. Let $\mathfrak{A}_i = \langle A_i, g_1^i, g_2^i, \ldots \rangle$ for $i \in I$ be algebras of type τ. The *direct product* $\prod \{\mathfrak{A}_i; i \in I\}$ is the algebra $\langle B, h_i, h_2, \ldots \rangle$, where B is the set of functions $e: I \to \sum \{A_i; i \in I\}$ such that $e(i) \in A_i$ for all $i \in I$ and for all $e_1, \ldots, e_n \in B$, $h_n(e_1, \ldots, e_m)$ is that member of B satisfying

$$h_n(e_1, \ldots, e_m)(i) = g_n^i(e_1(i), \ldots, e_n(i))$$

for all $i \in I$. Every identity which holds in $\mathfrak{A}_i$ for all $i \in I$ also holds in $\prod \{\mathfrak{A}_i; i \in I\}$; more generally, we have the following theorem.

THEOREM A1 (Birkhoff 1935). *A class K of algebras is an equational class iff K contains every subsystem and homomorphic image of algebras in K and every direct product of non-void families of algebras in K.*

Subdirect products. In the *direct product* $\prod \{\mathfrak{A}_i; i \in I\}$ (above) the mapping $\pi_i: B \to A_i$ defined by $\pi_i(e) =_{\text{def}} e(i)$ for all $e \in B$ is the *ith projection*. π_i is a homomorphism of $\prod \{\mathfrak{A}_i; i \in I\}$ into $\mathfrak{A}_i$. A subalgebra $\mathfrak{C}$ of $\prod \{\mathfrak{A}_i; i \in I\}$ is a *subdirect product* of $\{\mathfrak{A}_i; i \in I\}$ if each projection π_i maps $|\mathfrak{C}|$ on to A_i.

Subdirect products frequently occur in the following way.

THEOREM A2. *Let $\mathfrak{A}$ be an algebra and K be a family of congruences on $\mathfrak{A}$. Put $\Theta =_{\text{def}} \prod K$. Then $\mathfrak{A}/\Theta$ is isomorphic to a subdirect product of the family of quotient algebra $\{\mathfrak{A}/\theta; \theta \in \Theta\}$.*

Subdirect factorization An algebra $\mathfrak{A}$ is said to be *subdirectly reducible* if it has a separating family of congruences K which does not include $id_\mathfrak{A}$. $\mathfrak{A}$ is *subdirectly irreducible* if every separating family of congruences on $\mathfrak{A}$ contains $id_\mathfrak{A}$.

THEOREM A3 (Birkhoff 1944). *Every algebra is isomorphic to a subdirect product of subdirectly irreducible algebras.*

Note that every congruence on a two-element algebra is either the identity or it identifies the two elements. Hence every two-element algebra is subdirectly irreducible.

R Relational systems

A *relational system* or *structure* is a non-empty set A on which operations $g_1, g_2, \ldots$ are defined (as in an algebra) as well as relations $R_1, R_2, \ldots$. Comparison of such structures can be effected by augmenting the language used for the operations (as above) with a list P of *predicate letters*

$$P_1, P_2, \ldots$$

which denote the relations $R_1, R_2, \ldots$ respectively. The type-determining function τ must now be extended to define the number of arguments taken by the functions. Thus $\tau: F + P \to \mathbb{N}$, and for each n, f_n is a $\tau(f_n)$-place function symbol, $g_n: A^m \to A$ where $m = \tau(f_n)$, P_n is a $\tau(P_n)$-place predicate letter, and R_n is a $\tau(P_n)$-ary relation on A. Thus $\langle A, g_1, g_2, \ldots, R_1, R_2, \ldots \rangle$ is a structure (relational system) of *type* τ if, for all n, g_n is a $\tau(f_n)$-place operation on A and R_n is a $\tau(P_n)$-ary relation on A.

It is customary first to define an m-ary relation R on a set A as a subset of R^m. Then the *characteristic function* ϕ_R of R is defined as the function R^m with values in $\{0, 1\}$ such that $\phi_R(x_1, \ldots, x_m) = 1$ if $(x_1, \ldots, x_m) \in R$ and $\phi_R(x_1, \ldots, x_m) = 0$ if $(x_1, \ldots, x_m) \notin R$. It is useful in this book to reverse this process. We define an m-ary relation R in the first instance as a *function* $R: A^m \to \{0, 1\}$. Uniquely corresponding to this function is the *set* of m-tuples $\{(x_1, \ldots, x_m); R(x_1, \ldots, x_m) = 1\}$. Then R is the characteristic function of this set. The point of this inversion is that the set $\{0, 1\}$ can be considered as determining what can be called $\{0, 1\}$-valued relations. By replacing $\{0, 1\}$ with an arbitrary set L we now have a definition of L-valued relations.

T Topology

Only a few of the most basic definitions of topological concepts are employed in this book. There are many excellent books, such as Hocking and Young (1961), which expand these ideas.

Open sets Consider a set S. A subset $\mathbb{O}$ of $\mathbb{P}(S)$ is a *topology* on S and the members of $\mathbb{O}$ are *open* sets if

O1: the union of any subset of $\mathbb{O}$ is in $\mathbb{O}$,
O2: the intersection of any two members of $\mathbb{O}$ is in $\mathbb{O}$,
O3: $S, \emptyset \in \mathbb{O}$.

S, together with $\mathbb{O}$, is a *topological space*.

A subset X of S is a *closed set* if $S - X \in \mathbb{O}$. The *closure* $C(X)$ of a set X is the intersection of all closed sets which include X. The operator C satisfies the Kuratowski closure axioms (Example 3.2(iv)) (see Rasiowa and Sikorski 1963, Chapter 1).

The union of any two closed sets is a closed set. The intersection of any set of closed sets is also a closed set. Thus the open (closed) sets of a topological space form a ring of sets (Definition 5.9) and so are a distributive lattice (Chapter 5, §5.9.1). By **O1** these lattices are complete.

A subset X of S is said to be *dense in S* if $C(X) = S$ and it is said to be a *boundary set* if $S - X$ is dense in S.

A topological space is *separated* if it is the union of two disjoint open sets; otherwise it is *connected*.

A topological space S is a T_1-space if, given two points of S, each lies in an open set not containing the other. (This is equivalent to the condition that unit sets are closed.)

THEOREM T1. *In a T_1-space S, a closed set Y is dense in S if, and only if, $Y = S$, and $Y = S$ if, and only if, $Y \cdot Z \neq \emptyset$ for all closed sets Z.*

Compactness A topological space S is *compact* if for every subset C of open sets such that $\sum C = S$, there is a finite subset C' of C such that $\sum C' = S$.

S has the *finite intersection property* if for every collection K of closed sets

$$(L \in \mathbb{P}_\omega(K))(\prod L \neq \emptyset) \Rightarrow \prod K \neq \emptyset.$$

S has the finite intersection property iff S is compact.

W Words

Let A be an alphabet (set of objects called *letters*).

If $a_1, a_2, \ldots, a_n \in A$ then $\langle a_1, a_2, \ldots, a_n \rangle$ is the finite sequence whose first element is a_1, second element is $a_2, \ldots$, nth element is a_n. A *word* (on A) is a finite sequence of letters. The null sequence λ is counted as a word. A^* denotes the set of all words on A. If $U = \langle u_1, \ldots, u_n \rangle$ and $V = \langle v_1, \ldots, v_m \rangle$, then $U \cap V =_{\text{def}} \langle w_1, \ldots, w_{n+m} \rangle$ where $w_i = u_i$ for $1 \leq i \leq n$ and $w_{i+n} = v_i$ for $1 \leq i \leq m$.

Let $E_1, E_2, \ldots, E_n \in A^*$ and $y_1, y_2, \ldots, y_n$ be distinct letters in A. Then $(E_1, \ldots, E_n/y_1, \ldots, y_n)$ denotes the substitution operation which effects the simultaneous substitution of E_1 for y_1, E_2 for $y_2, \ldots$, and E_n for y_n. Thus if $W \in A^*$ then $W(E_1, \ldots, E_n/y_1, \ldots, y_n)$ is the result of simultaneously substituting E_1 for each occurrence of y_1 in $W, \ldots$, E_n for each occurrence of y_n in W. If σ, τ are two such substitution operators then the expression $(W\sigma)\tau$ (i.e. W is acted on first by σ and then by τ) is abbreviated to $W\sigma\tau$.

Let $\sigma = (E_1, \ldots, E_n/y_1, \ldots, y_n)$ and let $U, V \in A^*$. Then we have the following.

(i) $(U \cap V)\sigma = U\sigma \cap V\sigma$.

(ii) If the distinct letters $x_1, \ldots, x_n$ do not occur in U then
$$U\sigma = U(x_1, \ldots, x_n/y_1, \ldots, y_n)(E_1, \ldots, E_n/x_1, \ldots, x_n).$$

(iii) Let $F_1, \ldots, F_m \in A^*$ and suppose that the letters $y_1, \ldots, y_n$ do not occur in any of the words F_i. Let $z_1, \ldots, z_m$ be distinct letters which are distinct from the letters $y_1, \ldots, y_n$. Put $\sigma = (E_1, \ldots, E_n/x_1, \ldots, x_n)$ and $\tau = (F_1, \ldots, F_m/z_1, \ldots, z_m)$. Then
$$U\sigma\tau = U\tau(E_1\tau, \ldots, E_n\tau/y_1, \ldots, y_n).$$

Thus, if $z_1, \ldots, z_m$ do not occur in any E_i then
$$U\sigma\tau = U\tau\sigma.$$

References

Abraham, R. (1967). *Foundations of mechanics.* Benjamin, New York.
Alexander, H. G. (1956). (ed.) *The Leibniz–Clarke correspondence.* Manchester University Press, Barnes and Noble, New York.
Anderson, A. R. and Belnap, N. D. Jun. (1962). Tautological entailments. *Philosophical Studies,* **13**, 9–23.
Anderson, A. R. and Belnap, N. D. Jun. (1975). *Entailment: the logic of relevance and necessity,* Vol. 1. Princeton University Press.
Apt, K. R. (1981). Ten years of Hoare's logic: a survey—Part 1. *ACM Transactions in Programming Languages and Systems,* **3**, 431–83.
Aristotle (1936). *Physics.* (ed. W. Ross). Clarendon Press, Oxford.
Aristotle (1949). *Prior and posterior analytics* (ed. W. Ross). Clarendon Press, Oxford.
Arruda, A. I. (1977). On the imaginary logic of N. A. Vasil'ev. In *Non-classical logics, model theory and computability* (ed. A. I. Arruda *et al.*), pp. 3–24. North-Holland, Amsterdam.
Barnes, D. W. and Mackey, J. M. (1975). *An algebraic introduction to mathematical logic.* Springer-Verlag, New York.
Barwise, K. J. (1974). Axioms for abstract model theory. *Annals of Mathematical Logic,* **7**, 221–65.
Barwise, K. J. and Feferman, S. (ed.) (1985). *Model-theoretic logics.* Springer-Verlag, New York.
Beeson, N. (1979). Continuity in intuitionistic set theories. In *Logic Colloquium '78.* (ed. M. Boffa *et al.*) pp. 1–52. North-Holland, Amsterdam.
Bergson, H. (1911). *Creative evolution* (authorized translation by A. Mitchell). Macmillan, London.
Bernays, P. (1926). Axiomatische Untersuchung des Aussagenkalküls der 'Principia Mathematicae'. *Mathematische Zeitschrift,* **25**, 305–20.
Bernays, P. (1965). Betrachtung zum Sequenzen-Kalkül. In *Logic and methodology* (ed. A. T. Tymieniecka), Chapter I. North-Holland, Amsterdam.
Beth, E. W. (1959). *The foundations of mathematics.* North-Holland, Amsterdam.
Beth, E. W. (1962). *Formal methods.* Reidel, Dordrecht.
Białynicki-Birula, A. and Rasiowa, H. (1957). On the representation of quasi-Boolean algebras. *Bulletin de l'Académie Polonaise des Sciences, Classe III,* **5**, 259–61.
Birkhoff, G. (1933). On the combination of subalgebras. *Proceedings of the Cambridge Philosophical Society,* **29**, 441–61.
Birkhoff, G. (1935). On the structure of abstract algebras. *Proceedings of the Cambridge Philosophical Society,* **31**, 433–54.
Birkhoff, G. (1944). Subdirect unions in universal algebras. *Proceedings of the Cambridge Philosophical Society,* **50**, 746–68.
Birkhoff, G. (1948). *Lattice theory.* American Mathematical Society, Providence, RI.

Birkhoff, G. and von Neumann, J. (1936). The logic of quantum mechanics. *Annals of Mathematics*, **37**, 823–43.
Bishop, E. (1967). *Foundations of constructive analysis*. McGraw-Hill, New York.
Bishop, E. (1970). Mathematics as a numerical language. In *Intuitionism and proof theory* (ed. A. Kino et al.), pp. 53–71. North-Holland, Amsterdam.
Bishop, E. and Bridges, D. (1980). *Constructive Analysis*, Springer-Verlag, Berlin.
Blamey, S. (1986). Partial logic. In *Handbook of philosophical logic*, Vol. 3. (eds D. Gabbay and F. Guenthner), pp. 1–70. Reidel, Dordrecht.
Blau, U. (1978). *Die Dreiwertige Logik der Sprache*. de Gruyter, Berlin.
Bloor, D. (1983). *Wittgenstein: a social theory of knowledge*. Macmillan, London.
Bocheński, I. M. (1961). *A history of formal logic*. University of Notre Dame Press.
Bochvar, D. A(1939). On a three valued logical calculus and its application to the analysis of contradictions. (In Russian.) *Recueil Mathématiques*, N.S. **4**, 287–308.
Börger, E. (1986). *Berechenbarkeit, Komplexität, Logik*. Vieweg, Braunschweig.
Börger, E. (1989). *Computability, complexity, logic*. North-Holland, Amsterdam.
Bosanquet, B. (1911). *Logic or the morphology of knowledge*, Vol. I. Clarendon Press, Oxford (1st edn 1888).
Bourbaki, N. (1968). *Elements of mathematics: theory of sets*. Addison-Wesley, Reading, MA.
Bradley, F. H. (1922). *The principles of logic*, Vol. I. Oxford University Press (1st edn 1883).
Brouwer, L. E. J. (1905). *Leven, kunst en mystiek*. Waltman, Delft, 99 pp.
Brouwer, L. E. J. (1907) *Over de grondslagen der wiskunde*. Maas en van Suchtelen, Amsterdam.
Brouwer, L. E. J. (1908). De ontrouwbaarheid der logische princips. *Tijdschrift voor Wijsbegeerte*, **2**, 152–8.
Brouwer, L. E. J. (1909). *Het wezen der meetkunde* (Public lecture). Clausen, Amsterdam, 20 pp.
Brouwer, L. E. J. (1912). *Intuitisme en formalisme*. Clausen, Amsterdam (also published in *Wiskundig Tijdschrift*, **9** (1913), 180–8.)
Brouwer, L. E. J. (1918). Begründung der Mengenlehre unabhängig von logischen Satz vom ausgeschlossenen Dritte, Erste Teil, Allgemeine Mengenlehre. *Koninklijke Akademie van Wetenschappen te Amsterdam, Verhandlingen*, 1st Section, no. 5, 11–45.
Brouwer, L. E. J. (1923). Intuitionistische splitsing van mathematische grondbegrippen. *Koninklijke Akademie van Wetenschappen te Amsterdam*, **32**, 877–80.
Brouwer, L. E. J. (1925). Intuitionistische Zerlegung mathematischer Grundbegriffe. *Jahresbericht der deutschen Mathematiker Vereinigung*, **33**, 251–6.
Brouwer, L. E. J. (1927). Über Definitionsbereiche von Funktionen. *Mathematische Annalen*, **97**, 60–75.
Brouwer, L. E. J. (1928a). *Mathematik, Wissenschaft und Sprache*. Gistel, Vienna, 14 pp.
Brouwer, L. E. J. (1928b). *Die Struktur des Kontinuums*. Gistel, Vienna, 14 pp.
Brouwer, L. E. J. (1933). *De uitdrukkingswijze der wetenschap, kennistheoretische voordrachten door LEJ Brouwer e.a.* Gröningen, Batavia.
Brouwer, L. E. J. (1942). Zum freien Werden von Mengen und Funktionen. *Koninklijke Akademie van Wetenschappen te Amsterdam, Proceedings*, **45**, 322–3.
Brouwer, L. E. J. (1948). Essentiell negatieve eigenschappen. *Koninklijke Akademie*

van Wetenschappen te Amsterdam, Proceedings, **51**, 963–4. Also *Indagationes Mathematicae*, **10**, 322–3.

Brouwer, L. E. J. (1949a). Consciousness, philosophy and mathematics. In *Proceedings of the 10th International Congress of Philosophy*, pp. 1235–49. North-Holland, Amsterdam.

Brouwer, L. E. J. (1949b). De non-aequivalentie van de constructieve orderelatie in het continuum. *Indagationes Mathematicae*, **11**, 37–9.

Brouwer, L. E. J. (1952). Historical background, principles and methods of intuitionism. *South African Journal of Science*, **49**, 139–67.

Brouwer, L. E. J. (1953). Points and spaces. *Canadian Journal of Mathematics*, **6**, 1–7.

Büchi, J. R. (1952). Representation of complete lattices by sets. *Portugaliae Mathematicae*, **11**, 151–67.

Burstal, R. M. and Goguen, J. A. (1980). The semantics of CLEAR, a specification language. In *Abstract Software Specifications* (ed. D. Bjørner), Lecture notes in computer science, Vol. 86, pp. 292–332. Springer-Verlag, Berlin.

Burstal, R. M. and Goguen, J. A. (1984). Introducing institutions. In *Logic of Programs* (eds E. Clarke et al.), Lecture Notes in Computer Science, Vol. 164, pp. 221–56. Springer-Verlag, Berlin.

Campbell, A. D. (1943). Set coordinates for lattices. *Bulletin of the American Mathematical Society*, **49**, 395–8.

Campbell, N. R. (1920). *Physics—the elements*. Cambridge University Press.

Carnap, R. (1928). *Der logische Aufbau der Welt*, Weltkreis-Verlag, Berlin-Schlactensee. (Translated from the German by R. A. Wolf, Routledge and Kegan Paul, London, 1967.)

Carnap, R. (1935). Ein Gültigkeitskriterium für die Sätze der klassischen Mathematik. *Monatshefte für Mathematik und Physik*, **42**, 163–90.

Carnap, R. (1947). *Meaning and necessity*. University of Chicago Press.

Carnap, R. (1949). *Logical syntax of language*, Routledge and Kegan Paul, London.

Carnap, R. (1958). *Introduction to symbolic logic and its applications*. Dover Publications, New York.

Chang, C. C. (1974). Model theory 1945–1971. In *Proceedings of the Tarski Symposium* (eds L. Henkin et al.), pp. 173–86. American Mathematical Society, Providence, RI.

Chang, C. L. and Lee, R. (1973). *Symbolic logic and mechanical theorem proving*. Academic Press, New York.

Church, A. (1936). A note on the Entscheidungsproblem. *Journal of Symbolic Logic*, **1**, 40–1.

Church, A. (1956). *Introduction to mathematical logic*, Vol. 1. Princeton University Press.

Chwistek, L. (1948). *The limits of science*. Routledge and Kegan Paul, London.

Clarke, E. M. (1978). Programming language constructs for which it is impossible to obtain good Hoare-like axioms. *Journal of the Association for Computing Machinery*, **26**, 129–47.

Cleave, J. P. (1974a). An account of entailment based on classical semantics. *Analysis*, **34**, 118–22.

Cleave, J. P. (1974b). The axiomatization of material necessity. *Notre Dame Journal of Formal Logic*, **20**(1), 180–90.

Cleave, J. P. (1974c). The notion of logical consequence in the logic of inexact predicates. *Zeitschrift für mathematische Logik und Grundlagen der Mathematik*, **20**, 302–24.

Cleave, J. P. (1976). Quasi-Boolean algebras, empirical continuity and three-valued logic. *Zeitschrift für mathematische Logik und Grundlagen der Mathematik*, **22**, 481–500.

Cleave, J. P. (1980). Some remarks on the interpretation of 3-valued logics. *Ratio*, **22** (1), 52–60.

Cocchiarella, N. B. (1966). *Tense logic: a study in the topology of time*. Unpublished Ph.D. Thesis, University of California at Los Angeles.

Cohen, M. R. and Nagel, E. N. (1934). *An introduction to logic and scientific method*. Routledge and Kegan Paul, London.

Cohen, P. J. (1963). *The independence of the axiom of choice*. Mimeo, Stanford University.

Cohn, P. M. (1965). *Universal algebra*. Harper & Row, New York.

Cook, S. A. (1978). Soundness and completeness of an axiom system for program verification. *SIAM Journal of Computing*, **7**, 70–90.

Couturat, L. (1961). *La logique de Leibniz*. Georg Olms Verlagsbuchhandlung, Hildesheim.

Crawley, P. and Dilworth, R. P. (1973). *Algebraic theory of lattices*. Prentice-Hall, Englewood Cliffs, NJ.

Cresswell, M. L. and Hughes, G. E. (1968). *An introduction to modal logic*. Menthuen, London.

Cudia, D. F. and Singletary, W. E. (1968a). Degrees of unsolvability in formal grammars. *Journal of Symbolic Logic*, **15**, 680–92.

Cudia, D. F. and Singletary, W. E. (1968b). The Post correspondence problem. *Journal of Symbolic Logic*, **33**, 418–30.

Curry, H. B. (1963). *Foundations of mathematical logic*. McGraw-Hill, New York.

Curry, H. B. (1965). Inferential deduction. In *Contributions to logic and methodology in honour of J. M. Bocheński* (ed. A.-T. Tymieniecka), Chapter 2. North-Holland, Amsterdam.

van Dalen, D. and Troelstra, A. S. (1988). *Constructivism in mathematics, an introduction*, Vols I–II, North-Holland, Amsterdam.

Davis, M. (1958). *Computability and solvability*. McGraw-Hill, New York.

Dedekind, R. (1897). Über Zerlegung von Zahlen durch ihrer grössten gemeinsamen Teilen. *Festschrift der Technischen Hochschule zur Braunschweig bei Gelegenheit der 69. Versammlung Deutscher Naturforscher und Ärte*, 1–40. (Reprinted in *Gesammelte mathematische Werke*, II, pp. 103–47, 1931.)

Dedekind, R. (1900). Über die drei Moduln erzeugte Dualgruppe. *Mathematische Annalen*, **53**, 371–403.

Dirac, P. A. M. (1939). The relation between mathematics and physics. *Proceedings of the Royal Society of Edinburgh, Session 1938–39*, **59**, 122–9.

Dirac, P. A. M. (1963). The evolution of the physicist's picture of nature. *Scientific American* **208**(5), 45–53.

Doyle, J. and McDermott, D. (1980). Non-monotonic logic I. *Artificial Intelligence*, **13**, 27–39.

Doyle, J. and McDermott, D. (1982). Non-monotonic logic II. Non-monotonic modal theories. *Journal of the Association of Computing Machinery*, **29**, 33–57.

Duhem, P. (1954). *The aim and structure of physical theory* (transl. P. P. Wiener). Princeton University Press.

Dummett, M. (1977). *Elements of intuitionism*. Clarendon Press, Oxford.

van Emden, M. H. and Kowalski, R. A. (1976). The semantics of predicate logic as a programming language. *Journal of the Association of Computing Machinery*, **23**, 733–42.

Emerson, A. E. and Sistla, A. P. (1984). Deciding branching time logic: a triple exponential decision procedure for CTL*. In *Logic of programs* (eds E. Clarke et al.), Lecture notes in computer science, Vol. 86, pp. 176–92. Springer-Verlag, Berlin.

Fisch, M. and Turquette, A. R. (1966). Peirce's triadic logic. *Transactions of the Charles S. Peirce Society*, **2**, 81.

Fisher, M. J. and Ladner, R. E. (1979). Propositional dynamic logic of regular programs. *Journal of Computer and System Sciences*, **18**, 194–211.

Fitting, M. C. (1969). *Intuitionistic logic, model theory and forcing*. North-Holland, Amsterdam.

Floyd, R. M. (1967). Assigning meanings to programs. *Proceedings of the American Mathematical Society Symposia in Applied Mathematics*, **19**, 19–32.

Føllesdal, D. and Hilpinen, R. (1971). Deontic logic: an introduction. In *Deonetic logic: introductory and systematic readings* (ed. R. Hilpinen), pp. 1–35. Reidel, Dordrecht.

van Fraasen, B. C. (1975). The labyrinth of quantum logics. In *The logico-algebraic approach to quantum mechanics*, Vol. 1 (ed. C. A. Hooker), pp. 577–607. Reidel, Dordrecht.

Frayne, T., Morel, A., and Scott, D. (1962). Reduced direct products. *Fundamenta Mathematicae*, **51**, 195–228.

Frege, G. (1879). *Begriffschrift, eine der arithmetischen nachgebildete Formelsprache des reinen Denkens*. Verlag von Louis Nebert, Halle a/s.

Frege, G. (1969). *Gottlob Frege: Nachgelassenen Schriften*, Vol. 1. Felix Meiner Verlag, Hamburg.

Friedman, H. (1973). Some applications of Kleene's methods for intuitionistic systems. In *Proceedings of the Cambridge Summer School in Mathematical Logic*, lecture notes in mathematics, Vol. 337, pp. 113–70. Springer-Verlag, Berlin.

Fuhrken, G. (1964). Skølem-type normal forms for 1st order languages with generalised quantifier. *Fundamenta Mathematicae*, **54**, 291–302.

Gabbay, D., Pneuli, A., Shelah, S., and Stain, J. (1980). On the temporal logic of programs. *Proceeding of the 18th Symposium on the mathematical Foundations of Computer Science*, Springer lecture notes in computer science, Vol. 54, 278–92. Springer-Verlag, Berlin.

Garden, R. W. (1984). *Modern logic and quantum mechanics*. Adam Hilger, Bristol.

Geach, P. (1968). *A history of the corruptions of logic*. Inaugural address, University of Leeds.

Gentzen, G. (1932). Über die Existenz unabhänger Axiom System zu unendlichen Satz-systemen. *Mathematische Annalen*, **107**, 329–50.

Gentzen, G. (1934). Untersuchung über das logische Schliessen. *Mathematische Zeitschrift*, **39**, 176–210, 405–31.

Glivenko, V. (1929). Sur quelques points de la logique de Brouwer. *Académie Royale de Belgique, Bulletins de la Classe des Sciences*, Serie 5, **15**, 183–8.

Gödel, K. (1930). Die Vollständigkeit der Axiome des logischen Funktionenkalküls. *Monatshefte für Mathematik und Physik*, **37**, 349–60.

Gödel, K. (1931). Über formal unentscheidbare Sätze der Principia Mathematica und verwandter Systeme I. *Monatshefte für Mathematik und Physik*, **38**, 173–98.

Gödel, K. (1932–3). Zur intuitionistischen Arithmetik und Zahlentheorie. *Ergebnisse eines mathematischen Kolloquiums*, **4**, 34–8.

Gödel, K. (1933). Eine Interpretation des intuitionistischen Aussagenkalküls. *Ergebnisse eines mathematischer Kolloquiums*, **4**, 34–40.

Goodstein, R. L. (1964). *Recursive number theory*. North-Holland, Amsterdam.

Gödel, K. (1958). Über eine bisher noch nicht benützte Erweiterung des finiten Standpunktes. *Dialectica*, **12**, 280–7.

Grätzer, G. (1968). *Universal algebra*. Van Nostrand, Princeton, NJ.

Greechie, R. J. and Gudder, S. D. (1973). Quantum logics. In *Contemporary Research in the Foundations and Philosophy of Quantum Theory*, Vol. 2 (ed. C. A. Hooker), pp. 143–73. Reidel, Dordrecht.

Greechie, R. J. and Gudder, S. D. (1975). Quantum logic. In *The logico-algebraic approach to quantum mechanics*, Vol. 1 (ed. C. A. Hooker), pp. 546–70. Reidel, Dordrecht.

Grzegorczyk, A. (1964). A philosophically plausible formal interpretation of intuitionistic logic. *Indagationes Mathematicae*, **26**, 596–601.

Hájek, P., Benková, K., and Renc, Z. (1971). The GUHA method and the three-valued logic. *Kybernetica*, **7**, 421–35.

Hanf, W. (1974). Primitive Boolean algebras. In *Proceedings of the Tarski Symposium* (eds L. Henkin et al.), pp. 75–90. American Mathematical Society, Providence, RI.

Hardegree, G. M. (1975). The conditional in abstract and concrete quantum mechanics. In *The logico-algebraic approach to quantum mechanics*, Vol. 1 (ed. C. A. Hooker), pp. 49–108. Reidel, Dordrecht.

Hardegree, G. M. (1976). The conditional in quantum logic. In *Logic and probability in quantum mechanics* (ed. P. Suppes), pp. 55–72. Reidel, Dordrecht.

Hardy, G. H. (1969). *A mathematicians apology*. Cambridge University Press (1st edition 1940).

Hausdorff, F. (1914). *Grundzüge der Mengenlehre*. de Gruyter, Leipzig.

van Heijenoort, J. (1967a). Logic as language and logic as calculus. *Synthese*, **17**, 324–30.

van Heijenoort, J. (1967b). *From Frege to Gödel*. Harvard University Press.

Henkin, L. (1949a). Fragments of propositional calculus. *Journal of Symbolic Logic*, **14**, 42–8.

Henkin, L. (1949b). The completeness of the first order functional calculus. *Journal of Symbolic Logic*, **14**, 159–66.

Henkin, L. (1961). Some remarks on infinitely long formulas. In *Infinitistic Methods*, pp. 167–83. Pergamon, Oxford.

Herbrand, J. (1930). Recherches sur la théorie de la démonstration. *Travaux de la Société des Sciences et des Lettres de Varsovie, Classe III, Sciences mathématiques et physiques*, no. 33.

Hermes, H. (1965). *Enumerability, decidability, computability*. Springer-Verlag, Berlin.

Hermes, H. and Köthe, G. (1939). Die Theorie der Verbände. *Enzyklopädie der Mathematischen Wissenschaften I*, 2. Aufl., Vol. 5. Teubner, Leipzig.

Hertz, P. (1922). Über Axiomensysteme für beliebige Satzsysteme. I. Sätze der ersten Grades. *Mathematische Annalen*, **87**, 245–67.
Hertz, P. (1923). Über Axiomensysteme für beliebige Satzsysteme. II. Sätze höheren Grades. *Mathematische Annalen*, **89** (101), 457–514.
Heyting, A. (1930). Die formalen Regeln der intuitionistischen Logik. In *Sitzungsberichte der Preussischen Akademie der Wissenschaften, Physicalisch-mathematische Klasse*, 42–56. de Gruyter, Berlin.
Heuting, A. (1956). *Intuitionism*. North-Holland, Amsterdam.
Heyting, A. (1961). Infinitistic methods from a finitist point of view. In *Infinitistic methods*, pp. 185–92. Pergamon, Oxford.
Hilton, P. J. and Wylie, S. (1967). *Homology Theory*. Cambridge University Press.
Hindess, B. (1977). *Philosophy, methodology in the social sciences*. Harvester Press, Hassocks.
Hintikka, K. J. (1973). *Logic, language games and information*. Clarendon Press, Oxford.
Hoare, C. A. R. (1969). An axiomatic basis for computer programming. *Communications of the Association for Computing Machinery*, **12**, 576–83.
Hoare, C. A. R. (1971). Procedures and parameters: an axiomatic approach. In *Symposium on Semantics of Programming Languages* (ed. E. Engeler), Lecture notes in mathematics, Vol. 188, pp. 102–16. Springer-Verlag, Berlin.
Hocking, J. and Young, G. (1961). *Topology*. Addison-Wesley, Reading, MA.
Holland, S. S. (1975). The current interest in orthomodular lattices. In *The logico-algebraic approach to quantum mechanics, Vol. 1* (ed. C. A. Hooker), pp. 437–96. Reidel, Dordrecht.
Hoogewijs, A. (1979). A formalization of the non-definedness notion. *Zeitschrift für mathematische Logik und Grundlagen der Mathematik*, **25**, 213–21.
Horn, A. (1951). On sentences which are true of direct unions of algebras. *Journal of Symbolic Logic*, **16**, 14–21.
Huntington, E. V. (1904). Sets of independent postulates for the algebra of logic. *Transactions of the American Mathematical Society*, **5**, 288–309.
Husimi, K. (1937). Studies on the foundations of quantum mechanics I. *Proceedings of the Physical-Mathematical Society of Japan*, **19**, 766–89.
IEEE (1976). *Proceedings of the 6th Symposium on Multiple-valued Logic*. IEEE, New York.
Kalman, J. A. (1958). Lattices with involution. *Transactions of the American Mathematical Society*, **87**, 485–91.
Kalmár, L. (1936). Zurückführung des Entscheidungsproblems auf der Fall von Formeln mit einzigen binären Funktionsvariablen. *Compositio Mathematica*, **4**, 137–44.
Kalmbach, J. A. (1983). *Orthomodular lattices*. Academic Press, New York.
Kamlah, W. and Lorenzen, P. (1967). *Logische Propädeutik*. Bibliographisches Institut, Mannheim.
Kant, E. (1787). *Kritik der reinen Vernunft* (2nd edn), Johann Friedrich Hartknoch, Riga.
Kapp, E. (1942). *Greek foundations of traditional logic*. Columbia University Press.
Keat, R. and Urry, J. (1975). *Social theory as science*, Routledge and Kegan Paul, London (2nd edn 1982).

Keisler, H. J. (1970). Logic with the quantifier 'there exist uncountably many'. *Annals of Mathematical Logic*, **1**, 1–93.

Kleene, S. K. (1945). On the interpretation of intuitionistic number theory. *Journal of Symbolic Logic*, **10**, 109–24.

Kleene, S. C. (1962). *Introduction to metamathematics*. North-Holland, Amsterdam.

Kleene, S. C. and Vesley, R. E. (1965). *Foundations of intuitionistic mathematics*. North-Holland, Amsterdam.

Klein, F. (1932). Über einen Zerlegungsatz in der Theorie der abstrakten Verknüpfungen. *Mathematische Annalen*, **106**, 114–30.

Kline, G. L. (1965). N. A. Vasil'ev and the development of many-valued logics. In *Contributions to logic and methodology in honour of J. M. Bocheński* (ed. A.-T. Tymieniecka), pp. 315–26. North-Holland, Amsterdam.

Kneale, W. (1960). Modality de dicto and de re. In *1960 International congress on logic, methodology and philosophy of science* (eds E. Nagel et al.), pp. 622–33. North-Holland, Amsterdam.

Kneale, W. (1962). *The development of logic*. Clarendon Press, Oxford.

Kolmogoroff, A. (1932). Zur Deutung der Intuitionistische Logik. *Mathematische Zeitschrift*, **35**, 58–65.

König, D. (1926). Sur les correspondences multivoques des ensembles. *Fundamenta Mathematicae*, **8**, 114–34.

Körner, S. (1946–7). On entailment. *Proceedings of the Aristotelian Society*, **47**, 143–62.

Körner, S. (1955). *Conceptual thinking*, Cambridge University Press. (See also Dover, 1959).

Körner, S. (1966). *Experience and theory*. Routledge and Kegan Paul, London.

Körner, S. (1973). Material necessity. *Kant Studien*, **4**, 423–30.

Körner, S. (1976). *Experience and conduct*. Cambridge University Press.

Körner, S. (1980). Über Sprachspiele und rechtliche Institutionen. In *Ethics: foundations, problems and applications. Proceedings of the 5th International Wittgenstein Symposium* (Eds E. Morscher and R. Stranzinger), pp. 480–91, Hölder-Pichler-Tempsky, Vienna.

Körner, S. (1981). Intuition and formalization in mathematics. *Epistemologia*, **4**, 113–30.

Kowalski, R. A. (1974). Predicate logic as a programming language. In *Information Processing '74. Stockholm*, pp. 569–74. North-Holland, Amsterdam.

Kreisel, G. (1955). Models, translations and interpretations. In *Mathematical interpretation of formal systems* (eds Th. Skølem et al.), pp. 26–50. North-Holland, Amsterdam.

Kreisel, G. (1967). Informal rigour and completeness. *Problems in the philosophy of mathematics* (ed. I. Lakatos), pp. 138–85. North-Holland, Amsterdam.

Kreisel, G. and Krivine, J. L. (1967). *Elements of mathematical logic*. North-Holland, Amsterdam.

Kripke, S. (1965). Semantical analysis of intuitionistic logic. I. In *Formal systems and recursive functions*, pp. 92–130, North-Holland, Amsterdam.

Kuhn, T. (1962). *The structure of scientific revolutions*. University of Chicago Press.

Kuratowski, C. (1921). Sur la notion de lordre dans la théorie des ensembles. *Fundamenta Mathematicae*, **2**, 161–71.

Kuratowski, C. (1922). Sur l'operation $\bar{A}$ de l'analysis situs. *Fundamenta Mathematicae*, **3**, 182–99.
Kuratowski, C. (1961). *Introduction to set theory and topology*. Pergamon, Oxford.
Lakatoś, I. (1962). Infinite regress and the foundations of mathematics. *Aristotelian Society*, **36**, 155–94.
Lewis, C. I. (1918). *Survey of symbolic logic*. University of California Press, Berkeley, CA. (Reprinted Dover Publications, New York, 1960.)
Lewis, C. I. and Langford, C. H. (1932). *Symbolic logic*. Century Company, New York. (Reprinted Dover Publications, New York, 1951.)
Lloyd, J. W. (1984). *Foundations of logic programming*. Springer-Verlag, Berlin.
Löb, M. (1966). Extensional interpretation of modal logic. *Journal of Symbolic Logic*, **31**, 33–45.
Loemker, L. E. (1969). *Leibniz philosophical papers and letters*. Reidel, Dordrecht.
Lorenz, K. (1968). Dialogspiele als semantische Grundlagen von Logikkalkülen. *Archiv für mathematische Logik und Grundlagenforschung*, **11**, 32–55, 73–100.
Lorenz, K. (1973). Rules versus theorems. *Journal of Philosophical Logic*, **2**, 352–60.
Lorenzen, P. (1951). Algebaische und logistische Untersuchungen über freie Verbände. *Journal of Symbolic Logic*, **16**, 81–106.
Lorenzen, P. (1955). *Einführung in die operative Logik und Mathematik*. Springer-Verlag, Berlin.
Lorenzen, P. (1958). Logical reflection and formalism. *Journal of Symbolic Logic*, **23**, 240–9.
Lorenzen, P. (1959). Über die Begriffe 'Beweis' und 'Definition'. In *Constructivity in Mathematics* (ed. A. Heyting), pp. 169–77. North-Holland, Amsterdam.
Lorenzen, P. (1961a). Ein dialogisches Konstruktivitätskriterium. In *Infinitistic methods*, pp. 193–200, Pergamon, Oxford.
Lorenzen, P. (1961b). Das Begründungsproblem der Geometrie als Wissenschaft der Räumlichen Ordnung. *Philosophie Naturalis*, **4**, 415–31.
Lorenzen, P. (1962). *Metamathematik*. Bibliographisches Institut, Mannheim.
Lorenzen, P. (1971). *Differental and integral*. University of Texas Press.
Löwenheim, L. (1915). Über Möglichkeiten in Relativkalkül. *Mathematische Annalen*, **76**, 447–70.
Łukasiewicz, J. (1941). Die Logik und das Grundlagenproblem. In *Les entretiens de Zürich sur les fondements at la méthode des sciences mathématiques 6–9, 12th Dec. 1938*, (Ed. F. Gonseth), pp. 82–100. S. A. Leeman, Zürich.
McArthur, R. P. (1976). *Tense logic*. Reidel, Dordrecht.
McCarthy, C. (1984). Information systems, continuity and realizability. In *Logic of programs* Lecture notes in computer science, Vol. 164, pp. 341–59. Springer, Berlin.
MacColl, H. (1906). *Symbolic logic and its applications*. Longmans, London.
McDermott, D. (1982). A temporal logic for reasoning about plans and actions. *Cognitive Science*, **6**, 101–55.
McKinsey, J. C. C. (1939). Proof of the independence of the primitive symbols of Heyting's calculus of propositions. *Journal of Symbolic Logic*, **4**, 155–8.
Mal'cev, A. I. (1941). On a general method for obtaining local theorems in group theory. *Notices of the Pedagogical Institute of Ivanova, Physical-mathematical Sciences*, **1**, 3–9 (in Russian).
Manna, Z. and McCarthy, J. (1969). Properties of programs. Partial function logic. *Machine Intelligence*, **5**, 27–37.

Mendelson, E. (1964). *Introduction to mathematical logic*. van Nostrand, Princeton.
Meseguer, J. (1989). General logics. In *Logic Colloquium '87* (eds H.-D. Ebbinghaus et al.), pp. 275–329. North-Holland, Amsterdam.
Mitchell, O. H. (1883). On a new algebra of logic. In *Studies in logic by members of the John Hopkins University*, Johns Hopkins University Press.
Moisil, G. C. (1935). Recherches sur l'algèbre de la logique. *Annales Scientifique de l'Université de Jassy*, **22**, 1–117.
Montague, R. (1960). Logical necessity, physical necessity, ethics and quantifiers. *Inquiry*, **4**, 259–69.
Monteiro, A. (1960). Matrices de Morgan caractéristiques pour le calcul propositionnel classique. *Anais Academia Brasileira de Ciencias*, **32**, 1–7.
Moore, G. E. (1900). Necessity. *Mind NS.*, **35**, 289–304.
de Morgan, A. (1847). *Formal logic: or the calculus of inference, necessary and probable*. Taylor and Walton, London. (reprinted Chicago and London 1926, ed. A. E. Taylor).
de Morgan, A. (1856). On the symbols of logic, the theory of syllogism, and in particular of the copula and the application of the theory of probabilities to some questions of evidence. *Cambridge Philosophical Transactions*, **9**, 104: **10**, 346.
Mostowski, A. (1957). On a generalisation of quantifiers. *Fundamenta Mathematicae*, **44**, 12–36.
Mostowski, A. (1959). On various degrees of constructivism. In *Constructivity in Mathematics* (ed. A. Heyting), pp. 178–94. North-Holland, Amsterdam.
Mure, G. R. G. (1965). *The philosophy of Hegel*. Clarendon Press, Oxford.
Muškardin, V. (1978). *Quasi-Boolean algebras and semantics for logical entailment*. Unpublished Ph.D. Thesis, University of Bristol.
Myhill, J. (1975). Constructive set theory. *Journal of Symbolic Logic*, **40**, 347–82.
Nachbin, J. R. (1949). On a characterisation of the lattice of all ideals of a Boolean ring. *Fundamenta Mathematicae*, **36**, 137–42.
Nagel, N. E. (1972). *The structure of science: problems in the logic of scientific explanation*. Routledge and Kegan Paul, London.
Nelson, D. (1947). Recursive functions and intuitionistic number theory. *Transactions of the American Mathematical Society* **61**, 307–68.
Ore, O. (1938). Structures and group theory II. *Duke Journal*, **4**, 247–69.
Parkinson, G. H. R. (1966). *Leibniz logical papers*. Clarendon Press, Oxford.
Peano, G. (1889). *Formulaire de mathématiques* 2, §2. Logique mathématique. Bocca frères, Turin.
Peano, G. (1894). *Notations de logique mathématique*. La Rivista di matematica, Turin.
Pears, D. R. (1953). Incompatibilities of colours. In *Logic and language, second series* (ed. A. G. N. Flew), pp. 112–24. Blackwell, Oxford.
Peirce, C. S. (1870). Description of a notation for the logic of relatives resulting from an amplification of the conception of Boole's Calculus of Logic. *Memoirs of the American Academy*, **9**, 317–78.
Peirce, C. S. (1880). On the algebra of logic. *American Journal of Mathematics*, **3**, 15–57.
Peirce, C. S. (1885a). On the algebra of logic. *American Journal of Mathematics*, **7**, 80–202. (*Collected Papers*, Vol. 3, 210–38.)

Pivćević, E. (1974). Truth as structure. *Review of Metaphysics*, **28**, 311–27.
Pivćević, E. (1990). *Change and selves*. Clarendon Press, Oxford.
Pneuli, A. (1977). The temporal logic of programs. *Proceedings of the 18the Annual Symposium on the Foundations of Computer Science*, pp. 46–57, IEEE, New York.
Popper, K. (1947). New foundations for logic. *Mind*, **56**, 193–235.
Post, E. (1921). Introduction to a general theory of propositions. *American Journal of Mathematics*, **43**, 163–85.
Pratt, V. R. (1976). Semantical considerations in Floyd–Hoare logic. *17th IEEE Symposium on the Foundations of Computer Science, 1976*, pp. 109–21. IEEE, New York.
Prior, A. N. (1962). *Formal logic* (2nd edn), Oxford University Press.
Quine, W. V. (1945). On ordered pairs. *Journal of Symbolic Logic*, **10**, 95–6.
Quine, W. V. (1947). The problem of interpreting modal logic. *Journal of Symbolic Logic*, **12**, 43–8.
Quine, W. V. (1961). *From a logical point of view*. Harvard University Press Cambridge.
Quine, W. V. (1966). Reply to Professor Marcus. In *The ways of paradox and other essays*, pp. 3–20. Random House, New York.
Rasiowa, H. (1974). *An algebraic approach to non-classical logics*. North-Holland, Amsterdam.
Rasiowa, H. and Sikorski, R. (1963). *The mathematics of metamathematics*. Państwowe Wydanictwo Naukowe, Warsaw.
Reichenbach, H. (1935). *Wahrscheinlichkeitslehre*. Sijthoff, Leiden (1st edn 1891).
Reichenbach, H. (1975). Three-valued logic and the interpretation of quantum mechanics. In *The logico-algebraic approach to quantum mechanics*, Vol. 1 (ed. C. A. Hooker), pp. 53–98. Reidel, Dordrecht.
Rescher, N. (1969). *Many-valued logic*. McGraw-Hill, New York.
Rescher, N. R. and Urquhart, A. (1971). *Temporal logic*. Springer-Verlag, Berlin.
Robinson, A. (1951). *On the metamathematics of algebra*. North-Holland, Amsterdam.
Robinson, A. (1963). *Introduction to model theory and to the metamathematics of algebra*. North-Holland, Amsterdam.
Robinson, J. A. (1965). A machine-oriented logic based on the resolution principle. *Journal of the Association of Computing Machinery*. **12**(1), 23–41.
Rogers, H. (1967). *Theory of recursive functions and effective computability*. McGraw-Hill, New York.
Rosenbloom, P. C. (1950). *The elements of mathematical logic*. Dover Publications, New York.
Rosser, J. B. and Turquette, A. R. (1952). *Many-valued logics*. North-Holland, Amsterdam.
Russell, B. (1903). *Principles of mathematics*. Reprinted Allen and Unwin, London, 1948.
Russell, B. (1950). *Introduction to mathematical philosophy*. Allen and Unwin, London.
Russell, B. and Whitehead, A. N. (1910–13). *Principia Mathematica*, Vol. 1 (1910), Vol. 2 (1912), Vol. 3 (1913). Cambridge University Press.
Sacks, G. (1972). Differential closure of a differential field. *Bulletin of the American Mathematical Society*, **78**, 629–34.
Sasaki, U. (1954). On orthocomplemented lattices satisfying the exchange axiom. *Journal of Science of Hiroshima University*, Series A17, 293–302.

Ščedrov, A. (1985). Intuitionistic set theory. In *Harvey Friedman's research on the foundations of mathematics* (eds L. A. Harrington et al.), pp. 257–84. North-Holland, Amsterdam.

Schröder, E. (1890). *Algebra der Logik. Vol. I*, Leipzig. (2nd edn, Chelsea, New York, 1966.)

Schurz, G. (1983). Das deduktive Relevanzkriterium von Stephan Körner und seine Wissenschaftstheoretischen Anwendungen. *Grazer Philosophische Studien*, **20**, 149–77.

Schütte, K. (1961). *Beweistheorie*. Springer-Verlag, Belin. (English edition, *Proof theory*, 1977.)

Scott, D. (1970). *Outline of a mathematical theory of computation*. Technical Monograph PRG-2, Oxford University Computing Laboratory.

Scott, D. (1972). Background to formalisation. In *Truth, syntax and modality* (ed. H. Leblanc), pp. 244–73. North-Holland, Amsterdam.

Scott, D. (1973). On engendering an illusion of understanding. *Journal of Philosophy*, **68**, 787–807.

Scott, D. (1974a). Completeness and axiomatisability in many-valued logic. In *Proceedings of a Symposium in Pure Mathematics*, Vol. 25, (ed. L. Henkin), pp. 411–36. American Mathematical Society, Providence, RI.

Scott, D. (1974b). Rules and derived rules. In *Logical theory and semantic analysis*. (ed. S. Stenlund) pp. 147–61. Reidel, Dordrecht.

Scott, D. (1976a). Does many-valued logic have any use? In *Philosophy of logic* (ed. S. Körner), pp. 64–73. Blackwell, Oxford.

Scott, D. (1976b). *Data types as lattices*. Technical Monograph PRG-5, Oxford University Computing Laboratory.

Shepherdson, J. C. S. (1988). Negation in logic programming. In *Deductive databases and logic programing* (ed. H. Misker), Chapter 1. Morgan Kaufman, Los Altos, CA.

Šik, O. (1976). *The third way*. Wildewood House, London.

Skølem, Th. (1919). Untersuchungen über die Axiome des Klassenkalküls and über Produktions- und Summations-probleme, welche gewisse Klassen von Aussagen betreffen. *Skrifter, Vidensabsakademiet i Kristiana*, **3**, 37.

Skølem, Th. (1920). Logische-kombinatorische Untersuchungen über die Erfüllbarkeit und Beweisbarkeit mathematischer Sätze nebst einem Theoreme über dichte Mengen. *Skrifter, Vidensabsakademiet i Kristiana*, *I* **4**, 1–36.

Smiley, T. (1962). The independence of connectives. *Journal of Symbolic Logic*, **27**(4), 426–38.

Smiley, T. (1981). Frege and Russell. In *Un siècle dans la philosophie des mathématiques*, Archıvs de l'institute internationale des sciences théoretiques, Vol. 23, pp. 53–7. Office Internationale de Libraire, Brussels.

Smullyan, R. M. (1961). *Theory of formal systems*. Princeton University Press.

Stegmüller, W. (1964). Remarks on the completeness of logical systems relative to the validity concepts of P. Lorenzen and K. Lorenz. *Notre Dame Journal of Formal Logic*, **5**, 81–112.

van Stigt, W. (1971). *Brouwer's intuitionism*. Unpublished Ph.D. Thesis, University of London.

van Stigt, W. (1979). The rejected parts of Brouwer's dissertation on the foundations of mathematics. *Historia Mathematica*, **6**, 385–404.

Stoy, J. E. (1977). *The Scott–Strachey approach to programming language theory.* MIT Press, Cambridge, MA.

Szabo, W. E. (1969). The collected papers of Gerhard Gentzen. North-Holland, Amsterdam.

Tarski, A. (1930). Über einige fundamentalen Begriffe der Metamatematik. *Comptes rendus des séances de la société des sciences et des lettres de Varsovie, Classe III*, **23**, 22–9.

Tarski, A. (1948). *A decision method for elementary algebra and geometry.* Rand Corporation, Santa Monica, CA.

Tarski, A. (1952). Some notions on the borderline of algebra and metamathematics. *International Congress of mathematicians, Cambridge USA, 1950, Volume 1*, (ed. L. M. Graves et al.), pp. 705–20. American Mathematical Society, Providence.

Tarski, A. (1956). *Logic, semantics, metamathematics.* Clarendon Press, Oxford.

Tarski, A. (1965). *Introduction to logic.* Clarendon Press, Oxford.

Tarski, A. and Vaught, R. (1957). Arithmetical extensions of relational systems. *Compositio Mathematica*, **18**, 81–105.

Tarski, A., Robinson, R., and Mostowski, A. (1953). *Undecidable theories.* North-Holland, Amsterdam.

Thiel, C. (1981). Mathematics, logic and ontology. In *Un siècle dans la philosophie des mathématiques*, Archivs de l'institute internationale des sciences théoretiques, Vol. 23, pp. 95–112. Office Internationale de Libraire, Brussels.

Thom, R. (1967). Stabilité structurelle et morphogenèse. Benjamin, New York.

Thuraisingham, M. B. (1987). Reducibility relationships between decision problems for system functions. *Zeitschrift für mathematische Logik und Grundlagen der Mathematik*, **33**, 305–12.

Thuraisingham, M. B. (1990). *Recursion-theoretic properties of the inference problem in database security.* Report MTP 291, Mitre Corporation, Bedford, MA.

Toulmin, S. (1961). *Foresight and understanding.* Indiana University Press.

Urquhart, A. (1986). Many valued logic. In *Handbook of philosophical logic*, Vol. 3. (eds D. Gabbay and F. Guenthner), pp. 71–116. Reidel, Dordrecht.

Vasil'ev, N. A. (1924). Imaginary (non-Aristotelian) logic. In *Atti del Quinto Congresso Internationale di Filosofia, Napoli, 5–9 Maggio, 1924*, pp. 107–109.

Vaught, R. L. (1964). The completeness of logic with the added quantifier 'there are uncountably many'. *Fundamenta mathematicae*, **54**, 303–4.

Vaught, R. L. (1974). Model theory before 1945. In *Proceedings of the Tarski Symposium* (eds L. Henkin et al.), pp. 153–72, American Mathematical Society, Providence, RI.

Waismann, F. (1955). Are there alternative logics? *Proceedings of the Aristotelian Society*, **45**, 77–104.

Wajsberg, M. (1935). Beiträge zum Metaausagenkalkül I. *Monatshefte für Mathematik und Physik*, **42**.

Wajsberg, M. (1938). Untersuchung über den Aussagenkalkül von A. Heyting. *Wiadomości Mathemayczne*, **46**, 45–101.

Wang, H. (1961). The calculus of partial predicates and its extension to set theory. I. *Zeitschrift für mathematische Logik und Grundlagen der Mathematik*, **7**, 283–8.

Wang, H. (1963). *A survey of mathematical logic.* Science Press, Peking, and North-Holland, Amsterdam.

Weyl, H. (1946). Mathematics and logic. A brief survey serving as a preface to a review of the philosophy of Bertrand Russell. *American Mathematical Monthly*, **53**, 2–13.

Wiener, N. (1912). A simplification of the logic of relations. *Proceedings of the Cambridge Philosophical Society*, **17**, 387–90.

Wittgenstein, L. (1922). *Tractatus logico-philosophicus*. Routledge and Kegan Paul, London.

Wittgenstein, L. (1956). *Remarks on the foundations of mathematics*. Blackwell, Oxford.

Wolf, R. G. (1977). A survey of many-valued logic (1966–1974). In *Modern uses of multiple-valued logic* (eds J. M. Dunn *et al.*), pp. 167–328. Reidel, Dordrecht.

von Wright, G. H. (1951). *An essay in modal logic*. North-Holland, Amsterdam.

von Wright, G. H. (1968). *An essay in deontic logic and general theory of action*. North-Holland, Amsterdam.

Zermelo, E. (1930). Über Grenzzahlen und Mengenlehre. *Fundamenta Mathematicae*, **16**, 29–47.

Index

abstract calculus 6
abstraction §5.4
absurdity 19, 115, 348
activity 309, 357
addition 105
adequacy 385
admissible rule 93, 317
Albert of Saxony 129
algebra 390
   of formulas 20, 204, 263
   of truth-values 21, 199, 203, 306
   quotient- 144
all §4.4
analysable 280
analytic 362, 366, 382
Anderson and Belnap 286
antecedent 37, 101, 313
antisymmetric 141
apartness 113
approximation 162
arbitrary domain 17
Aristotelian connectives 203
Aristotle 1, 2, §1.2, 111, 114, 136, 142, 250, 254, 310, 360
Aristotle's theory of science §1.2
arithmetic 340, 358
artificial intelligence 61
ascending chain condition (ACC) 196
assignment 41, 213
assymetric 141
atom 40, 66, 191
   simple- 66
atomic formula 40, 366
axiom 6, 33
  general-/special- 313, 328, 378
axiomatizability 6, 373

BASIC 28
basic- 226
basic sequent 339
basic system of logical entailment 287
becoming 351
before/after 219
belonging 360
Bergson 351
Bernays 98, 194
Beth 195
binary
   operations 105, 120
   relation 136
binding 86
Birkhoff 114, 160, 168, 174, 187, 393

Bishop 118, §10.10
Bocheński 107, 360
body 27
Boole 3
Boolean
   algebra §5.10.6
   operations §9.2.2
Bosanquet 286
boundary 233
   elements §7.4
   sets 231, 395
bound variables 39, 86, 88
Bradley 286, 363
Brouwer 17, 114, chapter 10
Burleigh 117, 128

calculus of absurdity 348
calculus of classes 128, 147, 212
canonical
   assignment 46
   homomorphism 237
   mapping 83
   projection 142
   realization 45, 213
   valuation 46
Carnap 1, 6, 203, 364
causality §10.9.4
C-free 66
chain condition 196
chain-connected 250
chance 361
change 111, 344
chaos not mentioned in text
characteristic function 212, 244, 394
Church–Rosser property 146
Chwistek 195
classical imperative 11, §8.2
classical implicative lattice 177, 190
clause 26, 52
closed boundary set 395
closed formula 44
closed interval 229
closed set §3.6, 156, 161
closed set of top. space 395
   closed world assumption 113
closure 62, §9.2.2
   method 195
closure specification 29
closure operation 177
coboundary 233
code of conduct 385
combinatorial system §3.2

commute 187
compact §2.2.6, 51, 61, 197, §6.9.3, 395
  element 197
  kernel 60
  strongly- 84
compatibility 332
complement §5.10.3
complementation 246
completeness 17, §2.2.6, 34, 51, 81, 210, 220, §810
component 247
computation 354
computer science 5, 57, 219
conditional §4.2.1, 211
congruence 144
  natural- 148
connected 395
  element 249, §8.9
connected relation 141
connective §3.7.1, §4.1
consciousness §10.9.2
consequence 93, 112
  function 61
  junctor 377
consistent 22, 88
constant term 39, 87
construction 107, 340, §10.9.8
constructive 17, 28, §10.11
contact 250
context 306
contingent 129, 360
continuity §7.8.2, §8.5, §8.9
  principle §8.5
continuous 163
continuum 229, §10.9.10, §10.10.3
contradiction 19, 22, §4.2.2, §4.3.4
  law of- 115
converse 140
  domain 140
coordinate language 9
correlative definitions 7
couple 138
cover 148
creating subject 345
Curry 113, 136, 170, 311
cut rule 38, 100
cut condition 317

d-entails 303
d-irrelevant/slack/rigid 290
data-base 305
data-type §5.8.1
decidable 373
decisions 10, 254
de dicto/re 361
Dedekind 160, 168, 170
deduction theorem 108

deductive
  postulate 5
  science 5
  system 6
definite
  condition 244
  goal 55
  program 27
  program clause 27, 55
definition by abstraction 142
degenerate conjunction 268
deletion 289
dense 235
deontic formula/operator 386
derivation 33, 93
derived
  operation 392
  system 391
descending chain condition (DCC) 197
descriptions §8.3.3
descriptive counterpart 385
designated 193
deterministic 96
dialogue 107, 311, 328
difference 111
dilemma 169
Dingler 358
Diodorus Cronus 218
Dirac 12
direct
  derivation 33, 93
  power 209
  product 393
directly derivable 33, 93
disjunctive judgement/syllogism 286
distinguished occurrence 25
divisibility 229
DL = distributive lattice
dogma of structural stability 12, 365
duality 157
Duns Scotus 12

elementarization 98
elementary formal system 52
elementary theory of necessity 369
empirical
  continuity §7.8
  discourse 16, 243
  sentence 11
entailment 287, 302
equality 147, 244
equational class 391
equivalence
  class 142
  relation 141
Euclid 143
Euclidean geometry 13, 143, 305, 358
Eudoxus 143

Index 413

Euler 137
exact class 245
excluded middle 16, 253
existence 340
existential 44
  closure 45
  condition 256
  quantifier 38, 130
extensional 244
external world 357
extremal clause 28
extremity 233, 250

factor 247
fallacies 2
falsehood 19, 115
field of
  inexact sets 246
  relation 140
  of sets 156, 191
finitary set-to-set function 60
finite assignment 67
finite intersection property 30
finite subdivision 229
finite theory of necessity 369
finitely refutable 54, 113
finiteness theorem 30
first degree entailment 30
fixed point 80
  theorem 162
formalization 346, 350, 354
formal part 247
formal system §3.9
formula
  critical- 280
  minimal- §9.2.1
  modal- 368
  prenex- 44
  pure- 268
  effective quantifier- 311
  special- 381
  t-f-free 213
free product 230
free variable 39, 86
  Horn clause 52
Frege 3, 17, 86, 203, 363
formula 19, 40, 66
functionally free §6.7
function letters 39
future 220

game 311
Geach 2, 136
generalized cyclic group 169
generating function/specification 29
Gentzen system 37, 51, 277
geometry 143, 358, 365

glb 152
Glivenko 103, 349
God 354
Gödel 4, 103, 352, 356, 362
grade 318
graph 144
Greechie and Gudder 209
ground 286
  atom 40
  instance 47
group 85, 168
G-structure 98

Hájek 244
Hardy 16
Hauptsatz 51
Hausdorff 146
head 27
Hegel 111, 287
height 280
Herbrand model 47
Herbrand theorem §2.3.5
Heyting 5, 17, 107, 113, 115, 118, 347, 350
Hilbert 256, 352
Hilbert space 161, 188, 217
Hilbert system 33
Hoare 57, 363
Hobbes 142
Holland, Jr 210
homomorphic image 63
homomorphism 63, 392
hypothetico-deductive system 5

idealization 11, 113, 243
ideal of lattice 170, 231
identification 142
identity 140, 142, 391
  relation 140
I-lattice 177
implication 1, 19, §6.4.3, 141, 210
  material 106
  minimal- §9.3
  Philonian- 106
  relation 141
  strict- 361
inconsistent 22
index set 314
idealist 286, 353
individual 130
individual constants 39
inductive
  class 29
  definition §2.2.5, 113
incompatible properties 112
inequality 113

inexact 16, 246
  structure 275
infinite distributive laws 166
infinite
  formulas 25
  disjunction/conjunction 43
infinity lemma 31
initial
  elements 29
  sentences 37
instance (of identity) 230
integrated circuit 194
intersection 56, 60, 246
  of relations 140
intersubjective 342, 345
intuition §10.9.5, §10.10.6
intuitionism 17, 107, §10.9
intuitionistic analysis §10.9.10
involution §5.10.5, 228
irrelevant 290
iterated obligation 386
i-valid 277

Jevons 118, 189
Joachim Jungius 137
join 165
  homomorphism 170
  irreducible 172, 248
junctor 19, 204
  basic- 326
  regular- 326

Kant 2, §10.9.3
Kapp 310
Kleene 103, 244, 260, 317, 339
Klein, Felix 160
Klein, Fritz 160
Kneale 106, 361
knowledge 342, 363
Kolmogoroff 310
König's infinity lemma 31
Körner 247, 258, 288, 366, 385
Kreisel 4, 102, 345
Kripke 226, 340
Kuratowski 139

language §10.9.7
  of type τ 38, 390
lattice 80
  complemented- §5.10
  complete- §5.8
  distributive- §5.9.1
  implicative- 177
  involuted- 187
  of closed sets 156, 166
  orthocomplemented- 187
  orthomodular- 187, 210
  polynomial 205, 260
  pseudo-complemented- 177
  valued sets §6.8.1
law
  de Morgan- 128, 176, 275
  distributive- 160
  modular- 165
  Peirce- 169, 177
left rule 313
Leibniz 142, 143, 363
Lemmon 219
Lewis 361
Lindenbaum algebra 83, 177, 180, §8.8
linearity 220
linear tense structure 221
linguistic element 310
literal 20, 40, 268
Löb 386
logic §3.5, §6.3
  Arabian- 218
  classical- chapter 2, §10.8
  deontic- §11.8
  dynamic- 363
  effective- chapter 10
  effective modal- §11.7
  Greek- 309
  free variable- 87
  Indian 2, 111
  intuitionistic- chapter 10
  intuitionistic modal- 384
  maximum- 79
  many-valued 112, 116
  Megarian–Stoic- 106, 111, 117, 202, 218
  minimum- 79
  modal- chapter 11
  non-monotonic- 62
  of logics 77
  of truth-values §6.3, §8.4
  partial function- §8.3.2
  positive- §10.6
  propositional- 205
  program §2.3.8
  pure effective- 327
  pure effective modal- 383
  pure effective quantifier- 338
  pure implicational- 95
  quantifier- 88
  Scholastic- 117
  semantic- 193
  simple- 97
  standard deontic- 389
  strong word- 84
  subalgebra- 85
  subject/predicate- 2
  temporal- §6.9
  three-valued- 116, chapter 8
  word- 84
logical combination 20

logical consequence chapter 3
logical equivalence 22, §3.7.1, 115
logical identity 66, 205
logical necessity 364, 369
logical operation §3.7
logical particle/symbol 19
logically n-entails, n-equivalent 307
logistic thesis 3
Lorenz 107, 309
Lorenzen 3, 10, 339, 358
Löwenheim 17
lower
   bound 151
   semi-lattice 155
Łukasiewicz 128, 194, 203, 254

materially necessary 12, 365
matrix 44
maximum (maximal, minimal) element 149
Maxwell's equations 365
meaning 288
meet 156
mention 362
metaphysics 12, 225, 357, 361
Mill, J. S. 12
minimum
   system 219
   model §2.2.4
modal
   depth 368, 377
   formula (see formula) 368
   industry 360
   logic (see logic)
   operator 225, 366
   rule 374
   syllogism 360
modality
   real 366
   relative- §11.3
model 22, 203, 219
   minimum- §2.2.4
   minimum Herbrand- 56
   non-standard- 168
model structures §6.11
modus ponens 24, 33, 94, §4.2.1
monotone 61
de Morgan 137
multisubject semantics §9.4
Mure 111, 361
Muškardin 246, §9.4
mysticism 341, 343

Nagel 1, §1.2
name language 10
necessary belonging 360
necessity chapter 11
   general theory of- 368

junctor 367
   material- 365
   operator 366
   relative- §11.3
negation §4.2, §4.3, §10.6, §10.9.9
negative facts 111
New Nyāyā 2
von Neumann 209
neutral 10, 244
negative information 54
node 93
non-demonstrability 113
normal conjunction/disjunction §4.2.3, 116
normal form §2.2.3, §2.3.3, §8.7
normal implication §4.2.1, §6.4.3
normal QBA, see QBA
normative 358
NQBA = normal quasi-Boolean algebra
nucleus 235, §8.8

objective 16, 345
objects 356, 357
obligation §11.8
   operator 385
   relative- 385
observation 10
Ockham 128
order §5.5
   reversal 175
open set 177, 394
overlap 250

P-analytic 366
paradox §8.3.1
part 247
   principal- 249
partially ordered set (poset) 146
PASCAL 28
past/present/future 220
path 281
PCL = pseudo-complemented lattice
P-consequence 366
Peano 13, 141
Peirce 95, 160, 169, 253
PEM = principle of excluded middle
P-equivalent 369
permanence 344
Peter of Spain 117, 129
Philo 106
physical
   laws 365
   system 210
physics 365
Pivčevič 365
place 143
Plato 2, 111
plurality of minds 341

Poincaré 347
polynomial 205, 391
positive case 244
positive diagram 26
positivism §1.2, §1.3.1
possible 360, 377
  worlds 362
Post 194, 203, 254
power (of relation) 140
practical arithmetic 359
precision engineering 359
predicate calculus, three-valued §8.10
predicate letter 394
predication 111
prefix 44, 130
prelogic §3.2
prescription 359
primordial intuition §10.9.5
principal NQBA 230, 248
principle of excluded middle 115, 129, §10.9.9
principle of localization 30
principle of omniscience 355
principle of tolerance 1
Prior 389
procedural interpretation 27
product (of relations) 140
production rules 29
program 57, 363
projection map 142
proof 33, 93, 108
  tree 93, 280
proposition letter 19
proposition, subject/predicate- 137, 160
protocomplement 175
pseudo-complement 177, 183, 202
psychological criterion 288
pure entailment 289
P-valid 364

QBA = quasi-Boolean algebra
quantifier-free 44
quantifier §3.7.2
  weak- 216
quantum mechanics 189, §8.3.4
quasi-Boolean algebra 228
quasi-ordering 146
quasi-ordered set (qoset) 146
Quine 8, 10, 256, 362
Q_4-valued system 287

rank 318
ratio 143
r-congruence 144
real closed field 305
realism 353
reality §10.11.2
reality postulate 5

realization 213, 241
real numbers 113, 143
recursive function 372
recursively enumerable 53, 54, 373
reduced
  logic 83
  algebra 144
reducible 103
reductionism §1.3
refinement 259, 276
reflexive 141
refute 277
refutation procedure 34
refutability 113
regular 259, 326
relation §5.2
  recursively enumerable- 53
relational system 394
relative product 140
relative pseudo-complement 177
reliability 194
representable 53
representation 102, 172
  theorem 173
resolution 35, 55
resolvant 35
Rhenius 137
rigid
  formula §9.2
  implication §9.3.1
  tautology §9.3.2
ring of sets 156
root 94
rule
  cut, see cut rule
  right (left)- 313
  of inference 33, 93
  structural- 38
  thinning- 38
Russell 3, 111, 114, 138, 145, 256, 361

Sacks 4
satisfiable 22, 42
satisfies 42, 67, 112
science §1.2
scientific institutions 8
Scholastics 128, 360
Schröder 160, 168
Scott 162, 340, 363
SDPC = simple definite program clause
self-evidence 6
sentence 44
sequent 37, 100, 277
set theory 4, 13, 255, 340
set-to-set function 60
Šik 243

simple
  set 55
  definite program clause 66
simplicial complex 230
Skølem 176
Skølem extension 49
Skølem function §2.3.6
slack §9.2.1
Smiley 58, 195, 287
Smullyan 53
society §10.11.1
solipsism 17, §10.9
some §4.4
sound 81
space 143
spatial activity 359
special 318
  variable 332
specification, generating- 29
van Stigt §10.10
strong connective 83, 108, 109
strongly equivalent 103
strongly logical operation, see strong connective
subdirectly irreducible 393
subdirect product 393
subformula 20
subgoal 55
subject §9.4
subject/predicate 2
substitution 395
  instance 47
succedent 37, 101
syllogism 101, 137, 309
syllogistic 137, 169
symmetric 138, 141
  system 96
syntax 364
synthetic judgement 342

Tarski 59, 75, 203, 254, 373
tautology 22, 41, 194, 267, 277, 290
temporal
  logic, see logic
  operation 219
  precedence 219
tense structure 220
term 39, 66, 87
ternary opration 105
theorem 33, 80
theoremhood
theory 58, 80
  general—of necessity §11.6
  of logical necessity 369
  of necessity 267
  of necessity in groups 373
  of necessity in real closed fields 373
  of obligation 386

time 342
topological space, topology 394
transitive 138
  closure 140
tree 93, 264
triviality condition §10.4
truth 5, 10, 17, 19, §6.5, §10.11.3
  claims 357
  definite 9, 311
  table 21, 193, 202
  table, weak 260
  values 112, 203, §6.9.1, §8.3, 306
t-section 220
Turing-equivalent 372
type 0, 1 65
type τ 39

uniformity theorem 48
union 60, 246
unit clause 27
universal
  closure 45
  quantifier 130
universality 17
upper bound 151
upper semi-lattice 155
upper sequent 280

vagueness 10
valid 22, 42, 220, 316
  inference 1
valuation 22, 41, 204, 213, 263
  canonical- 46
  Herbrand- 46
variant 44
Valil'ev 112, 194, 254
verification 10
vertex sequent 281

Waissman 254
weak reducibility 103
Weyl 3, 351
Wiener 139
wisdom 344
witness 10, 111
Wittgenstein 16, 23, 112, 203, 364
world relation 225
von Wright 385

Y-instance 47

Zermelo 13

$\wedge$-closed 24
$\wedge$-formula 277
$\vee$-formula 277